Climate Change for Astronomers

Causes, consequences, and communication

Online at: https://doi.org/10.1088/2514-3433/acfcb6

AMERICAN ASTRONOMICAL SOCIETY IOP | ebooks™

AAS Editor in Chief

Ethan Vishniac, Johns Hopkins University, Maryland, USA

About the program:

AAS-IOP Astronomy ebooks is the official book program of the American Astronomical Society (AAS) and aims to share in depth the most fascinating areas of astronomy, astrophysics, solar physics, and planetary science. The program includes publications in the following topics:

- GALAXIES AND COSMOLOGY
- INTERSTELLAR MATTER AND THE LOCAL UNIVERSE
- STARS AND STELLAR PHYSICS
- EDUCATION, OUTREACH, AND HERITAGE
- HIGH-ENERGY PHENOMENA AND FUNDAMENTAL PHYSICS
- THE SUN AND THE HELIOSPHERE
- THE SOLAR SYSTEM, EXOPLANETS, AND ASTROBIOLOGY
- LABORATORY ASTROPHYSICS, INSTRUMENTATION, SOFTWARE, AND DATA

Books in the program range in level from short introductory texts on fast-moving areas, graduate and upper-level undergraduate textbooks, research monographs, and practical handbooks.

For a complete list of published and forthcoming titles, please visit iopscience.org/books/aas.

About the American Astronomical Society

The American Astronomical Society (aas.org), established 1899, is the major organization of professional astronomers in North America. The membership (~7,000) also includes physicists, mathematicians, geologists, engineers, and others whose research interests lie within the broad spectrum of subjects now comprising the contemporary astronomical sciences. The mission of the Society is to enhance and share humanity's scientific understanding of the universe.

Editorial Advisory Board

Climate Change for Astronomers

Causes, consequences, and communication

Edited by
Travis Rector
Department of Physics & Astronomy, University of Alaska Anchorage,
3211 Providence Dr., Anchorage, AK 99508, USA

IOP Publishing, Bristol, UK

ISBN 978-0-7503-3727-4 (ebook)
ISBN 978-0-7503-3725-0 (print)
ISBN 978-0-7503-3728-1 (myPrint)
ISBN 978-0-7503-3726-7 (mobi)

DOI 10.1088/2514-3433/acfcb6

Version: 20240501

AAS–IOP Astronomy
ISSN 2514-3433 (online)
ISSN 2515-141X (print)

British Library Cataloguing-in-Publication Data: A catalogue record for this book is available from the British Library.

Published by IOP Publishing, wholly owned by The Institute of Physics, London

IOP Publishing, No.2 The Distillery, Glassfields, Avon Street, Bristol, BS2 0GR, UK

US Office: IOP Publishing, Inc., 190 North Independence Mall West, Suite 601, Philadelphia, PA 19106, USA

This book is dedicated to
Claudia, mi hija maravillosa. Cada día que estamos juntos es un regalo especial.

Contents

Part VII Climate Change and Astronomy

Preface

T. A. Rector—University of Alaska Anchorage

> 'At every moment, we always have a choice, even if it feels as if we don't. Sometimes that choice may simply be to think a more positive thought.'
>
> —Tina Turner

The End of Normal

When I was approached in March 2020 by IOP Publications with the idea for this book, the world was on the verge of a massive transformation—in ways at that time we could barely imagine. The COVID-19 pandemic was sending us all into lockdown. Years later, the consequences of COVID on our lives cannot be overstated. Economies were upended. People lost their lives, and their livelihoods. Isolation took a terrible toll on our physical and mental health. Pictures of empty streets, idle airplanes, barren grocery stores, and mass graves revealed to us in stark contrast how fragile our lives—and our societal infrastructures—really are.

But during the pandemic we also saw resilience and adaptation. People sewed makeshift masks, choirs sang together online, and—where allowed—we spent more time outdoors. Because we weren't running around everywhere, many found themselves with extra time. The clock was no longer running our lives. With the hours we used to spend in cars and planes we could do more. People took up new hobbies. We learned new languages. Sales of bicycles skyrocketed. Taylor Swift released four albums.

Astronomy adapted too. COVID was an opportunity to rethink how we did things. Transformation of how we work accelerated. Remote observing became mainstream. Online conferences brought people together who, even in the best of times, would not have been able to travel. In our isolation we found ways to be together—to meet new colleagues and start new collaborations.

It turns out the delay from COVID was also fortuitous for this book. What we had originally planned then was quite different from what is in front of you now. It was to be focused on the physical science of climate change. While that is still of course important, the book now puts a greater emphasis on communication and advocacy. The science is established. What we need now more than anything is to take action based upon what we know.

The world in which this book resides is also rapidly changing. Not just from COVID, but in particular from the rapid growth of solutions to climate change, such as wind and solar energy. This growth hasn't been driven just by concerns about climate change, but from the simple fact that they are better. For example, the economic reality is that wind and solar are now as cheap or cheaper than fossil fuels. Five years ago we told audiences that we needed to switch to renewable energy even though they were more expensive. Now—despite what some say—there's little reason not to switch. We're transitioning from a message of austerity to one of opportunity. What could the future look like? And could it be something that we can look forward to?

The Way We Live Now

As I write this the east coast of the United States is engulfed in smoke from wildfires in Canada. It brings back memories of August 2019, when Anchorage was similarly shrouded so badly you couldn't go outside for several days. I remember the pain I could feel inside my lungs, and the anxiety I felt as I struggled to breathe. Living in Alaska I am often reminded that we are at the mercy of nature. We are of course the land of grizzly and polar bears. But this vulnerability felt different.

The experience I felt in 2019, and those on the east coast are feeling now, is a palatable reminder of the consequences of the use of fossil fuels–not only for climate change but also for air pollution. And while the smoke on the east coast will likely clear in a few days, Bill McKibben points out that this is what many people breathe every day.[1] In fact, air pollution is the world's leading killer—every year 1 in 5 die from it.

So what to do about it? The solutions are varied and complex, but also surprisingly simple. About 90% of climate change is from the extraction and use of fossil fuels. We need to stop. As Chapters 6 and 7 point out, this won't be easy—especially when fighting against industries that stand to lose trillions of dollars from the energy transition. But the rapid growth of wind and solar shows us that it's already happening. Our role is to help it happen even faster.

In writing and editing this book I have found hope. While the chapters on consequences and misinformation were emotionally difficult, the chapters on communications and solutions were uplifting—perhaps even moreso. I end this experience feeling better, more optimistic than when I started. That's not to say it's easy. Every day I feel the weight of what we face—and I'm sure you do too. On bad days it can be overwhelming. But on good days—which is most days now—I focus on the process: What can I do today to help? As Bernadette Rodgers points out in Chapter 15, there's a whole spectrum of advocacy—from teaching climate change in your classes to non-violent protest. It's for you to find where you belong. It is our hope that this book will help.

[1] https://billmckibben.substack.com/p/we-can-see-clearly-now

Foreword

Peter Kalmus

How many other technological civilizations exist in our Galaxy? This question has captivated me since I was a child. The answer, of course, depends on how long civilizations in the cosmos last, which depends on whether or not they learn to live in harmony with their planets before they collapse.

We are at a crossroads in the history of our 4.5-billion-year-old planet. These days in which we are alive are precious beyond measure, especially from the perspective of Earthlings who come after us. Every day the fossil fuel industry continues to exist makes our planet hotter, taking us more deeply into irreversible catastrophe. The only way out is to end the fossil fuel industry; the faster we do, the more we will save. And it would be tremendously helpful for astronomers to join in the fight.

We are currently losing a tremendous amount of biodiversity and habitability, such as the Amazon Rainforest, the Great Barrier Reef, boreal forests such as the Sierra Nevada with its giant sequoia trees, coastal cities, and hydrologic and agricultural stability. We are in the process of losing much more: habitable tropics, a large fraction of biodiversity, potentially civilization as we know it, which depends complexly on all of this. It is incredibly important to fight the fossil fuel industry, which has captured world leaders and international climate negotiations. We must stand up and be the adults.

I'm not capable of sugarcoating. Things seem worse now to me than they did even a few years ago; I am hoping that as our predicament becomes clearer and clearer, it will cause more people to fight. Yes, it's scary; but we need to fight. The summer of 2023, in particular, has forced me to revise my sense of how fast the Earth system is changing, and how well we climate scientists understand it. Our models predict global annual mean temperature increases well, at least for now, but they aren't very good at predicting localized extremes or nonlinear dynamic processes such as melting ice sheets and ocean currents.

Everything seems to be breaking down much faster than we thought it would. It's not too late to fight, and it never will be. But, it is late. I'd give anything to be back at the levels of global heating we had even ten years ago. And ten years from now, I'm sure we'd give anything to be back at today's levels of heating.

I started my scientific career in astronomy. As a kid, I loved Star Trek and Bradbury and Asimov. In college, I fell in love with cosmology, but in grad school I fell hard for an impossibly audacious new window on the universe. I'll never forget the first time I saw the LIGO Hanford gravitational wave detector: after a long path back to science from debilitating depression as an undergraduate, I felt deeply grateful for the opportunity to contribute to the grand project of human scientific understanding. Few things in life are as noble or as meaningful—or as much fun. I spent that summer rollerblading along the 4-kilometer-long arms of the world's most sensitive gravitational wave detector, capable of measuring differential lengths less than a ten thousandth of a proton, with expensive optical equipment carefully stowed in my backpack, pioneering its precision calibration via the momentum of a thin stream of sinusoidally modulated photons. As a student, and then as a postdoc, I led ultra-sensitive albeit optimistic searches for burst signals from supernovae and magnetars.

But all the while, I became more and more concerned about global heating. The more I learned, the more concerned I became. Soon I was so obsessed with climate science that I could no longer focus on astronomy, so in 2012 I made the big leap into climate science. I left Caltech—and all the career momentum I'd developed over seven years with LIGO—moving to the Jet Propulsion Laboratory a few miles up the Arroyo Seco in northeast Los Angeles to work on clouds and climate change. Three years later, my former colleagues discovered gravitational waves. I missed out on the discovery of a lifetime, but I don't regret my decision. I feel called to defend the Earth.

Astronomers, including former gravitational wave astronomers, tend to see Earth as a spaceship. A marvelous spaceship, on which all of human history has occurred; on which millions or perhaps billions of crazy species have flourished; on or slightly above which every human thought has been thunk; the air of which all of our friends and lovers, children and parents, colleagues and adversaries have breathed, connected at the molecular level. And now, a small number of people are intentionally and irreversibly damaging our spaceship's life support systems. For just a bit of money. It is a crime of cosmic ignorance.

Now is the time to stand up and stop this foolish, selfish, and short-sighted sabotage of Earth's livability. Please take risks to stop the irreversible breakdown of Earth. Please be a climate activist.

Many people will be moved to action when astronomers are moved to action. The climate movement needs you, desperately. You are the caretakers of the grand astronomical perspective, and it must be brought to bear in the public's climate discourse.

In case they are helpful, I've distilled three key principles for climate activism. 1. Fully embrace the emotions that come with the knowledge that our planet is breaking down irreversibly, and pointlessly. Transmit those emotions to others and allow them to drive your own action. 2. Work with other people. Join groups, make friends, cultivate mutual support. Find joy in activism, together. Urge each other to ever more audacious statements and actions. 3. Take risks. We need rapid transformation of very strong, solidified social norms, and our brains, finely attuned social instruments that they are, interpret pushing hard against social norms as risky and uncomfortable. If it doesn't feel risky, it probably isn't changing social norms very much. Be as creative and courageous as you can.

Tragically, we are rapidly discovering and understanding exoplanets even as our own planet dangerously overheats and spirals into mass extinction. I used to think this was a crazy coincidence, straight out of science fiction. But of course it is no coincidence: it is the result of technology without wisdom. Bringing wisdom is an even more necessary and noble project than adding to our growing scientific understanding. We need the wisdom to end the fossil fuel industry, end industrial animal agriculture, and change the goal of the human economic system of systems from profit for the few to flourishing for all.

The first time I was arrested for climate disobedience was the most cathartic experience of solidarity I have ever tasted. I want you to experience that. Raise your voice as loudly as you can. Join me and other climate activists on the right side of history—in rage, in love, and in solidarity.

Acknowledgments

The book is indebted to professionals inside and outside the field of astronomy who were kind enough to share their knowledge and insights, as well as fact check the book. In particular we wish to thank the following:

Janica Anderzén, Rosie Barnes, Robert Blum, Leo Burtscher, Will and Melissa Carpenter, Nelta Edwards, John Fasullo, Katherine Hayhoe, Jeremy Hoffman, K. Tabetha Hole, Katherine Honda, Knud Jahnke, Gunnar Knapp, Jonathan Izett, Jeremy Littell, David Lockard, Thomas Lucas, Michael Mann, Pierrick Martin, Theofanis Matsopoulos, Carrie McDougall, Ian McLennan, Joaquin Murrieta-Saldivar, Hilary Peddicord, Lucy Richardson, Amy Robertson, Sheryl Sotero, Luigi Tibaldo, Jackson Voelkel, Betsy Wilkening, and Ryan Wyatt.

Editor biography

Travis A. Rector

Travis A. Rector is an astrophysicist at the University of Alaska Anchorage. In recent years his focus has been on advocating for solutions to climate change. Living in Alaska, he has witnessed dramatic changes in his home state. He is also one of the founders of Astronomers for Planet Earth, an international grass-roots movement of astronomy students, educators, amateurs and scientists working to address the climate crisis from an astronomical perspective. He is currently serving as the chair of a task force for the American Astronomical Society, whose goal is to identify ways astronomy as a profession can reduce its carbon footprint on a scale commensurate with the terms of the Paris Agreement.

Department of Physics & Astronomy
3211 Providence Dr., Anchorage, AK 99508 USA
tarector@alaska.edu
+001–907–786–1242
Affiliation: University of Alaska Anchorage

List of Contributors

Anna Cabré Albos
University of Pennsylvania, USA

Paola Banchero
University of Alaska Anchorage, USA

Jeffrey Bennett
Big Kid Science; University of Colorado, Boulder, USA

Faustine Cantalloube
Aix Marseille University, France

Ron Ekers
CSIRO Space and Astronomy/Curtin University, Australia

David J. Helfand
Columbia University, USA

Aidan Hotan
CSIRO Space and Astronomy, Australia

Emily Kerrison
University of Sydney/CSIRO Space and Astronomy, Australia

Jürgen Knödlseder
Insitut de Recherche en Astrophysique et Planétologie (IRAP), France

Rika Kobayashi
Australian National University, Australia

Rachel Mason
Independent

Vanessa Moss
CSIRO Space and Astronomy/University of Sydney, Australia

Bob Raynolds
Denver Museum of Nature & Science, USA

Travis Rector
University of Alaska Anchorage

Glen Rees
TFOM

Bernadette Rodgers
Willamette University and Portland Community College, USA

Elizabeth Tasker
Japan Aerospace Exploration Agency (JAXA), Japan

Claire Trenham
CSIRO Environment, Australia

Maaike van Kooten
National Research Council Canada's Herzberg Institute of Astronomy and Astrophysics Research Centre (NRC-Herzberg), Canada

Kathryn Williamson
Kathryn Williamson Consulting LLC, USA

Ka Chun Yu
Denver Museum of Nature & Science, USA

Contributor biographies

Anna Cabré Albos

Anna Cabré Albos is a climate scientist, oceanographer, climate impact analyst, and policy advisor, with a background in physics, cosmology, and astronomy. She is associated with the University of Pennsylvania and has experience with global Earth system models as represented in the IPCC reports, especially on the analysis of ocean currents in the Southern Ocean and South Atlantic, oxygen levels in the ocean, biophysical modeling, and impacts on fisheries.

Affiliation: University of Pennsylvania

Paola Banchero

Paola Banchero is a professor of journalism and public communications at the University of Alaska Anchorage. Her research agenda includes media treatment of climate change. This year, she will guide undergraduate students as they interview survivors of climate change.

Department of Journalism and Public Communications
3211 Providence Dr., Anchorage, AK 99508 USA
Affiliation: University of Alaska Anchorage

Jeffrey Bennett

Jeffrey Bennett specializes in math and science education, writing for and speaking to audiences ranging from elementary age to college faculty. He has written two books on global warming: A Global Warming Primer and, for kids, The Wizard Who Saved the World. He is the founder of Big Kid Science and a recipient of numerous awards including the Science Communication Award from the American Institute of Physics and, most recently, the 2023 Klopsteg Memorial Lecture Award from the American Association of Physics Teachers.

680 Iris Ave., Boulder, CO 80304 USA
Affiliation: Big Kid Science; adjunct at University of Colorado, Boulder

Faustine Cantalloube

Faustine Cantalloube is an associate researcher at CNRS (France), working on ground-based instruments for imaging exoplanets and circumstellar disks. She characterises the performance of current instruments dedicated to exoplanet and disk imaging, develops specific post-processing methods and participates in designing future instruments.

Affiliation: Aix Marseille Univ, CNRS, CNES, LAM, Marseille, France

Ron Ekers

Ron Ekers is a radio astronomer and a CSIRO fellow (retired) based at CSIRO Space and Astronomy. During his career he has been director of radio observatories in the US and Australia and is a past president of the International Astronomical Union. He is an internationalist and has always been motivated by the use of technology innovations to pursue our science and facilitate our ability to collaborate.

CSIRO Space and Astronomy
PO Box 76, Epping NSW 1710, Australia
Affiliation: CSIRO/Curtin University

David J. Helfand

David J. Helfand, former chair of the Astronomy Department at Columbia University, has served on Columbia's faculty for nearly five decades. He was also president and vice chancellor of Quest University Canada. Helfand is the chair of the American Institute of Physics and past president of the American Astronomical Society. He is the author of *A Survival Guide to the Misinformation Age: Scientific Habits of Mind* (Columbia, 2016). and *The Universal Timekeepers: Reconstructing History Atom by Atom* (Columbia, 2023)

438 West 116th Street, Apt. 73, New York, NY 10027 USA
Affiliation: Columbia University

Aidan Hotan

Aidan Hotan is a senior system scientist working to support radio astronomy national facilities. He has worked on a wide range of science and instrumentation projects and has an interest in building effective collaborations across widely distributed teams.

CSIRO, 26 Dick Perry Ave, Kensington WA 6151, Australia
Affiliation: CSIRO Space and Astronomy

Emily Kerrison

Emily Kerrison is a PhD candidate at the University of Sydney where she is on the hunt for the youngest supermassive black holes. Through The Future of Meetings she has developed an interest in how more sustainable and inclusive meeting practices can improve opportunities for EMCRs.

Sydney Institute for Astronomy, Physics Building A28, Physics Road, The University of Sydney, Darlington 2006, NSW

Affiliation: University of Sydney/CSIRO Space and Astronomy

Jürgen Knödlseder

Jürgen Knödlseder is a staff scientist of the National Centre for Scientific Research (CNRS). He is working on instrumentation and observations in the field of gamma-ray astronomy. He is member of the astrophysics advisory board of the French Space Agency CNES, and he was chairing for nine years the Consortium Board of the Cherenkov Telescope Array (CTA). Besides doing research in the field of high-energy astrophysics he also works on sustainability issues related to astronomical research. He is member of Astronomers for Planet Earth, the French "Labos 1point5" initiative, and he is heading since 2022 the "Office for Environmental Footprint Reduction" of CTA.

Insitut de Recherche en Astrophysique et Planétologie (IRAP)
9, avenue Colonel Roche
31400 Toulouse, France

Affiliation: Insitut de Recherche en Astrophysique et Planétologie (IRAP)

Rika Kobayashi

Rika Kobayashi is a computational chemist at Australia's National Computational Infrastructure supercomputer facility in Canberra. Her specialization is in quantum chemical electronic structure theory—developing, implementing and applying novel methodology. As a sideline she has been exploring remote collaboration initiatives, especially for teaching.

Leonard Huxley Bldg 56, Australian National University, Mills Road, Acton, ACT, 2601, Australia

Affiliation: Australian National University

Rachel Mason

Rachel Mason is a broadly-trained scientist with particular interests in the role of animals in the food system. After several years as an astronomer at international observatories, she changed fields to work on problems occurring on our own planet. She has since published research into agriculture and climate change, meta-science, and alterations to the global nitrogen cycle, and now focuses on the complex consequences of industrial food animal production.

Affiliation: Independent

Vanessa Moss

Vanessa Moss is a Senior Experimental Scientist at CSIRO and oversees science operations for the ASKAP radio telescope. After chairing the original symposium in 2020, she has led "The Future of Meetings" community to advocate for improved online practices across astronomy and beyond, collaborating with like-minded groups and organizations worldwide.

CSIRO, PO Box 76, Epping NSW 1710, Australia

Sydney Institute for Astronomy, Physics Building A28, Physics Road, The University of Sydney, Darlington 2006, NSW

Affiliation: CSIRO Space and Astronomy/University of Sydney

Bob Raynolds

Bob Raynolds is a geologist based in Denver. He has taught at Peshawar University in Pakistan, Dartmouth College, the Colorado School of Mines and the Denver Museum of Nature & Science. Bob previously worked for the US Geological Survey, Exxon, and Amoco. Bob is currently a Research Associate at the Denver Museum of Nature & Science studying sediments in the Denver Basin that record the uplift of the Rocky Mountains. He is former president of the Friends of Dinosaur Ridge and of the Colorado Scientific Society. His recent lectures focus on the geological record and its role in helping to understand the impact of environmental changes on Colorado's ecology and water resources.

2001 Colorado Blvd.
Denver CO 80205
Affiliation: Denver Museum of Nature & Science

Glen Rees

Glen Rees is a Data Scientist doing machine learning and data analysis for clients in a wide range of industries. With broad expertise in technology and virtual reality applications for online interaction, he leads and coordinates many of the technical activities of TFOM.

Affiliation: TFOM

Bernadette Rodgers

Bernadette Rodgers is a full-time climate activist and part-time science educator. She currently teaches astronomy and earth science classes as adjunct faculty at the Willamette University Pacific Northwest College of Art, and at Portland Community College. She also serves as board chair of SustainUS, Inc, and is active with Climate Reality Portland, Scientist Rebellion, and Astronomers for Planet Earth.

5105 SW Richardson Dr, Portland, OR 97239

Affiliation: Willamette University and Portland Community College

Elizabeth Tasker

Elizabeth Tasker is an astrophysicist and science communicator at JAXA's Institute of Space and Astronautical Science. She is a member of the international outreach team for the Hayabusa2 asteroid exploration mission and the up-coming Martian Moons eXploration (MMX) mission, and is interested in using virtual space to explore environments that cannot be reached by humans.

Institute of Space and Astronautical Science, JAXA

3–1–1 Yoshinodai, Chuo-ku, Sagamihara City, Kanagawa Prefecture, 252–5210

Affiliation: Japan Aerospace Exploration Agency (JAXA)

Claire Trenham

Claire Trenham is a senior experimental scientist and data & digital lead in CSIRO's Climate Intelligence programme, and brings to "The Future of Meetings" a focus on using technology to improve inclusivity and reduce emissions associated with research.

Synergy Bldg 801, CSIRO BMSIP, Black Mountain, ACT, Australia, 2600.

Affiliation: CSIRO Environment, Canberra

Maaike van Kooten

Maaike van Kooten is an adaptive optics developer at the National Research Council Canada's Herzberg Institute of Astronomy and Astrophysics Research Centre. She works on adaptive optics research and development as well as helps build adaptive optics systems for telescopes around the world.

5071 W Saanich Rd, Victoria, BC V9E 2E7 Canada
Affiliation: National Research Council Canada's Herzberg Institute of Astronomy and Astrophysics Research Centre (NRC-Herzberg)

Kathryn Williamson

Kathryn Williamson is an independent Earth and Space Education and Outreach Consultant. She taught astronomy at West Virginia University from 2016–2022, where she implemented a variety of climate change educational efforts with students in the classroom, campus community members, K-12 teachers, and state legislators.

Affiliation: Kathryn Williamson Consulting LLC

Ka Chun Yu

Ka Chun Yu is an astronomer, and the Curator of Space Science at the Denver Museum of Nature & Science, which he joined as part of a team tasked to create planetarium software to visualize the known universe. He has helped produce movies for the digital dome; has created Earth educational programs for the planetarium; and has done research on how digital planetariums can be used to effectively teach astronomy. He is one of the founders of the Worldviews Network, a group using immersive visuals to place Earth within its cosmic context, and to connect public audiences with ecological, biodiversity, and climate issues. He participates in extensive education and public outreach including giving numerous talks to the public, and advising on science content in permanent and temporary museum exhibits.

2001 Colorado Blvd.
Denver, CO 80205
Affiliation: Denver Museum of Nature & Science

Part I

Introduction

Chapter 1

Why Astronomers?

T A Rector

> "It has been said that astronomy is a humbling and character-building experience. There is perhaps no better demonstration of the folly of human conceits than this distant image of our tiny world. To me, it underscores our responsibility to deal more kindly with one another, and to preserve and cherish the pale blue dot, the only home we've ever known."
>
> —Carl Sagan, *Pale Blue Dot*, 1994

We are living in a time where climate change is adversely affecting the habitability of the Earth. Humanity's response, particularly over the next decade, has critical consequences for what our future will hold. Fortunately astronomers are well positioned to make a difference. We offer a unique and important perspective that can help people understand the problem as well as the solutions. Our physics and astronomy classes, as well as our public outreach, are opportunities to teach climate change because they reach large numbers of people and cover related topics. But we need to recognize that climate change is different from other topics we teach. Communication about climate change has to connect with people's belief systems and cultural values. It has to help them overcome their fears as well as counter misinformation. Climate change is scary, but it is an opportunity to envision a better world for us all. And astronomers play an important role in creating that future.

1.1 My Story

Like many astronomers I started teaching climate change unintentionally. In 2004, during my first year at the University of Alaska, I taught our introductory course on the Solar System. During the planetary atmospheres lecture I explained how the surface of Venus was a raging hell because of its thick atmosphere of carbon dioxide. "Is that what global warming is going to do to the Earth?", asked a student. I assured her that the Earth wasn't at risk of turning into Venus. But I didn't have

doi:10.1088/2514-3433/acfcb6ch1

good answers for the follow up questions: "What *is* going to happen to the Earth?" "How bad will it be?" and "What can we do to stop it?" A few years later another student in one of my classes wrote an opinion piece for the campus newspaper about climate change. She said that learning about the geologic histories of the terrestrial planets opened her eyes to the threat we face. That was a formative moment for me, to realize the importance of teaching climate change in my astronomy and physics classes.

At first I was a little bit uncomfortable. I felt that I didn't know the science well enough, and I struggled to find resources that were useful. There was a large gap between materials written for the public—which were too basic—and the professional climate science journals—which used jargon I'd never heard before. David Helfand brought to my attention the excellent Princeton Primers on Climate series of books, which are useful for scientists in other fields. These and other resources helped me to develop a deeper understanding of climate science. But they also made it clear that I already had a strong enough foundation, from simple concepts like conservation of energy to more complex subjects like fluid dynamics and radiative transfer. Indeed, the physics of astronomy and of climate change greatly overlap.

What also became clear was that knowing the science wasn't enough. At first I was unconsciously assuming what is known as the "information deficit model"—the notion that apathy or skepticism about climate change was due to a lack of understanding about the problem. In other words, if we explain the science well enough people will be naturally motivated to take action. But knowledge isn't enough. Communication about climate change has to connect with people's belief systems and cultural values. It has to help them overcome their fears, and perhaps their guilt. In addition, it has to overcome misinformation—as there are individuals and organizations working to confuse and sow doubt about the causes and consequences of climate change. To be better prepared for "climate denial" talking points, I took the University of Queensland's EdX online course Making Sense of Climate Science Denial, which provided insights into the strategies deniers use and ways to counter them.

Eventually I also realized that we need to talk about solutions. Whenever I talked or taught about climate change the first question I'd get would invariably be along the lines of, "What can we do about it?" Like many scientists I was reluctant to advocate for particular solutions. But I've since found it possible to provide information and resources that enable people to see that there are a wealth of options that can minimize the consequences of climate change. And, in many other ways, they'll make our lives better too. Ultimately we want to build in people not just awareness and concern about climate change, but hope as well.

Living in Alaska over the last twenty years I've watched climate change unfold here in real time. Mostly due to the positive feedback loop ("vicious cycle") between ice melt and increased albedo, the Arctic is currently warming at about three times the rate as the rest of the world (Ballinger et al. 2021). I have pictures of myself standing in front of glaciers that—if you took the picture now from the same location—you can no longer see. Watching my beloved home literally melt has been heartbreaking. But Alaska is no longer unique in that way. Every place on Earth is now being noticeably affected. Just about everyone has experienced climate change in some way—whether it be through record heat waves, severe drought, or other extreme weather events.

1.2 Why Astronomers?

Why should astronomers teach climate change? Shouldn't we leave it up to the climate scientists and "stay in our lane"? There are multiple reasons why it is appropriate (and important!) that we teach climate change in our classes and talk about it at public speaking events, as well as with friends and family.

First and foremost, there is tremendous overlap between the physics of climate change and that of astronomy. We teach foundational concepts in our classes that are helpful for understanding the problem. This includes obvious topics like the nature of light and the greenhouse effect, but other subjects as well, such as the seasons and orbital eccentricity—which are part of Milankovitch cycles—and astronomical time-scales—which can lead to an appreciation of the difference between natural and anthropomorphic climate change. The geologic histories of Venus and Mars are powerful lessons that there is no guarantee that a planet like the Earth will be habitable, or remain so. And exoplanets give us a view into exotic worlds, some of which may lie within "habitable zones" but are clearly not habitable for us. More ways that astronomy connects to climate science are described in Chapter 8.

Astronomy also gives an important perspective. Known as the "overview effect," astronauts often talk about the viewpoint they gained when seeing the Earth from outer space. This effect was especially pronounced for the Apollo astronauts who went to the Moon, from where the Earth shrank to a small size against the blackness of space. Often described as steely eyed missile men, these astronauts described poetically the emotional impact of seeing the Earth from afar, and the awareness it created of its fragile beauty. Here are some examples of what they saw, and what they said (Figures 1.1–1.3)

While astronomers haven't experienced the overview effect in the same way, we too have that perspective. We also know that the Earth is special and unique. From looking at the terrestrial planets we know that our home could easily have turned out very differently. From our studies of exoplanets we know that the thousands of other worlds we've discovered are uninhabitable and—even if they were—they're too far away. We also know of the finiteness of the Earth, and the tremendous strain being put on its resources. Climate change is therefore ultimately an astronomical story. The phrase "There is no Planet B" has become a rallying cry for the climate change movement; and there's no better place to learn why than in an astronomy class.

With its spectacular images of planets, nebulae, and galaxies, astronomy also inspires a sense of awe and wonder. And people often find the immense size of the universe to be humbling. These experiences can help them to be more motivated to address climate change; e.g., Zhao et al. (2018) found that feelings of awe can reduce one's "social dominance orientation"—the belief that humans can dominate over nature—and enhance their willingness to engage in pro-environmental behavior.

Figure 1.1. "It suddenly struck me that that tiny pea, pretty and blue, was the Earth. I put up my thumb and shut one eye, and my thumb blotted out the planet Earth. I didn't feel like a giant. I felt very, very small."—Neil Armstrong (Apollo 11 Astronaut)[1]

Not only does astronomy have an important message about climate change, we have a wide reach. Fraknoi (2001) estimated that every year a quarter million students take an introductory astronomy class in the U.S. at the college level. That estimate has likely climbed to 300,000. Furthermore, for many students this is the last formal science course they'll take. From that point on what they learn about climate change will come from other sources, such as the news and social media. Informal education also provides opportunities to explain climate change. The Association of Science-Technology Centers estimates that there were 120 million visits to a science center worldwide in 2016,[2] and there were an estimated 150 million visits to a planetarium in 2019.[3] Museum attendees are also more receptive to learning about climate change (Stylinsky et al. 2017). Astronomy is popular among

[1] Photo: https://upload.wikimedia.org/wikipedia/commons/thumb/b/b1/AS11-44-6550.jpg/1024px-AS11-44-6550.jpg

[2] http://www.astc.org/wp-content/uploads/2017/09/ASTC_SCStats-2016.pdf

[3] https://www.lochnessproductions.com/reference/attendance/more_attend.html

Figure 1.2. "If somebody had said before the flight, "Are you going to get carried away looking at the Earth from the Moon?" I would have said, "No, no way." But yet when I first looked back at the Earth, standing on the Moon, I cried."—Alan Shepard (Apollo 14 Astronaut)[4]

most people and in general astronomers are well regarded. We can approach climate change from popular and less-controversial angles, such as exoplanets and astrobiology.

Finally, we are able to talk more freely about the science of climate change than climate scientists themselves. While overall scientists largely retain public trust (Funk 2017), some skeptics and deniers claim that climate scientists have an incentive to lie about or exaggerate the problem to attract funding for research. As astronomers we cannot be accused of such. Governmental agencies are major funders of climate research, and climate scientists are vulnerable to political pressure. Surveys in the U.S. (Donaghy et al. 2007) and Australia (Lewis 2020) reveal that many have felt pressure to downplay their results. Many climate scientists are reluctant to speak out because of the harassment some have endured

[4] Photo: https://www.lpi.usra.edu/resources/apollo/images/print/AS14/71/9845.jpg

Figure 1.3. "As we got further and further away, it diminished in size. Finally it shrank to the size of a marble, the most beautiful you can imagine. That beautiful, warm, living object looked so fragile, so delicate, that if you touched it with a finger it would crumble and fall apart. Seeing this has to change a man."—James Irwin (Apollo 15 Astronaut)[5]

(Valero 2023). We need to speak up on behalf of our climate science colleagues who cannot.

Like many scientists, astronomers have an aversion to "getting political." But it's important to keep in mind that climate scientists didn't politicize climate science. And epidemiologists didn't politicize COVID-19. These science-based issues were made political by other parties who have major financial interests at stake. Astronomers advocating for solutions to climate change is hardly new. For example, in 1985 Carl Sagan testified to the U.S. Congress about climate change. He explained the greenhouse effect—on Earth and on Venus—and then offered insights and recommendations on what to do about the problem.[6]

[5] Photo: https://nssdc.gsfc.nasa.gov/imgcat/html/object_page/a15_h_91_12343.html
[6] https://www.c-span.org/video/?125856-1/greenhouse-effect

1.3 Why the Urgency?

Especially in the era where we face problems such as the COVID-19 pandemic, social justice issues like diversity, equity, and inclusivity, as well as more astronomy-specific issues such as light pollution and radio-frequency interference from satellites, why should astronomers make climate change a priority? First, it's important to recognize that climate change isn't a stand-alone problem. It is often thought of as an environmental issue—and it is. But it is also a health issue, an economic issue, and a social justice issue. In many ways climate change can be thought of as a "threat multiplier"—it makes many of the other problems we face worse.

Second, time is of the essence. While we are already feeling the effects of climate change, the good news is that we can still avoid the worst consequences. But to do so we need to act quickly. The Paris Agreement sets as a goal to limit the global temperature rise to less than 2.0 °C above pre-industrial levels, and preferably less than 1.5 °C. The hope is to avoid triggering "tipping points"—defined as thresholds (temperature or otherwise) that, once crossed, lead to irreversible changes. And exceeding 1.5 °C of global warming could trigger multiple climate tipping points (Armstrong McKay et al. 2022). The IPCC AR6 Synthesis Report (IPCC 2023) makes it clear that to limit warming to 1.5 °C or 2 °C involves rapid and deep and, in most cases, immediate greenhouse gas emissions reductions *this decade*. Assuming carbon emissions continue at 2022 levels, the time remaining from the beginning of 2023 to limit global warming to 1.5 °C, 1.7 °C, or 2.0 °C are equivalent to 9, 18, and 30 years, respectively (Friedlingstein et al. 2022). In the equation for solving climate change, time is the most important variable.

Third, it's important to recognize that we aren't detached from the problem. Astronomy as a profession is being impacted by climate change—from changes in atmospheric conditions that affect data quality to forest fires that threaten our observatories. Astronomy is also contributing to climate change. We have one of the largest carbon footprints in the natural sciences (e.g., Blanchard et al. 2022). And while our profession is relatively small, on a per person basis our footprint is quite high—so we have as much obligation as anyone else to reduce our emissions. Let's be leaders and role models for what the future can bring.

1.4 About This Book

The book is written for astronomers, by astronomers. All of the chapters were penned by authors (or co-authors) who have backgrounds in astronomy and are now working in areas of climate science, education, and advocacy. We're hardly experts in all of the fields of study discussed in this book, so where possible we have brought in experts from other fields—e.g., climate scientists, engineers, and professional communicators—to help with the writing or to fact check. And the book has benefited greatly from their knowledge. While (hopefully!) this book will be useful for people who are not professional astronomers, it is assumed that you have some experience in astronomy and are comfortable with undergraduate-level physics (e.g., that you know what an MeV is).

What this book is: It is designed to help you be a better educator, communicator, and advocate regarding climate change and its solutions.

What this book is not: By and large it is not a deep dive into the details of climate science; e.g., how temperature anomalies are calculated. Other resources (e.g., the Princeton Primers) are already available. Nor does it include relevant astronomy topics such as astrobiology and exoplanets. The science content for these subjects can also be found elsewhere.

The book is divided into seven sections. The foreword and this chapter make the case for why astronomers should play a role in addressing climate change. Chapters 2 and 3 explain what we know about the Earth's past and how we are able to predict what our future may hold. Chapters 4–7 delve into why everyone should care about climate change, and what we can do about it. Chapters 8–11 give advice on how you can incorporate climate change into your formal (e.g., classroom) and informal (e.g., public talks and planetarium) education. Chapters 12 and 13 are intended to help you to be better prepared to have difficult conversations about what is a scary and highly emotional topic for many—including talking with those who are skeptical or in denial. Chapters 14–16 share examples and ideas of the role you can play to be a voice for change in your community and beyond. Finally, Chapters 17–19 explore how climate change is affecting astronomy, how astronomy is contributing to climate change, and what we can do to reduce our carbon footprint.

The subject of climate change is evolving rapidly—particularly in the context of solutions and public engagement. This book represents a snapshot of the topic in 2023. As described in Chapter 6, we have done our best to serve as "honest brokers" on the causes, consequences, and solutions of climate change. But in doing so we recognize our limits to understand all of these topics thoroughly; and we acknowledge that our perspectives are influenced by our incomplete knowledge and personal experiences. Even with the size of this book there is much that had to be left out. Hopefully this book (and the references therein) can provide a gateway into understanding each topic more deeply. That said, it is important to remember that you don't have to be an expert in all of these areas!

1.5 Your Story

As you read this book, reflect on your own journey of concern about climate change. If you are reading this, you are probably already pretty far down that path. When did you first learn about it? How have you been affected personally? Think about: When did you decide to devote time to understanding the problem more deeply? Have you made personal choices to reduce your carbon emissions? Do you talk about it with friends, family, and colleagues? Do you teach or give public talks about it? Are you advocating—or willing to advocate—for solutions? Every one of us is at a different point in this journey, and that's ok. Thinking about climate change is often overwhelming. It can feel that one person can't make a difference. But keep in mind that no gesture is too small. For example in Chapter 10 David Helfand talks about the impact an article by Wally Broecker had on him and how it guided his career. What impacts will you have on others?

Climate change is scary as hell, but it is an opportunity to envision a better world for us all. As Jeff Bennett discusses in Chapter 9, "What do we want the world to look like *after* we've solved climate change?" Hopefully this book helps you to play a role in creating that future.

References

Armstrong McKay, D. I., Staal, A., Abrams, J. F., et al. 2022, Sci, 377, eabn7950

Ballinger, T. J. 2021, Surface Air Temperature NOAA Technical Report OAR ARC, 21-02

Blanchard, M., Bouchet-Valat, M., Cartron, D., Greffion, J., & Gros, J. 2022, PLoSC, 1, e0000070

Donaghy, T., Freeman, J., Grifo, F., Kaufman, K., Maassarani, T., & Shultz, L. 2007, Atmosphere of Pressure: Political Interference in Federal Climate Science (Cambridge, MA: UCS Publications) https://www.ucsusa.org/resources/atmosphere-pressure

Fraknoi, A. 2001, AEdRv, 1, 121

Friedlingstein, P., et al. 2022, ESSD, 14, 4811

Funk, C. 2017, IST, https://issues.org/real-numbers-mixed-messages-about-public-trust-in-science/ (accessed 6.17.23)

Greenhouse Effect | C-SPAN.org https://www.c-span.org/video/?125856-1/greenhouse-effect (accessed 6.17.23)

IPCC 2023, Climate Change 2023: Synthesis Report. Contribution of Working Groups I, II and III to the Sixth Assessment Report of the Intergovernmental Panel on Climate Change, Core Writing Team ed. H. Lee, & J. Romero (Geneva, Switzerland: IPCC) 35

Lewis, D. 2020, Natur, 586, 19

More about Tallying The World's Planetarium Attendance, Loch Ness Productions. https://www.lochnessproductions.com/reference/attendance/more_attend.html (accessed 6.17.23)

Princeton Primers in Climate https://press.princeton.edu/series/princeton-primers-in-climate (accessed 6.17.23)

Sagan, C. 1994, Pale Blue Dot: A Vision of the Human Future in Space (1st ed.; New York: Random House)

Stylinski, C., Heimlich, J., Palmquist, S., Wasserman, D., & Youngs, R. 2017, ApEnEdC, 16, 234

Valero, M. V. 2023, Natur, 616, 421

Zhao, H., Zhang, H., Xu, Y., Lu, J., & He, W. 2018, FrPs, 9, 2367

Part II

Past and future

Chapter 2

Geologic Climate History of the Earth

Ka Chun Yu and Robert Raynolds

"We may observe in some of the abrupt grounds we meet with, sections of great masses of strata, where it is as easy to read the history of the sea, as it is to read the history of Man in the archives of any nation."

—Jean André Deluc, 1794

"We live in a strange world."

—Greta Thunberg, 2019

Just as Earth's surface has evolved over geological time, Earth's climate has also undergone radical changes. Secular changes in Earth's orbital parameters can result in modifications to solar insolation. Also playing important roles are the plate tectonics that alter the shape of the continents over millions of years, and the geophysical and geochemical interplay between the atmosphere, oceans, and land. Although the past amount of trace greenhouse gases in the atmosphere can be directly measured in air bubbles trapped in ice cores, a wide variety of proxies exist that can be used to indirectly infer temperatures much further in the past than we have an ice core record for. Although our knowledge of Earth's climate history becomes more uncertain the further back in time we investigate, studying this past record can help us understand the ways that Earth systems can be destabilized within short spans of time, and will let us better grasp the consequences of today's anthropogenic climate forcings.

2.1 Introduction

As we look back over the vast span of geologic time, it is clear that Earth's climate has undergone radical changes. Glacially polished rocks in the Sahara desert and in Manhattan's Central Park are indicators of drastic change over time. As in other observational sciences, the history of Earth's climate has been assembled by geoscientists examining the geological record. This record is like a rear view mirror with our recent past in greater focus than the distant past. Changes in climate are

doi:10.1088/2514-3433/acfcb6ch2

documented by increasingly higher resolution proxies, culminating in astronomical calculations, instrumental recordings, and modern observations.

Through the study of deep-time Earth history, we now understand that in the past billion years, our planet's climate has swung between two different climate states, one warm and one cool. The "hothouse" periods (pink-red in Figure 2.1) include some greenhouse warming conditions that are extreme enough for the polar regions to become ice-free and tropical. Earth's "icehouse" eras (light blue in Figure 2.1) make up a smaller proportion of geologic time, but they include periods when continent-wide ice sheets covered one or both poles (deep blue in Figure 2.1). The current icehouse period started with the buildup of Antarctic ice starting 34 million years ago (Ma), while ice cover became a permanent feature of the northern hemisphere starting 3 Ma.

Although the variations in climate and surface temperatures have generally been smooth on scales of millions of years, we have also discovered that the climate record has been punctuated by short deviations in the temperature record that occur on much shorter timescales. These include the Paleocene–Eocene Thermal Maximum at ~56 Ma (red bar in Figure 2.1; see Section 2.6), a period of elevated temperatures of as much as 8°C at the poles. This temperature anomaly was short, lasting for just $\sim 2 \times 10^5$ yrs, and was the result of greenhouse warming caused by staggering releases of carbon into the atmosphere and oceans. The existence of such short climatic aberrations, of which there are many that have been inferred from the geological record, shows that climatic change does not happen just gradually, but can occur rapidly.

The study of climate variations in the distant past has important consequences for us today. Such research can help address questions about how sensitive Earth's surface temperature is to abrupt increases in CO_2 levels. Researchers are also interested in understanding the impact of global warming on how rapidly ice sheets can decay, the subsequent effect on sea level rise, changes to precipitation patterns, and how acidified oceans and other modified environments impact ecosystems and

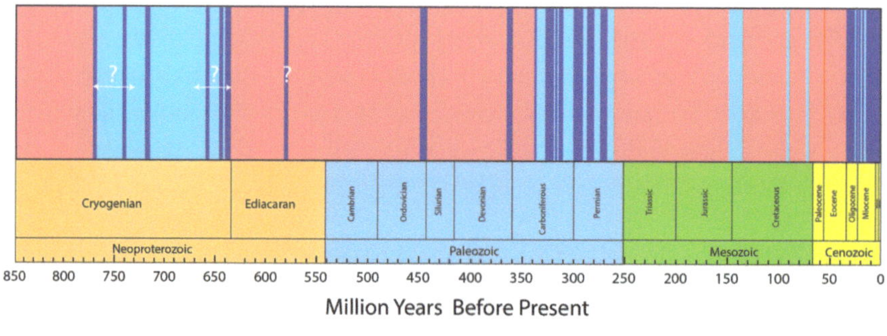

Figure 2.1. A timeline of Earth's "hothouse" (red-pink) and "icehouse" (light-blue) periods from the past 850 million years. Ice cover in the polar regions is denoted by the dark-blue bands, while the deep red shows punctuations in temperature such as the Paleocene–Eocene Thermal Maximum at ~56 Ma. The white question marks and arrows represent uncertainties about the extent of glaciation events in the Cryogenian. (After Figure 1.2 from National Research Council 2011).

the humans that live in them (National Research Council 2011). As anthropogenic global warming continues, it is important to learn from the past to find tipping points that can lead Earth's climate to move suddenly and permanently into new states. Such critical thresholds are not necessarily anticipated by current models that are tuned to records from the recent historical past.

In this chapter, we will first examine the various agents that continuously alter climate over time, driven by both external and internal factors (Zachos et al. 2001). The external forcing comes from minor variations in the parameters that describe Earth's orientation and orbit as a result of perturbations from the other planets, commonly referred to as the Milankovitch cycle. The internal drivers ultimately stem from plate tectonics, movements of Earth's crust driven by mantle convection that steadily rearrange the configuration of the continents. We will then review the history of Earth's climate, as reconstructed from a wide variety of lines of evidence. We will take a reverse "powers of ten" approach, starting with the largest timescales that also have the greatest uncertainties, and proceeding to subsequently smaller and more recent time periods, reviewing in each case the most important proxies for the temperature record. We will not be examining all of the different proxies or lines of evidence used to determine past climate, since they are too numerous to summarize even for this lengthy chapter. But for the ones we cover, we will review them in enough detail to show that our understanding of the climate record is robust, and follows different lines of evidence from multiple scientific disciplines.

2.2 Orbital Parameters

The orbital effects that affect climate are most familiar to astronomers. These processes affect insolation, the amount of solar radiation reaching Earth's surface (Figure 2.2). The three parameters first proposed by Milutin Milankovitch as astronomical contributions to climate are Earth's axial tilt, orbital eccentricity, and precession of its axis. The axial tilt ("obliquity") ε varies between 21.5° and 24.5° in a 41 thousand year (kyr) cycle. Changes in the eccentricity e of Earth's orbit happens on cycles at 400 kyr and 100 kyr. Precession of Earth's axis occurs with a 25.7 kyr period. A greater axial tilt results in greater summertime insolation at the poles, and less wintertime insolation at mid-latitudes. This leads to more extreme seasons at higher latitudes, with colder winters and hotter summers. The eccentricity varies between 0 and ~0.06, with the current value at e ~0.017 resulting in a 3% difference in aphelion and perihelion distance. At maximum eccentricity, the annual change in insolation is 24%.

The impact on Earth's climate from its orbital parameters do not occur in isolation, but the cycles can counteract or reinforce each other. Precession can amplify or mute seasonal extremes, by shifting the time of the year when Earth is closest to the Sun. Currently, perihelion passage occurs on or near January 3, meaning the longitude of perihelion, the angle between the vernal equinox and perihelion, is $\omega = 281°$. If the June solstice was to occur at perihelion passage, i.e., $\omega = 90°$, then the northern hemisphere will experience warmer summers by comparison, while at the same time, aphelion will occur when the southern

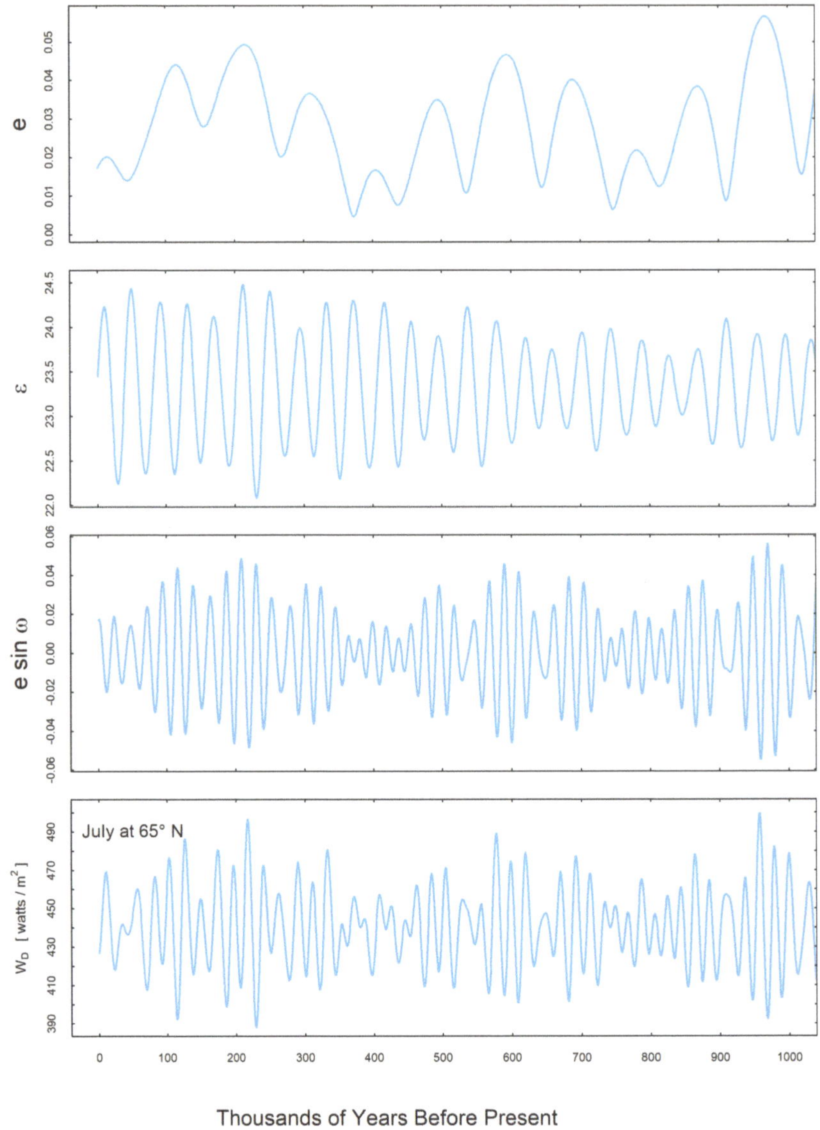

Figure 2.2. Variations in Earth's orbital eccentricity, obliquity, climate precession parameter, and daily insolation at 65° north latitude over the last million years. (Data from Berger & Loutre 1991.)

hemisphere is in its summer, leading to a cooler southern summer. Although precession of Earth's axis is on a 25.7 kyr cycle, there is also a "precession of the perihelion" that moves on a 112 kyr cycle. We can define a "climate precession" ϖ that is defined to be the angle between the vernal equinox and the perihelion. Since precession occurs on a 25.7 kyr cycle and the perihelion moves on a 112 kyr cycle, the climate precession is a combination of these two frequencies. Given that $1/25.7 + 1/112 \sim 1/21$, ϖ has a cycle of 21 kyr.

The orbital eccentricity also plays a role. If Earth's orbit is circular, $e = 0$, then there is no difference between the perihelion and aphelion distances. Conversely, as eccentricity increases, the effect of precession on climate becomes amplified. Thus, it makes sense to define a precession parameter $e \sin \tilde{\omega}$ that includes the effect of both the eccentricity and the climate precession.

If we define λ as the longitude of Earth in its orbit with respect to the vernal equinox (so that $\lambda = 180°$ is the location of the autumnal equinox), ϕ as the latitude of a location on Earth, and S as the solar constant, then the daily insolation can be defined as (Paillard 2010):

$$W_D = S\left(\frac{1 - e\cos(\lambda - \varpi)}{1 - e^2}\right)^2 \left(\frac{p \arccos\left(-\frac{p}{\sqrt{s + p^2}}\right) + \sqrt{s}}{\pi}\right),$$

where $s = \max(0; 1 - \sin^2\phi - \sin^2\varepsilon \sin^2\lambda),\ p = \sin\phi \sin\varepsilon \sin\lambda,$

and $W_D = 0$ if $s = p = 0$.

2.3 Surface Agents: Circulating Air, Water, and Plates

On time scales of hundreds to tens of million years, Earth's surface patterns of tectonics and atmospheric composition have modulated global water and atmospheric circulation as well as planetary surface temperatures (Berner 2004).

2.3.1 Circulating Air

Earth has special attributes that impact climate. First the planetary surface is mantled in fluids—air and water—that circulate according to temperature and density gradients caused by latitudinal differences in solar insolation. These result in *meridional flows* in the atmosphere that occur in the north–south direction, and *zonal flows* that run east–west. These flows are further modulated by the fictitious Coriolis force from planetary rotation on a parcel of gas or fluid, with the acceleration given $f \times v = 2\Omega \times \sin\phi \times v$, where f is the Coriolis parameter, v is the speed of the parcel in the Earth frame, Ω is the angular velocity of the Earth, and ϕ is the latitude. The accelerations in the zonal (east–west) and meridional (north–south) directions can be defined, respectively, as:

$$\frac{du}{dt} = f \times v,$$

$$\frac{dv}{dt} = f \times u.$$

Air rises due to solar heating along a low pressure band near the equator. The air is then pushed away towards mid-latitudes at high altitudes, creating a circulating meridional flow that forms the *Hadley cell*. This is the first of three meridional cells

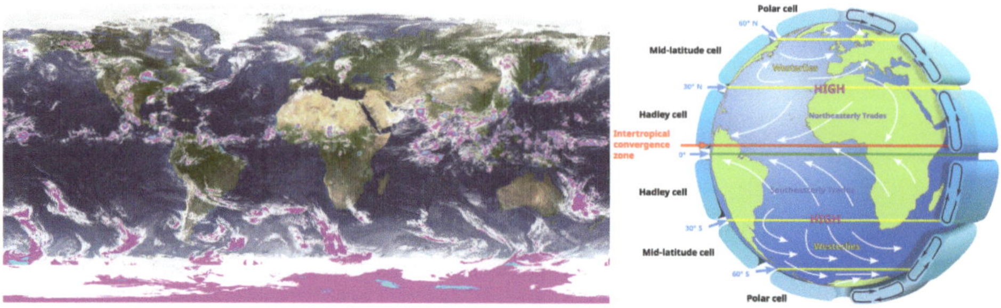

Figure 2.3. *Left* Cloud cover on Earth on 6 July 2022 from combined GOES, Meteosat, and MTSat infrared imaging. The colors represent elevation with white, pink, and teal representing low-, medium-, and high-altitude clouds. Not shown are thin clouds, the addition of which would result in ~70% or greater cloud cover over Earth on a typical day (Wylie et al. 2005, Warren, Eastman, & Hahn 2014). The Intertropical Convergence Zone (ITCZ) can be seen as a narrow band of clouds along the equator from the Pacific to Atlantic Oceans. Descending dry air from Hadley cell circulation partly explains the higher-albedo subtropical deserts with little cloud cover in the Tropics of Cancer and Capricorn between ±15°–30°. At mid-latitudes, storm tracks outline the meanders in the west-to-east flowing jet streams. (Credit: S. Albers, NOAA Aviation Weather Center.) *Right* Locations of the various meridional circulation cells in the atmosphere, along with arrows marking out the trade winds and westerlies. (Credit: NASA, Kaidor.)

(followed next by the Ferrel and Polar cells), each of which are divided by alternating high and low pressure zones. High evaporation rates from the oceans at the rising limb of the convective Hadley cell result in the Intertropical Convergence Zone (ITCZ), identified as a low-latitude band of storm clouds and heavy precipitation (Figure 2.3). At the sinking limb of the Hadley cells near 30°N and 30°S, dry air descends, creating a high pressure band with low precipitation. (By contrast, the Polar cell is driven by sinking cold air that creates a high pressure zone. The Ferrel cell is driven indirectly by the circulation of the Hadley and Polar cells.) This latitude corresponds to where the world's deserts are located. The temperature and pressure gradients at the boundaries between the Hadley–Ferrel cells and the Ferrel–Polar cells lead to the subtropical and polar west-to-east jet streams. Surface winds moving towards the equator as part of the Hadley cell circulation are deflected by the Coriolis force, leading to *trade winds* that blow toward the equator from the northeast in the Northern Hemisphere and from the southeast in the Southern Hemisphere. As part of the Ferrel cell circulation, air moving from the high pressure band towards the poles is also deflected by the Coriolis effect, leading to westerlies that blow to the east in the mid-latitudes at ~30–60°.

2.3.2 Circulating Water

Surface ocean circulation is driven by atmospheric winds, along with solar heating, evaporation, and precipitation. Friction from wind stress on the surface of the water plus Coriolis forces help drive the surface currents in the upper ocean. The varying topography of coastal regions further shape surface circulations. Water piling up along a coast leads to the formation of horizontal pressure gradients to push the water away. Like the air, water moves along high-to-low pressure gradients.

In addition to moving liquid water at a global rate of $\sim 10^7$ m^3 s^{-1}, ocean currents also transport heat, salt, and other dissolved matter. Meridional heat transport is on the order of 1 petawatt (10^{15} W = 1 PW) in each of the northern and southern hemispheres (Ganachaud & Wunsch 2000).

Large-scale surface ocean currents are dominated by the subtropical gyres between 20°–40° latitude, with the gyres circulating clockwise in the Northern hemisphere, and counter-clockwise in the Southern hemispheres (Figure 2.4). The western boundaries of gyres tend to consist of faster, narrower, and deeper currents (e.g., the Kuroshio in the Pacific Ocean and Gulf Stream in the Atlantic Ocean), while the eastern boundaries have slower, more diffuse currents. Gyre circulation helps move warm water from equatorial to higher latitudes. In the Southern hemisphere, the subtropical gyres are strongly influenced by the Antarctic Circumpolar Current. The ACC is driven by strong westerlies, which, unlike those in the north, pass over considerably less land mass which slows atmospheric winds via friction.

Water moving north from the tropics is also part of *meridional overturning circulation* (MOC). Along the western edge of the Atlantic Ocean, warm water moving to the northeast in the Gulf Stream gradually cools, while evaporation increases salinity. Both these effects make the water denser, leading it to sink and to contribute to the mass of cold water that makes up the North Atlantic Deep Water. The Antarctic Bottom Water (AABW) is an even colder and denser analog in the south, originating from the margins of the southern continent, including from the sinking of high salinity brine created as a result of salt rejection during sea-ice formation. Together, the effects of temperature and salt on water buoyancy collectively results in *thermohaline circulation*. Such temperature and salinity differences drive the Atlantic Meridional Overturning Circulation (AMOC), with the warm surface water cooling, becoming saltier, and sinking as it moves north. Up to

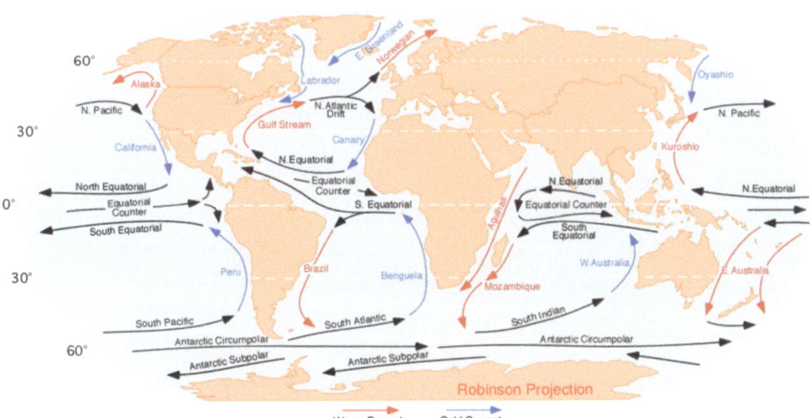

Figure 2.4. Schematic showing surface ocean currents. The currents are strongly variable over different timescales (e.g., seasonal, decadal, El Niño Southern Oscillation), so the arrows represent broad averages as opposed to persistent flows. (Image credit: https://commons.wikimedia.org/wiki/File:Corrientes-oceanicas.png, public domain.)

1.3 PW of heat is transported north, with its release keeping the climate of northwestern Europe relatively mild compared with the eastern coast of North America at comparable latitudes (Bryden & Imawaki 2001) The water forms a complete circuit after flowing south deep in the ocean. Salinity can decrease when there is an increased flux of freshwater into the ocean, which occurs when there is greater precipitation, river flow, or melting ice. As seawater migrates north, surface evaporation can also increase salinity.

2.3.3 Circulating Plates

The location and size of land masses affect the amount of heating of Earth's surface from the Sun, as well as how air and water are transported. Atmospheric winds and ocean currents are driven by the temperature gradients, with the equatorial regions receiving about three times the solar energy as the polar regions. The actual amount absorbed depends on the albedos of the various surface types. The averaged albedo of Earth is 0.3, meaning the ground and oceans absorb ~65% of incoming solar radiation. Because of the low albedo (typically ~0.07) and high specific heat of water (about 5× larger than soil and rock), and the well-mixing of the upper ~10–150 m of the oceans, sea surface temperatures (SSTs) are slower to respond to seasonal changes in solar insolation than land surface temperatures. For the same reasons, the seas, unlike the land, experience small swings in diurnal temperature.

Large temperature contrasts between the land and water lead to *monsoonal circulations*. In the summer, rapid heating of the land results in warming air rising, leading to a low pressure area at the surface. Air flowing towards the low pressure zone from the comparatively higher-pressure ocean carries with it water vapor, which condenses as monsoonal rainfall. In the winter, an opposite pattern occurs. The air over land cools faster, and descends, leading to higher pressures over the land. The oceans cool more slowly, so have comparatively lower pressures. The wind patterns are reversed, leading to cooler, dry air moving from land to sea. On top of this basic picture, the size and location of the continents are additional modifying factors of the seasonal flow of air and humidity. Northern hemisphere land masses are larger and higher in elevation (e.g., the Himalayan Mountains and Tibetan plateau) than those in the southern hemisphere. This leads to stronger summer monsoons in the north than in the south.

The shapes and positions of Earth's dynamic surface are not fixed in time, but are constantly being modified by drifting plates, erupting volcanoes, and active rock weathering in uplifted mountain belts and plateaus. These phenomena blur the past. Before 200 Ma, our understanding of plate geometries and thus the very shape and distribution of ocean basins and land masses is increasingly hypothetical. This is because the records of plate motion are recorded by sea floor magnetic anomalies and most ocean crust (and hence the records) older than 200 million years (Myr), have been destroyed by the process of subduction (see Cox [1973] and Oreskes & LeGrand [2001] for historical reviews).

At scales of tens to hundreds of million years, the ocean and landmass configurations shift over time, varying widely, and influencing the pattern and magnitude of ocean currents. When the continents merge to form one supercontinent, which happened most recently with Pangaea, 335–200 Ma, extreme temperature and aridity gradients between the interior and margins of the continent result in summer "mega-monsoons" (Kutzbach & Gallimore 1989). Development of mountains and high plateaus, such as when India collided with Asia 40–50 Ma, creating the Himalayan range and Tibetan plateau, also influence the large-scale patterns of precipitation. Solar heating of the Tibetan plateau pulls in moisture-laden air from the south, driving the South Asian monsoons to make it the strongest in the world. Mountain ranges also have an *orographic effect*: air flowing over them is forced to rise and cool, leading to condensation of the water vapor and precipitation, while air heats adiabatically as it descends down the other side, increasing the amount of water vapor it can carry before condensation occurs. This results in the establishment of regional rain catchments on one side of the mountains, and rain shadows on the other.

2.4 Internal Release of CO_2

As part of the carbon cycle, tectonic processes transport CO_2 from Earth's interior into the atmosphere by volcanic outgassing. In places where one plate subducts beneath a second plate, crustal material melts as it descends into the upper mantle. When the rising magma surfaces, it can form a belt of volcanic mountains (such as those of the Cascades in the Pacific Northwest of North America), which release CO_2 trapped in the rock when they erupt. In divergent margins on the seafloor, new crust is formed as plates spread apart, with the upwelling hot magma releasing CO_2 directly into the oceans. In this case, the older crust is pushed laterally away by new crust rising along the axis of the seafloor ridge. As the new crust cools, embedded magnetic minerals record the direction of Earth's magnetic field at the time of cooling. The seafloor therefore contains stripes of opposing magnetic polarity mirrored symmetrically on either side of the oceanic ridges. These magnetic lineations document the flipping of Earth's magnetic poles over the last 175 Myr (the period during which there is still existing oceanic crust that has not been subducted), with intervals of reversals happening from a few thousand years to several million years in length.

The rate of seafloor spreading can vary over time, as well as differ from one seafloor ridge to another. For instance, the East Pacific Rise spreads at $\gtrsim$70 mm yr^{-1}, nearly an order of magnitude faster than the Central Atlantic (Cande & Kent 1992). Faster spreading creates more new crust, but also results in faster subduction of older crust. When this older crust melts as it descends into the upper mantle, the CO_2 locked in the rock is released to be available for outgassing in volcanoes. In the last 20 My, the rate of seafloor spreading has dropped 30% to its current rate of $\sim 2.6 \times 10^6$ km^2 Myr^{-1}. This rate is correlated with the amount of CO_2 degassed from the formation of new crust, dropping from an estimated ~1000 ppm at 15 Ma to the pre-industrial level of 280 ppm. This variability of crust creation and CO_2 production directly accounts for the changes in global temperatures and the subsequent formation of ice sheets in the Northern Hemisphere in the last million years (Herbert et al. 2022; see Section 2.7).

Following this introduction to the basic mechanisms affecting climate on Earth, the remainder of this paper examines the record of climate change. We will follow time in scales ranging from billions of years through successively shorter time frames, with discussions of the variety of drivers impacting Earth's changing climate system at each time scale.

2.5 Four Billion Years of Change

The Earth's surface has had a remarkably stable climate for the last several billion years. It has grown neither too hot nor too cold to accommodate liquid water and life, despite the fact that stellar evolution models predict that solar energy output has increased through time, with the current luminosity ~48% greater than its initial zero-age main sequence value of 0.677 $L_\odot$ (Bahcall, Pinsonneault, & Basu 2001). The bolometric solar luminosity over time $L(t)$ can be approximated (Figure 2.5) by:

$$\frac{L(t)}{L_\odot} = \frac{1}{1 + \frac{2}{5}\left(1 - \frac{t}{t_\odot}\right)},$$

where $L_\odot = 3.85 \times 10^{26}$ W is the present-day luminosity, and $t_\odot = 4.57 \times 10^9$ yrs is the present-day age of the Sun (Gough 1981).

While early models for greenhouse gas content and composition are speculative, it is clear that the Earth has had liquid water on its surface for several billion years. Uranium-rich zircon minerals have been discovered from 4.2–4.4 billion years ago (Ga), with oxygen isotopic levels that require them to have been exposed to liquid water (Wilde et al. 2001; Mojzsis, Harrison, & Pidgeon 2001), suggesting water was present even before the end of the "Late Heavy Bombardment" period at 3.8 Ga. Sedimentary rock laid down between 3.5–3.8 Ga also provides evidence for liquid

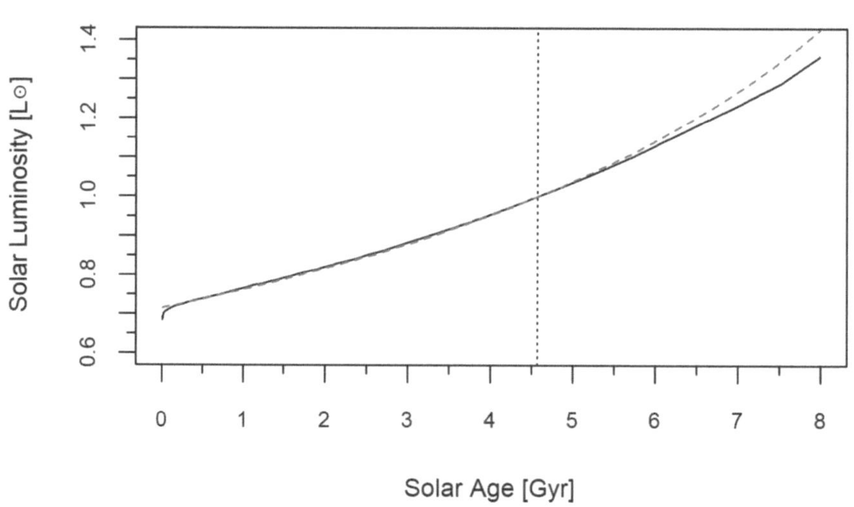

Figure 2.5. The Sun's luminosity over time as calculated by Bahcall, Pinsonneault, & Basu (2001) is shown as a black solid line, and Gough's (1981) analytic approximation in a blue dashed line. The current age of the Sun is marked by the vertical dotted line.

water (Fowler, Ebinger, & Hawkesworth 2002; Eriksson et al. 2004; Benn et al. 2006), including pillow basalt formed by lava extruded on the seafloor, and sedimentary rock with mud cracks or ripples from waves (Ramsay 1963). Sedimentary rocks, dating to 3.7 Ga and reflecting deposition by water, contain fossilized stromatolites that are created by layers of mat-building microorganisms living in sea water (Nutman et al. 2016).

The Sun was fainter by 25% at 3.8 Ga. Assuming a gray infrared opacity $\tau_{\rm IR}$, the mean surface temperature T_s of the atmosphere can be approximated by:

$$T_s^4 = \frac{1}{\varepsilon\,\sigma}\,\frac{r^2}{4}\,\frac{L_\odot}{4\pi d^2}\,(1 - A)\left(1 + \frac{3}{4}\,\tau_{IR}\right),$$

where ε is Earth's effective surface emissivity, σ is the Stefan-Boltzmann constant, r is Earth's radius, $L_\odot$ is the solar luminosity, A is Earth's albedo, and d is the mean Earth–Sun distance. Sagan & Mullen (1972) found that if the Earth had the same atmospheric composition and albedo as today, the global temperature should be below freezing at 2.3 Ga. They and others proposed increased greenhouse gases as the solution to this "Faint Young Sun Paradox." Early work, based on one-dimensional radiative convective models, suggested that the amount of CO_2 needed to keep the oceans liquid—as much as 300 mbar to maintain a surface temperature of 15°C (Kasting, Pollack, & Crisp 1984)—was inconsistent with the existence of minerals from this period that were precipitated out from ancient soils that require atmospheric CO_2 levels only one-tenth as high (Sheldon 2006). However, newer 3D atmospheric general circulation models (GCMs) have been developed that incorporate more realistic cloud feedback, lower albedo from smaller land fraction, shorter days, and other mechanisms. There are multiple scenarios in which the early Earth can remain temperate with CO_2 and methane levels that do not conflict with theoretical and observational constraints (Charnay et al. 2020).

2.6 Four Hundred Million Years of Change

The geological record suggests ancient periods of extensive glaciation that some have interpreted to be so widespread as to have covered the entire planet (Hoffman et al. 1998, Hoffman 2019). In this "Snowball Earth" model, polar continental plates may have hosted ice masses, increasing earth's albedo and creating a cooling feedback that caused the Earth to be shrouded in ice. While some have suggested this high albedo earth would have resulted in perpetual ice cover, others have speculated that the cooling would have been eventually broken by the accumulation of volcanogenic greenhouse gasses, resulting in greenhouse warming, and melting of the ice (Walker et al. 1981).

Over the past 400 Myr we can reconstruct patterns of plate motions well enough to see the influence of glaciation on land masses that lingered at the south pole. The Gondwanan land mass (the amalgamated southern continents) was glaciated in the time span from about 250–325 Ma as deduced from glacial deposits preserved today in Africa, South America, Australia, and the Indian subcontinent (Figure 2.6).

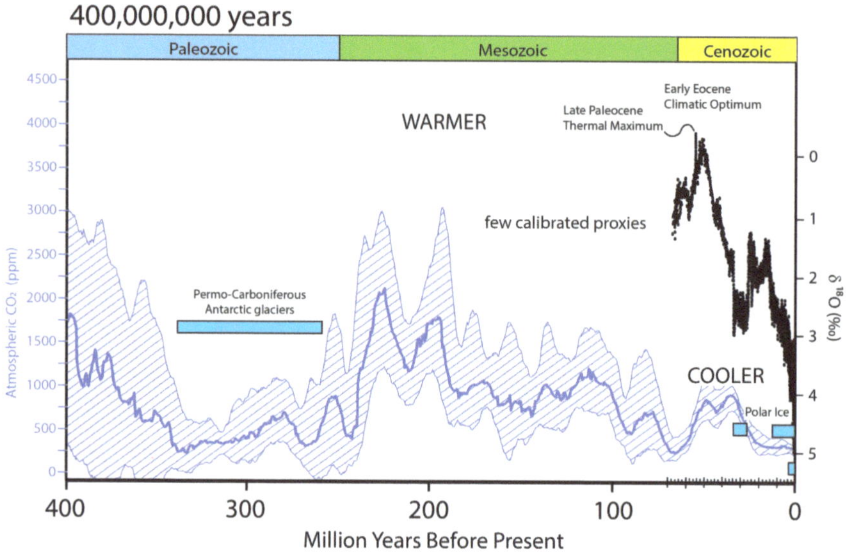

Figure 2.6. Last 400 Myr of climactic history. A fit of ~1500 different estimates of atmospheric CO_2 data from Foster, Royer, & Lunt (2017) using five independent techniques (size of leaf stomata, soil carbonates, boron isotopes in foraminifera, δ^{13}C in liverworts, and δ^{13}C from alkenones produced by phytoplankton) is shown as the dark blue line, with the cross-hatching representing 95% confidence intervals. The Cenozoic δ^{18}O record is from Hansen et al. (2008).

The record of atmospheric composition in time older than 100 Myr is deduced from evidence of ancient chemical reactions and from stable isotopes found in ancient sedimentary rocks. Early marine life photosynthesized and created oxygen that built up in our atmosphere to the point where it oxidized iron dissolved in seawater, creating global sedimentary units called banded iron formations. These are used as sources of iron ore and reflect a transition to an oxygenated atmosphere about 2.4 Ga (Ligrone 2019).

Understanding climate far in the past has been made possible primarily via paleoceanographic studies of fossilized single-celled ocean-dwelling microorganisms, and the minute changes in isotopic ratios of oxygen (^{18}O vs. ^{16}O) and carbon (^{13}C to ^{12}C) in their carbonate shells. When microscopic organisms die, they fall and settle onto the seafloor. Over time, the seafloor builds up a sedimentary record of past life, with rates of deposition occurring on the order of millimeters per millennium. Deep sea cores will contain a record of these biological populations over time, along with any geochemical signatures associated with changing environmental conditions they live in. The notation commonly used is defined as a fraction per mille (or per thousand):

$$\delta^{18}\,\mathrm{O} = \frac{\left({}^{18}\mathrm{O}/{}^{16}\mathrm{O}\right)_{\mathrm{sample}} - \left({}^{18}\mathrm{O}/{}^{16}\mathrm{O}\right)_{\mathrm{standard}}}{\left({}^{18}\mathrm{O}/{}^{16}\mathrm{O}\right)_{\mathrm{standard}}} \times 1000$$

For the isotopic concentration in current ocean water without any dissolved salts, $^{18}O/^{16}O = 2005.20$ ppm (Baertschi 1976). A $\delta^{18}O = 1.0‰$ means the sample has an $^{18}O/^{16}O$ ratio that is 0.1% greater than the standard.

Foraminifera are a class of fossils that is most commonly used to study ancient climates (Figure 2.7). These single-cell organisms, living in the top 200 m of the water column (planktic) or on the seafloor (benthic), build internal mineral shells, using HCO_3^{-1} ions dissolved in seawater to form hard calcium carbonate ($CaCO_3$).

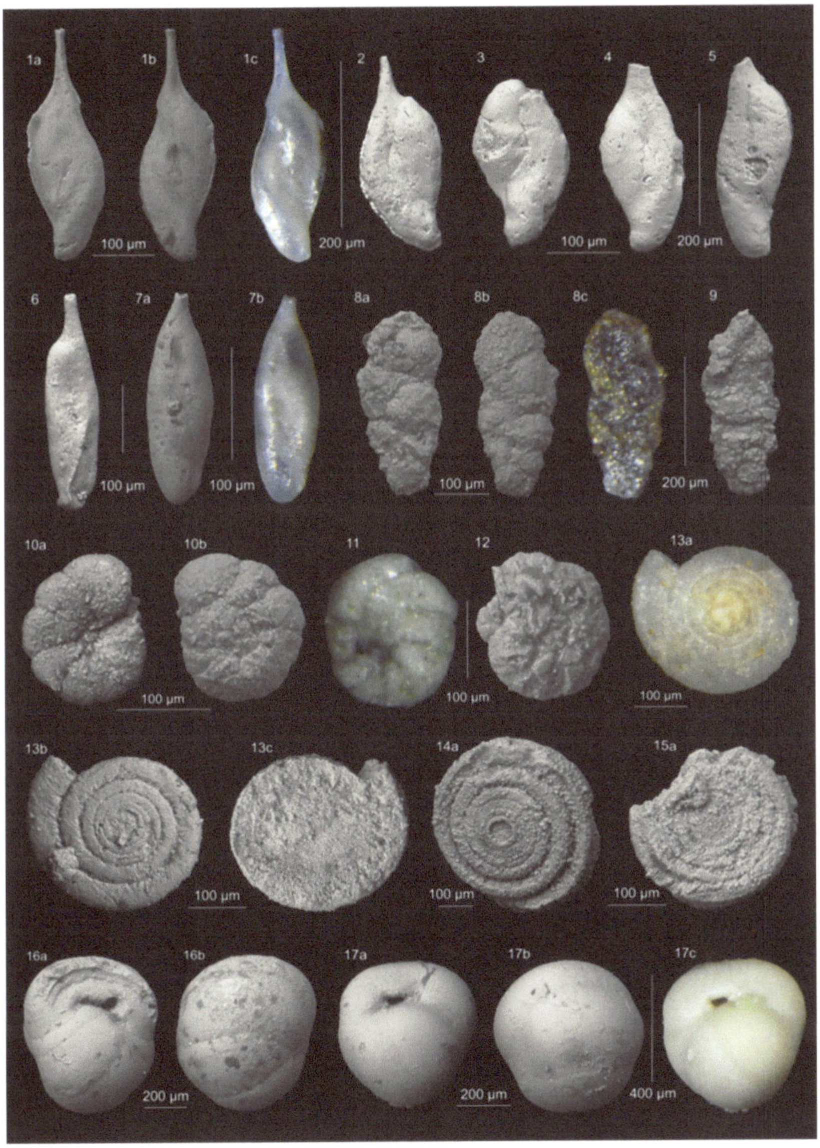

Figure 2.7. Examples of Mesozoic benthic foraminifera from the British Petroleum micropaleontology collection. (Image credit: Fox et al. 2018. Creative Commons Attribution-Share Alike 4.0 International license.)

The oxygen "in" HCO_3^{-1} comes from seawater, so the calcium carbonate precipitated by marine organisms contains a record of the ^{18}O fraction at a rate that is temperature-controlled and species-specific (De Vleeschouwer et al. 2017), with the concentration decreasing as the temperature increases. Hence a temperature estimate can be derived from isotopic measurements of the fossil shells (Marchitto et al. 2014). The isotopic fractionation dependence of temperature has been determined empirically from living forams to be of the form (Ravelo & Hillaire-Marcel 2007):

$$T = 16.9 - 4.38(\delta_c - \delta_w + 0.00027) + 0.10(\delta_c - \delta_w + 0.00027)^2,$$

where T is the water temperature during deposition, δ_c is the per mille difference between the sample carbonate with the standard, and δ_w is the per mille difference of ^{18}O of water with the standard. Although the modern value of δ_w can be directly measured, values for ancient samples need to be adjusted because of other confounding factors. For instance, during an ice age, ice sheets are built up preferentially from lighter water since water with ^{18}O evaporates less readily than water with ^{16}O, leading to an increase in the $^{18}O/^{16}O$ ratio in the oceans. Foram shells growing in warm water with high evaporation will be enriched with heavier ^{18}O. When water vapor precipitates as rain or snow, $H_2{}^{18}O$ tends to condense first. As water vapor is transported from the equator towards the poles, it becomes more and more depleted of ^{18}O as the lighter isotope precipitates out. The snow that built up the North American ice sheets were highly depleted of heavier water. Today, Greenland ice sheets have $\boldsymbol{\delta}^{18}O$ values of around −30‰, while the ratio in Antarctica is −55‰. When a glacial period ends, meltwater can decrease the ratio of heavier water in the oceans. Local changes can also occur as a result of greater evaporation (increasing $\boldsymbol{\delta}^{18}O$) or greater precipitation (decreasing $\boldsymbol{\delta}^{18}O$). And although most calcium carbonate is formed from carbonate precipitated out of the water, carbon may also originate from the organism itself. This metabolic (or "vital") source of carbon dioxide would not have the same $\boldsymbol{\delta}^{18}O$ as carbon precipitated as a result of thermodynamic isotopic equilibrium from seawater.

Based on the evidence from the coring of ocean sediments and in ice cores worldwide, the consensus is that the main ^{18}O signal correlates primarily with changes in ice volume. Other complicating factors, including changes in sea surface temperature, salinity, the habitats of foraminifera that change by depth and season, and dissolution of calcium carbonate at the seafloor due to undersaturation of the surrounding water, are secondary effects. Hence, the global isotopic signature found in layers of fossilized forams allow a series of glacial and interglacial periods to be established, with absolute ages set from radiocarbon, uranium (U-series), and other dating techniques. This standard paradigm has recently been reviewed and updated by Meckler et al. (2022), who point out the weaknesses of the accepted temperature curves, noting that variable groundwater storage of meteoric water (originating in the atmosphere) through time and variable interactions between seawater and oceanic crust—due for example to variations in rates of seafloor spreading (Herbert et al. 2022)—can be a significant modulator. But whether the proxy is that of $\boldsymbol{\delta}^{18}O$ measured from benthic foraminifera (Figure 2.7; Miller et al. 1987) or using the newer temperature-dependent clumping of rare isotopes (Meckler et al. 2022), the basic trend in the last 70 Myr is that of gradual cooling.

In addition to $\delta^{18}O$, five other proxies have been used for estimating CO_2 in the last half billion years of Earth history (Royer et al. 2004; Foster, Royer, & Lunt 2017): $\delta^{13}C$ fractionation in phytoplankton, which depends on the amount of dissolved CO_2 in water; carbon locked up in prehistoric soils and minerals as a result of carbon dioxide exchange with the atmosphere; the density of stomata (the microscopic pores where transpiration takes place) in fossil leaves that correlates with atmospheric CO_2 concentrations based on empirical relationships found in modern plants; a boron fractionation ratio that is dependent on the pH of the seawater, which can be used to infer the amount of dissolved CO_2; and a $\delta^{13}C$ signal in fossilized liverworts that is controlled by the atmospheric CO_2 concentration. Different magnitudes of errors are associated with the various proxies and at different geological epochs. Yet, combined together, they can still give a rough record of the evolution of the atmosphere. At the start of the Phanerozoic eon (543 Ma), the concentration of CO_2 was as high as 4000 ppm, and then decreased to modern day values by the Pennsylvanian period (~300 Ma). During this time, the most extensive glaciation of the past half billion years occurred when Antarctica and Australia drifted over the South Pole (Crowell 1999). This was followed by a CO_2 increase during the Mesozoic to 1000–3000 ppm; and then finally a decline to the present day.

At 56 Ma, an abrupt warming episode occurred that lasted 170,000 years. During this Paleocene–Eocene Thermal Maximum (PETM), sea surface temperatures increased by ~5°C. This drastic change was identified via a global negative carbon isotope excursion (CIE) in inorganic and organic sediments, caused by the release of an estimated ~1–1.5 × 10^4 gigatons of ^{12}C-enriched carbon into the atmosphere in as little as <500 yr (Zachos et al. 2007). One hypothesis suggested that methane clathrates, methane trapped in icy deep-sea mudrocks, were destabilized by hydrothermal vents on the ocean floor. However, fossilized forams do not show the $\delta^{13}C$ deficit expected from this source (Haynes & Honisch 2020). Volcanism can produce outgassing to match the isotopic signature, but normal volcanism cannot produce enough CO_2 to cause the observed warming. Coincident with the PETM was a much larger source of volcanic activity and carbon: the North Atlantic Igneous Province (NAIP), an enormous volcanic field covering Greenland and parts of the United Kingdom, which produced 1.8 × 10^6 km^3 of basalt (Gutjhar et al. 2017). The NAIP was associated with the seafloor spreading that opened up the Atlantic Ocean. As the lithosphere stretched, carbon-rich rock was brought up to the surface, where it released carbon as it melted (Gernon et al. 2022). Volcanism did not last through the onset of the PETM, but the sudden warming may have triggered other carbon reservoirs, such as methane from hydrothermal vents (Svensen et al. 2004), or carbon in permafrost or methane clathrates on the seabed (Kender et al. 2021).

A consequence of a large injection of CO_2 into the Earth system is greater acidification of the oceans, or a reduction in water pH. An observed decrease in $\delta^{11}B$ (comparing ^{11}B and ^{10}B) is associated with decreasing pH and reduced $CaCO_3$ content in fossilized shells (Muller et al. 2020), and correlated with an increase in anomalous CO_2 concentrations in the atmosphere during the PETM (Gutjahr et al. 2017).

2.7 Forty Million Years of Change

Over the past 40 Myr (an interval spanning the Middle Eocene to Present), we have robust seawater temperature proxy records derived from deep sea drilling from all over the world. Isotopic records from these cores can be compiled (stacked) to yield a high resolution curve showing past climate. The pattern of climate variability over this period reflects a gradual cooling from warm early Tertiary times to the "ice house" conditions of today (Figure 2.8). Superimposed on this is a relatively abrupt cooling episode, first near the Eocene-Oligocene boundary at 33.5 Ma, when a southern ice cap first developed (Liebrand et al. 2017). The broader cooling trend was interrupted between 27 and 15 Ma with warming that caused the ice sheets to retreat, although this was also punctuated with Antarctic ice build-up in the mid-Miocene at ~24 Ma. Although there were modest changes in the overall plate configuration, significant geometric changes in oceanic gateways affected ocean circulation. This time span saw the culmination of the collision of the Indian subcontinent with Asia, the formation of the Isthmus of Panama, and the opening of the Drake Passage at the Scotia Arc between Antarctica and South America. This latter event cleared the way for a circumpolar ocean circulation that isolated the landmass of Antarctica, promoting cooling and eventual extensive ice cap formation. Although there have been attempts to explain the gradual cooling to a single factor such as the widening ocean passageways around Antarctica, they do not explain the warming at the end of the Oligocene. Such large scale climatic shifts require additional factors such as decreasing atmospheric CO_2 concentrations, and tectonic activity in the form of mountain building and ocean gateways that modified

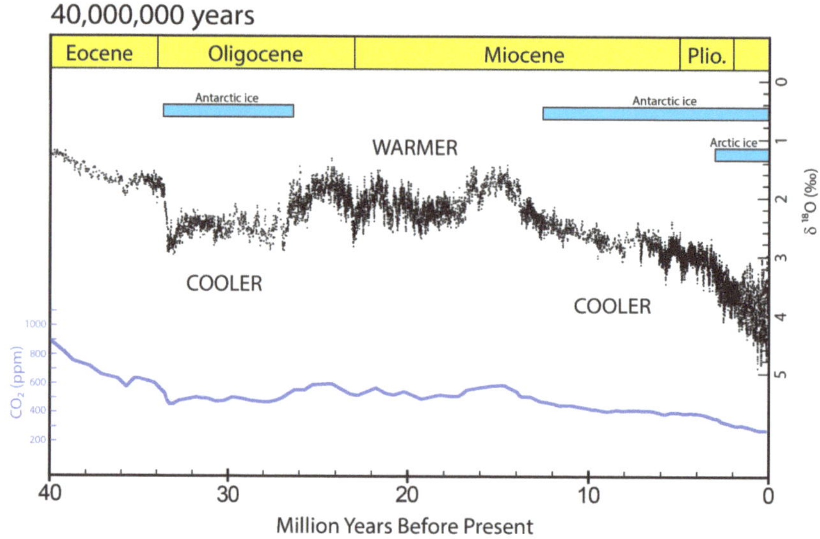

Figure 2.8. Last 40 Myr of climactic history with $\delta^{18}O$ compiled from 40 deep sea cores (Zachos et al. 2001) and a model of CO_2 concentration in the atmosphere calibrated by ice cores (Hansen et al. 2008).

atmospheric and oceanic circulations, and altered the flux of heat and moisture (Zachos et al. 2001).

Before 20–25 Ma, South America and Antarctica were connected. Kennett (1977) argued that this continental arrangement allowed warm water from low latitudes to be redirected south, preventing the glaciation of Antarctica. He speculated that after Drake's Passage opened up (preceded by the Tasmanian Gateway opening between Australia and Antarctica), a west-to-east circumpolar flow around Antarctica could now block warmer water from reaching the southern continent. However, subsequent simulations show that geography was not sufficient by itself to initiate cooling and the buildup of ice in Antarctica, but that atmospheric CO_2 concentration had to drop as well (DeConto & Pollard 2003; Pearson, Foster, & Wade 2009). Further support comes from the correlation of CO_2 and various temperature proxies (Doria et al. 2011; Pagani et al. 2011; Maxbauer, Royber, & LePage 2014).

Changes in topographic relief and the exposure of mountains of bedrock with associated weathering have been evoked as a way to diminish the greenhouse effect, with chemical reactions associated with rock weathering pulling CO_2 out of the atmosphere and sequestering it in secondary minerals. In hydrolysis-based weathering, CO_2 dissolved in rainwater makes the water weakly acidic with carbonic acid (H_2CO_3), which attacks granite and other silicate rocks (containing, e.g., $CaAl_2Si_2O_8$) in continental crust. Dissolved ions flow into the oceans, where they are incorporated into the shells of plankton as limestone ($CaCO_3$). When plankton die, they sink to the seafloor so that the weathered carbon is ultimately locked up in ocean sediments. The burial of these mineral byproducts in deep sedimentary basins depletes the atmospheric carbon reservoir, cooling the Earth. Over the course of tens of millions of years, this sediment is carried along the seafloor by plate tectonics, until it gets subducted into the Earth's interior. The carbon is released when the subducted material melts, and the CO_2 is returned back to the atmosphere through volcanic outgassing, completing the carbon cycle.

It has been suggested that the collision of India with Asia and the development of the Tibetan plateau, coupled with the regional rise of western North America over the past 50 Myr, may have exposed a steady supply of new rock for weathering, and abetted the long range cooling patterns we observe. In this uplift hypothesis, the rate of chemical weathering correlates with the amount of mineral surfaces exposed to the air. In mountains and plateaus that are formed as the result of crustal uplift, precipitation erodes the exposed rock; steep slopes result in rock falls; and glaciers pulverize the underlying ground to create fields of moraine debris. These active processes expose fresh, unweathered bedrock, where the steepest 10% of all terrain is responsible for roughly 40% of the chemical weathering (Larsen, Montgomery & Greenberg 2014). Increased rates of uplift have been recorded in the Himalayan range, the Tibetan plateau, and the Andes mountains over the last 5 My, which is also consistent with the greater transfer of sediment by rivers into the oceans (Raymo, Ruddiman, & Froelich 1988).

Newly refined models of the rates of seafloor crust production over the past 20 Myr (Herbert et al. 2022) suggest that a gradual though stepped diminution of seafloor spreading rates may be related to a diminishment of tectonic degassing at mid-ocean ridges and at subduction zones. This resulted in a decrease of CO_2 flux to the atmosphere and thus a reduction in the greenhouse effect. Together the effects of rates of plate tectonic processes and global weathering patterns are likely to explain the global cooling witnessed over the past 40 Myr.

2.8 Four Million Years of Change

In this time frame, our climate proxies are increasingly robust, now relying on composites of deep sea cores taken from all the world's oceans. The long term cooling trend discussed above is evident (Figure 2.9). From 3–2.5 Ma, the climate crossed a threshold, allowing accumulation of northern hemisphere polar ice and the onset of boreal glaciation, broadly defining the Pleistocene ice ages. However, up until the 1960s, geologists had only an incomplete land-based record of recent ice ages, leading them to believe that only four glaciations occurred (Imbrie 1982). With the advent of the coring and measurements of deep-sea marine sediments (Figures 2.10 and 2.11), the number of recognized glacial stages in the last 4 Myr increased by an order of magnitude.

In a seminal paper, Emiliani (1955) made the first attempt to measure $\delta^{18}O$ of a sediment core containing foraminifera from the Caribbean and reconstructed changes in SST from 400 ka. Hays, Imbrie, & Shackleton (1976) measured the $\delta^{18}O$ in planktic foraminifera from two deep sea cores from the southern Indian Ocean, with a continuous record going back 500 kyr. They showed that $\delta^{18}O$ correlated with changes in orbital insolation, with a lag of several thousand years from the summer insolation change to the change in ice volume. (The lag is predicted by Milankovitch's theory, since the rate of growth of ice is a maximum when

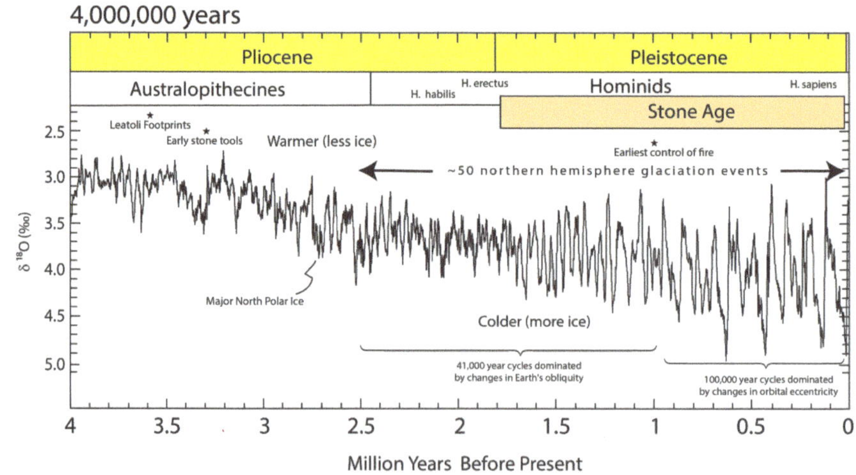

Figure 2.9. Last 4 Myr of climate history with $\delta^{18}O$ from Eastern Atlantic deep sea site 659 (Zachos et al. 2001). Greater $\delta^{18}O$ correlates with greater ice volume. The *Australopithecus afarensis* footprints at Laetoli are the oldest known hominin footprints at 3.6 Ma (Leakey 1981, Masao et al. 2016), while the earliest stone tools date to 3.36 Ma (Harmand et al. 2015), and the earliest controlled use of fire is at Wonderwerk Cave, South Africa at 1.0 Ma (Berna et al. 2012).

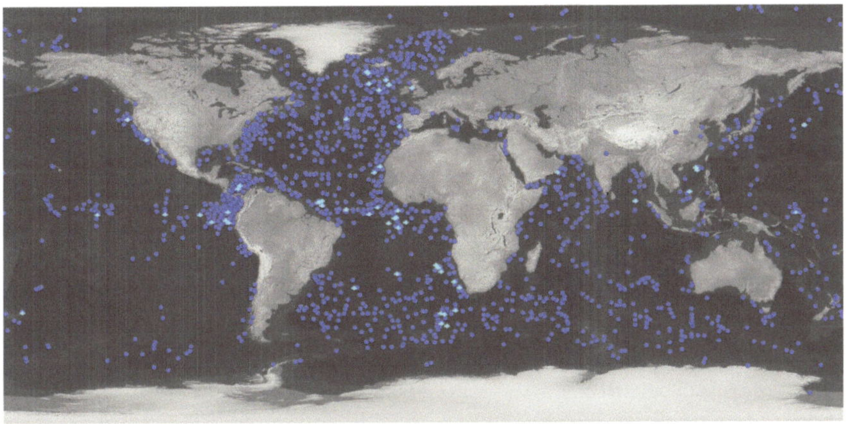

Figure 2.10. Locations of 2690 deep sea core sample sites (blue dots) from 514 different paleocean proxy studies, and 57 deep core sites used to build the LR04 stack (cyan crosses). Data from NOAA National Centers for Environmental Information (NCEI), Center for Weather and Climate (CWC), World Data Service (WDS) for Paleoclimatology, and Lisiecki & Raymo (2005).

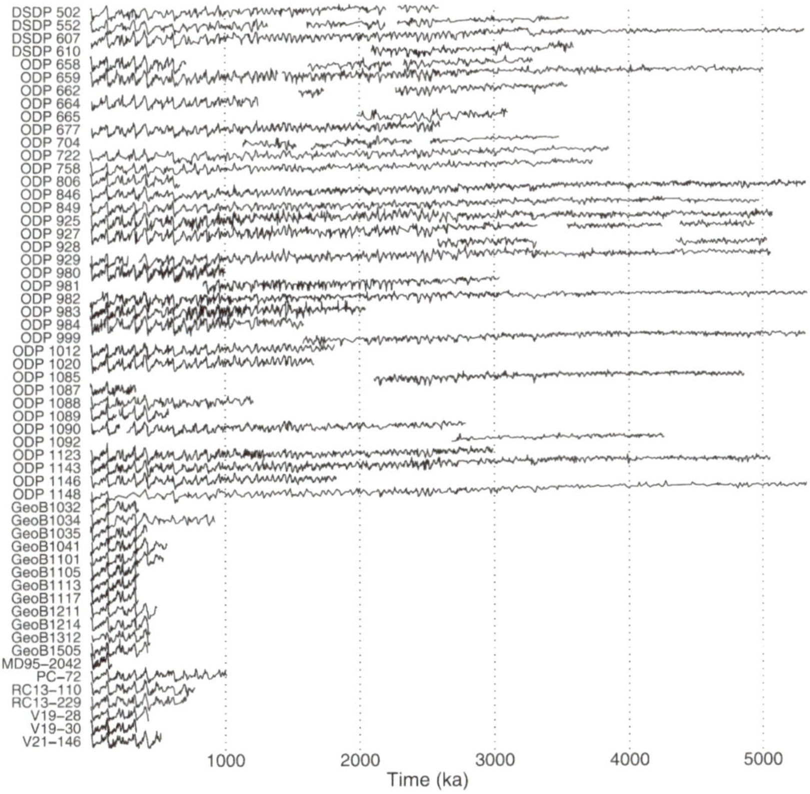

Figure 2.11. Aligned benthic $\delta^{18}O$ data from 57 deep-sea cores spanning the last 5 My used to generate the LR04 stack (figure 2 from Lisiecki & Raymo 2005; https://agupubs.onlinelibrary.wiley.com/doi/10.1029/2004PA001071; image credit: American Geophysical Union).

summer insolation is at a minimum, but when solar insolation begins to increase, ice volume continues to grow. Ice sheets reach their maximum size only after insolation is high enough for the rate of ice ablation to exceed that of accumulation.)

A $\delta^{18}O$ record of northern hemisphere glacial history going back 2.75 Myr (Raymo 1994) shows the long-term drift (part of the trend starting 50 Ma) to more positive $\delta^{18}O$ values, suggesting growing volumes of ice on land, and/or cooling deep-ocean temperatures. Multiple oscillations also occurred, with a 41 kyr frequency matching the obliquity cycle up until a transition between 1.2–0.9 Ma. After 0.9 Ma, the $\delta^{18}O$ variation changes to a signal with larger amplitudes, with a clear 100 kyr cycle starting at 0.6 Ma that correlates with orbital eccentricity changes (Figure 2.9). How a weaker insolation signal from the 100 kyr eccentricity cycle can dominate the stronger precession and obliquity forcing at the Mid-Pleistocene Transition (MPT) is still being debated. The proposal that the gradual decrease in CO_2 led to a greenhouse gas threshold being reached (Raymo, Oppo, & Curry 1997) is contradicted by boron isotope-based reconstructions showing the CO_2 level to be stable before the transition (Hönisch et al. 2009). By separating the contributions from ice volume and water temperature to the benthic $\delta^{18}O$ signal, Elderfield et al. (2012) found that the new 100 kyr cycle was initiated in Antarctica at 900 ka. A low Southern Hemisphere summer insolation suppressed the melting of the ice sheets during the interglacial period at the start of the MPT, with the subsequent growth of the southern ice sheets lowering sea levels by 120 m. Instead of being part of a long-term cooling trend driven by lowering CO_2 levels (Clark et al. 2006), the MPT originates from a reorganization of the climate system. What drives this reorganization is still under debate. The various possible mechanisms usually involve feedbacks for ice sheet growth. In these models, a commonly adopted framework is to suppose that the buildup and ablation of ice with changes in insolation occurs in different feedback regimes, and the climate system can drastically alter when it slips past different thresholds. For the smallest ice sheets, the ice volume changes linearly with insolation, but as the ice sheet grows, positive ice albedo or elevation-temperature feedbacks can allow the ice to survive insolation maxima. As the ice sheet grows even larger, it can cross yet another threshold where other physical mechanisms like glaciers calving into icebergs at the continental margins can lead to rapid termination of a glaciation episode (Berends et al. 2021).

Over the past 4 million years, there have been few appreciable changes in the large-scale continental and oceanic plate configuration. However, closure of ocean passageways such as the Isthmus of Panama (O'Dea et al. 2016) and the Indonesian Seaway (Cane & Molnar 2001) may have significantly affected marine circulation patterns. Based on geological, paleontological, and molecular biological evidence, the Central American Seaway closed between North and South America 3 Ma (O'Dea et al. 2016). The new isthmus of Panama blocked warm water in the Atlantic from being driven westward into the Pacific by trade winds. This water instead flowed north via the Gulf Stream, where the influx of saltier water inhibited sea ice formation (which requires water to precipitate salt out before it can freeze). Haug & Tiedemann (1998) argued that with less sea ice, greater evaporation from the North Atlantic led to increased precipitation and accumulation of snow over land, precipitating glacier formation. More recent research shows that establishment of the isthmus of Panama is not enough by itself to initiate glaciation in the Northern Hemisphere, but is only a small factor that must also be accompanied by a drawdown in atmospheric CO_2 combined with orbital changes (DeConto et al. 2008, Lunt et al. 2008). Brierley & Fedorov (2016) found that the opening of the Bering Strait (between Siberia and Alaska) had the opposite effect of the Central American closure and the narrowing of the Indonesian Seaway.

In the last 3 My, New Guinea and Australia shifted north by 3°, which constricted the Indonesian seaway, and restricted the flow of warm waters from the south Pacific into the Indian Ocean. From global circulation models, the reduced zonal flow cooled the central Indian Ocean waters by as much as 2–3°C, leading to aridification of East Africa (Cane & Molnar 2001).

2.9 Four Hundred Thousand Years of Change

Long ice cores have revealed both short-term climatic oscillations as well as longer orbital scale cycles. The δ^{18}O signal measured in the annual ice layers reflects ambient temperature, but is also impacted by other factors, such as how far the moisture has traveled before precipitating out, the elevation of the ice, and the season when precipitation occurred. Trapped air bubbles in the layers provide direct measurements of greenhouse gases like CO_2, methane (CH_4), and nitrous oxide (N_2O). The annual layers laid down in the ice sheets allow precise reconstructions of the climate record, with uncertainties growing with the deeper ice layers that have been thinned and stretched under the weight of the overlying layers. In the early 1990s, two cores were drilled into the summit of Greenland as part of the Greenland Ice Core Project (GRIP) and Greenland Ice Sheet Project (GISP2), with the latter reaching a depth of 3 km. The GRIP and GISP2 sites were selected to be over smooth bedrock to minimize distortion of the deepest layers from ice flowing over uneven ground. The two locations were separated by only 28 km, so that they would be sampling similar climate histories. When the two sequences were compared, they showed virtually the same climate record, reinforcing the idea that ice cores were reliable climate indicators (Grootes et al. 1993).

The reproducibility of stable isotope records between adjacent Greenland ice cores and between Greenland and Antarctica ice cores (Martrat et al. 2007, Spahni et al. 2005), and the marvelous detail of the ice core isotopic proxies (Steffensen et al. 2008) allow the ice record to be read in Greenland back 130,000 years and in Antarctica, going back in excess of a million years (Petit et al. 1999; Siegenthaler et al. 2005; Figure 2.12). The linkage and correlation between the climate history found in ice cores to the much longer, but lower resolution marine isotopic record, represents one of the greatest paleo-climatic breakthroughs. Together, the combined proxies offer the clearest record of climate over the past 60 Myr (see Sections 2.6–2.8).

Greenland ice cores with a continuous annual resolution going back >60 ky demonstrate that temperatures can vary abruptly over time spans of just a few years (Steffensen et al. 2008). Such rapid changes are remarkable as most climate modulators have longer response times. One potential explanation lies in the abrupt

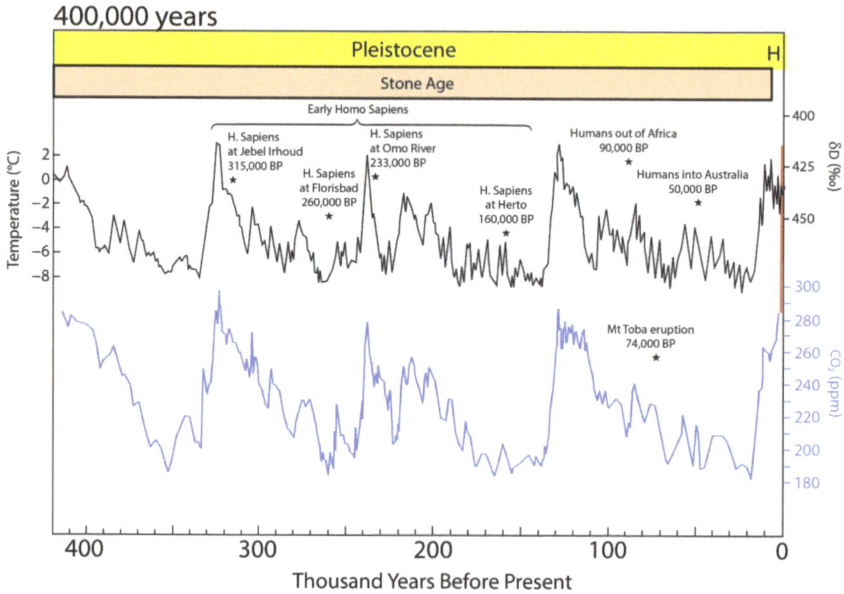

Figure 2.12. Last 400,000 years of climate history with temperature data from the Vostok ice cores, and CO_2 data from four separate ice cores (all from the Antarctic; Petit et al. 1999, Siegenthaler et al. 2005). δD represents the deuterium excess as measured in HDO compared to non-deuterated H_2O, and is another proxy for local temperature change. Ages are shown for the evidence of the earliest modern humans at Jebel Irhoud (Hublin et al. 2017), Omo River and Herto (Vidal et al. 2022). Australia was settled by humans dispersing from Africa 50 ka (Malaspinas et al. 2016). Sulfate aerosols from the Toba eruption likely resulted in a decade-long volcanic winter with maximum temperature drops of ~10–17°C, and reduced the genetic diversity in the surviving human population (Robock et al. 2009).

reorganizing of marine currents (thermohaline circulation) and linked atmospheric circulation patterns. This may be caused by changes in the amount of freshwater flowing into the North Atlantic, which impacts the warm water brought northward by the Gulf Stream (Lapointe & Bradley 2021; see more about this in Section 2.10).

The Last Interglacial occurred 129–116 ka (thousand years ago), when ice volume was at a minimum, and the atmospheric CO_2 content was similar to pre-industrial levels (Govin et al. 2015, Dutton et al. 2015). High eccentricity of Earth's orbit ($e = 0.039$ compared to 0.0167 today) combined with a perihelion near the June solstice and a high inclination resulted in a summer insolation in the Northern Hemisphere that was 30 W m^{-2} higher and sea surface temperatures 1–2°C warmer (Capron et al. 2014, 2017; Hoffman et al. 2017) and a sea surface level 6–9 m higher than today (Dutton et al. 2015). The Last Interglacial therefore provides a good test case to see how climate forcings of similar scale to ones today affected parameters like the extent and volume of the ice sheets, and sea level rise (Otto-Bliesner et al. 2017).

The ratio of $^{13}C/^{12}C$ has been used as another temperature proxy. Photosynthesis preferentially removes ^{12}C. Broecker (1982) argued that lower sea levels during ice ages would have exposed more continental shelf surface to erosion, leading to greater inputs of phosphates and other nutrients washing into the sea. This would drive greater productivity and growth in marine photosynthesizing organisms, resulting in increased removal of CO_2 at the surface. When planktic and benthic organisms were compared, Shackleton et al. (1983) observed a gradient in $\delta^{13}C$ between the surface and deep ocean. When calibrated with atmospheric gases trapped in bubbles in ice cores, the $\delta^{13}C$ gradient between upper and bottom ocean forams turns out to be a proxy of atmospheric CO_2 levels (Shackleton et al. 1992). However, while the CO_2 levels are in phase with orbital forcing, there is a phase lag between the ice volume which changes just behind the $\delta^{13}C$ gradient (Shackelton & Pisias 1985). This implies that the CO_2 is responsible for the climatic changes, and is not just passively responding to such changes. Furthermore, the washing of phosphates to fertilize the seas as proposed by Broecker could not be responsible for the drops in atmospheric CO_2, since the results from sea level changes would lag by $\sim 10^3$ yrs when compared with orbital forcing.

Stable isotope records from cave mineral deposits have also been used to determine changes in rainfall and climate patterns over orbital timescales. The majority of *speleothems* (from the Greek for "cave deposits") consist of carbonate depositions, and include stalagmites rising from the floor, stalactites hanging down, and flowstones created by flowing water. Carbonic acid formed from CO_2 in rainwater dissolves calcium carbonate as it flows through the ground. As drip water inside a cavern, this carbonate-rich water precipitates calcite or aragonite while releasing CO_2 into the cavern. Calcite and aragonite have the same chemical composition ($CaCO_3$) but in different crystalline forms, with precipitation favoring one or the other based on aridity of the region. Because uranium is soluble but thorium is not, measuring the latter in uranium series (U-series) dating, following the decay chain $^{238}U \rightarrow {}^{234}U \rightarrow {}^{230}Th$, can be used to determine precise ages of the accumulated layers with uncertainties typically <1% (Dorale et al. 2004). From such accurate dating, speleothems in caves in the Guadalupe Mountains in the southwest US have been shown to consist of annual bands that have been deposited at rates of 0.1 mm yr^{-1} (Polyak & Asmerom 2001). Because the drip water originates as surface precipitation, measuring $\delta^{18}O$ in calcite deposits measures indirectly the $\delta^{18}O$ in the source water of the precipitation that eventually finds its way into the cave (Hendy 1971). The exact composition of drip water in the cave is also modified by the surrounding climate conditions, with cooler regions having a closer relationship between $\delta^{18}O$ in cave formations with the $\delta^{18}O$ in rainfall (Baker et al. 2019).

Cave calcite deposits are often used to indicate the intensity of past monsoons. Stronger monsoons bring water that is more fractionated and rich in ^{16}O over time and distance. When monsoons are weak, precipitation is dominated by nearer sources of water, with more ^{18}O. The orbital monsoon hypothesis suggests that insolation variations caused by changes in precession are responsible for monsoonal strength (Kutzbach 1981). Higher summer insolation would lead to a stronger monsoon circulation, where greater heating of the land leads to a more vigorous inflow of moist air from the oceans. Measurements of $\delta^{18}O$ from caves in China and Brazil show changes that correlate with solar insolation on a 23 ky cycle over the past 160,000 years (Yuan et al. 2004, Cruz et al. 2005). The variations can be explained by a changing monsoon strength where negative $\delta^{18}O$ values represent strong monsoon flows with greater fractionation of oxygen isotopes, and positive $\delta^{18}O$ indicates weaker monsoons and less fractionation.

Evidence of climate change impacts on human evolution has also appeared within this period. Analysis of potassium depositions in lake sediment cores from Chew Bahir in Ethiopia, reveal the hydroclimate history of a region in East Africa where early humans were known to have lived. In the last 620 kyr, there were abrupt swings between wet and dry regimes, with the frequency and intensity of the swings increasing at 275 ka. These climatic impacts would have fragmented the human population, and could have driven cultural innovation and evolution (Foerster et al. 2022). The Toba volcanic eruption in northern Sumatra 74 ka was the largest in the last 100,000 years (Oppenheimer 2002), with the ejected dense rock equivalent to three orders of magnitude greater than the ejecta from the 1980 Mount St. Helens eruption. The amount of sulfate aerosols injected into the stratosphere would have led to a decade-long volcanic winter, with devastating impacts on plant and animal life including humans (Robock et al. 2009). Genetic evidence suggests that the diminished human population could have had as few as <10,000 individuals, resulting in a severe genetic bottleneck (Rampino & Ambrose 2000). However, there continues to be debate about whether the Toba eruption actually led to a short period of global cooling and a subsequent plunge in human numbers (Ge & Gao 2020).

2.10 Forty Thousand Years of Change

The last 40,000 years witnessed Earth's emergence from the last Ice Age. Since 20 ka, the Milankovitch cycle pointed to increasing Northern Hemisphere insolation. At 11 ka, the Earth's obliquity had reached a maximum, while the precession shifted the perihelion point to June 21. This resulted in maximum summer insolation in the Northern Hemisphere of 7% (30 W m^{-2}) greater than today, while the Southern Hemisphere winter insolation was reduced by 7% (Kutzbach et al. 1998), triggering melting of the northern ice sheets (Figure 2.2). In conjunction with this, very high resolution measurements of air bubbles trapped in the ice layers from Greenland ice cores show an increase in atmospheric CO_2 content from under 200 ppm at 20 ka to ~260 ppm at 10 ka (Luthi et al. 2008; Figure 2.13), as a result of CO_2 release from the deep southern ocean (Marcott et al. 2014).

The insolation-induced warming is supported by the fossil coral reef record in the Caribbean. Coral reefs grow just below the surface of the ocean, accumulating $CaCO_3$ in readily sampled annual layers that can indicate water temperatures and other ocean conditions at the time of calcification (Thompson 2022), including sea level at the time when the coral was alive. By adjusting for subsidence or uplift over time of the underlying bedrock, the elevation of fossil coral reefs can be used to infer not only the sea level, but also the volume of water locked up in ice. Radiometric dating of the reef structure can in turn give a record of sea levels and hence, ice volume, over time. Fairbanks (1989) took cores from fossil coral in Barbados extending back to the Last Glacial Maximum ~15 ka, when the sea level was 120 m

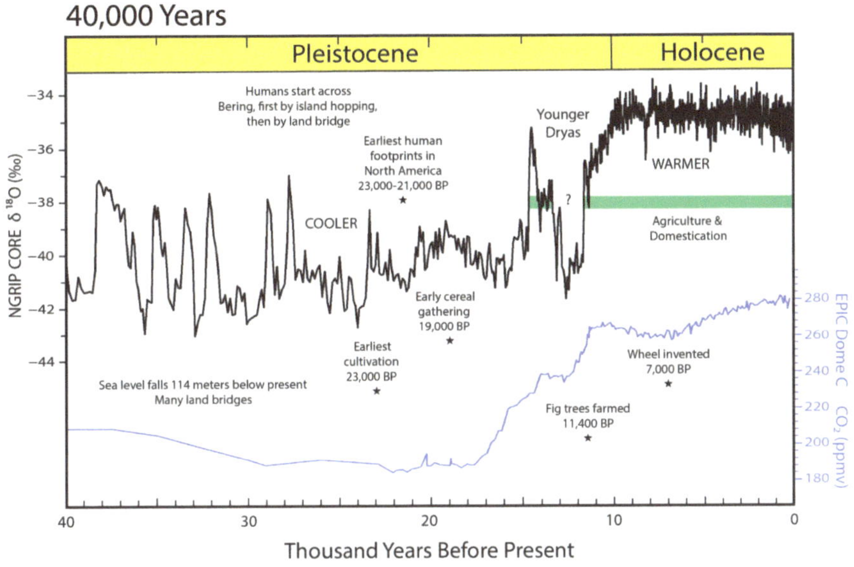

Figure 2.13. Last 40,000 years of climate history with $\delta^{18}O$ data from the GRIP Greenland ice core (Svensson et al. 2008) and CO_2 data from the Antarctic EPIC Dome C core (Luthi et al. 2008). Early cereal from Tanno & Willcox (2006) and first figs from Kislev, Hartmann & Bar-Yosef (2006); animal domestication (11,000 YBP) from Zeder (2008); founder crops from Weiss & Zohary (2011); and the earliest evidence for the cultivation of wild grains from Snir et al. (2015).

lower than it is today. Radiocarbon dating shows the trend of sea level rise due to deglaciation, but because the $^{14}C/^{12}C$ ratio has not been constant through time due to variations in Earth's magnetic field (which alters the incoming cosmic ray flux, and changes how much ^{14}C is added to the atmospheric reservoir), ^{234}U–^{230}Th mass spectrometry is used to give a more accurate indicator of age (Bard et al. 1990).

This warming period was interrupted by a brief, largely Northern Hemisphere cooling event called the Younger Dryas. At 12.9 ka, despite solar insolation being close to maximum, warming halted and temperatures in the North Atlantic dropped, bringing almost glacial cold and drier air to Europe. This 1200-year-long episode is named after the *Dryas* wildflower, which along with other Arctic vegetation extended their range south along with the cold air, with the most dramatic temperature drops occurring on timescales of a century or less.

The current consensus holds that the Younger Dryas cooling episode was due to changes in the direction of flow from the melting of the North American Laurentide ice sheet (Leydet et al. 2018). Waters from melting glaciers have as much as a 30‰ deficit of ^{18}O. The meltwater flow south down the Mississippi showed up as a negative $\delta^{18}O$ signal in Gulf of Mexico planktic foraminifera sediments. During the Younger Dryas, this current became disrupted, and the negative $\delta^{18}O$ pulse disappeared. Instead, water flowed east via the St. Lawrence River into the Arctic and North Atlantic seas. The freshwater reduced the salinity of the upper layer of water, disrupting the AMOC thermohaline circulation, reducing the flow of heat via the Gulf Stream, and interrupting the general warming trend. To this basic picture, other factors contributing to the cooling include additional periodic freshwater pulses elsewhere, dust in the atmosphere reducing insolation, changes in trace greenhouse gases like methane and nitrous oxide, and changes in atmospheric circulation (Renssen et al. 2015). The cooling was not global. While Greenland and Europe got colder, the Antarctic warmed. Because the circulation was halted in the North Atlantic, warm waters accumulated in the Southern Hemisphere, leading to a seesaw effect where cooling in the north is counterbalanced by warming in the south (Broecker 1998).

Based on cores from polar ice and the North Atlantic seafloor, the post-glacial Holocene Epoch has generally been thought to be a period of remarkable climate stability that followed the dynamic climate changes of the Pleistocene (Dansgaard et al. 1993). While other interglacial periods may have witnessed similar temperatures as the Holocene, none appear to have been as prolonged or as stable. This span of over 10,000 years of stable, equitable climate saw profound advances in human cultural development, including domestication of animals and plants, and settlement in urban settings allowing the development of specialized skills and increased social organization (Figure 2.13). However, in the tropics and subtropics (extending from the tropics to ~35° north and south), there is evidence for dramatic fluctuations in climate during the Holocene (Gasse 2000).

Some of the significant regional climate variations were related to orbital forcings, which reached a maximum at the beginning of the Holocene, raising the sea surface temperatures in the northern hemisphere (Koç, Jansen, & Halfidason 1993). The greater overall temperatures in the north led to a northward shift of the ITCZ and increases in the intensity of monsoons in Africa, with substantial moisture sweeping into the Sahara after the end of the Younger Dryas. Cores from the Mediterranean and in African lakes show that the Sahara was a savanna grassland filled with lakes interconnected by waterways (Drake et al. 2011), with abundant wildlife as seen in rock art (Guagnin 2014a, 2014b; Ben Nasr & Walsh 2020). This African Humid Period did not persist, but dried up by ~6–5 ka. The gradual decrease in solar insolation during the last 10 kyr is not likely to produce by itself the abrupt end to a green Sahara, so additional feedbacks have been proposed, such as albedo changes caused by the loss of plants, denuding of vegetation from grazing animals kept by pastoralists, positive feedback from the loss of lakes and wetlands, a decrease in SST at northern latitudes, and changes in ocean circulation (Tierney & deMenocal 2013, Wright 2017, Collins et al. 2017). The greening followed by aridification of the Sahara is part of a longer-term pattern of hydrological and climate instability in Africa observed at different times and regional scales during the Holocene.

From the start of the Holocene to today, obliquity varied from 24.23° to 23.44°, while the summer solstice shifted from perihelion to aphelion, resulting in a 48 W m^{-2} drop in summer insolation at 60°N. Among the atmospheric concentrations of greenhouse gases before the industrial era, methane decreased from 680 ppb at 10 ka to 570 ppb at 5 ka, before returning to the early Holocene value, and N_2O has varied between 250–270 ppb (Spahni et al. 2005). In the last millennium, the pre-industrial level of CO_2 has fluctuated between 278–285 ppm, or less than 3%

(Siegenthaler et al. 2005). Based on sunspot records and measurements of cosmogenic isotopes that correlate with sunspot numbers, it is estimated that solar luminosity has varied between <0.1% between the Late Maunder Minimum (1675–1715) and today, corresponding to a radiative forcing from total solar irradiance change of 0.0–0.10 W m^{-2} (Myhre et al. 2013). Large volcanic eruptions can inject sulfate aerosols into the stratosphere which can reflect sunlight back into space, contributing to cooling. Variations in solar activity and greenhouse gas concentrations have only altered the amount of radiative forcing in the atmosphere by small amounts (with stochastic spikes from volcanic activity) until the industrial era when anthropogenic greenhouse gas emissions gradually grow and then swamp the effect from natural variations (Jansen et al. 2007; Myhre et al. 2013).

As the glaciers retreated north, they left downwarped bedrock along their southern margins. Without the mass of ice weighing it down, the depressed bedrock rebounded via a mantle viscous response with a half life of 3000 yrs (i.e., the bedrock shifts up by 50% of its remaining vertical displacement every 3000 yrs). Because the response is slow, meltwater filling up these depressed areas formed *proglacial lakes*, whose overflows would cut new channels in the dynamic landscape. As the land rebounded, the southernmost edges would become too shallow to retain water, while the lakes' northern edges would grow to follow the receding glacier. The largest of the deglacial lakes was Lake Agassiz, which over several thousand years flooded ~500,000 km^2 of western Canada, although its maximum extent was less than half that, and at any given time, its actual area was much smaller. A brief cooling event at 8.2 ka is attributed to a catastrophic outburst of Lake Agassiz in western Canada (Teller et al. 2002, Clarke et al. 2004, Flesche Kleiven et al. 2008) that poured a vast volume of freshwater into the North Atlantic. Similar to the freshwater influx that led to the Younger Dryas, this new lid of freshwater substantially diminished the downward flow of cold saline water, disrupting the AMOC and the northward transport of warm waters. This led to an abrupt cooling of 2–6°C recorded in the Greenland ice records.

2.11 Four Thousand Years of Change

Modern civilization arose during the stable climate regime of the Holocene (Figure 2.14). As a generalization, early societies developed along river systems where perennial water supplies allowed stable lifestyles and a gradual transition from hunter gathering to sedentary lifestyles with domesticated animals and plants. Examples include the Nile, Tigris-Euphrates, Indus, and Yellow river systems. Settled populations allowed craft specialization, and ultimately urbanization.

The Medieval Warm Period (MWP) has been inferred for 1080–1200 (Le Roy Ladurie 1971), followed by a cool period from 1400 to roughly 1850 termed the Little Ice Age (LIA; Lamb 1977). However, both these episodes are based on temperature records from Europe and the North Atlantic, and may just be regional phenomena as opposed to representing truly global or even hemispheric changes (Hughes & Diaz 1994; Bradley & Jones 1993). Even when regions experienced

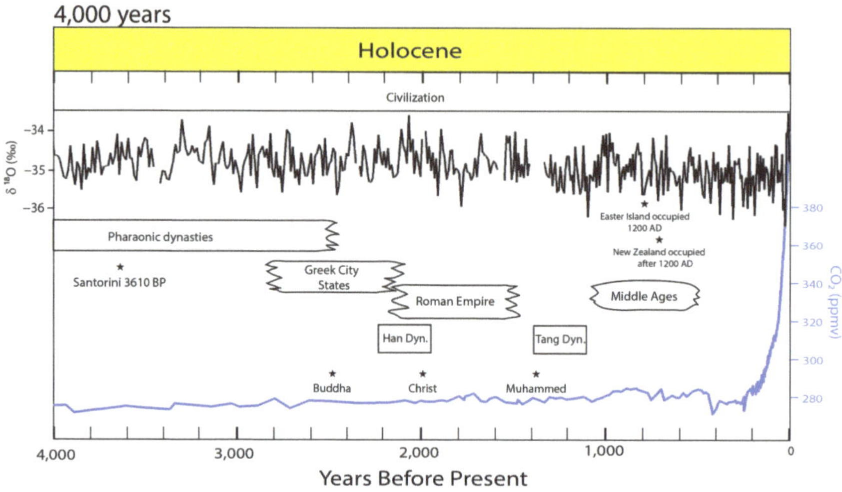

Figure 2.14. Last 4,000 years of climate history with the $\delta^{18}O$ from the GISP2 Greenland ice core (Grootes et al. 1993; Stuiver et al. 1998) and CO_2 from Our World in Data and Friedlingstein et al. (2022). Easter Island arrival date from Hunt & Lipo (2006).

warming during the MWP, or cooling during the LIA, they did not do so synchronously (Jones & Mann 2004). Different regions experienced cooling at different times during the Little Ice Age, whose onset has been tied to changes in North Atlantic circulation (Clarke et al. 2004; Lapointe & Bradley 2021). These authors suggest a precursor event of strong polar heat transport was followed by a period of extreme glacier melt and calving and a modification of AMOC circulation patterns that, coupled with other feedbacks, resulted in the cold period in Europe.

The definition and timing of onset of the *Anthropocene*, the age of humans (Crutzen 2002; Steffen et al. 2015), is still in flux (Waters et al. 2016). Some suggest a very early onset associated with anomalous increases in atmospheric CH_4 and CO_2 as a result of early agriculture and clearcutting practices (Ruddiman 2003, 2007). From a paleontological perspective, the extirpation of megafauna associated with arrival of humans (Martin & Klein 1984; Smith et al. 2019) marks a biological change that will be clear in the paleontological record as a faunal boundary. Others suggest that it should start with radionuclides associated with atmospheric nuclear tests that peaked in the early 1960s (Steffen et al. 2015; Waters et al. 2015). In any case, the term underlines the concept that our civilization has altered our global ecosystem to a degree that has been manifested as a fundamental change in Earth history that has been recorded in the geological and paleontological records.

The last several thousand years of climatic records can now rely on the precise biological record of tree rings (Figure 2.15). The modern field of dendrochronology was founded by astronomer A.E. Douglass, who pioneered many of the techniques that are still in use today. Douglass hoped to use tree rings to indirectly study the solar cycle (which he believed to affect precipitation). Each annual growth ring consists of "earlywood" produced in spring, consisting of large thin-walled cells, and is lighter in color and less dense than the "latewood" grown at the end of the season

Figure 2.15. A segment from a bristlecone pine (*Pinus longaeva* D. K. Bailey) from the White Mountains, California, approximately 50 mm thick. This is one of many specimens from this area that have been used to build chronologies to help reconstruct climate over multi-millennial timescales. (Image credit: Peter Brewer.)

made up of smaller, thick-walled cells. Rings are narrower when moisture is low, or when temperatures are cold. Instead of searching for trees standing in the middle of their ideal ecological niche, where the local growth conditions overwhelm any climatic signals, dendrochronologists select trees in isolated groups located at the edges of their range, where they are especially sensitive to environmental conditions (Fritts 1976). For instance, trees located near the arid edge of a forest will have its growth limited by the availability of moisture, while a tree at the highest elevation or latitude of its natural range will be most stressed by low temperatures. Although changes in climate will reveal themselves in ring widths, studies of modern trees show that the maximum density of rings (found in the latewoods) is a better indicator of temperature (Schweingruber, Briffa, & Nogler 1993). Today, a combination of density and width data are used to reconstruct the climatic record (Briffa et al. 1995). Measurements from tree rings are correlated with the available meteorological records, and the derived relationships from recent times are used to estimate past temperature and precipitation where the instrument record does not exist.

Crossdating, by synchronizing cores from multiple trees, is used to validate the ring record not only across a region, but overlapping ring records from different trees can help reconstruct a chronology further back in time. Often, radiocarbon dating is used to approximately date a piece of wood before crossdating is used to date to within a year's precision. Modern day ^{14}C dating uses cyclotrons and mass spectrometers to count the number of ions of ^{12}C, ^{13}C, and ^{14}C directly in sample masses of $\sim$1 mg (Muller 1977). However, plants do not assimilate carbon isotopes in the same proportions found in the atmosphere. Because ^{12}C is fixed more easily than ^{14}C during photosynthesis, a fractionation can occur with ^{14}C depletion at rates

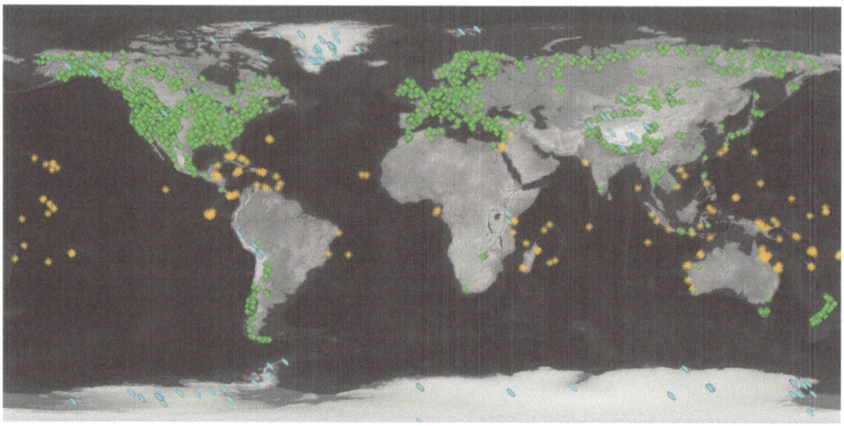

Figure 2.16. Locations where coral samples (orange asterisks), tree rings (green diamonds), and ice cores (cyan ellipses) have been collected to gather paleoclimate proxy data. The three categories of climate proxies correspond roughly to the tropical, mid-latitude, and polar regions. (Image credit: NOAA Paleoclimatology Program.)

as much as −5%, while the amount of fractionation can also vary depending on plant species by a factor of 2–3 (Olsson 1974). Long-term secular variations in the atmospheric reservoir of ^{14}C up to 2% are thought to be caused by factors related to the amount of incoming cosmic rays, including increases in flux from supernovae, as well as modulation by solar activity and changes in the Earth's magnetic field (Damon, Lerman, & Long 1978). In the modern industrial era, there is a sharp drop in ^{14}C caused by the burning of fossil fuels, which releases fossil carbon that is depleted of the heavier isotope (Suess 1965). This phenomena offers one of the key pieces of evidence that the documented increase in atmospheric CO_2 is derived from the combustion of fossil fuels and has been termed the Suess effect (Keeling 1979).

Trees from different parts of the world (Figure 2.16) preserve the regional climate record, showing that wide climate variability existed across different parts of the globe. Although the Little Ice Age represented substantial cooling in the Northern Hemisphere, the temperature anomaly is missing from tree ring records from the Southern Hemisphere (Cook et al. 1996).

Tree ring records from extremely long-lived bristlecone pine trees from the White Mountains, California (Ferguson 1969), and from Irish and German oaks (Becker 1993) have been used to establish a chronology going back 11.4 ka, that can be used to calibrate the ^{14}C dates to calendar dates. The resulting radiocarbon calibration curves have been extended to 24 ka (predating the Holocene) using marine sediments and corals dated with U-Th series (Stuiver et al. 1998). Regional effects from coastal upwelling, latitudinal gradients, and thawing of frozen earth limits the accuracy of the ^{14}C time scale to ±50 yr.

2.12 Four Hundred Years of Change

The last 400 years have seen the advent of direct instrumental measurements. In 1593, Galileo Galilei invented the thermoscope, a precursor to the modern thermometer, which he and his students used to observe temperature differences at different locations over time. However, during his Inquisition trial and even after his death in 1642, making meteorological measurements was politically dangerous enough that one of his collaborators destroyed his instruments and records lest he be discovered by the authorities (Camuffo 2002). Galileo's supporters and followers persevered to establish the first international meteorological network in the mid-1650s, which grew to include cities far outside the Italian peninsula, including Paris and Warsaw. Observers used identical instruments made by the same Florentine instrument maker, and followed the same uniform procedure for observations, including having the instruments mounted outside on a north-facing wall. Observational records were sent back to the Accademia del Cimento (Academy of Experiment) so that they could be compared with one another. Although this network shut down after only a decade because of pressure from the church, it influenced others to establish weather networks throughout Europe over the next two centuries (Frisinger 1977).

Written records can also record other proxies as indirect climate records of phenomena from natural and artificial causes. The latter includes recorded data related to the dates of harvests, the sizes of agricultural yields, when waterways freeze or become ice-free, and the timing of ceremonies meant to break droughts (Brazdil et al. 2005). Such datasets have limitations. For example, often only extremes are recorded; records have spatial and temporal inhomogeneities; qualitative descriptions can be weighted with personal biases; and linguistic meanings of words can change over time. As a result, such records must be interpreted with caution, and require independent calibration with the instrumental record in order to be useful (Jones & Mann 2004). For instance, the date of harvesting of grapes in vineyards is typically 30 days after the date of the onset of ripening, which in turn is affected by temperature and rainfall. Harvest date records going back 600 years have been used to infer maximum August temperature anomalies after calibration with contemporary reference data to give a rate of 6 to 10 days earlier harvest for every degree Celsius in warmer temperatures (Daux 2021). Climate change has induced a trend in recent decades (1981–2007) of average harvest dates occurring 10.2 days earlier (and greater than one standard deviation) when compared to the baseline average from 1600–1900 (Cook & Wolkovich 2016).

Reconstructions of the temperature anomaly record from the last thousand years show quite stable temperatures in the northern hemisphere until the last century. As a result of anthropogenic forcing, a positive temperature anomaly appears in the past century which would otherwise not have occurred if there were only natural sources of radiative forcing. The industrial period saw a rapid increase in CO_2, N_2O, and CH_4 to levels higher than ever measured in the ice core records, and at rates faster than recorded in the past two millennia, to provide the radiative forcing (Jansen et al. 2007). This is punctuated by short events, such as the solar irradiance "Maunder minimum" that occurred in the middle of the 17th century that led to cooling in the northern hemisphere.

The largest driver of change over the past 400 years has been the exponential growth of the human population, which further drives a "Great Acceleration" in

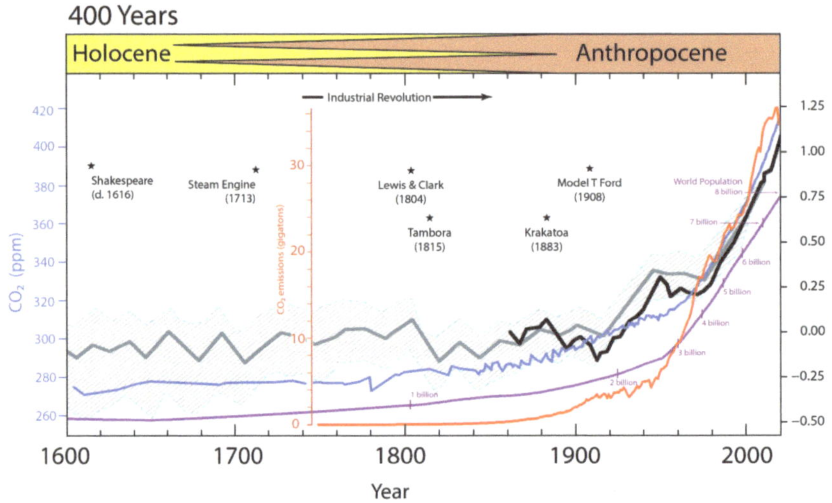

Figure 2.17. The "Great Acceleration" (Steffen et al. 2015) in Earth systems trends that is visible in the last 400 years of climate history. Shown are the global temperature anomaly as a decadal average relative to 1850–1900 (observed is black line; reconstructed is in gray) from IPCC (2021), CO_2 concentration (blue) from Our World in Data using NOAA data tabulated in https://gml.noaa.gov/ccgg/trends/global.html); annual CO_2 emissions (red) from Our World in Data and Friedlington et al. 2022, and global population (purple) from US Census Bureau and McEvedy & Jones (1978). The definition of the start of the human-dominated geological epoch, the Anthropocene, is under considerable debate (Smith & Zeder 2013). One possible boundary is represented here based on the arrival of Europeans in the Americas, and the subsequent transfer of plant species between the New and Old Worlds, with their pollen showing up in the marine sedimentary record (Lewis & Maslin 2015).

human-driven impacts on Earth systems (Steffen et al. 2015; Figure 2.17). Estimated to have reached 8 billion in 2022 (UN DESA 2022), the population has doubled in the past ~40 years. While the global rate of population growth peaked at 2.2% in 1963, the annual rate is now ~0.8% (UN Population Division), with people living longer lives contributing to our growing total population. Our collective impact on Earth is not just limited to climate as a result of greenhouse gas emission, but also involves land use changes, such as from deforestation and overgrazing; biodiversity loss from degradation of ecosystems; and changes to the global biogeochemical cycles of carbon, nitrogen, and water (Steffen et al. 2005). The planetary boundaries framework (Rockström et al. 2009) is one way to group and categorize these and other critical processes regulating the Earth system. The boundaries are defined as the points where human-caused perturbations of these subsystems could destabilize the Holocene-like state that has allowed modern human societies to develop and flourish. Gaps in our knowledge about the functioning of the Earth system leads to uncertainties about what the "safe" boundaries are. For climate change, the reference boundary is to an atmospheric CO_2 concentration of 350 ppm with a radiative forcing at the top-of-the-atmosphere of +1.0 W m^{-2} relative to preindustrial levels, while the uncertainty for maintaining a stable Earth system is in the range 350–450 ppm for CO_2 and +1.0–1.5 W m^{-2} for radiative forcing (Steffen et al. 2015).

2.13 Forty Years of Change

In the past 40 years, atmospheric CO_2 concentration has been rising for $\sim$ 2 ppm yr^{-1} (Figure 2.18), and the increased emitted carbon sourced from fossil fuels means that there has been a decrease of $\delta^{13}C$ from the pre-industrial level of −6.5‰ to −8.5‰ (Rubino et al. 2013). Current measurements of climate systems are increasingly sophisticated, especially with sensor networks in the oceans and from orbit that track the magnitude of glacial melt, sea level rise, and fluxes of heat and water mass in the oceans.

Measurements of ice mass in Greenland and Antarctica by the Gravity Recovery and Climate Experiment (GRACE; 2002–2017) and GRACE Follow-On (2018–present) satellites offer monthly data on ice loss rates. Micron-precision measurements of the distance between pairs of co-orbiting satellites using microwave ranging (and laser interferometry in GRACE-FO) provide an estimate of Earth's gravity field, and changes over time of ice sheet and groundwater mass (Landerer et al. 2020). The analysis of the changing gravity fields indicate that between 2002 and 2019, Greenland has been losing ~260 Gt yr^{-1} of ice and Antarctica ~110 Gt yr^{-1} (Velicogna et al. 2020). The continuous long-term measurements have revealed large year-to-year variability, with the Greenland ice sheet mass loss slowing to ~100 Gt yr^{-1} in 2017–2018, followed by a record loss of 532 Gt yr^{-1} in 2019 (Sasgen et al.

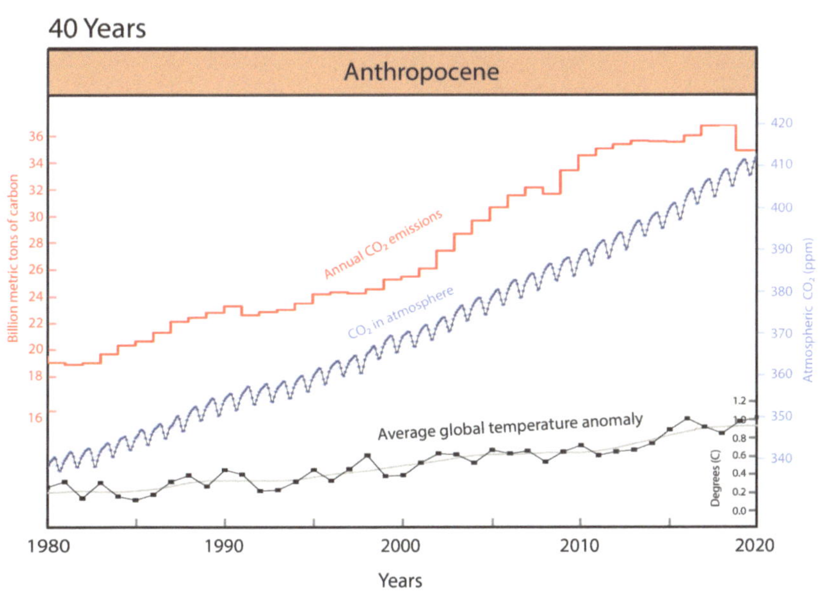

Figure 2.18. Last 40 years of climate history. The average global temperature anomaly is with respect to the 1951–1980 average, and uses data from the Goddard Institute for Space Studies (GISS) Surface Temperature Analysis (v4; https://data.giss.nasa.gov/gistemp/graphs_v4/). The data for the amount of CO_2 in the atmosphere is from the US Earth System Research Laboratory, Global Monitoring Division, sourcing Mauna Loa Observatory series and a Global Average series over marine surface sites (https://datahub.io/core/co2-ppm). The annual CO_2 emissions is from the Global Carbon Project (https://www.globalcarbonproject.org/carbonbudget/21/data.htm).

2020). This mass of meltwater entering the seas accounts for 1.2 mm yr^{-1} of sea level rise or about a third of the total (with another third from thermal expansion of seawater, and the remainder from melting mountain glaciers and groundwater extraction).

Measurements of sea level rise are done through the radar altimetry from a series of satellites providing continuous unbroken measurements, starting with TOPEX/Poseidon in 1992, and continuing with the Jason satellites (first launched in 2001). Sentinel 6, the latest of the Jason series, was lofted into orbit Nov 2020, and like its predecessors, deploys a radar altimeter as the primary instrument, with a microwave radiometer used to quantify the radio signal delay caused by water vapor in the troposphere (Donlon et al. 2021). In addition to thermal expansion of seawater, a smaller factor affecting the interpretation of sea level rise is the post-glacial rebound of the continents that is still occurring since the end of the last ice age. The weight of the ice sheets pushed underlying bedrock horizontally as well as vertically. With the melting of the continental ice sheets, glacial isostatic adjustment means that mantle underneath the oceans is shifting back towards land, leading the ocean basins at the continental margins to get deeper, and counteracting sea level rise at a rate of ~0.3 mm yr^{-1} (Tamisiea 2011). Since 1993, the global mean sea level has been rising at a rate of 3.1 mm yr^{-1}, with an acceleration in the past quarter century at 0.10 mm yr^{-2} (Cazenave, Palanisamy, & Ablain 2018). The acceleration is also confirmed by the GRACE satellite data, which suggests that meltwater inflow into the oceans contributes 2.5 mm yr^{-1} out of the 3.8 mm yr^{-1} of total mean sea level rise observed between 2005 and 2017 (Tapley et al. 2019).

As noted previously, Northern Hemisphere cooling episodes like the Younger Dryas (Section 2.10) and the Little Ice Age (Section 2.11) can be explained by large influxes of freshwater that disrupt meridional circulation in the Atlantic Ocean. Paleoclimate records show that a reorganization of AMOC leading to abrupt climate change can occur on extremely short timescales—years and decades instead of millennia (Broecker 1997, Henry et al. 2016). There is thus a need to understand how AMOC would slow or collapse with global warming, how a slowdown would affect climate in Europe, and how much cooling we would have if a shutdown was to occur (NRC 2002). Measurements of the AMOC are conducted through a variety of methods, including via submarine cables, automated submersibles (gliders and the robotic Argo floats), moored sensor arrays, and satellite altimetry and gravimetry (Perez et al. 2015). One such system is the line of RAPID monitors running from Florida to Morocco at 26.5°N, constructed to monitor where the Gulf Stream flow is restricted to the narrow Florida Straits where it is more economical to measure; the flow is still near maximum; and before heat is released northeast towards Europe. AMOC at this latitude is derived from three observed components: the mass flux of the northward flowing Florida Current is estimated by the voltage induced in submerged telephone lines between Florida and the Bahamas, created by charged ions in seawater passing through Earth's magnetic field; satellite measurements of east–west wind-driven stress which when combined with the Coriolis force produces a south-to-north meridional flow (Ekman transport) in the top 50 m of the ocean; and the transport of seawater measured directly along vertical profiles by a series of

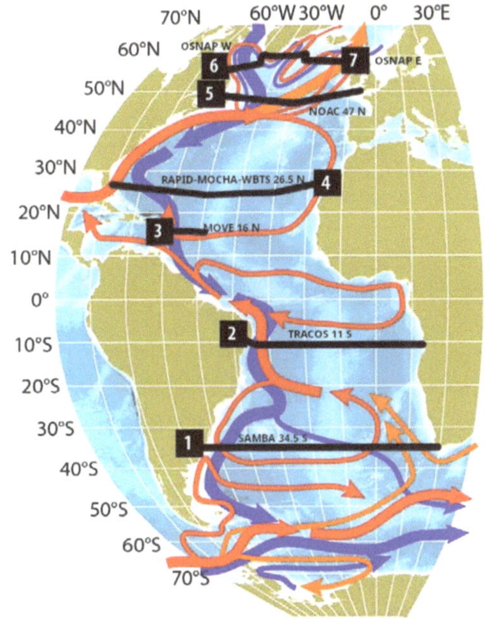

Figure 2.19. A diagram showing the high salinity, lower density, warmer (red and orange arrows) and low salinity, higher density, cooler (blue arrows) flows of the Atlantic Meridional Overturning Circulation. Black lines represent different ocean transport observational arrays, in order from 1 to 7: SAMBA 34.5°S, TRACOS 11°S, MOVE 16°N, RAPID-MOCHA-WBTS 26.5°N, NOAC 47°N, OSNAP West, and OSNAP East. Image credit: Berx et al. 2022; https://doi.org/10.5670/oceanog.2021.supplement.02-04; Creative Commons Attribution 4.0 International License (CC BY 4.0).

moored arrays at different longitudes along the 26.5°N line (Strokosz 2004, Bryden et al. 2009, McCarthy et al. 2015). RAPID and other ocean transport observation systems (Figure 2.19) can currently provide direct measurements of seasonal and interannual variability in AMOC (Frajka-Williams et al. 2019). Data from tethered buoy arrays can be further combined with satellite altimetry to monitor large scale ocean circulation patterns (Sanchez-Franks et al. 2021).

2.14 Concluding Thoughts

This short review shows the scope of the evidence needed to describe how Earth's climate has evolved and changed over time up to the fossil fuel-driven global warming of the present day. Anthropogenic climate change poses one of the greatest challenges to human civilization, since it adversely impacts a wide variety of human systems, including freshwater access, food security, infrastructure, and general health and wellbeing (IPCC et al. 2022). We are also reminded of the magnitude of the task ahead of us to bend the rising CO_2 curve. During the COVID 19-driven shutdown of global activities in spring 2020, the reduction in global commerce and mobility resulted in a 7% drop in fossil fuel emissions in 2020 compared to 2019. This decrease in the anthropogenic CO_2 signal was swamped by natural variability, resulting in no measurable global decrease in atmospheric carbon (Lovenduski et al. 2021), while in 2021, CO_2 emissions from fossil fuels generally rebounded to pre-pandemic 2019 levels (Jackson et al. 2022, Crippa et al. 2022, Davis et al. 2022).

The Earth climate story in this chapter alludes to the scale of the global scientific research effort underlying the narrative in this review. A vast army of researchers across multiple scientific disciplines has collected, analyzed, and synthesized the data that describes how Earth's climate has changed in the past, and contributes to our understanding of how it might continue to do so in the future. This collaborative effort includes not just the scientists themselves, but also the large support teams that are necessary to keep such scientific programs going. This would include, for example, the seafaring crews operating the vessels that deploy and maintain oceanic sensor arrays; the refrigeration specialists who support the room-sized freezers that warehouse Greenland and Antarctic ice cores; and the librarians and archivists who curate and track down handwritten manuscripts and other physical documents containing centuries-old weather records. Although climate change is driven by growing atmospheric greenhouse gases, understanding climate change requires the involvement of many more scientific disciplines than just atmospheric science. The endeavor to understand Earth's climate may be the largest global scientific collaboration in human history.

No matter how quickly society decarbonizes, global temperatures will continue to rise through the mid-century and likely beyond (UNEP 2021). As extreme weather events become more common, the need to research strategies and pathways for adaptation to and mitigation of climate impacts will become more pressing (Corell et al. 2014). Since climate impacts on society are broad, and the scale and types of the effects vary widely depending on the geographic region and the socioeconomic conditions of the impacted country, multidisciplinary cooperation is even more important. For instance, the cost of adaptation in developing countries could grow to US $500 billion annually by 2050 (Pauw et al. 2020). Financing climate adaptation will require researchers to join together to determine vulnerabilities, assess risk, and target aid, and for collaboration between the worlds of science and policy (Currie-Alder et al. 2021). This work will need to be done in conjunction with all levels of government, from national to community levels, plus engaging a broad range of stakeholders that include private firms, non-government organizations, and other practitioners (Corell et al. 2014). The collaboration that gave us our current understanding of Earth's climate history, incomplete as it is, is an example of the type of global partnerships needed to not only understand our future climate trajectory, but as a model for the type of cooperation needed moving forward. Despite the vast challenges facing society as a result of the climate crisis, we can end on a faintly positive note of reflection of what we must do to determine humanity's climate future.

References

Baertschi, P. 1976, E&PSL, 31, 341

Bahcall, J. N., Pinsonneault, M. H., & Basu, S. 2001, ApJ, 555, 990

Baker, A., et al. 2019, NatCo, 10, 2984

Bard, E., Hamelin, B., Fairbanks, R. G., & Zindler, A. 1990, Natur, 345, 405
Becker, B. 1993, Radcb, 35, 201
Ben Nasr, J., & Walsh, K. J. 2020, JMA, 33, 3
ed. Benn, K. Mareschal, J.-C. & Condie K. C. (ed) 2006, Archean Geodynamics and Environments (Geophysical Monograph) (164,; Washington, DC: American Geophysical Union)
Berends, C. J., Köhler, P., Lourens, L. J., & van de Wal, R. S. W. 2021, RvGeo, 59, e2020RG000727
Berger, A., & Loutre, M. F. 1991, QSRv, 10, 297
Berna, F., Goldberg, P., Horwitz, L. K., Brink, J., Holt, S., Bamford, M., & Chazan, M. 2012, PNAS, 109, E1215
Berner, R. A. 2004, The Phanerozoic Carbon Cycle: CO2 and O2 (New York: Oxford Univ. Press)
Bradley, R. S., & Jonest, P. D. 1993, Holoc, 3, 367
Brázdil, R., Pfister, C., Wanner, H., Storch, H. V., & Luterbacher, J. 2005, ClCh, 70, 363
Brierley, C. M., & Fedorov, A. V. 2016, E&PSL, 444, 116
Briffa, K. R., Jones, P. D., Schweingruber, F. H., Shiyatov, S. G., & Cook, E. R. 1995, Natur, 376, 156
Broecker, W. S. 1982, PrOce, 11, 151
Broecker, W. S. 1997, Sci, 278, 1582
Broecker, W. S. 1998, PalOc, 13, 119
Bryden, H. L., & Imawaki, S. 2001, InGeo, 77, 455
Bryden, H. L., Mujahid, A., Cunningham, S. A., & Kanzow, T. 2009, OcSci, 5, 421
Camuffo, D. 2002, ClCh, 53, 7
Cande, S. C., & Kent, D. V. 1992, JGRB, 97, 13917
Cane, M. A., & Molnar, P. 2001, Natur, 411, 157
Capron, E., et al. 2014, QSRv, 103, 116
Capron, E., Govin, A., Feng, R., Otto-Bliesner, B. L., & Wolff, E. W. 2017, QSRv, 168, 137
Cazenave, A., Palanisamy, H., & Ablain, M. 2018, AdSpR, 62, 1639
Charnay, B., Wolf, E. T., Marty, B., & Forget, F. 2020, SSRv, 216, 1
Clarke, G. K., Leverington, D. W., Teller, J. T., & Dyke, A. S. 2004, QSRv, 23, 389
Clark, P. U., Archer, D., Pollard, D., Blum, J. D., Rial, J. A., Brovkin, V., Mix, A. C., Pisias, N. G., & Roy, M. 2006, QSRv, 25, 3150
Clarke, G. K., Leverington, D. W., Teller, J. T., & Dyke, A. S. 2004, QSRv, 23, 389
Collins, J. A., Prange, M., Caley, T., Gimeno, L., Beckmann, B., Mulitza, S., Skonieczny, C., Roche, D., & Schefuß, E. 2017, NatCo, 8, 1372
Cook, E. R., Buckley, B. M., & D'Arrigo, R. D. 1996, Climatic Variations and Forcing Mechanisms of the Last 2000 Years, ed. P. D. Jones, R. S. Bradley, & J. Juzel (Berlin: Springer) 141
Cook, B. I., & Wolkovich, E. M. 2016, NatCC, 6, 715
Corell, R. W., Liverman, D., Dow, K., Ebi, K. L., Kunkel, K., Mearns, L. O., & Melillo, J. 2014, Climate Change Impacts in the United States: The Third National Climate Assessment, ed. J. M. Melillo, T. C. Richmond, & G. W. YoheU.S. Global Change Research Program (Washington, DC: US Government Printing Office) 707
Crowell, J. C. 1999, Pre-Mesozoic Ice Ages: Their Bearing on Understanding the Climate System (Geological Society of America Memoir (192,; Boulder, CO: GSA)

ed. Cox A. (ed) 1973, Plate Tectonics and Geomagnetic Reversals (San Francisco, CA: WH Freeman)
Crippa, M., et al. 2022, CO2 Emissions of All World Countries (Luxembourg: Publications Office of the European Union)
Crutzen, P. J. 2002, Natur, 415, 23
Cruz, F. W., Burns, S. J., Karmann, I., Sharp, W. D., Vuille, M., Cardoso, A. O., Ferrari, J. A., Silva Dias, P. L., & Viana, O. 2005, Natur, 434, 63
Currie-Alder, B., Rosenzweig, C., Chen, M., Nalau, J., Patwardhan, A., & Wang, Y. 2021, ComEE, 2, 220
Damon, P. E., Lerman, J. C., & Long, A. 1978, AREPS, 6, 457
Dansgaard, W., et al. 1993, Natur, 364, 218
Daux, V. 2021, Paleoclimatology (Frontiers in Earth Sciences), ed. G. Ramstein, et al. (Cham: Springer Nature) 205
Davis, S. J., et al. 2022, NatCC, 12, 412
De Vleeschouwer, D., Vahlenkamp, M., Crucifix, M., & Pälike, H. 2017, Geo, 45, 375
DeConto, R. M., & Pollard, D. 2003, Natur, 421, 245
DeConto, R. M., Pollard, D., Wilson, P. A., Pälike, H., Lear, C. H., & Pagani, M. 2008, Natur, 455, 652
Donlon, C. J., et al. 2021, RSEnv, 258, 112395
Dorale, J. A., Edwards, R. L., Alexander, E. C., Shen, C. C., Richards, D. A., & Cheng, H. 2004, Studies of Cave Sediments: Physical and Chemical Records of Paleoclimate, ed. I. D. Sasowsky, & J. E. Mylroie (New York: Kluwer Academic/Plenum) 177
Doria, G., Royer, D. L., Wolfe, A. P., Fox, A., Westgate, J. A., & Beerling, D. J. 2011, AmJS, 311, 63
Drake, N. A., Blench, R. M., Armitage, S. J., Bristow, C. S., & White, K. H. 2011, PNAS, 108, 458
Dutton, A., Carlson, A. E., Long, A. J., Milne, G. A., Clark, P. U., DeConto, R., Horton, B. P., Rahmstorf, S., & Raymo, M. E. 2015, Sci, 349, aaa4019
Elderfield, H., Ferretti, P., Greaves, M., Crowhurst, S., McCave, I. N., Hodell, D., & Piotrowski, A. M. 2012, Sci, 337, 704
Emiliani, C. 1955, JG, 63, 538
ed. Eriksson, P. G. Altermann, W. Nelson, D. R. Mueller, W. U. Catuneanu, O. & Catuneanu O. (ed) 2004, The Precambrian Earth: Tempos and Events (Developments in Precambrian Geology vol 12) (Amsterdam: Elsevier)
Fairbanks, R. G. 1989, Natur, 342, 637
Ferguson, C. W. 1969, Tree-Ring Bull, 29, 1
Frisinger, H. H. 1977, History of Meteorology: To 1800 (Historical Monograph Series) (Boston, MA: American Meteorological Society)
Fritts, H. C. 1976, Tree Rings and Climate (London: Academic)
Foerster, V., et al. 2022, NatGe, 15, 805
Foster, G. L., Royer, D. L., & Lunt, D. J. 2017, NatCo, 8, 14845
ed. Fowler, C. M. R. Ebinger, C. J. & Hawkesworth C. J. (ed) 2002, The Early Earth: Physical, Chemical and Biological Development (Geological Society Special Publication no 199) (London: Geological Society)
Fox, L. R., Stukins, S., Hill, T., & Bailey, H. 2018, JMicP, 37, 11
Frajka-Williams, E., et al. 2019, FrMS, 6, 260

Friedlingstein, P., et al. 2022, ESSD, 14, 1917
Ganachaud, A., & Wunsch, C. 2000, Natur, 408, 453
Gasse, F. 2000, QSRv, 19, 189
Ge, Y., & Gao, X. 2020, QuInt, 559, 24
Gernon, T. M., Barr, R., Fitton, J. G., Hincks, T. K., Keir, D., Longman, J., Merdith, A. S., Mitchell, R. N., & Palmer, M. R. 2022, NatGe, 15, 573
Gough, D. O. 1981, SoPh, 74, 21
Govin, A., et al. 2015, QSRv, 129, 1
Grootes, P. M., Stuiver, M., White, J. W. C., Johnsen, S., & Jouzel, J. 1993, Natur, 366, 552
Guagnin, M. 2014a, AfrArRv, 31, 407
Guagnin, M. 2014b, EnAr, 20, 52
Gutjahr, M., Ridgwell, A., Sexton, P. F., et al. 2017, Natur, 548, 573
Hansen, J., Sato, M., Kharecha, P., et al. 2008, OASJ, 2, 217
Harmand, S., Lewis, J. E., Feibel, C. S., Lepre, C. J., Prat, S., Lenoble, A., & Roche, H. 2015, Natur, 521, 310
Haug, G. H., & Tiedemann, R. 1998, Natur, 393, 673
Hays, J. D., Imbrie, J., & Shackleton, N. J. 1976, Sci, 194, 1121
Haynes, L. L., & Hönisch, B. 2020, PNAS, 117, 24088
Hendy, C. H. 1971, GeCoA, 35, 801
Henry, L. G., McManus, J. F., Curry, W. B., et al. 2016, Sci, 353, 470
Herbert, T. D., Dalton, C. A., Liu, Z., et al. 2022, Sci, 377, 116
Hoffman, P. F. 2019, AREPS, 47, 1
Hoffman, P. F., Kaufman, A. J., Halverson, G. P., & Schrag, D. P. 1998, Sci, 281, 1342
Hoffman, J. S., Clark, P. U., Parnell, A. C., & He, F. 2017, Sci, 355, 276
Hönisch, B., Hemming, N. G., Archer, D., Siddall, M., & McManus, J. F. 2009, Sci, 324, 1551
Hublin, J. J., et al. 2017, Natur, 546, 289
Hughes, M. K., & Diaz, H. F. 1994, ClCh, 26, 109
Hunt, T. L., & Lipo, C. P. 2006, Sci, 311, 1603
Imbrie, J. 1982, Icar, 50, 408
IPCC 2022, IPCC Climate Change 2022: Impacts, Adaptation and Vulnerability. Contribution of Working Group II to the Sixth Assessment Report of the Intergovernmental Panel on Climate Change, ed. H.-O. Portner, D. C. Roberts, M. Tignor, et al. (Cambridge: Cambridge Univ. Press)
Jackson, R. B., et al. 2022, ERL, 17, 031001
Jansen, E. J., et al. 2007, The Physical Science Basis Contribution of Working Group I to the Fourth Assessment Report of the Intergovernmental Panel on Climate Change, ed. S. Solomon, D. Qin, M. Manning, Z. Chen, M. Marquis, K. B. Averyt, M. Tignor, & H. L. Miller (Cambridge: Cambridge Univ. Press)
Jones, P. D., & Mann, M. E. 2004, RvGeo, 42,
Kasting, J. F., Pollack, J. B., & Crisp, D. 1984, JAtC, 1, 403
Keeling, C. D. 1979, EnInt, 2, 229
Kender, S., et al. 2021, NatCo, 12, 5186
Kennett, J. P. 1977, JGR, 82, 3843
Kislev, M. E., Hartmann, A., & Bar-Yosef, O. 2006, Sci, 312, 1372
Kleiven, H. K. F., Kissel, C., Laj, C., et al. 2008, Sci, 319, 60
Koç, N., Jansen, E., & Haflidason, H. 1993, QSRv, 12, 115

Kutzbach, J. E. 1981, Sci, 214, 59
Kutzbach, J. E., & Gallimore, R. G. 1989, JGRD, 94, 3341
Kutzbach, J., Gallimore, R., Harrison, S., Behling, P., Selin, R., & Laarif, F. 1998, QSRv, 17, 473
Lamb, H. H. 1977, Climate: Present Past and Future Volume.2: Climatic History and the Future (New York: Routledge)
Landerer, F. W., et al. 2020, GeoRL, 47, e2020GL088306
Lapointe, F., & Bradley, R. S. 2021, SciA, 7, eabi8230
Larsen, I. J., Montgomery, D. R., & Greenberg, H. M. 2014, Geo, 42, 527
Le Roy Ladurie, L. E. 1971, Times of Feast, Times of Famine: A History of Climate Since the Year 1000, ed. B. Bray (Garden City, NY: Doubleday)
Leakey, M. D. 1981, RSPTB, 292, 95
Lewis, S. L., & Maslin, M. A. 2015, Natur, 519, 171
Leydet, D. J., Carlson, A. E., Teller, J. T., Breckenridge, A., Barth, A. M., Ullman, D. J., Sinclair, G., Milne, G. A., Cuzzone, J. K., & Caffee, M. W. 2018, Geo, 46, 155
Liebrand, D., et al. 2017, PNAS, 114, 3867
Lisiecki, L. E., & Raymo, M. E. 2005, PalOc, 20, PA1003
Ligrone, R. 2019, Biological Innovations That Built the World: A Four-Billion-Year Journey Through Life and Earth History (Cham: Springer) 129
Lovenduski, N. S., Chatterjee, A., Swart, N. C., Fyfe, J. C., Keeling, R. F., & Schimel, D. 2021, GeoRL, 48, e2021GL095396,
Lunt, D. J., Valdes, P. J., Haywood, A., & Rutt, I. C. 2008, ClDy, 30, 1
Lüthi, D., et al. 2008, Natur, 453, 379
Malaspinas, A. S., et al. 2016, Natur, 538, 207
Marchitto, T. M., Curry, W. B., Lynch-Stieglitz, J., Bryan, S. P., Cobb, K. M., & Lund, D. C. 2014, GeCoA, 130, 1
Marcott, S. A., et al. 2014, Natur, 514, 616
ed. Martin, P. S. & Klein R. G. (ed) 1984, Quaternary Extinctions: A Prehistoric Revolution (Tucson: Univ. Arizona Press)
Martrat, B., Grimalt, J. O., Shackleton, N. J., et al. 2007, Sci, 317, 502
Masao, F. T., Ichumbaki, E. B., Cherin, M., Barili, A., Boschian, G., Iurino, D. A., Menconero, S., Moggi-cecchi, J., & Manzi, G. 2016, Elife, 5, e19568
Maxbauer, D. P., Royer, D. L., & LePage, B. A. 2014, Geo, 42, 1027
McCarthy, G. D., Smeed, D. A., Johns, W. E., et al. 2015, PrOce, 130, 91
McEvedy, C., & Jones, R. 1978, Atlas of World Population History (London: Penguin Books)
Meckler, A. N., et al. 2022, Sci, 377, 86
Miller, K. G., Fairbanks, R. G., & Mountain, G. S. 1987, PalOc, 2, 1
Mojzsis, S. J., Harrison, T. M., & Pidgeon, R. T. 2001, Natur, 409, 178
Muller, R. A. 1977, Sci, 196, 489
Müller, T., et al. 2020, Geo, 48, 1184
Myhre, G. D., et al. 2013, The Physical Science Basis. Contribution of Working Group I to the Fifth Assessment Report of the Intergovernmental Panel on Climate Change, ed. T. F. Stocker, D. Qin, G.-K. Plattner, M. Tignor, S. K. Allen, J. Boschung, A. Nauels, Y. Xia, V. Bex, & P. M. Midgley (Cambridge: Cambridge Univ. Press)
National Research Council 2002, Abrupt Climate Change: Inevitable Surprises (Washington, DC: National Academies Press)

National Research Council 2011, Understanding Earth's Deep Past: Lessons for Our Climate Future (Washington, DC: National Academies Press)
Nutman, A. P., Bennett, V. C., Friend, C. R., Van Kranendonk, M. J., & Chivas, A. R. 2016, Natur, 537, 535
O'Dea, A., et al. 2016, SciA, 2, e1600883
Olsson, I. U. 1974, GFF, 96, 311
Oppenheimer, C. 2002, QSRv, 21, 1593
ed. Oreskes, N. & LeGrand H. E. (ed) 2001, Plate Tectonics: An Insider's History of the Modern Theory of the Earth (Cambridge: Westview Press)
Otto-Bliesner, B. L., et al. 2017, GMD, 10, 3979
Pagani, M., Huber, M., Liu, Z., Bohaty, S. M., Henderiks, J., Sijp, W., Krishnan, S., & DeConto, R. M. 2011, Sci, 334, 1261
Paillard, D. 2010, CRGeo, 342, 273
Pauw, P, Weikmans, R, Watson, C, Jahns, H, Prowse, M, Quevedo, A, & Puri, J 2020, 2021 Global progress on financing for adaptation, Adaptation Gap Report (Nairobi: United Nations) pp 23–31
Pearson, P. N., Foster, G. L., & Wade, B. S. 2009, Natur, 461, 1110
Perez, R. C., Baringer, M. O., Dong, S., et al. 2015, MTSJ, 49, 167
Petit, J. R., et al. 1999, Natur, 399, 429
Polyak, V. J., & Asmerom, Y. 2001, Sci, 294, 148
Rampino, M. R., & Ambrose, S. H. 2000, Volcanic Hazards and Disasters in Human Antiquity, ed. F. W. McCoy, & G. Heiken, (Geological Society of America Special Paper 345) 71
Ramsay, J. G. 1963, JG, 66, 353
Ravelo, A. C., & Hillaire-Marcel, C. 2007, Proxies in Late Cenozoic Paleoceanography, ed. C. Hillaire-Marcel, & A. De Vernal (Amsterdam: Elsevier) 735
Raymo, M. E. 1994, AREPS, 22, 353
Raymo, M. E., Oppo, D. W., & Curry, W. 1997, PalOc, 12, 546
Raymo, M. E., Ruddiman, W. F., & Froelich, P. N. 1988, Geo, 16, 649
Renssen, H., Mairesse, A., Goosse, H., et al. 2015, NatGe, 8, 946
Robock, A., Ammann, C. M., Oman, L., et al. 2009, JGRD, 114, (D10)
Rockström, J., Steffen, W., Noone, K., et al. 2009, EcoS, 14, 32
Royer, D. L., Berner, R. A., Montañez, I. P., Tabor, N. J., & Beerling, D. J. 2004, GSAT, 14, 4
Rubino, M., et al. 2013, JGRD, 118, 8482
Ruddiman, W. F. 2003, ClCh, 61, 261
Ruddiman, W. F. 2007, RvGeo, 45,
Sagan, C., & Mullen, G. 1972, Sci, 177, 52
Sanchez-Franks, A., Frajka-Williams, E., Moat, B. I., & Smeed, D. A. 2021, OcSci, 17, 1321
Sasgen, I., Wouters, B., Gardner, A. S., et al. 2020, ComEE, 1, 1
Schweingruber, F. H., Briffa, K. R., & Nogler, P. 1993, IJBm, 37, 151
Shackleton, N. J., & Pisias, S. N. 1985, The Carbon Cycle and Atmospheric CO2: Natural Variations Archean to Present, Geophysical Monograph Series, (Vol. 32, ed. E. T. Sundquist, & W. S. Broecker; Washington, DC: American Geophysical Union) 303
Shackleton, N. J., Hall, M. A., Line, J., & Shuxi, C. 1983, Natur, 306, 319
Shackleton, N. J., Le, J., Mix, A., & Hall, M. A. 1992, QSRv, 11, 387
Sheldon, N. D. 2006, PreR, 147, 148
Siegenthaler, U., et al. 2005, Sci, 310, 1313

Smith, B. D., & Zeder, M. A. 2013, Anthr, 4, 8
Smith, F. A., Smith, R. E. E., Lyons, S. K., Payne, J. L., & Villaseñor, A. 2019, QSRv, 211, 1
Snir, A., Nadel, D., Groman-Yaroslavski, I., Melamed, Y., et al. 2015, PLoSO, 10, e0131422
Spahni, R., et al. 2005, Sci, 310, 1317
Steffen, W., et al. 2005, Global Change and the Earth System: A Planet Under Pressure (Berlin: Springer)
Steffen, W., Broadgate, W., Deutsch, L., Gaffney, O., & Ludwig, C. 2015, AnthRv, 2, 81
Steffensen, J. P., et al. 2008, Sci, 321, 680
Srokosz, M. 2004, EOSTrAGU, 85, 78
Stuiver, M., et al. 1998, Radcb, 40, 1041
Suess, H. E. 1965, JGR, 70, 5937
Svensson, A., et al. 2008, CliPa, 4, 47
Svensen, H., Planke, S., Malthe-Sørenssen, A., Jamtveit, B., Myklebust, R., Rasmussen Eidem, T., & Rey, S. S. 2004, Natur, 429, 542
Tamisiea, M. E. 2011, GeoJI, 186, 1036
Tanno, K. I., & Willcox, G. 2006, Sci, 311, 1886
Tapley, B. D., et al. 2019, NatCC, 9, 358
Teller, J. T., Leverington, D. W., & Mann, J. D. 2002, QSRv, 21, 879
Thompson, D. M. 2022, WIREsCC, 13, e745
Tierney, J. E., & deMenocal, P. B. 2013, Sci, 342, 843
UN Dept Economic and Social Affairs Population Division 2022, World Population Prospects 2022: Summary of Results UN DESA/POP/2022/TR/NO 3 (New York: United Nations)
UN Environment Programme 2021, Emissions Gap Report 2021: The Heat Is On—A World of Climate Promises Not Yet Delivered (Nairobi: United Nations)
UN Population Division 2022, World Population Prospects, https://population.un.org/wpp/Download/ [accessed 12 April 2023]
Velicogna, I., et al. 2020, GeoRL, 47, e2020GL087291
Vidal, C. M., et al. 2022, Natur, 601, 579
Walker, J. C., Hays, P. B., & Kasting, J. F. 1981, JGRC, 86, 9776
Warren, S., Eastman, R., & Hahn, C. J. 2014, Encyclopedia of Atmospheric Sciences, (Vol. 2, (2nd ed. ed. G. R. North, J. Pyle, & F. Zhang; Amsterdam: Elsevier) 161
Waters, C. N., et al. 2015, BAS, 71, 46
Waters, C. N., et al. 2016, Sci, 351, aad2622
Weiss, E., & Zohary, D. 2011, CAn, 52, S237
Wilde, S. A., Valley, J. W., Peck, W. H., & Graham, C. M. 2001, Natur, 409, 175
Wright, D. K. 2017, FrEaS, 5, 4
Wylie, D., Jackson, D. L., Menzel, W. P., & Bates, J. J. 2005, JCli, 18, 3021
Yuan, D., et al. 2004, Sci, 304, 575
Zachos, J., Pagani, M., Sloan, L., Thomas, E., & Billups, K. 2001, Sci, 292, 686
Zachos, J. C., Bohaty, S. M., John, C. M., McCarren, H., Kelly, D. C., & Nielsen, T. 2007, RSPTA, 365, 1829
Zeder, M. A. 2008, PNAS, 105, 11597

Chapter 3

Predictions of Our Future Global Climate: Models, Scenarios, and Projections

Anna Cabré Albós

> "Climate change is more than statistics. It's more than data points. It's more than net-zero targets. It's about the people. It's about the people who are being impacted right now."
>
> —Vanessa Nakate

Climate change poses an existential threat to humanity, evident through its current effects on both natural and human systems. Climate action requires cooperation on a global and local scale, while also tackling other pressing global crises. To achieve meaningful progress, we must approach climate action with a comprehensive, interconnected, and far-reaching perspective that places long-term human and ecological wellbeing at the forefront. How do we embark on the necessary transformations? Climate models and projections decades ahead serve as the building blocks of our strategy, guiding our actions to limit the impacts of climate change and simultaneously address social inequalities. In this chapter, we explore the history and state-of-the-art of Earth System models, and how the projected hazards and impacts, as well as the pathways for mitigation and adaptation, are communicated by scientists to other sectors via the United Nations Intergovernmental Panel on Climate Change with the aim to aid high-level global negotiations, economical decisions, and whole-of-society actions.

3.1 Why Do We Need Climate Models?

A general circulation climate model is used to simulate the climate on Earth or another planet over a long period of time. Unlike weather forecasts, which can only predict weather as far as 10 days into the future, climate models are created to learn about the long-term average spatial and temporal patterns in our world's climate, to answer questions such as: What drives the seasonality of this place? What are the

doi:10.1088/2514-3433/acfcb6ch3

interactions and feedback mechanisms between the different Earth elements, e.g. atmosphere, ocean, ice, and land? What are the main mechanisms behind climate natural variability? What is the climate predicted to be in 2050 or 2100? What was the climate on Earth before humans?

Models provide valuable information for times and places with scarce observations. Thanks to the combination of climate models and observations, we can now have a better idea about what has happened over geological timescales, what happens in remote parts of the world where there are few observations—such as the deep ocean—and what might happen in the future due to the anthropogenic emission of greenhouse gases.

Models help us understand the mechanisms behind well-observed climate patterns that are too complex to understand from the observations alone. Models are used to isolate the role of distinct mechanisms by creating parallel simulated worlds with different external forcings or different configurations, or by testing a set of hypotheses that might be relevant for a climate process. Hence, models are also used to answer questions such as: How was the climate when Antarctica was still attached to South America? How would the climate be in a world with a number of times more active volcanoes (as it was millions of years ago)? How would a world without mountains be? Or how would the Earth be if it had a different arrangement of continents? An ocean without biology? And more relevant for our immediate future, what will our world be like when the summer Arctic sea ice is completely gone (as is poised to happen in a decade or two)?

Moreover, models for the Earth help us strategize observation campaigns by targeting places and seasons most prone to significant variations. As an example, SOCCOM[1] (Southern Ocean Carbon and Climate Observations and Modeling) is an initiative that has put around two hundred robotic floats in the Southern Ocean, a crucial part of the ocean and the global climate that is comparatively unexplored. Increasing observations in remote locations is important to improve the reliability of models.

Climate models are not only used to understand Earth, but also to simulate climate on other planets, both inside and outside of our Solar System. Starting from basic information such as planet size, mass, tilt, and distance from the most nearby star, scientists can put constraints on possible climate scenarios within a range of possible atmospheric compositions, or hypothesize the presence or the absence of an ocean (Shields 2019). These models help identify the most promising planets for study and hence maximize the benefit of costly space missions.

Finally, models allow us to make predictions on seasonal, decadal, centennial, and millennial time scales. Of course models are extremely useful for climate change, as these allow us to plan the best mitigation pathways (reduction of greenhouse gas emissions), adapt to upcoming changes, and understand the most important feedbacks which are crucial to determine the feared tipping points. Models allow us to distinguish between natural and anthropogenic forcings and therefore quantify the role that humans play in climate, even in isolated extreme events in studies of

[1] https://soccom.princeton.edu/

attribution. Most climate scientists agree that we have entered a new geological epoch called the Anthropocene where our (human, e.g. *anthropos* in Greek) activities on Earth—i.e., the use of natural resources and emission of greenhouse gases to the atmosphere—represent the greatest contribution to changes in the Earth's climate. Geologically, we are living in an interglacial epoch, which would be followed by a glacial epoch 50,000 years from now if there was no human influence. Scientists predict that this transition has been delayed by at least another 50,000 years due to human-made global warming (Ganopolski et al. 2016); our activities are therefore altering the climate on very long time scales. We are now the biggest force in the climate and also the biggest source of uncertainty to the future of our climate.

3.2 How Do Climate Models Work?

The earliest and most basic numerical climate models were called Energy Balance Models (EBMs). These models examine the balance between the incoming solar radiation and the outgoing radiation to analytically determine the surface temperature of a planet. The first numerical simulations (i.e., not possible to solve analytically)—called General Circulation Models (GCMs)—were created later on, with the arrival and expansion of computers in the 1960s. These models consist of a complex set of mathematical equations that typically simulate at least the physics of the atmosphere, the ocean, or the atmosphere-ocean coupled system, and have become more sophisticated with time, incorporating more elaborate coupling between systems. Nowadays, climate models also include ice, land, vegetation, chemical tracers, and carbon cycling, and are usually called Earth System Models (ESMs), as they are trying to simulate the full complexity of the Earth System. See a timeline of the history of climate modeling (Hickman 2018)

The operation of these sophisticated models is based on physical and chemical principles. All models include fundamental physics such as the conservation of energy, momentum, and mass in fluid dynamics, or the mechanics of the orbits of the Earth around the Sun. Some equations need to be approximated to account for the discretization of continuous equations, e.g. Navier-Stokes equations of fluid motion, which describes the speed, density, pressure, and temperature of the gases in the atmosphere and the water in the ocean. Models also include parameterized equations to describe empirically well-known processes that are not driven from fundamental principles, e.g. evaporation as a function of wind speed and humidity, or to explicitly describe complex processes, e.g. biogeochemical processes in vegetation on land or phytoplankton growth in the ocean. Climate scientists divide the world in a 3-dimensional spatial grid (e.g. latitude-longitude-height or depth), where they can apply the basic equations into a temporal grid with numerical methods on a computer. The outputs usually include hourly, daily, and monthly timeframes, and variables such as temperature, humidity, wind speed, rainfall, ice cover, radiation fluxes, or salinity and pH in the oceans, among many others, typically stored in the NetCDF data format.[2]

[2] https://www.unidata.ucar.edu/software/netcdf/

3.3 What Are the Components of Climate Models?

Most complex Earth system climate models consist of four components (Stocker, 2022) (see Figure 3.1):

1. Atmosphere: Includes Solar radiation energy balance, formation of clouds, precipitation and evaporation, atmospheric heat and water vapor transport, heat exchange with oceans and land. It also accounts for atmospheric gases, dust, aerosols, and chemical tracers.
2. Ocean: Includes global currents and mixing, inflow exchange between the different ocean basins, transport of water masses, transport of heat in the ocean, exchange of water vapor, CO_2 and other gases with the atmosphere, exchange of heat between the ocean and the atmosphere, and chemical and biological carbon cycle, as the ocean is the dominant reservoir of carbon in the global earth system.
3. Land Surface: Includes the structure and position of the continents as a determining factor of the climatic zones and the ocean currents, changes in sea level, transformation of short-wave to long-wave radiation, albedo of the Earth's surface (e.g. reflectivity), dust reservoir, transfer of momentum and energy, and also vegetation, carbon cycle and storage, snow cover, soil moisture, and rivers.
4. Sea ice: As ice melts and expands, the model includes the changes in absorption and reflection of solar radiation (albedo) (e.g. ice reflects 80% of the incoming radiation), a crucial feedback mechanism for cooling and warming the poles. It also simulates the release of salt to the ocean when sea ice is created every winter, crucial for the creation of deep water currents.

In the present, the most sophisticated models also include ice shelves and glacial dynamics, the thawing of permafrost, and more complex biological processes, both

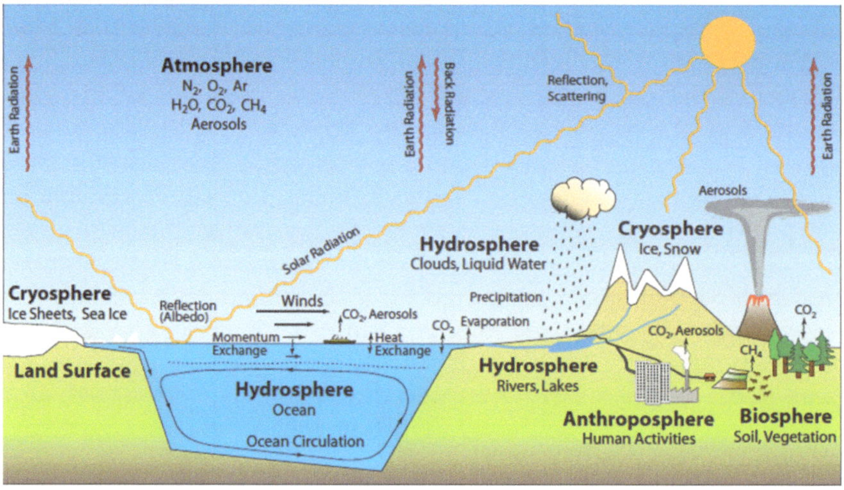

Figure 3.1. Main components and associated processes of the climate system on a global scale (from Stocker 2022, available from www.climate.unibe.ch/stocker).

on land and in the ocean. It is worth noting that scientists include in models the basics of the climate system, but it is the model itself that creates storms, droughts, or sea ice.

3.4 Limitations of Models and Sources of Uncertainty

No model is perfect. These are some of the factors that contribute to their limitations (e.g, McSweeney & Hausfather 2018; The emergent patterns of climate change by Schmidt[3]):

3.4.1 Computing Power

The amount of calculations that a model can handle is constrained by computing power, which limits the level of complexity as well as the spatial and time resolution that can be coded in a model. Note that a typical climate model has around a million lines of code. Nowadays, a coarse resolution global climate model has grid cells of around 100 km wide and tens of vertical levels. The most precise global models have a resolution of around 1 km. Since in general models are three-dimensional, increasing the resolution by a factor of two requires about 8 times the computing power. Regional models overcome the spatial limitation by downscaling global models to a particular region, so they can increase spatial resolution and have a finer structure detail of processes at a local level. In general, the accuracy of simulations improves at higher resolution because a lot of processes depend on it. For example, mountains are important for the formation and location of clouds, rain, and winds. At low resolution, models cannot distinguish well between peaks and valleys in mountains, but once resolution is improved, these types of inaccuracies are quickly reduced. The same happens with bottom-sea topography, crucial for the formation of ocean currents and eddies.

3.4.2 Parametrization

Many processes occur on smaller scales and/or faster than the model spatial/temporal resolution, which means that the model is not able to capture them well, and therefore these processes need to be parametrized. Parametrization is also used to describe processes that are not fully understood or that do not derive from fundamental physics; and it is one of the main sources of uncertainty. In order to fine tune the model parameters, a system of feedback is used to optimize and recalibrate the parameters by comparing the processes and output with observational data. Even after calibration, the modeled Earth system is not completely in balance, e.g. the radiation balance at the top of the atmosphere is usually tuned *a posteriori*. This particular adjustment is necessary mostly because of clouds and albedo uncertainties. Climate modelers are working to reduce biases by correcting the underlying causes of bias instead of overcorrecting with calibration.

One of the negative consequences of prioritizing precision is that most models will only run tens of ensembles with different initial conditions, which is short of the ideal

[3] https://www.ted.com/talks/gavin_schmidt_the_emergent_patterns_of_climate_change

number needed to calculate model uncertainty. However, single-model uncertainty is usually lower than multi-model uncertainty, which tells us that most uncertainty in modeling comes from the set of parameters used, not from initial conditions within a model, and that in general multi-model averages perform better at reproducing observations than single-model ensemble averages.

3.4.3 Spatial and Temporal Coverage

Another limitation in climate models is the paucity of observations as well as their sparsity, specifically their lack of continuity in time and space. Models do better where there are plenty of observations because that allows long-term calibration and a better understanding of the underlying processes driving climate. Luckily observations and especially remote sensing have increased exponentially since the era of satellites began in the 1970s, and most recently with robotic observations. Still, most observations cluster at the surface and focus more on physical indices such as temperature, salinity, and less on indices describing chemistry and biology. Moreover, the biological empirical relationships do not derive from fundamental principles, vary across regions, and are more difficult to incorporate successfully into models. Ice sheets, fundamental for centennial climate change, are only starting to be added to Earth System models now, and land modeling is a relatively new niche.

However, it is important to note that climate is mostly driven by the ocean and the atmosphere on decadal and centennial timescales. The addition of ice and land is not going to change climate predictions in a fundamental way, but still need to be incorporated especially because melting of the Antarctica and Greenland glaciers is going to play an increasing role in the future.

3.4.4 Fundamental Knowledge

The other fundamental limitation is the level of understanding. Some processes are intrinsically very complex and therefore difficult to describe with equations—and this is reflected in models. One of the main limitations is the representation of clouds. Clouds cover around two thirds of the Earth at any given time—and can warm or cool the atmosphere, depending on the type of cloud and time of day. Clouds form and disperse in minutes so there are not many observational records. Yet, they are a crucial factor for climate so their uncomplete description represents one of the greatest sources of uncertainty in models. Related to the uncertainty in clouds, global models show a persistent double Intertropical Convergence Zone (ITCZ). The ITCZ is a low pressure system where most of the tropical rain occurs, and is located north of the equator most of the year, although it moves by a few degrees throughout the year and it splits into two ITCZs for a brief period every year. Current climate models exaggerate this split into two ITCZs, leading to the well-known double-ITCZ bias of the models throughout the year. Jet streams also appear to be too strong and straight in climate models. In the ocean, deep water formation is mostly recreated but sometimes via the wrong mechanism. In Antarctic waters, most models cannot simulate the continental shelf sinking of heavy waters mostly due to a lack of resolution. Instead they create deep water formation through mixing events further from the coast. As expected, as models increase in resolution, these processes are better captured.

3.4.5 Natural Variability

Natural variability is the greatest source of uncertainty when attributing observed climate changes to anthropogenic causes; natural variability can hinder or amplify climate change. Considerable research goes into understanding the natural sources of variability so that we can model them accurately and predict the moment of emergence of climate change as a signal significantly different from a natural cycle. At the beginning of the 21st century, the rise in global temperatures went through a brief hiatus which was widely used by climate skeptics as proof of no global warming. The truth is that a strong El Niño in 1998 that generated above-average warming was followed by La Niña and other natural processes that cooled the planet. This cooling made it harder to see the ongoing anthropogenic warming for about a decade. However, since then the warming has increased at a higher rate than ever; and it is clear that the warming that we observe is due to human activities. Other events—such as a weakening of the thermohaline circulation—cannot yet be attributed to humans as the natural variability is too large and slow (low frequency) for the signal to be significantly detectable yet. This natural variability is especially important when assessing climate and carbon feedback and tipping points, and this is why climate predictions are so important. We might not see it yet, but thanks to models we have a good idea of the timeline of emergence of different impacts. Finally, while it is still difficult to attribute specific extreme climate events, such as hurricanes, to anthropogenic climate change, we can affirm that extreme weather events such as torrential downpour, floods, and prolonged droughts are now significantly more intense and frequent than before, and we can estimate how much more likely these events are in a warmer world. As an example, the World Weather Attribution[4] "uses weather observations and climate models to understand how climate change influences the intensity and likelihood of extreme weather events."

3.4.6 Validation

In order to validate climate models, scientists compare them with past observations in what's called hindcast modeling. Climate models are run over the historical period, from around 1850 to near-present. They run the simulation with known climate forcing factors, such as greenhouse gas emissions, variations in the solar incoming radiation, aerosols from recorded volcanic eruptions and human activity, and land-use changes. Then they compare the output temperature, rain, snow, sea ice extent, or winter to summer differences, to recorded observations. It has been shown that models respond well to big climate forcings such as volcanic eruptions that significantly modify temperature and humidity in the following years. Climate models also do well in terms of mean climatological state, as they are able to

[4] https://www.worldweatherattribution.org/

reproduce the changes between winter and summer. In general, the ensemble average of several models is more accurate and reliable than any model alone. Intercomparison projects are used to evaluate models with the same exact climate forcings. In some cases, models are optimized to perform better within a specific realm, for example sea ice extent, and used specifically for this purpose. Models since the 1970s have also done a good job predicting future (now present) global warming so far.

3.4.7 Aspects That We Can Trust

Overall the strong correspondence between modeled and observed temperatures increases scientists' confidence that models are accurately capturing both the factors driving climate change and the level of short-term natural variability in the Earth's climate. Most findings from models have been proven robust and not varied much for decades, such as the warming effect of greenhouse gases emissions and the ocean role in climate. Models have continuously improved with time due to an increased understanding of climate processes, better parametrization of the processes, better resolution, and an expansion in the extent of observations.

ESMs (e.g. Earth System Models) reproduce very well the observed historical temperatures on average, capture the level of natural variability and the main factors driving climate change, and so the hypothesis is that they can reproduce well what happens in the future. The biggest uncertainty now is our collective behavior. Are we going to shift successfully to sustainable energy? How long is it going to take? Are we going to keep burning fossil fuels until the end of their supply? Are we going to shift the transport systems? Are we going to change how we manage land and our food systems? Are we going to find scalable carbon sequestration solutions?

Ultimately, models are subject to some trade-offs. We want as much predictive power as possible. At the same time, we also want the model to be as simple as possible; otherwise we run into overcomplexity and lack of CPU power. In order to improve models, observational, instrumental, and theoretical science is advancing in parallel so that all this new knowledge can be fed into modeled equations. The future of modeling is trending towards still higher resolution as well as the inclusion of artificial intelligence and deep learning to complement the current models and help solve some of the fundamental problems.

3.5 History of IPCC and CMIP Models

In 1988, the United Nations and the World Meteorological Organization formed the Intergovernmental Panel on Climate Change[5] (IPCC). This is the world leading initiative publishing reports on climate change, with the aim to "provide the world with a clear scientific view on the current state of knowledge in climate change and its potential environmental and socio-economic impacts."

Shortly after, in 1995, the World Climate Research Programme (WCRP) started the Coupled Model Intercomparison Project (CMIP), which aims to create a

[5] https://www.ipcc.ch/

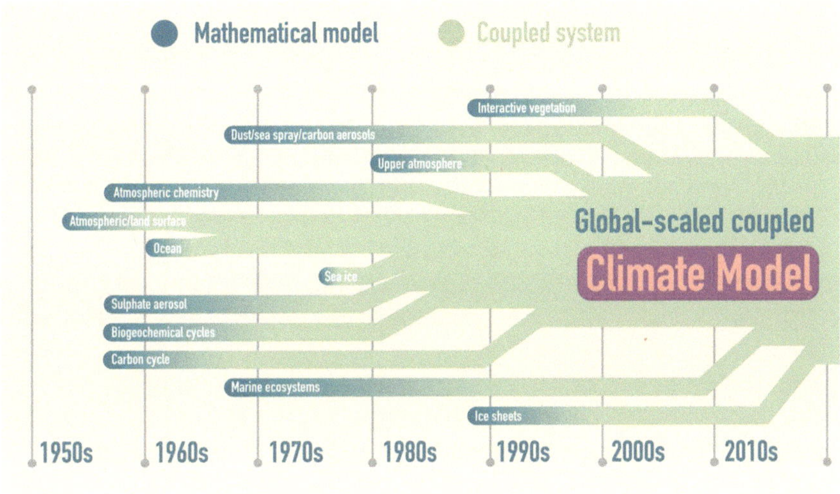

Figure 3.2. Evolution of climate modeling since the 1950s by Ichiko Sugiyama (based on original figure from Carbon Brief https://www.carbonbrief.org/qa-how-do-climate-models-work/)

"standard experimental protocol" for designing and studying the climate change projected scenarios of Earth System Models in the decades ahead.

Since then and every 6–7 years, climate scientists around the globe have joined efforts to publish a new IPCC Assessment Report, in what is very likely the most scrutinizing referee process in the scientific world. The report updates the trends in historical observations and also the projections into the future using climate models that research institutions submit following the standard protocol defined by CMIP. These IPCC reports are used by governments, institutions, companies and are the go-to reference for climate decision-making all over the world. The models used have evolved a lot in the last few decades, from including solely physics in the atmosphere and oceans, to including biology, interactions, land and ice dynamics, and more (see Figure 3.2). As CPU power is enhanced, climate modelers can increase the spatial and temporal resolution used. For these fully coupled ESMs, the spatial resolution has gone from 500 km in the 2000s to better than 100 km in the most recent CMIP6 models.

Climate models are run for hundreds or thousands of (simulated) years to ensure that they reach equilibrium before adding greenhouse gas emissions. Once a model is assumed to be in equilibrium, it is run from pre-industrial conditions (1850) until now with prescribed observed greenhouse emissions in what's called historical simulations. These simulations are only forced by greenhouse gases but the rest is free running. In order to continue the simulation into the future, we need to include a model of human behavior, beyond climate. Integrated Assessment Models are socio-economic models that simulate how population, economic growth, and energy use interact with climate, and help inform policy through indices such as the social cost of carbon. These models are used to predict the levels of greenhouse gas emissions in the future given a range of scenarios that then climate modelers can include in their simulations.

The fifth published Assessment Report (IPCC AR5) was published in 2013 with the output from 50 CMIP5 models from over 30 climate groups, and used the projections from four Representative Concentration Pathways (RCP) scenarios (IPCC 2013, Climate Change 2013: The Physical Science Basis). These projected scenarios describe a variety of trajectories of carbon dioxide concentrations from 2000 to 2100 based on different assumptions that come from Integrated Assessment Models. These scenarios do not portray specific pathways to the future, but they are useful to encompass a wide range of possible climate outcomes for the 21st century. The scenario numbers (e.g., RCP8.5) refer to the radiative forcing by 2100 (in $W{\cdot}m^{-2}$). Note that these scenarios input into the models the prescribed concentration, derived but different from the prescribed emissions. This facilitates the comparison among models with the disadvantage that it ignores some climate feedbacks.

RCP scenarios have been widely criticized for not corresponding to a particular future scenario in the socioeconomic domain. This is why the new 'Shared Socioeconomic Pathways' (SSPs) were created in time for the most recent IPCC report (The Physical Science Basis, IPCC 2021), in which around 50 global scientific institutions produced the next generation of around 100 climate models. These pathways now explore the emissions related to different socioeconomic theories. They include: a world of sustainability-focused growth and equality (SSP1); a "middle of the road" world where trends broadly follow their historical patterns (SSP2); a fragmented world of "resurgent nationalism" (SSP3); a world of ever-increasing inequality (SSP4); and a world of rapid and unconstrained growth in economic output and energy use (SSP5).

Apart from the historical and future scenario, a pre-industrial extended simulation is also run from 1860 onwards. This control simulation describes how the world would be if greenhouse emissions had remained at pre-industrial levels. Control runs are useful to understand natural variability separated from other changes, especially the ones due to increasing greenhouse gas emissions, and to diagnose and correct for long-term model drifts unrelated to external climate forcings.

It is crucial to communicate that the IPCC reports are trustworthy due to the strict referee process, but also conservative, due to the need for consensus and overcaution. Hence, we should take the results very seriously and expect that the impacts might even be worse than stated. The last synthesis report (IPCC 6th Assessment Report) says: "For any given future warming level, many climate-related risks are higher than assessed in AR5" (Assessment Report 5).

3.6 Other Types of Models

In order to research the Earth's geological climate patterns, scientists use **paleo-climate models**, which are also useful to understand the rapid changes we are experiencing now in comparison to past sudden climate events, and also to predict the long-term extent effects of the Anthropocene (Haywood et al. 2019). These models are simpler than the most complex ESMs because they are modeling much longer time scales and therefore only incorporate processes that matter at centennial–millennial scales.

Reanalysis models aim to mimic the recent past climate as accurately as possible, with the objective to provide complete maps that fill the gaps in observations and therefore help scientists understand climate processes and interannual variability. Reanalysis models combine observations with past short-range weather forecasts in a process called data assimilation. They use the same technology as weather forecasts, but at lower resolution and applied to the recent past climate. These models are used by scientists as almost-real representations of observations. These are different from **hindcast models**, which also model the near past but unlike reanalysis models, only include observed climate forcings (e.g. volcanic eruptions, solar radiation, greenhouse gases) and run freely without data assimilation.

As we have said, global models represent well the overall patterns of climate, but have several biases especially at regional scales. **Regional climate models** are a better option when the objective of the analysis is to focus on a particular part of the world. These models are derived from global models, downscaled to a particular area. Since they cover a smaller area, they can run quicker and at a higher resolution than global models, which fixes some biases due to low resolution. The Coordinated Regional Climate Downscaling Experiment[6] CORDEX is the largest intercomparison project on regional models.

3.7 Climate Change Predictions: What All of This Tells Us about Our Future

There is compelling evidence supporting the reality of a warming planet, derived from a wide array of independent indicators that span from the upper atmosphere to the depths of the ocean (e.g. Figure 3.3). These indicators include the rise in surface-land, atmosphere, and ocean-temperatures, the increase in ocean heat content at all depths, the retreat of glaciers and the reduction in snow cover and sea ice, the sea level rise, and the increase in atmospheric water vapor, among others.

The most recent IPCC reports (e.g. IPCC 2019a, IPCC 2019b, IPCC 2021) provide a summary: "Observed increases in well-mixed greenhouse gas (GHG) concentrations since around 1750 are unequivocally caused by human activities." As a consequence, global surface temperature was 1.1°C higher in the second decade of the 21st century compared to the preindustrial period (1850–1900). This warming is a direct consequence of the long-time accumulation of greenhouse gases in the atmosphere, and is predicted to increase by as much as 4.4°C by the end of the century (2080–2100) under the worst case scenario, the SSP5-8.5. All the modeled possible scenarios predict a gradual increase in global temperature that persists long after the greenhouse gas emissions peak. Warming on land is going to exceed warming over the ocean on average by a factor of 1.4 to 1.7 at the end of the 21st century. This temperature change will not be uniform across regions. The observed level of warming is unprecedented in at least the last 2000 years.

The Arctic region is warming at least twice as fast as the rest of the world, due in part to the increased melting of summer sea ice, which significantly decreases albedo

[6] https://cordex.org/

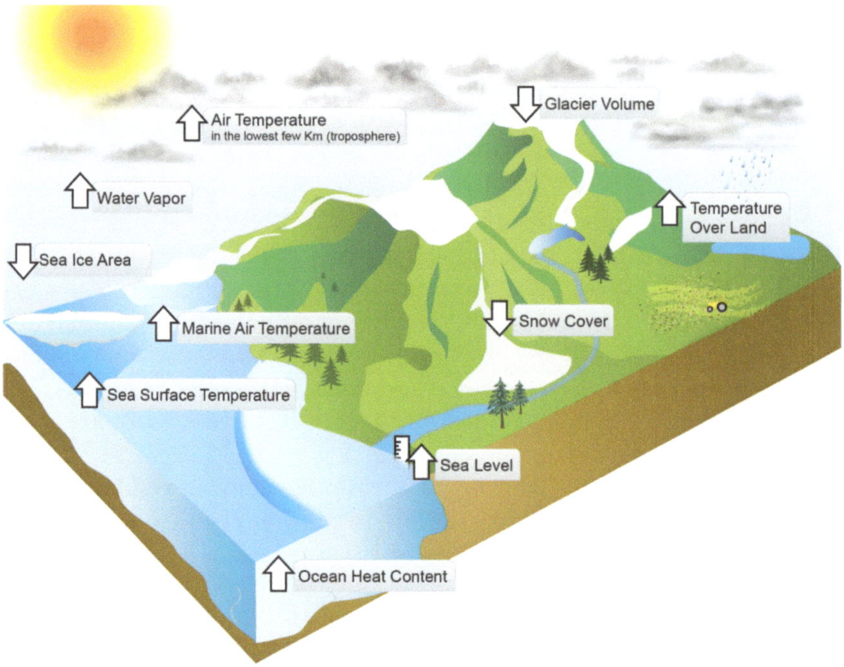

Figure 3.3. Independent analyses of many components of the climate system that would be expected to change in a warming world exhibit trends consistent with warming (arrow direction denotes the sign of the change) (from IPCC 2013, Chapter 2, FAQ 2.1, Figure 1).

as the dark ocean mirrors back less light than the white ice. The absorption by the oceans feeds back into further warming and further melting of sea ice (albedo positive feedback). Part of the melting is also due to changing ocean currents and weakening of the polar jet stream, whose intensity depends on the temperature difference between the Arctic and the tropics. This polar amplification is also intense in West Antarctica, but not as much in East Antarctica. Global warming has already shrunk all the components of the cryosphere. Besides the Arctic reduction in sea ice extent and thickness, global warming is shrinking glaciers, snow cover, and permafrost extension, and is predicted to accelerate the ongoing melting in Greenland and Antarctica. Mountain glaciers and ice sheets are melting and currently contribute about two thirds to sea level rise (the other third is due to the thermal expansion of the ocean under warming). In the future, the contribution from Greenland and Antarctica will dominate the sea level rise.

The increase in temperatures is intensifying the water cycle and is the main cause for the increase of extreme weather events—hurricanes, intense storms, flooding, droughts—which are the earliest to be noticed and to impact human lives. Warm air can hold more water vapor before it rains it out (Clausius—Clapeyron law), which in general, translates into an increase in global precipitation that is, however, highly dependent on location. The intensified hydrological cycle is predicted to increase the contrast between wet and dry regions and seasons. In the atmosphere, mean sea level pressure is projected to decrease in high latitudes and increase in the mid-latitudes as global temperatures rise. In the tropics, the tropical Hadley and Walker Circulations are likely to slow down (Baker 2022). Poleward shifts in the mid-latitude jets of about 1 to 2 degrees latitude are predicted by the end of the 21st century under the business-as-usual scenario in both hemispheres, with weaker shifts in the NH. However, the recovery of the ozone hole during austral summer in the Southern Hemisphere opposes the changes due to greenhouse gas emissions.

Most of the anthropogenic heat (91%) and approximately 25% of the anthropogenic CO_2 has been absorbed by the ocean, mitigating the amount of climate change that we observe, or feel immediately, on land. The rest of the heat goes into warming the land (5%), ice (3%), and the atmosphere (1%). As a consequence, the surface ocean has already warmed by 0.88°C and the heat has reached the deepest regions of the ocean. Ocean warming contributes to a third of the current sea level rise, as warm water dilates. Warmer water is the main contributor to an intensified water cycle and the intensification in frequency and intensity of extreme events such as storms and flooding in coastal areas. Warm water combined with fresh water from intense melting of ice is predicted to stabilize the water column and modify the world currents driven by changes in temperature and salinity. This also has consequences for biological activity in the ocean as the main producers, phytoplankton, live near the nutrient-depleted sunlit surface, and therefore need strong mixing from waters below to get the necessary nutrients. The biological interplay between nutrients and organic carbon production is predicted to change due to climate change, in general terms benefiting phytoplankton growth in the high latitudes, but hindering their growth in low latitudes. A warmer ocean also decreases in average the oxygen levels necessary for life in the ocean. Poleward migrations towards colder regions are predicted for most fish stocks. The warming of the ocean has already led to massive coral bleaching and the death of half of the shallow-water corals on Australia's Great Barrier Reef. With a 2°C global rise in temperatures, 99% of tropical coral reefs are projected to die. The physical and chemical oceanic carbon uptake is acidifying the ocean and corroding shell organisms, among other effects in the biological and chemical cycles, and reducing the efficiency of CO_2 absorption into the ocean.

On land, the predictions are increased warming and increased frequency, intensity, and duration of heat events. Droughts are projected to increase particularly in the Mediterranean region and sub-Saharan Africa, while extreme rainfall events are going to increase in many regions. Climate zones are projected to shift poleward in the middle and high latitudes. Current and predicted levels of warming are predicted to increasingly impact boreal forests with droughts, wildfires, and pest outbreaks, and also increase dryland water scarcity, soil erosion, vegetation loss, permafrost thawing, and tropical crop yield decline. All of these changes are already affecting food and water security, the access to a healthy life, while exacerbating social inequalities among countries, and within societies.

How to communicate these changes broadly? A rapid growth in human prosperity has only been possible in the last 10,000 years thanks to a very stable climate period, the Holocene, compared to other periods on Earth history. We are rapidly deviating from this climate stability upon which our survival depends. Global warming is nearly proportional to cumulative CO_2 emissions, which means that the level of current warming is already accounting for most cumulative past emissions, or that, perhaps more importantly, *future warming depends mostly on future emissions*. This is important to note when fighting discourses of climate delay or doomism. There is a bit of delayed response in the Earth system temperature, but it is relatively short compared to the timescale of human inertia determined by our dependence on fossil fuels. Dvorak et al. (2022) show that "following abrupt cessation of anthropogenic emissions"—an unrealistic experiment in the socio-economic realm—"decreases in short-lived aerosols would lead to a warming peak within a decade, followed by slow cooling as GHG concentrations decline."

3.8 Abrupt Changes and Tipping Points

Although most countries have pledged to reduce their greenhouse-gas emissions, the current national commitments would result in global warming by at least 2.6°C by 2100 (Hausfather & Forster 2021).[7] In order to keep the warming below a maximum of 1.5°C, the internationally agreed preferred pathway, the global and national ambitions need to be much more transformational and need to be implemented faster. The hope is that advancements in these initial pledges will push governments, companies, and society to strive for better commitments, and a cascading effect across nations and sectors will finally transform society in the time scale that we need to achieve the Paris mitigation goal.

Now, one may ask: what happens if the temperature keeps increasing? Climate tipping points triggered by anthropogenic forcing are relatively sudden changes in terms of geological timescales, and can push the Earth system into a different state and feed back dangerously into further temperature rise. While most changes are gradual and will eventually cease if we stop emitting greenhouse gases, the Earth system is not linear, which means that certain forcings are going to transform the Earth irreversibly for very long timescales from centuries to millennia. The changes can self-perpetuate and feature hysteresis and/or the climate can settle into alternative stable states. This is true especially for recovering from ice loss or ocean heat accumulation.

Most recently, Armstrong McKay et al. (2022) build on previous research (e.g. Lenton et al. 2019), and identify nine global tipping elements "which contribute substantially to Earth system functioning" and seven regional tipping elements "which contribute substantially to human welfare or have great value as unique features of the Earth system," five of which might have already been crossed (Figure 3.4). Most of these tipping points are related to the melting of ice under global warming and thus caused by changes at high latitudes or high mountains, but they have a global effect. The Arctic is rapidly losing sea ice. The chance of becoming an ice-free region in summer increases substantially at 2°C warming, with consequences for climate all over the world and local ecosystems, as an ice-free

[7] https://www.carbonbrief.org/analysis-do-cop26-promises-keep-global-warming-below-2c/

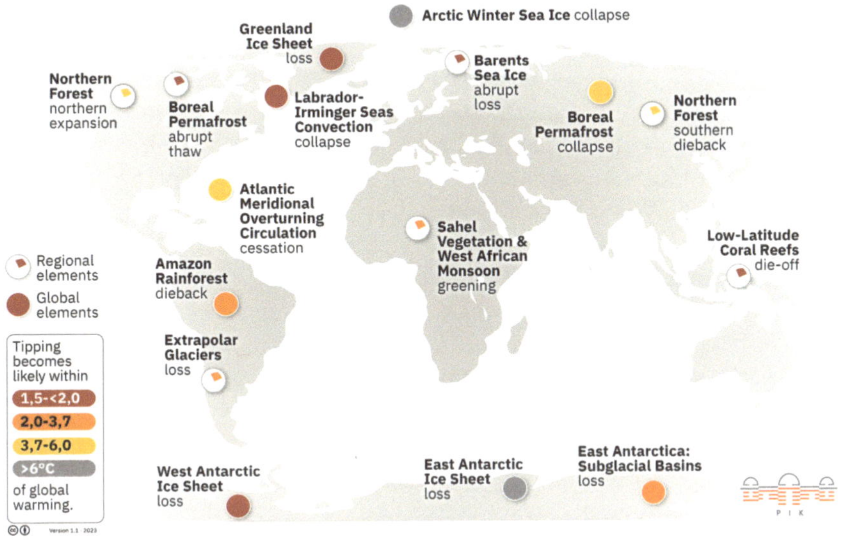

Figure 3.4. The geographical distribution of global and regional tipping elements, color-coded according to the best estimate for their temperature thresholds, beyond which the element would likely be "tipped." Figure designed at PIK, based on Armstrong McKay et al., Science (2022).

region reflects much less solar radiation and so it warms even quicker, even though this is not considered a tipping point as it does not self-perpetuate. The permafrost across the Arctic, a deep layer of frozen underground, is warming and starting to irreversibly thaw. This reactivates the biological cycle in the soil, which releases methane on top of CO_2—methane is a greenhouse gas 30 times more potent than CO_2 over a 100-year scale—which causes further warming and thawing. The Greenland ice sheet is also melting rapidly and will continue doing so for centuries. Some models suggest that the irreversible tipping point with no return could happen at only 1.5°C warming, or that perhaps we have already passed it. As part of the subpolar gyre, convection in the Labrador Sea in the North Atlantic might have already collapsed according to models as stratification increases.

In Antarctica, the paleontological records and recent observations point to a widespread collapse of the West Antarctic ice sheet, leading to 3 meters sea-level rise on timescales of centuries to millennia. The Amundsen Sea embayment of West Antarctica is retreating rapidly and might have already passed a tipping point, which could destabilize the rest of the West Antarctic ice sheet. On the other side of Antarctica, the Wilkes Basin, part of the East Antarctic ice sheet, is also becoming unstable, which could add another 3 or more meters to long-time sea level rise.

Arctic region warming and melting are increasing the flux of fresh water into the North Atlantic, hindering the formation of deep waters contributing to slowing down the Meridional Overturning Circulation (AMOC), the global conveyor belt that transports heat and salt across the oceans. The slowdown of AMOC (possibly detected from the 1950s) is one of the most speculated and feared tipping points, as it could destabilize the West African and the East Asian monsoons that control where and when it rains in the region of Sahel or the Amazon, and could cause an increased warming of the Southern Ocean, increasing the Antarctic ice sheet melting.

In the biosphere, we might have already passed a tipping point in the tropical oceans and forests. Ocean heating has led to mass coral bleaching events. Around half of the surface water corals on Australia's Barrier Reef are already dead, and the prediction jumps to 99% in the event of a 2°C temperature rise. Coral reefs host a big proportion of the global marine biodiversity, and feed millions of people. Finally, the other worrying tipping point is the drying out of the world's largest rainforest, the Amazon, which has its own climate system and hosts 16,000 different tree species. The forest plays an important role in recycling water. Scientists believe that this feedback between the ecosystem and the water cycle can eventually break if the area of the forest is significantly reduced via deforestation, fires, and warming, and becomes an ecosystem more like a savannah. We might be nearer than predicted to this dangerous tipping point, which would feed back into climate warming and the carbon cycle.

In summary, we are dangerously increasing the chances of a West Antarctic and Greenland ice sheet collapse, the die-off of warm-water corals, and abrupt thawing of permafrost. Labrador convection has probably already collapsed. Above 1.5°C warming, high-latitudes melting becomes more widespread and mountain glacier loss becomes very significant. See this Carbon Brief summary on tipping points.[8]

3.9 Meeting the 1.5°C Threshold

During the Conference of the Parties (COP) 21 in Paris in 2015, 196 parties adopted the Paris Agreement,[9] a legally binding international treaty on climate change. Its goal is to "limit global warming to well below 2, preferably to 1.5 degrees Celsius, compared to pre-industrial levels." Despite numerous well-founded criticisms, it has been a major diplomatic success, able to plant the initial seed in companies and governments to design a roadmap to a sustainable future. But where does this 1.5°C threshold come from?

Since 2008, a group of Least Developed Countries, especially Small Island Developing States, and others have been calling for limiting the global temperature rise to 1.5°C above pre-industrial levels to prevent the worst of climate change. The level of warming that we are experiencing today is already hurting everyone, but especially people living on low islands and in vulnerable countries and communities. Warming by 1.5°C (again, with respect to pre-industrial levels) will trigger lasting changes in many natural systems that will irreversibly transform life as we know it, but the assumption is that the human species can adapt under this optimistic scenario while it would be much more difficult and costly once we pass this threshold. Climate and life systems are nonlinear and some changes might soon become irreversible, so it is crucial to communicate effectively that any degree of

[8] https://www.carbonbrief.org/explainer-nine-tipping-points-that-could-be-triggered-by-climate-change/

[9] https://unfccc.int/process-and-meetings/the-paris-agreement/the-paris-agreement

warming that we can avoid is key for survival and wellbeing on Earth now and for generations to come.

The IPCC special report "Global warming of 1.5°C" came after the Paris Agreement, (IPCC 2018), and reports on the impacts of global warming of 1.5°C, exploring "related global greenhouse gas emission pathways, in the context of strengthening the global response to the threat of climate change, sustainable development, and efforts to eradicate poverty." The report concludes once more that the climate-associated risks for natural and human systems are higher for a global warming of 1.5°C than at present. A 1.5°C world is not safe for all, but the risks associated with this warming are much lower than, for instance, 2°C. Global warming is *likely* to reach 1.5°C between 2030 and 2052 if it continues to increase at the current rate, although we could sporadically reach this level in the following 5 years. To avoid the 1.5°C threshold, the world would have to halve emissions by 2030 (by 45% from 2010 levels) and reach net-zero emissions by 2050. There are different ways to get there, but all the scenarios require major and immediate transformation, a low-carbon unprecedented transition that affects all the sectors, from electricity to the food sector, and that must be accompanied with a just transition. It is also important to notice that all the possible pathways to a 1.5°C world, including the most optimistic, rely upon additional carbon removal on scales that do not exist yet. The report is also a call to action of all sectors of our society necessary to achieve transformative social change.

The good news is that the inclusion of a 1.5°C temperature limit in the 2015 Paris Agreement was a major victory for vulnerable countries after a long fight, and that this goal is still alive after the coronavirus pandemic and growing geopolitical turmoil. We need to communicate that the 1.5°C goal is achievable if a massive transformation happens across all sectors of our society. The problem is that this temperature threshold represents only a 0.3°C increase from the global temperature in 2022, already 1.2°C above pre-industrial levels. Therefore, many fear that it is too ambitious given the current state of the socio-economic system. In such a crucial moment, one way of communicating the 1.5°C goal is through the social justice lens, describing this temperature threshold as a lifeline for the most vulnerable countries, acknowledging that it is a very ambitious goal that needs to be found by equally ambitious actions and systemic social change, a reason for hope for many.

After the release of the 6th IPCC report, intergovernmental negotiations are at a crucial point as the first global stocktake (GST) occurs (2022–2023)—"a process for taking stock of the implementation of the Paris Agreement with the aim to assess the world's collective progress towards achieving the purpose of the agreement and its long-term goals." The good news is that mitigation and adaptation efforts are already occurring everywhere and there are indications of ongoing social change. We could be heading in the right direction, but again, we need changes across all sectors and levels of society to happen much quicker and in a more transformative way (e.g. Emissions Gap Report 2022,[10] State of Climate Action 2022[11]).

[10] https://www.unep.org/resources/emissions-gap-report-2022

[11] https://www.wri.org/research/state-climate-action-2022

References

Armstrong McKay, D. I., et al. 2022, Sci, 377, 7950
Baker, A. J. 2022, NatCC, 12, 615
Dvorak, M. T., Armour, K. C., Frierson, D. M. W., Proistosescu, C., Baker, M. B., & Smith, C. J. 2022, NatCC, 12, 547
Ganopolski, A., Winkelmann, R., & Schellnhuber, H. J. 2016, Natur, 529, 200
Hausfather, Z., & Forster, P. 2021, Analysis: Do COP26 promises keep global warming below 2C? [WWW Document]. Carbon Brief. https://www.carbonbrief.org/analysis-do-cop26-promises-keep-global-warming-below-2c/ (accessed 6.15.23)
Haywood, A. M., Valdes, P. J., Aze, T., et al. 2019, Earth Syst Environ, 3, 1
Hickman, L. 2018, Timeline: The history of climate modelling (Carbon Brief) https://www.carbonbrief.org/timeline-history-climate-modelling/ (accessed 6.15.23)
IPCC 2013, Climate Change 2013: The Physical Science Basis. Contribution of Working Group I to the Fifth Assessment Report of the Intergovernmental Panel on Climate Change (Cambridge: Cambridge Univ. Press) 1535 https://www.ipcc.ch/report/ar5/wg1/
IPCC 2018, Global Warming of 1.5°C. An IPCC Special Report on the impacts of global warming of 1.5°C above pre-industrial levels and related global greenhouse gas emission pathways, in the context of strengthening the global response to the threat of climate change, sustainable development, and efforts to eradicate poverty, ed. V. Masson-Delmotte, P. Zhai, H.-O. Pörtner, et al. (Cambridge: Cambridge Univ. Press) 3 https://www.ipcc.ch/sr15/chapter/spm/
IPCC 2019a, IPCC Special Report on the Ocean and Cryosphere in a Changing Climate, ed. H.-O. Pörtner, D. C. Roberts, V. Masson-Delmotte, et al. https://www.ipcc.ch/srocc/chapter/summary-for-policymakers/
IPCC 2019b, Climate Change and Land: an IPCC special report on climate change, desertification, land degradation, sustainable land management, food security, and greenhouse gas fluxes in terrestrial ecosystems, ed. P. R. Shukla, J. Skea, E. Calvo Buendia, et al. In press. https://www.ipcc.ch/srccl/chapter/summary-for-policymakers/
IPCC 2021, Climate Change 2021: The Physical Science Basis. Contribution of Working Group I to the Sixth Assessment Report of the Intergovernmental Panel on Climate Change, ed. V. Masson-Delmotte, P. Zhai, A. Pirani, et al. (Cambridge: Cambridge Univ. Press) https://www.ipcc.ch/report/ar6/wg1/
Lenton, T. M., Rockström, J., Gaffney, O., et al. 2019, Natur, 575, 592
McSweeney, R., & Hausfather, Z. 2018, Q&A: How do climate models work? (Carbon Brief) https://www.carbonbrief.org/qa-how-do-climate-models-work/ (accessed 6.15.23)
Shields, A. L. 2019, ApJS, 243, 30
Stocker, T. F. 2022, Introduction to Climate Modelling (Bern: Univ. of Bern) 199 Available from www.climate.unibe.ch/stocker
Wayne, G. P. SkSci, URL https://skepticalscience.com/rcp.php?t=1 (accessed 6.15.23)

Part III

Consequences and Solutions

Chapter 4

Consequences of Climate Change

T A Rector

> "Do the best you can until you know better. Then when you know better, do better."
>
> —Maya Angelou

Why does it matter if the Earth warms by a few degrees? Why should we care? And could it possibly be good for us? These are commonly asked questions. Climate change is negatively affecting the wellbeing of people around the world. It is impacting our planet's already strained ability to provide food, water, and shelter for its 8 billion inhabitants. It is more than an environmental problem—it is a health problem, an economic problem, and a social justice problem. This chapter explores some of the most consequential effects of climate change, as well as those most connected to the science of astronomy. It explores how and why extreme weather and climate events are becoming more common. It discusses impacts on the Earth's hydrosphere and cryosphere. And it explores "tipping points" that—once crossed—lead to irreversible changes. The consequences of exceeding 1.5°C and 2.0°C temperature anomaly thresholds are also discussed.

4.1 Introduction

The consequences of climate change are far reaching, and it is impossible to cover them all. Needless to say, all of the consequences discussed here are important because they have, directly or indirectly, ramifications for the physical health of people. The mental health outcomes of climate change are serious as well, ranging from minimal stress and distress symptoms to clinical disorders—including anxiety and sleep disturbances to depression, post-traumatic stress, and suicidal thoughts (Cianconi et al. 2020).

In this chapter we're going to focus on the direct and indirect effects of climate change. The list of consequences discussed is far from complete. It includes those that are most likely to be connected to your teaching and outreach, as well as ones

doi:10.1088/2514-3433/acfcb6ch4

that are commonly asked about. As mentioned elsewhere, 90% of climate change is the result of fossil fuels, and they also have other deleterious health effects. In particular they are a major source of airborne fine particulate matter ($PM_{2.5}$) that can enter the lungs and irritate our respiratory linings and are a key contributor to the global burden of mortality and disease. Vohra et al. (2021) found that more than 8 million people died in 2018 from fossil fuel pollution—which is about 1 in 5 deaths worldwide. So it is important to emphasize that solutions to climate change can also address other existing problems.

When talking about consequences it's important to note that not all places are affected to the same level at the same time. For example, the Arctic is currently warming at about three times the rate as the rest of the world (Ballinger et al. 2021). Also, not all times of day or season are affected the same. Temperature increases are higher for nighttime lows than daytime highs—and winters are warming faster than summers—because radiative forcing from greenhouse gases play a larger role in the energy budget when the Sun is down or not as high in the sky.

Also, not everyone in an area is affected in the same way. In particular, the poor and marginalized suffer disproportionately from climate change; e.g., in the U.S., Black and African American individuals are 40% more likely to live in areas with the highest projected increases in extreme temperature related deaths (EPA 2021). Worldwide, the "Global South" refers to the regions of Latin America, Asia, Africa, and Oceania that are lower income and often politically or culturally marginalized when compared to "developed" countries like those in North America and Europe. Carbon emission rates per capita for residents of the Global South are typically less than 10% of those of developed countries. They are the least responsible for climate change but most vulnerable and under-prepared for its impacts (e.g., Sen Roy 2018). Climate change is therefore not just a health and economic issue, but a social justice issue as well.

It is also a local issue. When discussing climate change it is important to focus on local impacts and risks. Talking about, e.g., polar bears or future generations,[1] reinforces a common misconception that climate change is a problem elsewhere—in space as well as in time. Talking about local impacts that are happening now is more motivational (van der Linden 2015). Therefore it is important to understand how climate change is affecting people locally. Fortunately there are resources available to explain this for many parts of the world. For example, in the U.S., the Fourth National Climate Assessment report (USGCRP 2017) assesses the impacts, risks, and adaptation by region.

While not everyone is being affected equally or in the same way, no one is immune to the effects of climate change. Climate risks are "transboundary," i.e., events triggered or exacerbated by climate change in one country or region can cross borders, continents, and oceans to affect communities on the other side of the world. So are the consequences of some adaptation actions. In a world that is increasingly interconnected, these risks are transmitted through shared natural resources and

[1] Unless you are Inuit, in which case both polar bears and future generations figure prominently in your cultural dialogue!

ecosystems, trade links, finance, and human mobility (Anisimov & Magnan 2023). As eloquently put by Damian Barr, "We are not all in the same boat. We are all in the same storm. Some of us are on super-yachts. Some have just the one oar."[2]

4.2 A Matter of Degree

The Paris Agreement has set a goal to limit global temperature increases to well below 2.0°C (3.6°F) compared to pre-industrial levels, ideally keeping the increase below 1.5°C (2.7°F). Half of a degree may sound inconsequential but the effects increase nonlinearly. An IPCC Special Report on the impacts of global warming of 1.5°C (SR15)(IPCC 2018) outlines the implications of overshooting 1.5°C. Predictions of when the 1.5°C and 2.0°C milestones would be reached of course depend on model assumptions. Ebi et al. (2018) find that warming of 1.5°C is projected to be reached in about the 2030s for all multi-model means under all scenarios. Warming of 2°C is projected to be reached in about the 2050s under most scenarios. And warming as high as 4°C by 2080 is not ruled out (IPCC 2021). What is clear is that unless there are deep reductions in global greenhouse gas emissions, the goal of limiting warming well below 2°C and close to 1.5°C will soon be out of reach. Chapter 3 describes how climate models work. Some of the consequences of this increased warming are discussed below.

4.3 Global Weirding

Described as "global weirding" by Katherine Hayhoe, climate change is causing extreme climate and weather events to be more likely. For example, heat waves (Habeeb, Vargo, & Stone 2015), air turbulence (Williams 2017), heavy precipitation (Schär et al. 2016), and river flooding (Mallakpour & Villarini 2015) are now more common. USGCRP (2017) gives a thorough breakdown of the increased likelihood for such events, particularly for the United States. According to the World Meteorological Association, extreme weather, climate, and water-related events contributed to nearly 12,000 disasters from 1970–2021. Reported economic losses are US$4.3 trillion; and the death toll is at 2 million, 90% of which are in developing countries.[3]

When there is a major weather event, such as a heat wave or a hurricane, don't ask, "Did climate change cause this?" Instead ask, "How did climate change make it worse?" Climate change attribution science is now an established field of study, such that it is now possible to attribute the impact of climate change on individual events; e.g., Stott, Stone & Allen (2004) determined that the unusually strong heat waves that hit Europe in 2003 were twice as likely because of climate change. As of 2015 the World Weather Attribution group studies extreme weather events shortly after they occur to determine to what extent climate change contributed to its severity.

[2] https://www.damianbarr.com/latest/https/we-are-not-all-in-the-same-boat

[3] https://public.wmo.int/en/resources/atlas-of-mortality

4.3.1 Changes in Precipitation

The water cycle is the movement of water as it makes a circuit from the oceans to the atmosphere and back. Some of the warm water in the oceans evaporates into the atmosphere, eventually cooling and condensing into cloud particles. Over time these particles grow and eventually precipitate as rain or snow. Some precipitation accumulates as ice caps, glaciers, and snowpack. In warmer climates, snow melts in the spring and flows into streams. Most precipitation falls back into the oceans or onto land, where it flows over the ground as surface runoff. A portion of runoff enters rivers and continues toward the oceans, where the cycle repeats. Another important factor in the water cycle is transpiration—the flow of water through plants. Several major impacts of climate change are associated with the change in water balance from changes in how water is transferred from the land to the atmosphere by evaporation and by transpiration from plants (e.g., Crausbay et al. 2020).

The Earth's atmosphere can be thought of as a sponge because it holds water vapor. Warmer air can hold more water vapor, as many of us have experienced in the wintertime when cold air leads to dry, cracked skin. As described by the Clausius–Clapeyron equation, air can hold 4% more water vapor when its temperature increases by 1°F (or 7% per °C). This water vapor will eventually be released as precipitation, either as rain or snow.

The water cycle, and all of its elements, are therefore sensitive to temperature. Current climate models indicate that rising temperatures will intensify the Earth's water cycle, as increasing evaporation of the oceans adds more water vapor to the atmosphere. Increased ocean evaporation will result in more frequent and intense storms. However, increased evaporation on land will result in the drying of some land areas. As a result, storm-affected regions are likely to experience increases in precipitation and increased risk of flooding, while areas located far away from storm tracks are likely to experience less precipitation and increased risk of drought.

4.3.2 Tropical Cyclones, Heavy Precipitation Events, and Flooding

Climate change is causing tropical cyclones (i.e., hurricanes and typhoons) to be bigger, wetter, and slower. Climate change does not appear to be changing the frequency of these storms in total—about 80 occur each year (Ramsay 2017). However, more powerful storms are becoming more common (Holland & Bruyère 2014). Known as "rapid intensification," storms now also form more quickly (Balaguru, Foltz, & Leung 2018). The primary source of energy for tropical storms is warm ocean water. Warmer oceans from climate change therefore mean more energy is available to power stronger storms. Because warm air holds more water, these storms are also getting wetter. And, for reasons not yet clear, tropical storms also appear to be moving more slowly (Kossin 2018). This means that a storm can remain over an area for a longer period of time, dumping more rain. For example, Hurricane Harvey, which hit the coast of Texas in 2017, released 1.5m (60") of rainfall (Blake & Zelinsky 2017) and caused massive flooding. Extreme rainfall events equal in magnitude to Hurricane Maria, a Category 5 hurricane that devastated the northeastern Caribbean in September 2017, are twice as likely to occur under 2.0°C as compared to 1.5°C (Vosper, Mitchell, & Emanuel 2020).

The polar vortices are low-pressure regions of cold air that encircle the Arctic and Antarctic. The circulation of the Earth's jet stream normally traps cold air in the polar regions. The rapidly warming Arctic has weakened the temperature gradient between it and sub-Arctic regions. This has caused more disruptions of the Arctic polar vortex, wherein cold air can "escape" and cause extreme cold temperatures at mid-latitudes (Cohen et al. 2021). When these cold snaps collide with warm, wet air it can produce extreme snowfall events (i.e., "blizzards"). Extreme snow events are already more common in the Northeastern United States (Chen et al. 2019), and it is anticipated these events will continue to become more common (Quante et al. 2021).

4.3.3 Drought and Extreme Fires

Drought is generally defined as a prolonged period (usually a season or more) with abnormally low levels of precipitation, resulting in a water shortage that causes a serious hydrological imbalance (Crausbay et al. 2020). Studies based on observations and model simulations have demonstrated increased frequency and severity of drought over global land areas as a result of climate change, along with an increase in the frequency and severity of heat related extremes; e.g. heat waves, warm spells, and number of hot days (Hao et al. 2018). Hotter temperatures increase evaporation rates and decrease soil moisture, creating conditions suitable for a "flash drought"—a rapid onset or intensification of drought conditions caused by lower-than-normal rates of precipitation combined with abnormally high temperatures, winds, and radiation.

Limiting global warming to 1.5°C is expected to substantially reduce the probability of extreme drought in some regions. In Europe, the Mediterranean, Amazon, and southern Africa, the 1.5°C scenario has significantly lower probability of drought than the 2.0°C scenario. However the drought risk does not change significantly over the U.S. Central Plains and Southwest for both of these scenarios (Lehner et al. 2017).

Droughts have significant impacts on the flora and fauna of a region, as well as on its people. Drought can reduce the availability and quality of water needed for productive farms, ranches, and grazing lands (Arora 2019). They can lead to shortages of drinking water as well as reducing its quality (Misra 2014). Because of the increased risk of dust storms during a drought, air quality can be reduced from increased particulate matter in the air from these events (Achakulwisut, Mickley, & Anenberg 2018). The particulate matter can irritate bronchial passages and lungs. Allergen patterns are also changing, and air pollution can modify the allergenic potential of pollen (M. De Sario, Katsouyanni, & Michelozzi 2013). Drought can also result in the spread of disease, such as those carried by mosquitoes that breed in stagnant water; however the overall effect of climate change on such diseases is unclear (Levy et al. 2016).

Because of climate change and induced drought, fire seasons are longer, drier, hotter, and windier. Dried out (and more flammable) vegetation is increasing the probability of large-scale wildfires (Littell et al. 2016). Warmer air temperatures also mean more lightning. Romps et al. (2014) find that the number of lightning strikes increases by about 12% for every degree of warming. As a result, in recent decades there has been a profound increase in the number of large forest fires as well as the acreage burned (Dennison et al. 2014). For example, since 1984 climate change has doubled the cumulative forest fire area in the western United States (Abatzoglou & Williams 2016). Fires are also getting larger. Megafires (greater than 100,000 acres) and gigafires (more than a million acres) are occurring with increasing frequency (e.g., Canadell et al. 2021). The risk of wildfires in North America increases significantly in the 2.0°C scenario as compared to 1.5°C (Romero-Lankao et al. 2014). Unfortunately these fires also release CO_2 in significant amounts. For example, the 2020 wildfires in California released an estimated 127 million tons of CO_2, which was nearly as much as passenger vehicles in the state generate in a typical year (Jerrett et al. 2022).

Wildfires affect health with thermal injuries, exposure to smoke, loss of physical infrastructure, as well as impacts on mental health and wellbeing. Like many other consequences of climate change, people of color are also the most vulnerable to wildfire (Davies et al. 2018). It is important to note that smoke from wildfires is a growing public health risk worldwide due to the enormous amount of smoke-related pollution that is produced which can travel thousands of kilometers from its source (Eisenman & Galway 2022).

4.3.4 Heat Waves and Heat Domes

While it doesn't get the same media attention as hurricanes and fires, heat is one of the largest weather-related causes of death, a problem that is worsening due to climate change (Shindell et al. 2020). Heat waves have increased in intensity, frequency and duration, with these trends projected to worsen (Perkins-Kirkpatrick & Lewis 2020). A "heat dome" occurs when hot air gets trapped over a large area, resulting in dangerously high temperatures for several days to several weeks. Climate change is making these events more common as well. For example, the heat dome that occurred in the Pacific Northwest in 2021, which killed 800 people, would occur just once every 10,000 years without the enhancing influence of climate change (McKinnon & Simpson 2022).

Heat stress occurs in the human body when it is unable to regulate its temperature in hot conditions. A "wet-bulb"[4] temperature of 35°C (95°F) has been theorized to be the limit to human adaptability to extreme heat, however Vellinga et al. (2002) found that heat stress can occur in young, healthy adults at temperatures significantly lower than this. Exposure to extreme heat is also associated with acute kidney injury, adverse pregnancy outcomes, worsened sleep patterns, worsening of underlying cardiovascular and respiratory disease, and increases in non-accidental and injury-related deaths

[4] "Wet bulb" temperature measurements include the effects of evaporative cooling. If the relative humidity is 100%, no evaporative cooling occurs and wet and dry bulb temperatures will be the same.

(Romanello et al. 2022 and references therein). Heat waves also impact mental health; e.g., an increased risk of suicide in high temperatures (Thompson et al. 2018).

Vicedo-Cabrera et al. (2021) find that 37% of heat-related deaths can be attributed to climate change and that increased mortality is evident on every continent. In the worst-case scenarios heat-related deaths could be of the same order as cancers or all infectious diseases (Carleton et al. 2020). Heatwaves will make parts of the world, especially in Africa and Asia, uninhabitable within decades (IFRC 2023).

Compared to a 1.5°C world, under 2.0°C warming the frequency of extreme heat waves would double over most of the globe. In the 1.5°C scenario 13.8% of the world population will be exposed to severe heat waves at least once every 5 years. This fraction becomes nearly three times larger (36.9%) under 2.0°C warming, i.e. a difference of around 1.7 billion people (Dosio et al. 2018).

Even within a region, heat doesn't affect everyone equally. The elderly are particularly vulnerable (e.g., Kenney, Craighead & Alexander 2014). Hispanics and Latinos in the U.S. have high participation in weather-exposed industries, such as construction and agriculture, which are especially vulnerable to the effects of extreme temperatures (EPA 2021). Globally, extreme heat is having the greatest impact economically in the poor tropical regions least culpable for warming (Callahan & Mankin 2022). Likewise, larger cities feel a greater effect. Known as the "urban heat island effect," temperatures in cities are about 1–4°C higher than in nearby rural areas because man-made structures are overall darker and absorb more energy than natural vegetation (Hibbard et al. 2017).

4.4 Tipping Points

A "tipping point" can be defined as a threshold (temperature or otherwise) that, once crossed, leads to irreversible changes (Lenton et al. 2008). Triggering one tipping point can also lead to others (Lenton et al. 2021).

Extreme fire events are an example of a tipping point. These events are causing the desertification of forest ecosystems due to the resultant erosion and changes in vegetation (Neary 2021), a tipping point from which the ecosystem may not recover. Other tipping points include deforestation of the Amazon rainforest (Lovejoy & Nobre 2018), the melting of the Antarctic ice sheets (DeConto et al. 2021), and the Atlantic Meridional Overturning Circulation (Kim et al. 2022). Arctic sea ice loss is often cited as an example of a tipping point, but there is evidence that it is reversible (Serreze 2011; Tietsche et al. 2021).

The goal of the Paris Agreement is to keep the global temperature rise to less than 2.0°C above pre-industrial levels. However research (e.g., Armstrong McKay et al. indicates that exceeding 1.5°C could trigger multiple climate tipping points, and that some of these tipping points may have already been passed. There is evidence that this may be the case for the melting of the Greenland ice sheet (King et al. 2020; Boers & Rypdal 2021).

Tipping points are often described as "points of no return," and are often the basis of the "climate doomism" mindset, wherein it is argued that it is already too late to address climate change and by logical extension no effort should be made to

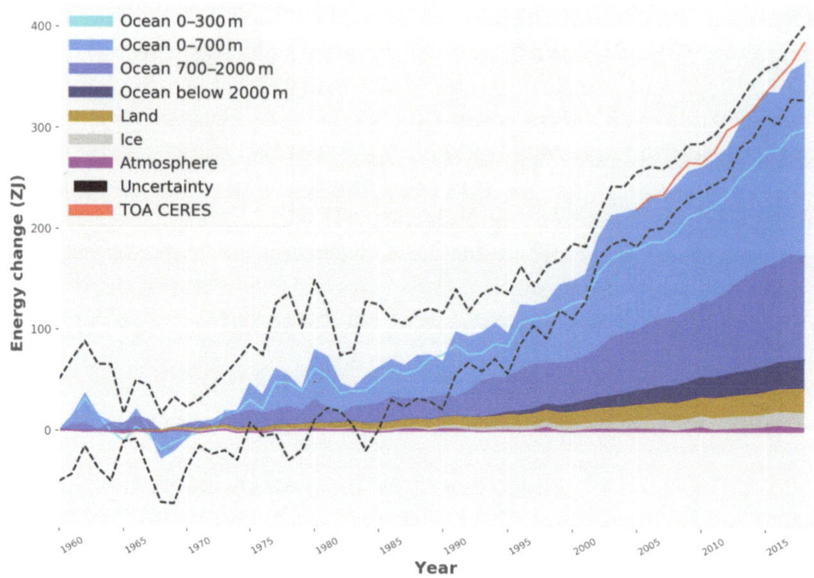

Figure 4.1. [5]This graph shows the buildup of excess heat (thermal energy) in the ocean, land, ice, and atmospheric components of the Earth system since 1960 in ZetaJoules (10^{21} Joules). The heat propagation to greater ocean depths with time is evident. Dashed lines bracket the total uncertainty of the data. Red line compares the trend of CERES satellite-based observations of the energy flow imbalance at top of atmosphere (TOA) (von Schuckmann et al. 2020).

do so. However these should not be interpreted as an excuse for inaction, as any reduction in greenhouse gas emissions is beneficial; e.g, 1.8°C is not ideal but it is still better than 1.9°C.

4.5 Impacts on Water and Ice

When we think of global warming we usually focus on the increase in the atmospheric temperature, which is natural as that is the temperature we feel. However, only 1–3% of the excess thermal energy due to climate change has gone into the atmosphere. Ninety percent has been absorbed by the oceans (Figure 4.1), with the balance absorbed by land and ice (von Schuckmann et al. 2020). As with the case of the land and atmosphere, the oceans have not warmed equally (Figure 4.2).

Rising ocean temperatures have several consequences. As mentioned before, the primary source of energy for tropical storms is warm ocean water. Generally speaking, warmer water provides more heat energy for tropical storms to develop.

Ocean currents (Figure 4.3) are the movement of ocean water as a result of wind, tides, and density differences due to temperature (thermo) and salinity (haline) variations. These in turn are driven by solar isolation, the Coriolis effect, and the

[5] https://essd.copernicus.org/articles/12/2013/2020/essd-12-2013-2020-f06-high-res.png

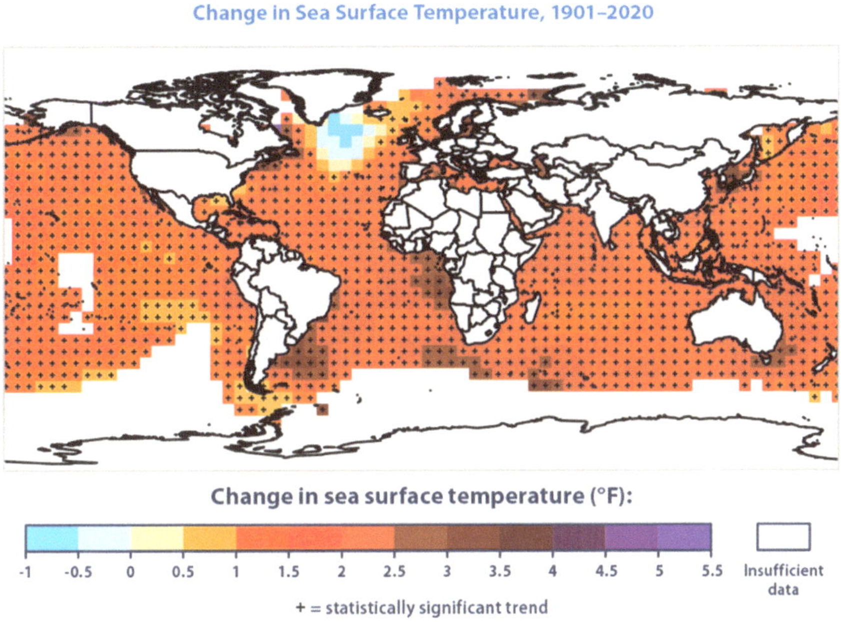

Figure 4.2. This map[6] shows how average sea surface temperature around the world changed between 1901 and 2020. It is based on a combination of direct and satellite measurements (Huang et al. 2020). A black "+" symbol in the middle of a square on the map means the trend shown is statistically significant. White areas did not have enough data to calculate reliable long-term trends. An exception to this warming is the subpolar North Atlantic. This exception has been hypothesized to result from a slowdown of the Atlantic Meridional Overturning Circulation (AMOC), but may also be due to atmosphere-induced cooling from a northward migration of the jet stream (Li et al. 2022).

tidal interaction with the Moon and Sun. Ocean currents have been described as "conveyor belts" that distribute energy throughout the Earth's climate system. As the Earth's climate changes, so are ocean currents. For example, the Atlantic meridional overturning circulation (AMOC) consists of a northward flow of warm, salty water in the upper layers of the Atlantic Ocean and a southward flow of colder, deep waters. Because of climate change the AMOC is at risk of a transition to a weaker circulation mode (Boers 2021). The adding of freshwater to the Atlantic from increased rainfall and the melting of the Greenland ice sheet is diluting the surface salt water and, due to its increased buoyancy, is slowing down the circulation. The AMOC brings energy to the North Atlantic, including North America and Europe, so a slowdown would result in a cooling that would at least partially offset global warming in the region, if not globally. It would also cause changes in precipitation (Jackson et al. 2015) that would have agricultural (Vellinga & Wood 2002) and economic (Link & Tol 2004) impacts.

[6] https://www.epa.gov/climate-indicators/climate-change-indicators-sea-surface-temperature

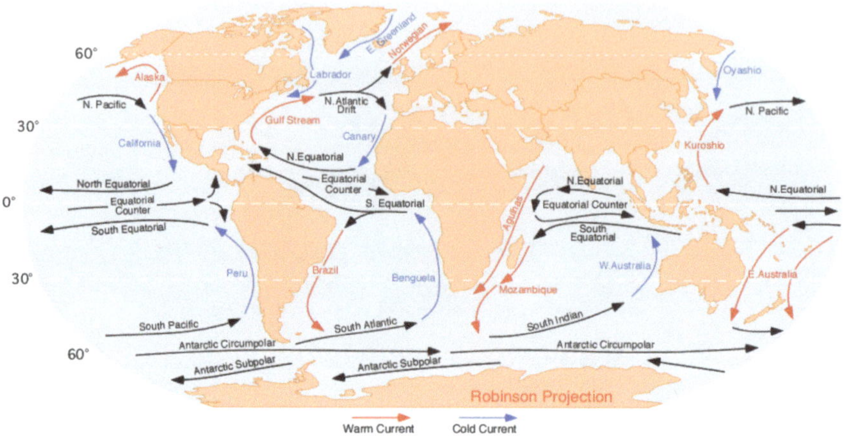

Figure 4.3. [7]The major ocean currents. Notice that many of them are parts of rotating ocean currents known as "gyres." There are five major gyres. Also note that ocean currents are three dimensional. (Credit: Michael Pidwirny)

4.5.1 Ocean Levels

Rising atmospheric and oceanic temperatures are also affecting sea levels. Globally this is driven by two factors: As land ice (primarily in Greenland and Antarctica) melts, it is increasing the volume of water in the oceans. Secondly, ocean levels have risen due to the thermal expansion of seawater (Fasullo & Gent 2017). Ice-mass loss has caused twice as much sea-level rise since 1900 as has thermal expansion (Frederikse et al. 2020). It is important to note that decreases in sea ice do not affect sea levels, just as the melting of ice cubes in a glass of water does not cause it to overflow.

The ice sheets of Greenland and Antarctica store about two-thirds of all the freshwater, and 99% of the frozen freshwater, on Earth. Warming atmospheric and oceanic temperatures are causing these ice sheets to melt from above and below the surface (Figure 4.4). The Antarctic ice sheet is the largest body of ice, with a volume equivalent to more than 50m global mean sea level rise (Vaughan et al. 1999). The Greenland Ice Sheet holds about 7m of sea level equivalent (Aschwanden et al. 2019). Antarctica is losing ice mass at an average rate of about 150 billion tons per year, with these rates accelerating (Paolo et al. 2015). Greenland is losing ice mass at about twice that rate because warmer air over Greenland, as compared to Antarctica, has led to greater changes in precipitation, ice flow discharge and meltwater run-off. This has made the Greenland ice sheet the largest contributor to sea level rise (Box et al. 2022).

Globally, sea levels are expected to rise 0.24 to 0.32 m by the year 2050 (Oppenheimer et al. 2019). Longer-term sea level rise estimates are less certain, but could be as high as two meters by 2100 (Bamber et al. 2019). Locally, coastal sea level changes depend on additional factors. For example, the accumulation of sediment or geological uplift from melting glaciers can cause the local ground level to rise. Land levels can also lower due to factors such as erosion or decreases in

[7] https://en.wikipedia.org/wiki/Ocean_current#/media/File:Corrientes-oceanicas.png

Figure 4.4. [8]Data from NASA's GRACE (2002–2017) and GRACE Follow-On (2018–2022) satellites show that the land ice sheets in both Antarctica (upper) and Greenland (lower) have been losing mass since 2002. The ice sheets of Greenland and Antarctica store about two-thirds of all the freshwater on Earth. Meltwater coming from these ice sheets is responsible for about one-third of the global average rise in sea level since 1993.

groundwater. Changes in atmospheric and ocean currents can also affect sea levels by pushing more water against some coastlines and pulling it away from others.

Increasing sea levels are causing increased incidents of flooding in coastal areas, eroding shorelines, and increasing the flow of salt water into agricultural soil and nearby groundwater aquifers, and the loss of habitat for local flora and fauna. Higher sea levels also make coastal infrastructure more vulnerable to damage from

[8] https://climate.nasa.gov/vital-signs/ice-sheets/

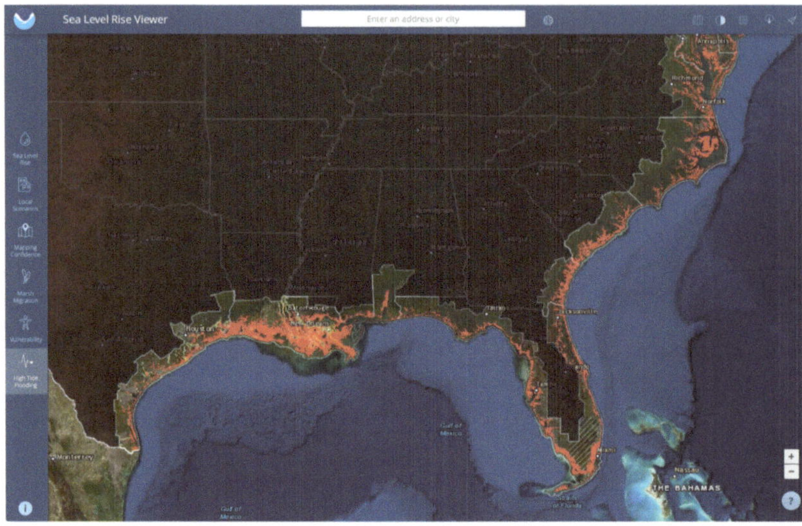

Figure 4.5. The NOAA Sea Level Rise Viewer is an interactive tool that allows the user to view sea level rise and potential coastal flooding impact areas. In the above map, areas in the southern United States currently subject to tidal flooding are shown in red.

tidal flooding (which is also known as "sunny day," "recurrent," or "nuisance" flooding). For example, incidents of tidal floods have increased 5- to 10-fold or more since the 1960s along the U.S. coastlines (Jacobs et al. 2018). Naturally, higher sea levels also make coastal regions more vulnerable to flooding due to storms (i.e., "storm surges"). Change has caused the volume and extent of storm surge inundation to increase (Camelo, Mayo, & Gutmann 2020). Tebaldi et al. (2021) estimate that by 2100 roughly half of the 7,000+ locations they considered will experience present-day 100 yr extreme-sea-level events at least once a year, even under 1.5°C of warming. The tropics appear more sensitive than the Northern high latitudes. Impacts on regions can be explored using the NOAA Sea Level Rise viewer (Figure 4.5).

Jevrejeva et al. (2018) estimate a median global sea level rise by 2100 of up to 52cm for a temperature rise of 1.5°C and up to 63cm for the 2.0°C scenario. They also find that a difference of 11cm global sea level rise could result in additional losses of US$ 1.4 trillion per year (0.25% of global GDP) if no additional adaptation is assumed.

4.5.2 Ocean Ecosystems

Climate change is a dire threat to ocean ecosystems such as coral reefs. Rising ocean temperatures and ocean acidification are causing mass "coral bleaching" events and infectious disease outbreaks to occur more frequently (Hoegh-Guldberg et al. 2007). When water is too warm coral will expel the algae living within them, causing the coral to turn white. Ocean acidification contributes to the problem, as it reduces the calcification rates in reef-building and reef-associated organisms (Anthony et al. 2008).

Other effects of climate change, such as sea level rise, changes to the frequency and intensity of tropical storms, and altered ocean circulation patterns, are also a threat to reefs.

Coral reefs are among the most productive and biologically diverse ecosystems on Earth. Despite their small size (0.1% of the ocean surface) they provide habitat for at least 25% of all marine species, with estimates of over one million species living in and around coral reefs (Reaka-Kudla & Wilson 1997). They also provide about US $375 billion annually in global goods and services that include coastal protection, building materials, fisheries, and tourism (Costanza et al. 1997). Chen et al. (2015) estimate that the lost value in terms of the global coral reef value under climate change scenarios ranges from US$3.95 to US$23.78 billion annually. Bleaching events can occur locally as well as globally. For example, in 2005 the U.S. lost half of its coral reefs in one year due to a massive bleaching event in the Caribbean (Eakin et al. 2010). During the 2014–2017 global coral bleaching event, unusually warm waters (partially associated with a strong El Niño) affected 70% of coral reef ecosystems worldwide, which is more than three times as many reefs as the previous global bleaching event in 1998 (Skirving et al. 2019). Coral reefs have therefore already shown their vulnerability to the global warming that has occurred to date (Lough, Anderson, & Hughes 2018). However it is possible for a coral reef to survive a bleaching event, even if up to 90% of the reef is killed (Gilmour et al. 2013). It is expected that the majority (70–90%) of coral reefs will be degraded or destroyed even if global warming is constrained to 1.5°C, with these values increasing to 99% at 2.0°C (Frieler et al. 2013).

4.5.3 Ocean Acidification

As mentioned earlier, 25%–30% of anthropogenic CO_2 emissions are absorbed by the oceans (Friedlingstein et al. 2022). When CO_2 dissolves in water it produces carbonic acid (H_2CO_3), which dissociates into a bicarbonate ion (HCO_3^-) and a hydrogen ion (H^+). The free hydrogen ions decrease the ocean pH of the ocean, causing it to become more acidic. While the oceans still have an alkaline pH, they are now about 30% more acidic than before industrial times (Jacobson 2005).

Described as global warming's "evil twin brother," ocean acidification has numerous consequences. For example, the increased acidity impedes the calcification process used by many marine organisms, such as the "framework builders" of coral reefs (Anthony et al. 2008), shellfish such as Red King and Tanner crabs (Long et al. 2013), and sea plankton such as pteropods (Bednaršek et al. 2019). Ocean acidification is contributing to coral bleaching and die-off as well as disrupting other marine ecosystems. It is therefore a threat to the food sources and livelihoods for over three billion people (FAO 2016).

Like global warming, ocean acidification isn't happening at the same rate in all parts of the world (Figure 4.6). Carbon dioxide is more soluble in cold water, meaning that colder bodies of water found at higher latitudes are overall more affected.

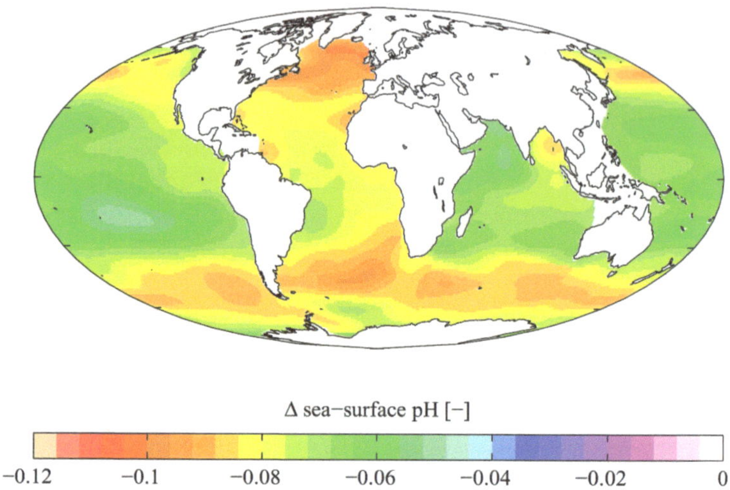

Figure 4.6. [9]Estimated change in annual mean sea surface pH between the pre-industrial period (1700s) and the present day. Calculated from fields of dissolved inorganic carbon and alkalinity from the Global Ocean Data Analysis Project (GLODAP) climatology and temperature and salinity from the World Ocean Atlas (2005). Land masses and bodies of water missing data are shown in white. Colder and higher latitude waters have the capacity to absorb more CO_2 and therefore become more acidic.

4.5.4 Decline of Sea Ice

Sea ice is an important element of the Earth's climate system. In the Arctic, sea ice expands and contracts with the seasons, achieving maximum levels in March and minimum levels in September. About 15% of the ocean contains sea ice for at least part of the year. As mentioned before, sea ice plays an important role in the overall climate of the Earth. In particular, the high reflectivity of sea ice affects the climate energy budget.

As one would expect, increasing ocean and atmospheric temperatures have negatively impacted sea ice (Figure 4.7). From 1979 to 2018, the Arctic lost 11% of its sea ice as measured during the maximum extent in March and 45% as measured during the minimum extent in September (Stroeve & Notz 2018). The Arctic ice season has also shortened at an average rate of more than 5 days per decade between 1979 and 2013 (Parkinson 2014). The amount of ice that survives year round has also dropped. The extent of multi-year[10] ice declined by more than 50% over 1999–2017; and first-year ice now makes up more than two-thirds of the Arctic sea ice cover (Kwok 2018). This loss of sea ice is driving the ice-albedo positive feedback loop. There has been concern that the transition from perennial to seasonal ice cover would trigger a tipping point, however Tietsche et al. (2011) find that the Arctic Sea can recover from ice-free summer conditions within just a few years. Under the 1.5°C scenario, Arctic summer sea ice is likely to be

[9] https://commons.wikimedia.org/wiki/File:AYool_GLODAP_del_pH.png

[10] "Multi-year ice" lasts for more than one winter cycle.

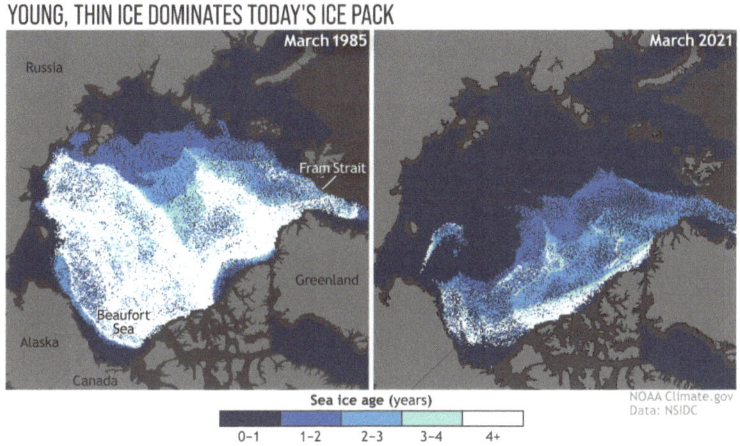

Figure 4.7. [11]In mid-March 1985 (left), the winter maximum ice pack was dominated by ice at least 4 years old (white). In 2021 (right), only a small strip of very old ice remained tucked up against the islands of the Canadian Arctic. More than half of the winter ice pack was less than a year old (dark blue). NOAA Climate.gov image, based on data from the National Snow and Ice Data Center.

maintained, whereas under the 2.0°C scenario the risk of an ice-free Arctic in summer is much worse. The likelihood of any September to be completely ice-free under a 2.0°C scenario is 35%, whereas this likelihood is only a few percent in the 1.5°C scenario (Jahn 2018).

The overall reduction of sea ice, and the shortening of the ice season, has consequences for the different elements of the Arctic food chain—triggering a cascade of changes that may be good for some species and harmful to others. Single-celled algae that rely on sunlight to survive form the basis of the food chain throughout the world's oceans. In the Arctic, these algae can live on the underside of sea ice (ice algae) and in water (phytoplankton). The decline of sea ice, and resultant increase in exposed ocean water, has therefore caused an increase in the productivity of phytoplankton and decrease of ice algae (Søreide et al. 2010). The increase of phytoplankton has led to growth of the bowhead whale population in the Pacific Arctic (Druckenmiller et al. 2018). At the same time, warmer waters mean that algae species capable of producing deadly neurotoxins are spreading poleward (Frey et al. 2020) (Figure 4.8).

Polar bears are an iconic symbol of climate change. They are an apex predator, and as such their health is reflective of the health of the Arctic food chain as a whole. The primary source of nutrition for polar bears is ringed and bearded seals, which they can only hunt while on sea ice. Coastal bear populations, such as those along the southern Beaufort Sea and southern Hudson Bay, are forced onto land during summer months—during which time they are unable to feed. An increase in this fasting period due to the decrease in the length of the ice season has dire long-term consequences for the polar bears (Molnár et al. 2020). It is important to note that the

[11] https://www.climate.gov/media/13890

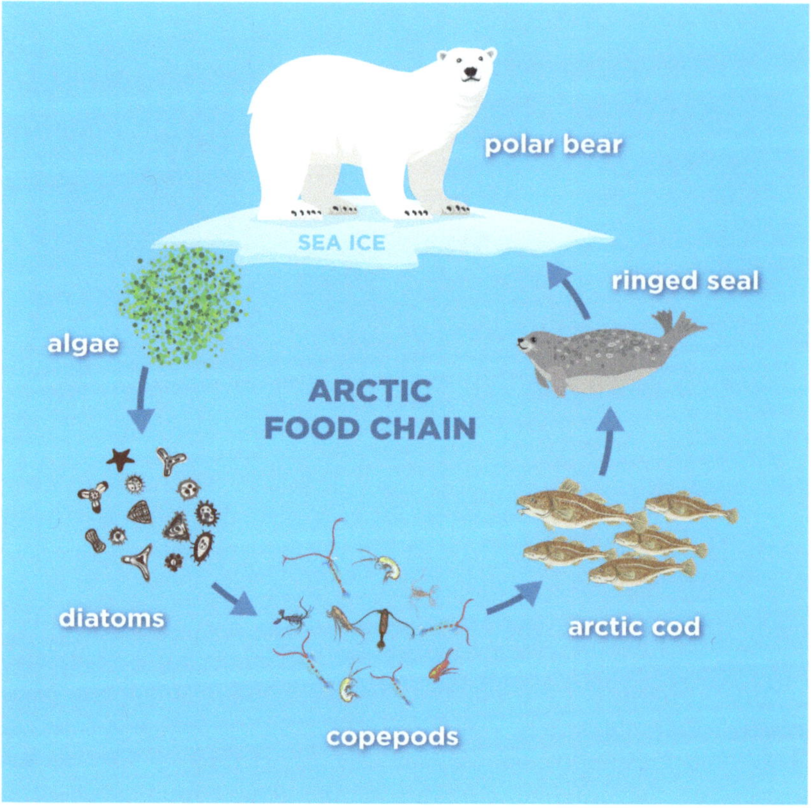

Figure 4.8. The Arctic food chain begins in part on algae that is attached to the underside of sea ice. This algae serves as the start of a food chain that eventually leads to ringed and bearded seals, the primary source of nutrition for polar bears (Copyright: Peppermint Narwhal/Polar Bears International).

global population of about 26,000 bears in the Arctic is divided into nineteen subpopulations, each of which is being affected differently. While most populations are being negatively impacted, some been reported as stable or productive in the short term (Regehr et al. 2016). This is sometimes "cherry picked" by climate skeptics as evidence that polar bear populations are actually fine and/or climate change isn't really happening.

4.5.5 Decline of Snow and Glacier Ice

Mountains are sometimes called the "water towers" of the world because they store and supply a significant fraction of the water needed by humans for drinking, irrigation, hydropower, and industrial purposes. Nearly two billion people rely upon mountainous water sources (Immerzeel et al. 2020).

Glaciers and snowpack act as sponges, stockpiling water (in snow and ice form) during winter months and releasing it gradually as meltwater in the summer. They offset the intrinsic variability of regional precipitation, acting as a buffer against droughts. Warming can create drought simply by preventing the accumulation of mountain snowpack. For example, in the mountainous western United States the total snowpack has declined 15–30% over 1955–2016 (Mote et al. 2018).

Similarly, glaciers have endured greater rates of melting, leading to cascading impacts on downstream systems (Milner et al. 2017). For example, glaciers in the Himalaya–Karakoram mountain ranges, which modulate the flow of freshwater to almost 900 million people, are losing mass at an accelerating rate (Nie et al. 2019).

Many animals rely on the meltwater from snowpack and glaciers. One example is Pacific salmon, an important species that supports subsistence, recreational, and commercial fisheries worth billions of dollars annually. Pacific salmon spawn in rivers and streams that are fed by snowmelt and glacial meltwater. The cold water provided by this snow and ice melt helps to regulate the temperature and water levels of rivers and streams. Less snowpack therefore means lower water levels as well as warmer water. Rising water temperatures are thermally stressing salmon (Mantua, Tohver, & Hamlet 2010).

4.6 Population Explosion and Mass Extinction

Homo sapiens have existed on Earth for two to four hundred thousand years. At the onset of agriculture roughly 10,000 years ago, an estimated one to ten million people lived on Earth (Darchen 2009). Since then growth has been exponential, with roughly a billion people by 1800. The population growth rate peaked at 2.1% in 1970, and has declined slightly since (Horiuchi 1993). Nonetheless, the population of the Earth has doubled to 8 billion since the iconic "blue marble" photo was taken in 1972 (Figure 4.9)—an astonishing rate of growth that is clearly unsustainable.

While not officially designated as such (yet), the "Anthropocene" is defined as the most recent period in Earth's history when human activity has had a significant impact on the planet's climate and ecosystems—through climate change and other human activity. Most of this impact has occurred since World War 2, where scientific and technological advances have led to rapid growth in many areas, including population and life expectancy, as well as energy and materials consumption. During the rapid post-war economic growth, energy, mobility, and food systems became increasingly reliant on oil and less on coal (Johnstone & McLeish This growth has come at the expense of other species. The Earth is currently undergoing a mass-extinction event, wherein species are going extinct at 100–1000 times the normal rate (Pimm et al. 2014).

A common question is: can species successfully adapt to climate change? To survive, a species will (1) stay within an area if the climate remains within its tolerance limits, (2) migrate to regions where the climate is currently within the species' tolerance limits, and/or (3) evolve to adapt to the new climate parameters (Davis, Shaw, & Etterson 2005). Failing these it will go extinct. For many species, it will be the rate of climate change, rather than the overall change in temperature and other factors, that will prove to be critical to its survival. For most species adaptation by evolution is not realistic on the timescale of climate change. For

Figure 4.9. Known as the "Blue Marble," this iconic photograph of the Earth was taken on December 7th, 1972, by the crew of the final Apollo mission, Apollo 17, as they traveled toward the Moon. Since the time of this photograph, the Earth's population has doubled. (Source: NASA).[12]

example, Quintero and Wiens (2013) studied over 500 species and found that they would need to evolve at rates more than 10,000 times faster than typically observed in order to adapt to our rapidly changing climate.

The timing of biological events, such as reproduction and migration (known as "phenological traits") are changing for many species, but they may be adapting too slowly to be able to persist in the long term (Radchuk et al. 2019). Global warming is also a threat to species with temperature-dependent sex determination, e.g., all species of sea turtle. During embryonic development, rising temperatures might lead to the overproduction of one sex and, in turn, could bias populations' sex ratios to an extent that threatens their persistence (Lockley & Eizaguirre 2021).

Climate change also complicates the way we perceive and consider invasive species (defined as an organism that is not indigenous, or native, to a particular region) as impacts to some will change and others will remain unaffected; other

[12] https://www.nasa.gov/image-feature/apollo-17-astronauts-capture-iconic-blue-marble-50-years-ago

nonnative species are likely to become invasive; and native species are likely to shift their geographic ranges into novel habitats. Under the 1.5°C scenario, 6% of insects, 8% of plants, and 4% of vertebrates are projected to lose more than half of their climatically determined geographic range by 2100. In the 2.0°C scenario, those percentages double or triple to 18% of insects, 16% of plants, and 8% of vertebrates (Warren et al. 2018).

It is important to note that a species' response to climate change can be complicated. In the case of Pacific salmon the increased meltwater output from glacier retreat will, in the short term, can create new habitat—but only for populations that spawn in systems with a high proportion of glacierized watersheds (Pitman et al. 2021). This can be helpful for these populations to survive other climate-change induced stressors. However in the long term the glaciers feeding these rivers and streams will melt away entirely, destroying these new habitats. As discussed in the chapter on misinformation, climate skeptics often point to a particular species, or a population of species, that is currently doing well and argue that this is evidence that climate change either isn't happening or isn't a concern.

4.7 Is More CO_2 Good?

It has been proposed that increased concentrations of CO_2 in the atmosphere could be beneficial to plants, especially for those that use the C_3 carbon fixation pathway. Indeed, this is a common climate skeptic/denier talking point, that "carbon dioxide is good for plants." However, during a 20-year field experiment Reich et al. (2018) found that some plant species benefited from the extra CO_2 in the short term but not in the long term. This is likely due to induced imbalances in the availability of soil nitrogen. Known as Liebig's law of the minimum, agricultural growth is dictated not by total resources available, but by the scarcest resource. So increased CO_2 can be beneficial to plant productivity if nothing else is limiting—nitrogen, water, thermal regime, etc. But in reality these other things do limit growth—and some of them (water, notably) will do so even more in a climate changed future. And while that extra CO_2 may allow plants to grow bigger their nutrients are more dilute and therefore of less nutritional value (Semba et al. 2022). Chapter 5 discusses this issue in more detail.

4.8 Economic Impact

So far we have focused on "environmental" consequences, but these also have economic impact. Climate change is causing additional health-care costs, loss of workforce productivity, and economic losses through the damage caused by climate-related extreme events. These costs and losses subsequently affect household incomes and national economies.

Burke, Hsiang, and Miguel (2015) find that overall economic productivity is highly sensitive to temperature in all countries, with productivity peaking at an annual average temperature of 13°C and declining strongly at higher temperatures. Already today, the cumulative effect of heat on economic output is commensurate with other health-related impediments (Day et al. 2019). Two thirds of all labor

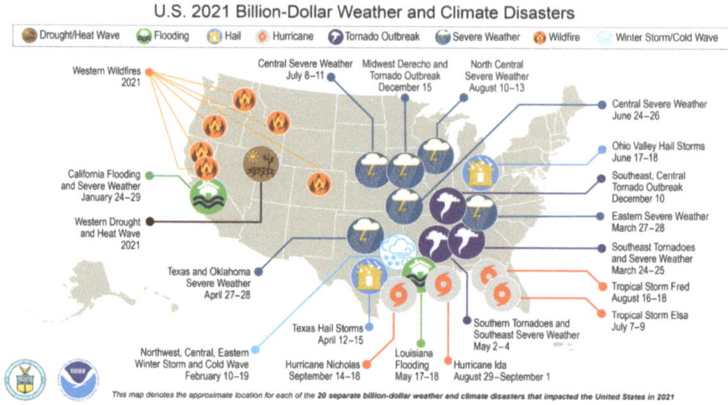

Figure 4.10. [14]In 2021, the United States experienced a record 20 weather or climate disasters that each resulted in at least $1 billion in damages.

hours lost globally in 2021 were in the agricultural sector (Romanello et al. 2022). Likewise, there are monetized costs of premature mortality due to air pollution. Goodkind et al. (2019) estimated that in the U.S. anthropogenic $PM_{2.5}$ was responsible for 107,000 premature deaths in 2011, at a cost to society of $886 billion.

In 2021 climate-related extreme events induced measurable economic losses of US $253 billion (Romanello et al. 2022). As a proportion of Gross Domestic Product (GDP), losses in countries with a high Human Development Index are double the global average. However losses in low-income countries are more deeply felt because most of these losses were uninsured. By 2050, it is estimated that annual financing needed to address the damage caused by climate change may reach up to US $1,132–1,741 billion (WFPHA 2022).[13]

The number of expensive, extreme weather events is increasing. In the 1980s the United States suffered on average three major weather/climate disaster events, each with losses exceeding $1 billion, every year. In 2021 there were twenty such events (Figure 4.10). Since 1980, the U.S. has sustained 323 weather and climate disasters where overall damages/costs reached or exceeded $1 billion—including Consumer Price Index adjustments to the year 2022 (Figure 4.11). The total cost of these 323 events exceeded $2.195 trillion (Smith 2020). These disasters of course also result in a loss of life. The twenty events of 2021 caused at least 688 direct or indirect fatalities. The costliest, and deadliest, single event in the United States was Hurricane Katrina, which harmed the city of New Orleans and the surrounding areas. It caused over 1,800 fatalities and $125 billion in damage (NOAA 2017).[15]

---

[13] https://www.wfpha.org/statement-on-the-public-health-imperative-for-financing-reparative-action-to-address-climate-change-loss-damage/

[14] https://www.climate.gov/media/13976

[15] https://www.nhc.noaa.gov/news/UpdatedCostliest.pdf

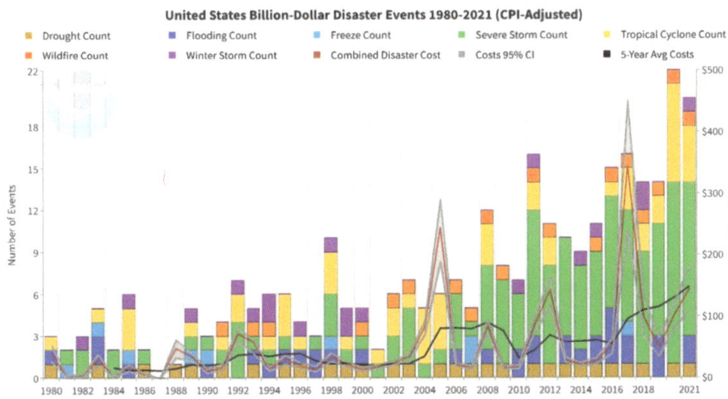

Figure 4.11. [16]The history of billion-dollar disasters in the United States each year from 1980 to 2021, showing event type (colors), frequency (left-hand vertical axis), and cost (right-hand vertical axis.) The number and cost of weather and climate disasters is rising due to a combination of population growth and development along with the influence of human-caused climate change on some types of extreme events that lead to billion-dollar disasters.

The rapid intensification of storms is also making it harder for communities to properly respond (e.g., to initiate timely evacuations). Drought conditions impact flow rates on large rivers that are economically vital. For example, in 2022 the Mississippi River in the United States, the Rhine and Danube in Europe, and the Yangtze in China all fell to historically low levels, impacting transportation of goods via barge, tourism via river cruises, as well as hydropower.

Carleton et al. (2020) estimate that the mean global increase in mortality risk due to climate change, accounting for adaptation benefits and costs, is valued at roughly 3.2% of global GDP in 2100 under the high-emissions (RCP 8.5) scenario. Burke, Davis, and Diffenbaugh (2018) find that limiting warming to 1.5°C would reduce economic damages relative to 2°C, saving the world economy upwards of US$20 trillion.

The cost of preventing these events is also rapidly increasing. For example, the annual fire suppression budget in California increased from less than $30 million in the 1980s to approximately $640 million during 2015–2019.

Neumann et al. (2015) estimate that the costs for U.S. coastal areas from sea level rise and storm surge, including adaptation, will total $930 billion to $1.1 trillion by 2100, not including additional damages to business activity, infrastructure such as roads and power grids, and natural resources and wildlife.

4.9 Effects on Civilization and Agriculture

Because civilization cannot exist in unlivable or uninhabitable places, climate change poses a threat to cause civilizations to collapse (or "climate collapse"), defined as the loss of a society's capacity to maintain essential governance functions,

[16] https://www.climate.gov/media/13979

such as the rule of law, and the provision of basic necessities such as food and water. This could first occur in specific, vulnerable locations (such as Syria) and then spread worldwide. It might not be an abrupt event but rather an extended process that occurs over a century or more (Steel et al. 2022). The severity and duration of the recent Syrian drought, which is implicated in the current conflict, was more than twice as likely because of climate change (Kelley et al. 2015).

The fundamental question is: How will the Earth continue to provide the necessary nutrition for humanity? The Earth is already straining to provide sufficient nutrition. According to the United Nations World Health Organization (UN WHO 2021), in 2021 more than 2.3 billion people (or 30% of the global population) lacked year-round access to adequate food.[17] Without adaptation, by 2050 climate change will reduce global agriculture yields up to 30% (Global Center on Adaptation 2019)[18] while food demand goes up 50–100% (Valin et al. 2014) as the population of the Earth approaches 10 billion people (United Nations 2022). While it may be possible to genetically modify some crops for climate changed conditions it is not without risk (Wolfenbarger & Phifer 2000). This topic is so important Chapter 5 is devoted to it.

4.10 Conclusions

This chapter provides a brief overview of some of many of the ways in which climate change is impacting us. However, when talking about climate change resist the urge to try to talk about them all! The goal here is to help you to be informed so that you can help others understand why we should care. In particular, focus on local impacts.

But this information alone doesn't necessarily help. *In fact it can backfire*, reinforcing fears that nothing can be done. If you are going to talk about consequences—and you should—you also need to talk about solutions. The solutions and communications chapters (e.g., Chapters 6 and 12) talk about how to do this.

References

Abatzoglou, J. T., & Williams, A. P. 2016, PNAS, 113, 11770

Achakulwisut, P., Mickley, L. J., & Anenberg, S. C. 2018, ERL, 13, 054025

Anthony, K. R. N., Kline, D. I., Diaz-Pulido, G., Dove, S., & Hoegh-Guldberg, O. 2008, PNAS, 105, 17442

Armstrong McKay, D. I., et al. 2022, Sci, 377, eabn7950

Arora, N. K. 2019, ES, 2, 95

Aschwanden, A., et al. 2019, SciA, 5, eaav9396

Balaguru, K., Foltz, G. R., & Leung, L. R. 2018, GeoRL, 45, 4238

Ballinger, T. J. 2021, Surface Air Temperature NOAA technical report OAR ARC ID# 34475, *National Oceanic and Atmospheric Administration*

[17] https://www.who.int/news/item/12-07-2021-un-report-pandemic-year-marked-by-spike-in-world-hunger

[18] https://gca.org/reports/adapt-now-a-global-call-for-leadership-on-climate-resilience/

Bamber, J. L., Oppenheimer, M., Kopp, R. E., Aspinall, W. P., & Cooke, R. M. 2019, PNAS, 116, 11195

Bednaršek, N., et al. 2019, FrMS, 6,

Blake, E.S., & Zelinsky, D.A. 2017, https://www.nhc.noaa.gov/data/tcr/AL092017_Harvey.pdf

Boers, N. 2021, NatCC, 11, 680

Boers, N., & Rypdal, M. 2021, PNAS, 118, e2024192118

Box, J. E., et al. 2022, NatCC, 12, 808

Burke, M., Davis, W. M., & Diffenbaugh, N. S. 2018, Natur, 557, 549

Burke, M., Hsiang, S. M., & Miguel, E. 2015, Natur, 527, 235

Callahan, C. W., & Mankin, J. S. 2022, SciA, 8, eadd3726

Camelo, J., Mayo, T. L., & Gutmann, E. D. 2020, FrBuEn, 6,

Canadell, J. G., Meyer, C. P., (Mick), Cook, G. D., Dowdy, A., Briggs, P. R., Knauer, J., Pepler, A., & Haverd, V. 2021, NatCo, 12, 6921

Carleton, T. A., et al. 2020, NBER Working Paper Series No. 27599 National Bureau of Economic Research

Chen, P.-Y., Chen, C.-C., Chu, L., & McCarl, B. 2015, GEC, 30, 12

Chen, G., et al. 2019, JCli, 32, 7561

Cianconi, P., Betrò, S., & Janiri, L. 2020, FrPs, 11,

Cohen, J., Agel, L., Barlow, M., Garfinkel, C.I., & White, I. 2021, Sci, 373, 1116

Costanza, R., d'Arge, R., de Groot, R., et al. 1997, Natur, 387, 253

Crausbay, S. D., et al. 2020, OEart, 3, 337

Darchen, S. 2009, CJRS, 32, 315

Davies, I. P., Haugo, R. D., Robertson, J. C., & Levin, P. S. 2018, PLoSO, 13, e0205825

Davis, M. B., Shaw, R. G., & Etterson, J. R. 2005, Eco, 86, 1704

Day, E., Fankhauser, S., Kingsmill, N., Costa, H., & Mavrogianni, A. 2019, CliPo, 19, 367

DeConto, R. M., et al. 2021, Natur, 593, 83

Dennison, P. E., Brewer, S. C., Arnold, J. D., & Moritz, M. A. 2014, GeoRL, 41, 2928

Dosio, A., Mentaschi, L., Fischer, E. M., & Wyser, K. 2018, ERL, 13, 054006

Druckenmiller, M. L., Citta, J. J., Ferguson, M. C., Clarke, J. T., George, J. C., & Quakenbush, L. 2018, DSRII, 152, 95

Eakin, C. M., et al. 2010, PLoSO, 5, e13969

Ebi, K. L., Hasegawa, T., Hayes, K., Monaghan, A., Paz, S., & Berry, P. 2018, ERL, 13, 063007

Eisenman, D. P., & Galway, L. P. 2022, BMCPH, 22, 2274

EPA 2016, Climate Change Indicators: Sea Surface Temperature https://www.epa.gov/climate-indicators/climate-change-indicators-sea-surface-temperature (accessed 6.14.23)

EPA 2021, Climate Change and Social Vulnerability in the United States—A focus on Six Impacts https://www.epa.gov/system/files/documents/2021-09/climate-vulnerability_september-2021_508.pdf

FAO 2016, The State of World Fisheries and Aquaculture 2016. Contributing to food security and nutrition for all, Report (Food and Agriculture Organization) https://www.fao.org/3/i5555e/i5555e.pdf

Fasullo, J. T., & Gent, P. R. 2017, JCli, 30, 9195

Frederikse, T., et al. 2020, Natur, 584, 393

Frey, K. E., Comiso, J. C., Cooper, L. W., Grebmeier, J. M., & Stock, L. V. 2020, Arctic Report Card 2020: Arctic Ocean Primary Productivity: The Response of Marine Algae to Climate Warming and Sea Ice Decline,

Friedlingstein, P., et al. 2022, ESSD, 14, 4811
Frieler, K., Meinshausen, M., Golly, A., Mengel, M., Lebek, K., Donner, S. D., & Hoegh-Guldberg, O. 2013, NatCC, 3, 165
Gilmour, J. P., Smith, L. D., Heyward, A. J., Baird, A. H., & Pratchett, M. S. 2013, Sci, 340, 69
Global Center on Adaptation, 2019, Adapt now: a global call for leadership on climate resilience https://gca.org/reports/adapt-now-a-global-call-for-leadership-on-climate-resilience/ (accessed 6.14.23)
Goodkind, A. L., Tessum, C. W., Coggins, J. S., Hill, J. D., & Marshall, J. D. 2019, PNAS, 116, 8775
Habeeb, D., Vargo, J., & Stone, B. 2015, NatHa, 76, 1651
Hao, Z., Hao, F., Singh, V. P., & Zhang, X. 2018, ERL, 13, 124022
Hibbard, K.A., Hoffman, F.M., Huntzinger, D., & West, T.O. 2017, In *Climate Science Special Report: Fourth National Climate Assessment, Volume I*, ed. D. J. Wuebbles et al. (Washington, DC: U.S. Global Change Research Program) 277
Hoegh-Guldberg, O., et al. 2007, Sci, 318, 1737
Holland, G., & Bruyère, C. L. 2014, ClDy, 42, 617
Horiuchi, S. 1993, PT, 21, 6
Huang, B., et al. 2020, JCli, 33, 1351
Immerzeel, W. W., et al. 2020, Natur, 577, 364
IFRC 2023, Extreme heat: Preparing for the heat waves of the future | IFRC https://www.ifrc.org/document/extreme-heat-preparing-heat-waves-future (accessed 6.14.23)
IPCC 2018, Global Warming of 1.5°C. An IPCC Special Report on the impacts of global warming of 1.5°C above pre-industrial levels and related global greenhouse gas emission pathways, in the context of strengthening the global response to the threat of climate change, sustainable development, and efforts to eradicate poverty, ed. V Masson-Delmotte, et al. (Cambridge: Cambridge Univ. Press) https://doi.org/10.1017/9781009157940
IPCC 2021, Climate Change 2021: The Physical Science Basis. Contribution of Working Group I to the Sixth Assessment Report of the Intergovernmental Panel on Climate Change, ed. V. Masson-Delmotte, P. Zhai, A. Pirani, et al. (Cambridge: Cambridge Univ. Press) 2391
Jackson, L. C., et al. 2015, ClDy, 45, 3299
Jacobs, J. M., Cattaneo, L. R., Sweet, W., & Mansfield, T. 2018, TRR, 2672, 1
Jacobson, M. Z. 2005, JGRD, 110,
Jahn, A. 2018, NatCC, 8, 409
Jerrett, M., Jina, A. S., & Marlier, M. E. 2022, EnP, 310, 119888
Jevrejeva, S., Jackson, L. P., Grinsted, A., Lincke, D., & Marzeion, B. 2018, ERL, 13, 074014
Johnstone, P., & McLeish, C. 2020, ERSS, 69, 101732
Kelley, C. P., Mohtadi, S., Cane, M. A., Seager, R., & Kushnir, Y. 2015, PNAS, 112, 3241
Kenney, W. L., Craighead, D. H., & Alexander, L. M. 2014, MedSciSpEx, 46, 1891
Kim, S.-K., Kim, H.-J., Dijkstra, H. A., & An, S.-I. 2022, npjCAtS, 5, 1
King, M. D., et al. 2020, ComEE, 1, 1
Kossin, J. P. 2018, Natur, 558, 104
Kwok, R. 2018, ERL, 13, 105005
Lehner, F., et al. 2017, GeoRL, 44, 7419
Lenton, T. M., et al. 2008, PNAS, 105, 1786
Lenton, T. M., et al. 2021, NRvEE, 2, 91
Levy, K., Woster, A. P., Goldstein, R. S., & Carlton, E. J. 2016, EST, 50, 4905

Li, L., Lozier, M. S., & Li, F. 2022, ClDy, 58, 2249
Link, P. M., & Tol, R. S. J. 2004, PoEcJ, 3, 99
Littell, J. S., Peterson, D. L., Riley, K. L., Liu, Y., & Luce, C. H. 2016, GCBio, 22, 2353
Lockley, E. C., & Eizaguirre, C. 2021, EvAp, 14, 2361
Long, W. C., Swiney, K. M., Harris, C., Page, H. N., & Foy, R. J. 2013, PLoSO, 8, e60959
Lough, J. M., Anderson, K. D., & Hughes, T. P. 2018, NatSR, 8, 6079
Lovejoy, T. E., & Nobre, C. 2018, SciA, 4, eaat2340
Mallakpour, I., & Villarini, G. 2015, NatCC, 5, 250
Mantua, N., Tohver, I., & Hamlet, A. 2010, ClCh, 102, 187
McKinnon, K. A., & Simpson, I. R. 2022, GeoRL, 49, e2022GL100380
Milner, A. M., et al. 2017, PNAS, 114, 9770
Misra, A. K. 2014, IJBEn, 3, 153
Molnár, P. K., et al. 2020, NatCC, 10, 732
Mote, P. W., Li, S., Lettenmaier, D. P., Xiao, M., & Engel, R. 2018, npjClAtSci, 1, 1
Neary, D. G. 2021, Conifers—Recent Advances (IntechOpen)
Neumann, J. E., Emanuel, K., Ravela, S., Ludwig, L., Kirshen, P., Bosma, K., & Martinich, J. 2015, ClCh, 129, 337
Nie, Y., Pritchard, H.D., Liu, Q., et al. 2021, NRvEE, 2, 91
NOAA 2017, Costliest U.S. tropical cyclones tables updated. https://www.nhc.noaa.gov/news/UpdatedCostliest.pdf
Oppenheimer, M., et al. 2019, Sea Level Rise and Implications for Low-Lying Islands, Coasts and Communities. *IPCC Special Report on the Ocean and Cryosphere in a Changing Climate*, ed. H.-O. Pörtner, et al. (Cambridge: Cambridge Univ. Press) 321
Paolo, F. S., Fricker, H. A., & Padman, L. 2015, Sci, 348, 327
Parkinson, C. L. 2014, GeoRL, 41, 4316
Perkins-Kirkpatrick, S. E., & Lewis, S. C. 2020, NatCo, 11, 3357
Pimm, S. L., Jenkins, C. N., Abell, R., et al. 2014, Sci, 344, 1246752
Pitman, K. J., Moore, J. W., Huss, M., et al. 2021, NatCo, 12, 6816
Quante, L., Willner, S. N., Middelanis, R., & Levermann, A. 2021, NatSR, 11, 16621
Quintero, I., & Wiens, J. J. 2013, EcoL, 16, 1095
Radchuk, V., et al. 2019, NatCo, 10, 3109
Ramsay, H. 2017, Oxford Research Encyclopedia of Natural Hazard Science (Oxford: Oxford Univ. Press)
Reaka-Kudla, M. L., & Wilson, D. E. 1997, Biodiversity II: Understanding and Protecting Our Biological Resources (Washington, DC: Joseph Henry Press)
Regehr, E. V., Laidre, K. L., Akçakaya, H. R., et al. 2016, BioL, 12, 20160556
Reich, P. B., Hobbie, S. E., Lee, T. D., & Pastore, M. A. 2018, Sci, 360, 317
Romanello, M., et al. 2022, Lancet, 400, 1619
Romero-Lankao, P., et al. 2014, Climate Change 2014: Impacts, Adaptation, and Vulnerability. Part B: Regional Aspects. Contribution of Working Group II to the Fifth Assessment Report of the Intergovernmental Panel on Climate Change, ed. V. R. Barros, C. B. Field, D. J. Dokken, et al. (Cambridge: Cambridge Univ. Press) 1439
Romps, D. M., Seeley, J. T., Vollaro, D., & Molinari, J. 2014, Sci, 346, 851
Sario, M. D., Katsouyanni, K., & Michelozzi, P. 2013, ERJ, 42, 826
Schär, C., et al. 2016, ClCh, 137, 201
Semba, R. D., Askari, S., Gibson, S., Bloem, M. W., & Kraemer, K. 2022, AdN, 13, 80

Sen Roy, S. 2018, ed. S. Sen Roy Linking Gender to Climate Change Impacts in the Global South, Springer Climate (Cham: Springer International Publishing) 1
Serreze, M. C. 2011, Natur, 471, 47
Shindell, D., Zhang, Y., Scott, M., Ru, M., Stark, K., & Ebi, K. L. 2020, GeoH, 4, e2019GH000234
Skirving, W. J., Heron, S. F., Marsh, B. L., Liu, G., De La Cour, J. L., Geiger, E. F., & Eakin, C. M. 2019, CorRe, 38, 547
Smith, A.B. 2020, U.S. Billion-dollar Weather and Climate Disasters, 1980 - present (NCEI Accession 0209268). NOAA National Centers for Environmental Information https://doi.org/10.25921/stkw-7w73. Accessed 06/01/2023
Søreide, J. E., Leu, E., Berge, J., Graeve, M., & Falk-Petersen, S. 2010, GCBio, 16, 3154
Steel, D., DesRoches, C. T., & Mintz-Woo, K. 2022, PNAS, 119, e2210525119
Stott, P. A., Stone, D. A., & Allen, M. R. 2004, Natur, 432, 610
Stroeve, J., & Notz, D. 2018, ERL, 13, 103001
Tebaldi, C., et al. 2021, NatCC, 11, 746
Thompson, R., Hornigold, R., Page, L., & Waite, T. 2018, PH, 161, 171
Tietsche, S., Notz, D., Jungclaus, J. H., & Marotzke, J. 2011, GeoRL, 38, L02707
UN WHO 2021, Pandemic year marked by spike in world hunger https://www.who.int/news/item/12-07-2021-un-report-pandemic-year-marked-by-spike-in-world-hunger (accessed 6.14.23)
United Nations, Department of Economic and Social Affairs, Population Division 2022, World Population Prospects 2022: Ten Key Messages (US EPA, O) 2021. Social Vulnerability Report https://www.epa.gov/cira/social-vulnerability-report (accessed 6.14.23)
USGCRP 2017, Climate Science Special Report: Fourth National Climate Assessment, Volume I, ed. D. J. Wuebbles, et al. (Washington, DC: U.S. Global Change Research Program) doi: 10.7930/J0J964J6
Valin, H., et al. 2014, AgEc, 45, 51
van der Linden, S., Maibach, E., & Leiserowitz, A. 2015, PPsSci, 10, 758
Vaughan, D. G., Bamber, J. L., Giovinetto, M., Russell, J., & Cooper, A. P. R. 1999, JCli, 12, 933
Vecellio, D. J., Wolf, S. T., Cottle, R. M., & Kenney, W. L. 2022, JAPhy, 132, 340
Vellinga, M., & Wood, R. A. 2002, ClCh, 54, 251
Vicedo-Cabrera, A. M., et al. 2021, NatCC, 11, 492
Vohra, K., Vodonos, A., Schwartz, J., Marais, E. A., Sulprizio, M. P., & Mickley, L. J. 2021, ER, 195, 110754
von Schuckmann, K., et al. 2020, ESSD, 12, 2013
Vosper, E. L., Mitchell, D. M., & Emanuel, K. 2020, ERL, 15, 104053
Warren, R., Price, J., Graham, E., Forstenhaeusler, N., & VanDerWal, J. 2018, Sci, 360, 791
Williams, P. D. 2017, AdvAtSci, 34, 576
Williamson, K. 2023, The Global Transboundary Climate Risk Report, ed. A. Anisimov, & A. K. Magnan (Stockholm: The Institute for Sustainable Development and International Relations and Adaptation Without Borders) https://adaptationwithoutborders.org/knowledge-base/adaptation-without-borders/the-global-transboundary-climate-risk-report
Wolfenbarger, L. L., & Phifer, P. R. 2000, Sci, 290, 2088
World Meteorological Association, Atlas of Mortality and Economic Losses from Weather, Climate, and Water related Hazards. 2023. https://library.wmo.int/idurl/4/57564

Chapter 5

Agriculture and Climate Change

Rachel Mason

"A nation that destroys its soils destroys itself. Forests are the lungs of our land, purifying the air and giving fresh strength to our people."

—Franklin D. Roosevelt

Agriculture began to develop roughly 12,000 years ago, around the time that the erratic climate of the Pleistocene epoch gave way to the relative stability of the Holocene (Richerson et al. 2001). Subsistence and smallholder agriculture is now practiced by about 2 billion people on 475 million small farms in the developing world (Rapsomanikis 2015). At the same time, much of the world's food comes from "industrial" farms and livestock facilities optimized for high yields of uniform products. All of these production systems are being affected in some way by climate change. Conversely, the world's food system is responsible for a substantial fraction of anthropogenic greenhouse gas emissions. These emissions can be reduced by changing what we eat—e.g., reducing meat and dairy consumption—as well as simply wasting less food. As climate change worsens, adaptation or transformation will be necessary throughout the food system.

5.1 Effects of Climate Change on Agriculture

Higher temperatures and changes in the amount, frequency, intensity, and seasonality of rainfall have direct effects on crop yields and livestock production. (Box 1 discusses the direct effects of elevated atmospheric CO_2 levels on crops.) Heat stress has numerous negative effects on plants and animals, leading to outcomes ranging from poor germination of rice and wheat (Jha et al. 2014) to lower milk yields from dairy cows (Polsky & von Keyserlingk 2017). Some crops, such as apples and pomegranates, may no longer experience the cold winter temperatures they require in order to set fruit (Luedeling et al. 2009). At the same time, however, in some areas warmer temperatures are increasing yields and enabling the introduction of new crops.

doi:10.1088/2514-3433/acfcb6ch5

Droughts are well known for their profound effects on agriculture, but intense rain and floods can also destroy crops.

Beyond these direct impacts, climate change is having many secondary effects on agriculture. Warmer temperatures have allowed weeds, pathogens, and insect pests to move to higher latitudes and elevations (Bebber et al. 2013). Smoke from fires exacerbated by drier, hotter conditions affects the taste of wine (not to mention the health of farmworkers). Extreme rain events and flooding can erode productive land, deposit sand and weed seeds on fields, damage infrastructure, and delay farm operations such as planting crops. Some crops tend to build up toxic substances during droughts, intense rains, or heatwaves.

Clearly, the impacts of climate change can be highly localized and specific to particular crops, regions, and production systems (Box 2 illustrates the effects on one particular crop and region). Nonetheless, at the global scale, climate change may already have reduced staple crop yields by ~1% (Ray et al. 2019), and models suggest further reductions, such as a 4–6% reduction in global wheat yields for every 1°C of warming (Liu et al. 2016). The fraction of yields lost to insect pests is projected to rise by 10–25% per 1°C of warming as insects grow and reproduce faster (Deutsch et al. 2018), while the risk of simultaneous failures of major crop-growing regions appears to be increasing (Gaupp et al. 2020). Some researchers argue that an increasingly unstable climate will make agriculture impossible within 1–2 centuries (Gowdy 2020). However, agricultural technologies are advancing rapidly, and others are more optimistic about humanity's ability to maintain or improve crop yields (Aggarwal et al. 2019).

Ultimately, agriculture is part of a complex social-ecological system. Agronomists can model how a crop will respond to rising temperatures, but predicting the combined effects of future temperatures, precipitation patterns, insect populations, etc. is much more difficult, and a farmer's or country's options for adaptation are shaped by the social and economic conditions in which they operate. Further complicating matters, the effects of climate change on agriculture may reshape global geopolitics, by pushing people to migrate from food-insecure regions and by shifting food production towards cooler countries such as Canada and Russia. Changes such as these could in turn affect the greenhouse gas (GHG) emissions from agriculture itself (Schierhorn et al. 2019)—the subject of the next section.

Box 1: Is CO_2 good for crops?

Atmospheric CO_2 has increased from pre-industrial levels of ~280 ppm to ~410 ppm at present, and is now at its highest level in at least a million years. This means that plants now have access to ~50% more of a resource that is essential for photosynthesis. In experiments that expose plants to elevated CO_2 (eCO_2; e.g. Free Air CO_2 Enrichment studies) higher crop yields are often observed. It has even been suggested that the positive effects of eCO_2 will help humanity achieve food security.

That simplistic conclusion ignores many important factors. To give just a few examples (Ainsworth & Long 2021): (1) eCO_2 only benefits plants that use the "C_3"

photosynthetic mechanism; so-called "C_4" crops, such as maize and sorghum, are unable to take advantage of eCO_2 in most conditions. (2) Even C_3 plants can only support increased growth if they have access to adequate nutrients (nitrogen phosphorus, etc.), which is not always the case for resource-constrained farmers. (3) Concentrations of protein, zinc, iron, and other nutrients in crops decline under eCO_2 (Myers et al. 2014). (4) In some cases, weeds benefit more than crop plants from eCO_2 (Ziska et al. 2010). Rather than concluding that "CO_2 increases yields," a more accurate interpretation would be "rising CO_2 may increase yields of some crops in some circumstances while yields will probably decrease in others. Crop nutritional quality is also likely to decline. We need to understand how to mitigate and manage these effects in ways that are tailored to local environments, needs, and resources."

5.2 Agriculture's Contribution to Climate Change

On the other side of the equation, agriculture has a profound influence on the climate. Agriculture-related emissions comprise roughly 25% of total anthropogenic GHG emissions, and reducing these emissions is essential for limiting warming to 2°C (Hedenus et al. 2014; Wollenberg et al. 2016).

Emissions from agriculture arise throughout the chain of production (Poore & Nemecek 2018). Land use change (such as deforestation to make way for crops and pastures) releases CO_2 from biomass and soil, and can also produce N_2O. The manufacture, transport, and use of fertilizers on crops and pastures relies on fossil fuels, then applying nitrogen fertilizer to land releases N_2O. The global food system involves very large numbers of animals, and ruminant animals (cattle, sheep, goats, etc.) emit CH_4 as the bacteria in their rumens digest plant matter. Microbial processing of animal waste stored in liquid form emits CH_4 and/or N_2O; when manure and urine are applied to land (by humans or animals), N_2O is emitted. Transporting, processing, and refrigerating crops and animal products generally relies on fossil energy, while decomposing food waste in landfills is another source of CH_4.

The relative contributions of each stage (and therefore opportunities for mitigation) vary widely by sector. Land use change can account for a significant fraction of emissions, especially from tropical products such as chocolate, coffee, palm oil, and beef. On-farm processes are the biggest emissions source for many fruits and vegetables, although total emissions are relatively low for these crops. Perhaps surprisingly, emissions from transportation ("food miles") comprise only a small fraction of GHG emissions for most crops.

Globally, emissions related to animal production account for >50% of total agricultural emissions, with CH_4 produced by cattle and sheep being about 40% of that figure (Gerber et al. 2013). At the same time, however, grazing cattle, sheep, and other ruminants can potentially help to sequester carbon (C). Grazing and trampling can stimulate plant growth, and when that plant material dies and decomposes, some of its carbon may remain in the soil. The amount of C that can be sequestered in this manner depends on environmental factors and grazing practices in ways that are not yet thoroughly understood (Mcsherry & Ritchie 2013). Soil C sequestration also relies on continued good management, can be reversed by poor management, and eventually reaches a limit (Godde et al. 2020; Garnett et al. 2017). Good grazing practices can be part of the response to climate change, but calculations suggest that globally, the amount of C sequestered only partially offsets emissions from grass-based meat production (Garnett et al. 2017).

5.3 What Should We Do?

Humanity needs to find ways of producing food in a changed and changing climate. On the farm, practices such as cover cropping, agroforestry, or floating gardens can provide resilience to climate change effects and/or help sequester carbon, while also having many environmental co-benefits. Scientists and farmers can breed plants that resist heat, drought, and pests (although introducing these characteristics while retaining flavor and other desirable qualities can be challenging). Researching climate-friendly practices, plants, and technologies, and enabling farmers to adopt them, requires funding and a supportive policy environment.

We also need to produce food without further disrupting the climate—and the Earth system as a whole. While global principles have been proposed, disagreement remains about the desirability of particular actions, technologies, etc. (Lavania et al. 2015; Mercer et al. 2012). The subject of emissions from livestock production, and how to reduce them, is especially contentious. Cattle feedlots can be efficient in terms of GHG emissions per unit of meat or milk produced (Capper 2012), but increasing demand for meat could mean that emissions rise regardless (Popp et al. 2010). In contrast to feedlots, grazing systems can sequester some carbon, but grass-based systems alone cannot satisfy the current level of demand for animal protein in developed countries (Van Zanten et al. 2016). A shift from beef towards pig meat, poultry, and fish could reduce GHG emissions, but potentially at the expense of other environmental outcomes (Gerber et al. 2007), public health (Manyi-Loh et al. 2018), and animal welfare (Bessei 2006).

That researchers and organizations come to different conclusions is a consequence of several factors. These include (1) the difficulty of the basic biophysical questions to be answered (e.g. N_2O emissions from fertilized soil and manure are produced by a complex set of interacting biological processes that are not yet completely understood; Butterbach-Bahl et al. 2013), (2) the uncertainties involved in drawing conclusions about large-scale social-ecological systems (e.g. removing UK dairy cows from pasture and afforesting the spared land can reduce emissions in the UK, but potentially displaces beef production to the Amazon; Styles et al. 2018), and (3) values-based disagreements about how the food system does and should work (are market-based solutions preferable? Who benefits from biotechnology? What should be expected of people who have had little role in causing the climate crisis? etc.) that may or may not be explicitly recognized by the researchers themselves (Garnett 2014).

Nonetheless, there is fairly broad agreement about a handful of things. For example, further deforestation must be avoided (Foley et al. 2011). Producing less animal-based food would reduce pressure on natural systems, including the climate system (Hedenus et al. 2014; Scarborough et al. 2023; Wollenberg et al. 2016; Wilson et al. 2019). We also need to reduce the fraction of food that is wasted (Hall et al. 2009). Implementing such changes on a meaningful scale will require concerted efforts from government and civil society. However, individuals can still take actions to reduce their personal impact, and perhaps contribute to changing what is considered normal and acceptable.

5.3.1 Replace Meat and Dairy with Plant Products

Levels of meat consumption vary widely around the world, scaling with levels of affluence (Sans & Combris 2015). Those of us who are currently near the top of the global meat league have the opportunity to reduce our impact by replacing animal products with plant-based proteins. This does not have to mean strict veganism: eating meatless meals just one day per week reduces the carbon footprint of a meat-heavy diet by around 5% (and your pet can help, too; Alexander et al. 2020). Beyond personal consumption, our roles as scientists may provide opportunities to make plant-based food the default. For example, when planning in-person scientific meetings we should consider the environmental impact of any food that is served (Sanz-Cobena et al. 2020).

5.3.2 Waste Less Food

"If food waste were a country, it would be the third largest [GHG] emitter." Roughly a third of food goes to waste at some point on the journey from farm to fork, representing an unnecessary use of resources and a source of avoidable emissions. While much waste occurs upstream of the consumer, in higher-income countries ~30–40% of food waste occurs in households, restaurants, etc. (FAO 2013). In the UK, eliminating avoidable food waste in a household would reduce GHG emissions as much as insulating the attic, or cutting out a return flight to Europe for the whole family each year (Quested et al. 2013). There are many ways of reducing food waste in the home; see here for ideas that might work for you.

5.3.3 What about Buying Local, Eating Organic, Etc.?

The main factor determining agricultural GHG emissions is the type of food (animal vs plant), not whether it's organic or how far away it is produced. Buying organic produce from the local farmers' market, or milk direct from your neighbor's herd, can have many other benefits (perhaps reducing plastic waste, building community connections, or preserving wildlife habitat). It's not going to fix climate change, but climate change isn't the only problem in the world. Just be sure to eat those vegetables before they go to waste.

Box 2: Case Study: Climate change and coffee in Central America
Although native to Ethiopia, coffee is now grown commercially in more than 50 countries. Most of the world's coffee is grown by small farmers: roughly 60% of each year's crop comes from 12.5 million smallholders farming on less than 5 ha (12 acres).

Coffee prices are notoriously volatile, and often drop below the cost of production. Combined with the fact that small producers only receive a few percent of the price of a cup of coffee sold in Europe or the USA, making a living from small-scale coffee production is a difficult task. Nonetheless, the industry employs millions of people, and many communities take pride in their coffee-growing heritage.

In Latin America the main coffee species is *Coffea arabica*, known for its rich, smooth flavor. *C. arabica* has very specific requirements for good yields of high-quality beans. Optimal temperatures range from 18–21°C, while higher temperatures lead to poor-quality fruit and eventually damage the plant. The amount and timing of rainfall are also important. *C. arabica* needs ~1800 mm of rain each year, distributed fairly evenly except for a 3–4 month dry period that induces uniform flowering and makes the crop easier to harvest.

Climate change in the coffee-producing regions of Latin America is characterized by rising temperatures, increasingly unpredictable and intense rainfall, and more frequent, longer droughts (Imbach et al. 2018; Depsky et al. 2020). Farmers are certainly noticing these changes (Harvey et al. 2018); as one coffee grower put it, "We can no longer trust the weather. We no longer know what day it rains, nor what day it does not rain" (Ruiz-de-Oña et al. 2019). In some regions, warmer spring temperatures have been causing coffee plants to flower at the beginning of the dry season, which leads to rapid ripening that lowers the quality and market price of the resulting crop. Droughts lead to low yields of poor-quality beans, while intense rainfall, flash flooding, and uprooted trees during hurricanes have a record of damaging crops and infrastructure. Modeling suggests that half of the coffee-producing land in Central America will transition from "excellent" or "good" suitability to "moderate" or "marginal" by 2050 (Lara-Estrada et al. 2021).

As well as direct effects on coffee plants, climate change may also have created favorable conditions for coffee's most economically damaging disease, leaf rust. Coffee leaf rust had long been present in Central America at a manageable level, but the severe rust epidemic that started in 2008 hit farmers hard, impacting the livelihoods and food security of thousands of growers and laborers. Warmer temperatures appear to reduce the period between the germination of the leaf rust fungus (*Hemileia vastatrix)* and its sporulation and dispersal to infect other plants. This, combined with factors like low coffee prices that may have left farmers unable to optimally manage their crops, increases the intensity of a leaf rust epidemic (Avelino et al. 2015).

A degree of adaptation to all of these effects is possible. Coffee could be grown on land at higher elevations (albeit potentially increasing deforestation and its corresponding CO_2 emissions), or farmers at low latitudes and altitudes could introduce new crops. New hybrid varieties, more tolerant of warmer conditions and more resistant to rust, are being developed (highlighting the importance of preserving wild crop relatives as a genetic reservoir). Agroecological adaptations, such as interplanting shade trees to regulate temperatures, or using mulch and cover crops to conserve soil moisture and reduce erosion, can also help (Perfecto & Vandermeer, 2015). However, investing in these adaptation strategies requires capital, credit, and technical assistance, which smallholder farmers often struggle to access (Harvey et al. 2018).

Coffee production is becoming increasingly difficult in Central America, as is smallholder and subsistence agriculture in general. This has already led to migration to more prosperous countries, influencing the social and political climate of those countries. For all these troubles, though, this probably isn't the end of the road for coffee. Avid coffee drinkers may be comforted to read that scientists are now developing coffee brewed from plant cell cultures in the lab.

References

Ainsworth, E. A., & Long, S. P. 2021, GCBio, 27, 27
Aggarwal, P., Vyas, S., Thornton, P., Campbell, B. M., & Kropff, M. 2019, GFSec, 23, 41
Alexander, P., Berri, A., Moran, D., Reay, D., & Rounsevell, M. D. A. 2020, GlEnCh, 65, 102153
Avelino, J., et al. 2015, FS, 7, 303
Bebber, D. P., Ramotowski, M. A. T., & Gurr, S. J. 2013, NatCC, 3, 985
Bessei, W. 2006, WPScJ, 62, 455
Butterbach-Bahl, K., Baggs, E. M., Dannenmann, M., Kiese, R., & Zechmeister-Boltenstern, S. 2013, PTRSB, 368,
Capper, J. L. 2012, Animals, 2, 127
Depsky, N., & Pons, D. 2020, ERL, 16,
Deutsch, C. A., et al. 2018, Sci, 361, 916
FAO 2013, Food Wastage Footprint, www.fao.org/publications
Foley, J. A., Ramankutty, N., Brauman, K. A., et al. 2011, Natur, 478, 337
Garnett, T. 2014, JCP, 73, 10
Garnett, T., Godde, C., Muller, A., et al. 2017, Grazed and confused? Ruminating on Cattle, Grazing Systems, Methane, Nitrous Oxide, the Soil Carbon Sequestration Question-and what it All Means for Greenhouse Gas Emissions https://www.oxfordmartin.ox.ac.uk/publications/grazed-and-confused/
Gaupp, F., Hall, J., Hochrainer-Stigler, S., & Dadson, S. 2020, NatCC, 10, 54
Gerber, P., Opio, C., & Steinfeld, H. 2007, Poultry production and the environment-A review, (available at http://www.fao.org/ag/againfo/home/events/bangkok2007/docs/part2/2_2.pdf)
Gerber, P. J., Steinfeld, H., Henderson, B., Mottet, A., Opio, C., Dijkman, J., Falcucci, A., & Tempio, G. 2013, Tackling Climate Change Through Livestock—A Global Assessment of Emissions and Mitigation Opportunities (Rome: Food and Agriculture Organization of the United Nations (FAO))
Godde, C. M., de Boer, I. J. M., zu Ermgassen, E., et al. 2020, CC, 161, 385
Gowdy, J. 2020, Fut, 115, 102488
Hall, K. D., Guo, J., Dore, M., & Chow, C. C. 2009, PLoSO, 4, 9
Harvey, C. A., Saborio-Rodríguez, M., Martinez-Rodríguez, M. R., et al. 2018, AgFSe, 7, 1
Hedenus, F., Wirsenius, S., & Johansson, D. J. A. 2014, CC, 124, 79
Imbach, P., Chou, S. C., Lyra, A., Rodrigues, D., Rodriguez, D., Latinovic, D., Siqueira, G., Silva, A., Garofolo, L., & Georgiou, S. 2018, PLoSO, 13, 1
Jha, U. C., Bohra, A., & Singh, N. P. 2014, PB, 133, 679
Lara-Estrada, L., Rasche, L., & Schneider, U. A. 2021, REnvCh, 21,
Lavania, D., Dhingra, A., Siddiqui, M. H., Al-Whaibi, M. H., & Grover, A. 2015, PlPhB, 86, 100
Liu, B., et al. 2016, NatCC, 6, 1130
Luedeling, E., Gebauer, J., & Buerkert, A. 2009, CC, 96, 219
Manyi-Loh, C., Mamphweli, S., Meyer, E., & Okoh, A. 2018, Mol, 23, 795
Mcsherry, M. E., & Ritchie, M. E. 2013, GCBio, 19, 1347
Mercer, K. L., Perales, H. R., & Wainwright, J. D. 2012, GlEC, 22, 495
Myers, S. S., et al. 2014, Natur, 510, 139
Perfecto, I., & Vandermeer, J. 2015, Coffee Agroecology: A New Approach to Understanding Agricultural Biodiversity, Ecosystem Services and Sustainable Development (London: Routledge)
Polsky, L., & von Keyserlingk, M. A. G. 2017, JDSci, 100, 8645

Poore, J., & Nemecek, T. 2018, Sci, 360, 987
Popp, A., Lotze-Campen, H., & Bodirsky, B. 2010, GEnvC, 20, 451
Quested, T. E., Marsh, E., Stunell, D., & Parry, A. D. 2013, ReCRe, 79, 43
Rapsomanikis, G. 2015, The Economic Lives of Smallholder Farmers: An Analysis Based on Household Data from Nine Countries (Rome: Food Agric. Organ. United Nations)
Ray, D. K., West, P. C., Clark, M., Gerber, J. S., Prishchepov, A. V., & Chatterjee, S. 2019, PLoSO, 14, 1
Richerson, P. J., Boyd, R., & Bettinger, R. L. 2001, AmAnt, 66, 387
Ruiz-de-Oña, C., Rivera-Castañeda, P., & Merlín-Uribe, Y. 2019, SS, 8, 1
Sans, P., & Combris, P. 2015, MSci, 109, 106
Sanz-Cobena, A., Alessandrini, R., Bodirsky, L., et al. 2020, NatF, 1, 187
Scarborough, P., Clark, M., Cobiac, L., et al. 2023, NatF, 4, 565
Schierhorn, F., Kastner, T., Kuemmerle, T., et al. 2019, ERL, 14,
Styles, D., Gonzalez-Mejia, A., Moorby, J., Foskolos, A., & Gibbons, J. 2018, GCBio, 24, 681
Van Zanten, H. H. E., Meerburg, B. G., Bikker, P., Herrero, M., & De Boer, I. J. M. 2016, Animal, 10, 547
Wilson, N., Cleghorn, C. L., Cobiac, L. J., Mizdrak, A., & Nghiem, N. 2019, AdN, 10, S389
Wollenberg, E., Richards, M., Smith, P., et al. 2016, GCBio, 22, 3859
Ziska, L. H., Tomecek, M. B., & Gealy, D. R. 2010, AgJ, 102, 118

Chapter 6

How To Think about Solutions

T A Rector and Ka Chun Yu

"To be hopeless is to be uninformed."

—Elin Kelsey (2020)[1]

We know what's causing climate change, and we understand the wide-ranging consequences. But what can we do about it? Are there solutions? Are the solutions as bad or worse than the problem? Or is it too late? The bad news is that years of inaction mean that we are now feeling the effects of climate change. But the good news is that the worst consequences can still be avoided—and for the most part we have the resources and technology to do so. But we have to act fast. It's tempting to look at climate change from purely a technological viewpoint—from which the problem looks quite solvable. But we also need to consider the sociological, political, and economic factors. After all, having the solutions is irrelevant if there isn't the will to adopt them. This chapter explores hurdles beyond the technical that must be overcome for us to address climate change.

6.1 Introduction

When teaching or talking about climate change, most of the time the first question is, "What can we do about it?" Most people are beyond needing to be convinced that it is a problem. They want to know if anything can be done—and if so, what? When I (T.A.R.) first started teaching climate change, I didn't have any answers. I felt that it wasn't my role as a scientist to advocate. The result was that many students in my class left feeling at best unfulfilled, and at worst demoralized. Because I didn't provide insights to solutions I fear it reinforced a common misconception that nothing can be done. Many Americans who accept that climate change is happening cannot express specific reasons to be hopeful that we can address the problem, and instead find it easier to identify doubts (Marlon et al. 2019). As discussed in more

[1] Hope Matters: Why Changing the Way We Think is Critical to Solving the Environmental Crisis. United States: Greystone Books.

doi:10.1088/2514-3433/acfcb6ch6

detail in Chapter 13, this is known as "climate doomism"; and it is a real problem that we need to address.

If you're discussing only the problem but not its solutions, you're only telling half of the story. We need to help people understand the solutions available so that they can have hope. Talking about solutions does not mean downplaying the dangers or urging a particular outcome. It means helping people understand the situation so they can make informed choices. Talking about solutions is also what people want (e.g., Koepfler et al. 2010). They want to know, for example, if electric vehicles (EVs) are less polluting, if wind and solar are viable, and—most importantly—if it is possible for these solutions to save us from the worst consequences. The short answer to these questions is "yes." And there are reasons beyond climate change why we should want to adopt many of these solutions. But that doesn't mean you should cheerlead. You can give information about the strengths and weaknesses of each solution so that people can be better informed about what can be done, so they can work on and advocate for solutions that align with their values. The purpose of this chapter is to help you be better prepared for those conversations.

6.1.1 Should Scientists Advocate?

Scientists often feel uncomfortable advocating for solutions. That is to say, we frequently feel that our role is to use the process of science to determine the causes and consequences, but leave it up to others to determine how to advocate for solutions. We fear engaging in advocacy might jeopardize our objectivity, or our stature in the public eye. But being objective is not the same thing as being indifferent. As Stephen Schneider said, "Staying out of the fray is not taking the high ground; it is just passing the buck."

In general people want scientists to be part of the solution process. About half of Americans believe it should be a high priority for scientists to interact with policymakers and the public (ScienceCounts 2020),[2] an increase since 2019. Cologna et al. (2021) find that 70% of German and 74% of American respondents think that scientists should actively advocate for climate-related policies in general. They note that this support decreases when asked about specific policies. For example, only half of Germans and 62% of Americans support scientists advocating for carbon taxes. This is likely a result of the support for carbon taxes in general; i.e., people don't want *anyone* advocating for policies that they don't support. Only half of the public supported scientists endorsing protests that call for action on climate change. So scientists advocating for unpopular policy measures or actions might thus face greater challenges in legitimizing their advocacy.

Pielke (2007) outlines four roles that scientists can take. One role is the "Pure Scientist," who shows no interest in advocacy. The second is that of the "Science Arbiter," who serves as a resource—providing information only as requested. The third is as an "Issue Advocate," who uses their knowledge to push for a single, or

[2] https://sciencecounts.org/wp-content/uploads/2021/03/Important-Shifts-in-Americans-Attitudes-Towards-Science-2020.pdf

limited set, of solutions. Finally, the "Honest Broker" is one who provides information on a wide range of solutions so that policy makers can make appropriate decisions based upon this information and other considerations. These are of course simplistic delineations, but they do illustrate the range of options available to scientists. Naturally it should be our goal to serve as honest brokers and provide as much information as possible on the range of solutions available, with the understanding that as individuals we cannot fully know all of the strengths and weaknesses of all available solutions. We also need to be aware that we have biases towards certain solutions based upon our own personal preferences and values.

If you feel the need, you can distinguish your roles. For example, you can say, "As a scientist, I understand the causes of climate change and the severity of the problem. As a concerned citizen/parent/etc, I have decided to get involved and work for the following solutions…"

6.1.2 Why the Urgency?

As discussed in the consequences section, to avoid the worst consequences of climate change we want to keep the global temperature anomaly to below 2.0°C, with a strong preference to keep the increase at or below 1.5°C. To keep warming below 2°C, the cumulative carbon emissions between 2011 and 2050 need to be limited to around 1100 gigatonnes of carbon dioxide (Meinshausen et al. 2009). Accomplishing this goal calls for a rapid reduction of carbon emissions—to about half of current (2020) levels by 2030 and to "net zero" by 2050. Figure 6.1 shows four example model pathways to accomplish this goal. All of these pathways require after 2050 some level of "carbon negative" emissions (i.e., more CO_2 is absorbed from the atmosphere than emitted), with a greater need the longer we delay.

But, as discussed in Chapter 12, there is nothing particularly magical about these limits; and we need to be careful of how we talk about them. Climate change is often perceived as having only two outcomes—either we survive it or we don't. But climate change truly is a matter of degree. While the goal is to keep the temperature anomaly below 2.0°C, and preferably 1.5°C, it is not as though the world will be destroyed if we fail to achieve either benchmark. A lower temperature anomaly will always be better, no matter what the scenario. So rather than focusing on outcomes we should focus on the process—i.e., what can we do to minimize emissions as rapidly as possible?

6.2 A Framework

The science is clear on the severity of the problem. As discussed in Chapters 4 and 5, climate change has dramatic and scary social, environmental, and economic impacts. Their scale and significance vary dramatically for different countries and groups within their societies. The science is also clear on the causes. Approximately 90% of climate change is from the extraction and use of fossil fuels. The rest is mostly due to changes in land use; e.g., clearcutting of forests for agricultural purposes (Friedlingstein et al. 2022). The impacts of climate change can therefore be

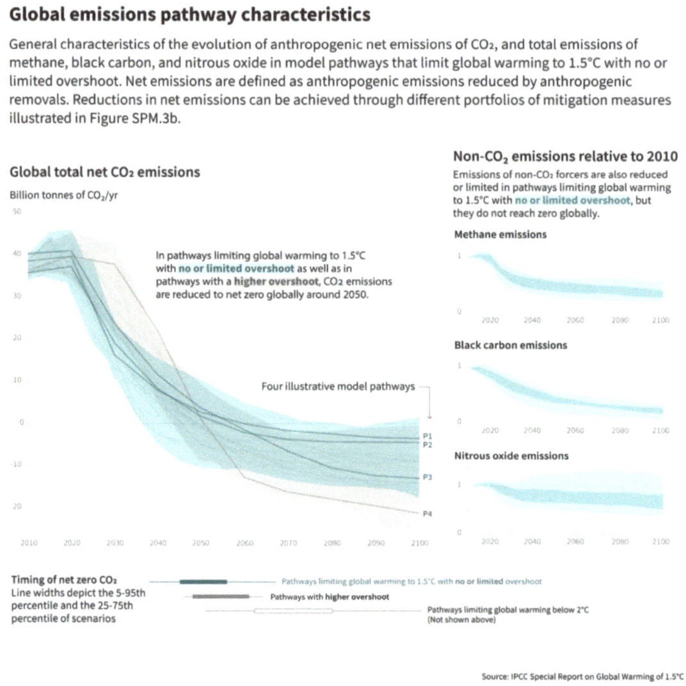

Figure 6.1. [3,4,5]The main panel shows global net anthropogenic CO_2 emissions in pathways limiting global warming to 1.5°C with no or limited (less than 0.1°C) overshoot and pathways with higher overshoot. The shaded area shows the full range for pathways analyzed in this Report. The panels on the right show non-CO_2 emissions ranges for three compounds with large historical forcing and a substantial portion of emissions coming from sources distinct from those central to CO_2 mitigation. Shaded areas in these panels show the 5–95% (light shading) and interquartile (dark shading) ranges of pathways limiting global warming to 1.5°C with no or limited overshoot. Box and whiskers at the bottom of the figure show the timing of pathways reaching global net zero CO_2 emission levels, and a comparison with pathways limiting global warming to 2°C with at least 66% probability. Four illustrative model pathways are highlighted in the main panel and are labeled P1, P2, P3 and P4, corresponding to the LED, S1, S2, and S5 pathways assessed in Chapter 2 of the IPCC SR15 report. Descriptions and characteristics of these pathways are available in figure SPM.3b.

reduced by decreasing carbon emissions and improving land use. The science is also clear on how quickly we need to adopt solutions to avoid the worst consequences. And it is clear we will need to adapt to the climate change that is already inevitable; e.g., protect coastlines from sea level rise and develop community-level emergency plans for extreme weather events. Both mitigation and adaptation are difficult, with widely different implications for different groups. It is not at all obvious what is the "right" global balance between reducing greenhouse gas emissions, other mitigation efforts, and adaptation to climate change and its impacts—or the best way to achieve that balance.

---

[3] https://www.ipcc.ch/sr15/graphics/

[4] https://www.ipcc.ch/site/assets/uploads/sites/2/2019/02/SPM3a.png

[5] SPM.3a from IPCC 2018

So how do we get there?

The goal for this chapter is to establish a framework for thinking about solutions. Climate change can be seen through different perspectives, e.g.:

- Technical: what steps—now and in the near future—are feasible to minimize climate change?
- Social/Moral: what do individuals and societies feel is the right thing to do based upon their values?
- Political: what policies should leaders enact, and how can they build support for them?
- Economic: what impacts do climate change and its solutions have on local and global commerce and industry?

These frames of course overlap and feedback upon one another; e.g. in principle democracies elect politicians that reflect people's values, who then enact policies that have economic consequences. But of course commerce and industry also influence elections; and many people look to leadership for guidance for what is important to them.

As scientists we have a tendency to focus on the technical framework, as it is most closely aligned with our knowledge and expertise. Chapter 7 therefore largely focuses on the technical side. But we also need to be cognizant that the other perspectives mentioned above play a role—and are just as important, if not moreso. Ideally any systematic comparison of solutions to climate change should consider all of these frames, and be done with respect to clearly stated objectives and metrics. These include but are not limited to the following:

- Significance: how much is the solution expected to contribute to solving the problem; e.g., what is the potential reduction to fossil fuel emissions that might result?
- Efficiency: how costly is the solution relative to others with similar benefits?
- Simplicity: how straightforward is the solution to adopt? To what extent does it require specialized expertise? What kind of infrastructure might it require?
- Adaptability: to what extent is the solution broadly suited (e.g., EVs, better home insulation) vs. geographically specific (e.g., geothermal and wind energy)
- Fairness: how equitable is the solution to different groups? How feasible is it to compensate groups that are negatively impacted?
- Political viability: what kind of support or opposition does the solution have? How might this support or opposition change over time?
- Timescale: if not now, how soon will the solution be available? And how quickly can it be implemented?
- Future improvement: to what extent might the efficacy or cost of the solution improve as technology advances.
- Integration: how well does the solution work with other solutions?

6.3 Technical Solutions

There are a variety of potential technical approaches for reducing carbon emissions. These include but aren't limited to:

- Switching to other forms of energy production (e.g., solar, wind, hydro-electric, and nuclear)
- Increasing efficiency of energy use (e.g., better insulation, more fuel efficient transportation)
- Short-term reductions in energy consumption (e.g., driving less, flying less, changing thermostat settings)
- Long-term reductions in energy consumption (e.g., restructuring communities to reduce or remove the necessity for driving)
- Carbon sequestration through technical or natural means (e.g., carbon capture from flue gases, forestry).

There are also a wide range of steps (from simple to highly disruptive) for mitigating the effects of climate change, such as building seawalls, changing the mix of crops grown in agriculture, and moving populations to locations less adversely affected.

Since about 90% of climate change is caused by the use of fossil fuels, Chapter 7 focuses on the technical viability of different energy sources, particularly in the U.S. But the technical feasibility of these potential mitigation and adaptation approaches are irrelevant if they aren't adopted. So we need to consider the different social, political, and economic implications for different countries, geographic regions, economies, and socioeconomic groups. In the following sections of this chapter we discuss how these other framings contribute to the discussion about solutions.

6.4 Sociology of Climate Change

In the study of climate change most of the focus has been on the natural sciences, to understand the physical nature of the problem. But knowing the science isn't enough if it doesn't lead to appropriate social action. It is similar to the COVID-19 pandemic: Billions were spent to develop a vaccine, but that money was wasted if the vaccine doesn't make it into people's arms. Well-organized social structures to produce, distribute, and dispense the vaccine are just as important as good communication about its safety and effectiveness. This communication is especially important when people are exposed to misinformation, as they are for both COVID-19 and climate change.

How we as individuals think and feel about climate change is important. But it is also relevant to see how social and historical forces shape what we think and feel. And critically, how those thoughts and feelings cause us—and others—to behave. Sociologists call this the *sociological imagination*. It's your ability to see yourself at this particular historical moment, and how you fit in with your social surroundings. For example, a person can be highly concerned about climate change but if they live in a location that doesn't have sufficient public transportation and is not suitable for alternative transportation (e.g., unsafe for walking or biking), they are most likely going to drive to and from work even though they understand that they are contributing to the problem. It helps to understand the history of the physical and sociological changes that occurred to accommodate individual cars so that they became the predominant form of transportation in the U.S. but not in other industrialized countries (Norton 2008). Understanding this history can help people to see the changes that need to be made to enable viable low-carbon transportation. It's also important to consider the roles that social structures—such as patriarchy, racism, and classism—play in the process of decision making; e.g., construction of the Interstate Highway System in the U.S. in the 1950s predominantly impacted poor and minority communities (Karas 2015).

It's important to acknowledge that climate change is also a social justice issue, as not everyone is contributing to the problem—nor are they feeling the effects—equally. Chancel (2022) estimates that in 2019 the bottom 50% of the world population in wealth emitted 12% of global CO_2 emissions, whereas the top 10% emitted about half; and each of the richest 1% is emitting close to 100 times more than the members of the poorest 10% (Chancel & Piketty 2015). At the same time low-income communities have the least resources to adapt and respond to the threats posed by climate change. As individuals we have some influence over our housing, transportation, and food. However our influence over these factors is proportional to our wealth; i.e., what options are available to you depend on what you can afford. Thus, the people most responsible for climate change are also the most able to respond, and of course have the highest obligation to do so.

We therefore need to consider the sociological aspects of not only the causes and consequences of climate change, but also the solutions. The sociological causes and implications of climate change are too many to explore here, but there are a few key concepts worth understanding as they can inform how we communicate: social norms and social change. For a deeper understanding, Islam & Kieu (2021) provide a review of the sociological perspectives on climate change. Szasz and Kiehl also provide useful teaching resources[6].

6.4.1 Social Norms

Imagine that you're in a crowded theater and you see what appears to be a fire burning in a far corner. If you perceive that no one else is responding to the smoke and flames, you might conclude that there isn't really a problem and choose to do nothing—even though there is ample physical evidence of a problem. Now imagine you're in the same theater, and while you see no indication of a fire everyone starts running for the exits. Without thinking about it you probably would too! After all, there must be a reason for everyone fleeing. This is an extreme example[7] of responding to social norms, defined as a group's rules of behavior. Norms more

[6] https://teachingclimate-2021.sites.ucsc.edu

[7] Extreme but this situation isn't purely hypothetical. A similar situation played out during a fire at a Great White rock concert in 2003. https://en.wikipedia.org/wiki/The_Station_nightclub_fire

subtle than this example are continuously at play in our daily lives, as we interact with different social groups. Social norms are important because people often choose a position on topics such as climate change that reinforces their connection to others with whom they share important ties rather than based on the available evidence (Kahan 2010). It is worth noting that people are part of multiple social groups—and they come in a range of sizes, from family units to national identity—and that each group can have different rules for appropriate behavior.

Social norms can be divided into two categories: "Injunctive" norms that tell us what we *should* do, and "descriptive" norms that specify what we *actually* do. When these norms are in conflict, people are often demotivated to change their behavior (Fielding & Louis 2020). For example, people often hear about climate change and the dire threat it poses, but in their daily lives they don't see individuals or larger social institutions doing much to address it. The most common result is that people implicitly assume that it's "not that bad" or it's a problem for the future—and they focus on other, more pressing issues. People are more likely to engage in a behavior if they see others in their peer group do the same (Goldstein et al. 2008). Likewise they are unlikely to engage if climate skepticism is perceived to be a group value. Personalizing and localizing the threat of climate change, and enhancing the norm that most people support action, can therefore be effective in overcoming doubt and inaction (Ballew et al. 2022).

6.4.2 Social Change

Often people ask whether addressing climate change should be based on personal decisions or collective action. The answer is "yes." Both are important. It's not people's fault that they live in a society that requires the use of fossil fuels for just about everything. Nor can you blame people for not engaging in greener behavior if the existing infrastructure is prohibitive; e.g., riding a bike to work on busy roads that lack safe bike lanes. However individual steps can normalize concern about climate change, which can be an important step toward widespread social change. The challenge is to transition from "What can I do?" to asking your peer groups "What can we do?" (Corbett 2021).

Social norms can be changed by individuals groups as well as institutions. This can be done on large scales; e.g., the "flight shame"[8] movement that started in Sweden in 2018 has drawn attention to the high carbon footprint of air travel, and in particular on its use for short-haul travel when lower-emission alternatives such as trains are available. Social norms can also be changed on a personal scale; e.g., conversing with a respected neighbor who has installed solar panels on their home or purchased an EV can increase a person's awareness level of such solutions and motivate them to take similar actions. In fact, the most likely indicator of whether or not someone will install solar panels on their home is if their neighbors have already done so (Barton-Henry et al. 2021). Social institutions can also encourage people to

[8] "Flugscham" in German or "flygskam" in Swedish.

make changes; e.g., government programs that offer rebates or tax reductions to install solar panels.

While the causes of climate change are well delineated, the solutions are not. That's because they're based not just on science and technology but social perceptions, values, and norms. Problems of social policy have been labeled as "wicked" because their requirements are poorly defined and can be contradictory (Rittel & Webber 1973). Climate change has been characterized as a "super wicked" problem consisting of four key features: time is of the essence; those who are causing the problem also need to find solutions; the central authority needed to address the problem is weak; and, partly as a result, policy responses irrationally discount the value of the future (Levin et al. 2012).

A major problem is that individuals and institutions are naturally averse to change. In sociology, "social inertia" depicts the resistance to change in societies or social groups (Bourdieu 1985). Brulle & Norgaard (2019) argue that climate change constitutes a "cultural trauma" that is met with resistance and attempts to maintain the status quo. Efforts to avoid large-scale social changes to address climate change constitute an effort to avoid cultural trauma, and result in social inertia at individual, institutional, and societal levels.

Part of the problem is that unhealthy behavior may not be seen as such once it becomes normalized. Often in our lives we have adjusted to existing problems to the point where they aren't noticed; e.g., we don't think about the health problems caused by air pollution even though we have all felt nausea after being exposed to exhaust from an internal-combustion engine. It all seems normal because, as far as we can recall, that's how it's always been. And when contemplating a change in behavior people tend to undervalue existing, known problems and overly worry about novel ones; e.g., in a phenomenon known as "range anxiety," people fixate on running out of charge in EVs while overlooking the inherent benefits of charging their vehicle at home. What's often not clear is how things can be better.

That of course doesn't mean change is impossible. There are many examples of changed social norms; e.g., it wasn't that long ago that smoking was deemed acceptable in most any venue, including enclosed spaces such as restaurants and airplanes. While some changes, such as the gradual phase out of public smoking, took many decades. Other changes can occur quite rapidly (e.g., LGBTQ+ rights) and some happen dramatically (e.g., the fall of the Berlin Wall). As discussed in the consequences section, a "tipping point" can be defined as a threshold (temperature or otherwise) that, once crossed, leads to irreversible changes (Lenton et al. 2008). Triggering one tipping point can also lead to others (Lenton et al. 2019). These are usually thought of in the context of physical systems, but there are also social tipping points where a small change in the underlying behavior of community members triggers an abrupt and irreversible change in a social system (Juhola et al. 2022). Tipping points can be negative, such as threats to societal structures from environmental impacts (Steel et al. 2022). But they can also be a good thing. Otto et al. (2020) describe several beneficial social tipping points, including when renewable energy production yields higher financial returns than fossil fuels as well as when investors perceive fossil fuel investments as too risky.

Tipping points can also occur in social movements. Chenoweth & Stephan (2011) conducted a study of protests for social change during 1900–2006. They found that nonviolent protests are twice as likely to succeed than violent protests, and that all nonviolent protests succeeded when at least 3.5% of the population actively participated. While this number is a small minority, it suggests there is a much larger population that passively supports the cause. It argues that it is less important to convince skeptics and more important to help those who are passively concerned to transition to actively engaged. We are already close to this threshold. Of the Six Americas discussed in Chapter 12, half of the Alarmed and a quarter of the Concerned said they would be willing to support an organization that engages in nonviolent civil disobedience against corporate or government activities that make global warming worse (Campbell et al. 2022).

6.5 Politics

Policy—at local, national, and international levels—is critical to the feasibility and effectiveness of climate change solutions. Considering that the science is clear on the importance of taking action, why are we moving so slowly on implementing policy changes? There are several factors to consider.

First, climate change is truly a global problem. Unlike other forms of air pollution, a CO_2 molecule affects the Earth's climate in the same way regardless of from where it is emitted. No country on its own can solve the problem; and no country is immune to its consequences. Thus it is necessary for worldwide cooperation. Extremely challenging international negotiations have led to agreements that commit countries to reduce greenhouse gas emissions as well as for wealthy, developed countries to help poorer countries (i.e., the "global south") with the costs of mitigation and adaptation. For the most part these agreements have set goals for greenhouse gas (GHG) emissions reductions for each country—known as a nationally determined contribution (NDC)—but have left it to individual countries to decide how to do so. Of course if countries don't meet their own commitments, other countries are less likely to meet theirs—devaluing the impact of the agreement.

Ratified in 1992 by 197 countries, the UN Framework Convention on Climate Change (UNFCCC) was the first global treaty. It established an annual meeting known as the Conference of the Parties (COP) to discuss climate change policy. This led to the Kyoto Protocol, which was signed in 1997 by 192 countries but not ratified by the U.S. It was the first legally binding climate treaty, with specific GHG reduction goals during a commitment period of 2008–2012, during which 36 countries achieved their goals. However this did not include major emitters such as the U.S., China, and India. And overall global emissions continued to rise during that period.

The most important treaty so far is the Paris Agreement, which was ratified in 2015 by 197 countries, including the U.S. It has set the goal to keep the temperature anomaly from rising above 2.0°C, and preferably below 1.5°C. It also sets the goal for global emissions to be neutral by 2050. The U.S. briefly withdrew from the agreement at the end of 2020, but reentered a few months later. Many consider the

Paris Agreement to be insufficient (e.g., Falkner 2016). Currently the world is not on pace to meet its goals (e.g., Climate Action Tracker 2022).

How can countries achieve their NDC goals? There are a wide variety of potential economic and non-economic strategies for incentivizing or mandating the reduction of emissions. These may be divided into three broad groups, with many different strategies within each:

- Government regulations and mandates, such as:
    - Banning production or consumption of fossil fuels for particular uses and/or locations
    - Mandating production of non-fossil fuels for particular uses or locations
    - Nationalizing the fossil fuel industry
- Government economic incentives, such as:
    - Taxes on fossil fuel consumption or production
    - Subsidies for non-fossil fuel consumption or production
    - Cap and trade mechanisms
    - Supporting research into non-renewable energy sources
- Moral incentives, such as encouraging individuals to voluntarily reduce fossil fuel consumption.

What strategies are feasible are critically limited by politics. This is particularly the case in democracies, but it matters even for authoritarian societies. No matter how much sense a given policy may make, governments are limited in their ability to enforce it if people don't want to do it. Policy change can also have unintended consequences; e.g., in 2003 the European Union declared biofuels to be carbon-neutral, triggering a surge in wood use. However projected growth in wood harvest for bioenergy could increase atmospheric CO_2 for at least a century before declining (Sterman et al. 2018). Relatedly, policy can be used to resist change; anti-protest laws have been passed to quell opposition to government policy and fossil fuels; e.g., in South Australia in 2023.[9] And politics have been used to silence climate scientists (e.g., Donaghy et al. 2007) and government officials; e.g., in 2011 officials at the Florida Department of Environmental Protection were ordered not to use the terms "climate change" or "global warming" in any official communications (Korten 2015).

A major part of the problem is that the timescales of politics and climate change do not align. Elections occur every few years whereas policies can take decades to be effective, often incurring short-term challenges in the process. These can be exploited by opposition parties; e.g., in 2012 a carbon tax[10] was enacted in Australia by the Labor Party and then revoked in 2014 after the Liberal–National coalition came to power. As discussed in Chapter 12, climate change often lags behind other issues (e.g., health and social justice issues) even among those who are the most concerned.

[9] https://www.theguardian.com/australia-news/2023/may/31/south-australia-passes-laws-to-crack-down-on-protest-after-disruption-of-oil-and-gas-conference

[10] https://www.cleanenergyregulator.gov.au/Infohub/CPM/Pages/About-the-mechanism.aspx

And this concern varies; e.g., it can peak shortly after an extreme weather event but then fade.

The conflict between good politics and good policy leads to difficult—and often contradictory—decisions. For example, economists would argue that high prices are perhaps the best incentive for reducing fossil fuel consumption. Yet almost nothing is more unpopular to the public. So when the start of the Russian–Ukraine war in 2022 led to a steep rise in energy prices, rather than welcoming the high prices as a way to encourage switching to electric vehicles or more energy-efficient gas vehicles, the Biden administration actively encouraged Saudi Arabia and other OPEC countries to produce more oil. They also released oil from the Strategic Petroleum Reserve in an effort to bring down U.S. gas prices. More dramatically, the "gilets jaunes" (yellow vests) protesters brought major French cities to a standstill because of their objection to proposed carbon taxes which would have increased gas prices.

Perhaps the largest hurdle is the political influence of industries that stand to lose if fossil fuel use is reduced. Colgan et al. (2021) argue that climate politics can be best understood as a contest between owners of assets that accelerate climate change, such as fossil fuel plants, and owners of assets vulnerable to climate change, such as coastal property. As discussed in Chapter 13, fossil fuel and related industries spend billions on advertising, lobbying, and political contributions in an effort to deter climate legislation; and these efforts have increased at times when politicians are considering climate policy (Coan et al. 2021). Fossil fuel interest groups work to deter policy on local, national, and global levels; e.g. over 600 fossil fuel lobbyists attended the COP27 meeting in 2022.[11] Stokes (2020) offers a detailed examination of how over the years fossil fuel companies and electric utilities have fought policy changes that promote clean energies in the U.S. Considering the relatively low priority of climate change among an often-distracted electorate and the staunch resistance by the fossil fuel industry, politicians are understandably reluctant to take major steps forward in addressing the problem—especially as it could cost them in the next election. However as the renewable energy industry grows, so will its popularity and its political influence.

Because of these challenges, it has been very difficult to pass laws on a national level in the U.S. However there have been some successes. For example, the 2022 U.S. Inflation Reduction Act (IRA) updates tax credits to make it more affordable to operate carbon capture and storage (CCS) projects. The credit is now $85/ton for CO_2 sequestered geologically (compared to $60/ton for industrial utilization including enhanced oil recovery or EOR), while direct-air capture (DAC) projects with geological storage are given a $180/ton credit (Clean Air Task Force 2022). The $85/ton tax credit helps defray the cost of CCS, but is still not enough to completely cover its new deployment in most industries (Energy Futures Initiative 2023), so additional subsidies or investments will help. In May 2023 the U.S.

[11] https://www.reuters.com/business/cop/cop27-fossil-fuel-lobbyists-criticised-africa-energy-debate-simmers-2022-11-10/

Environmental Protection Agency (EPA) proposed performance standards that limit the amount of carbon emissions from fossil fuel plants.[12] Although the EPA does not have the legal authority to compel utility companies to install CCS, the new standards are strict enough that adopting CCS is one of the ways that existing coal and gas power plants can comply. Models show that the new rules plus the IRA tax credits combined can lead to a 77% reduction in carbon emissions in power generation by 2030 relative to 2005 levels (Lynch & Doinger 2023). This gets very close to the goal of creating electrical power that is 80% free of carbon emissions by 2030, a necessary step towards net-zero energy emissions by 2050 (Esposito et al. 2021). Of course these policy initiatives could be changed by future administrations.

While national-level policy has been slow to develop, there are many examples of successful state-level decarbonization legislative efforts. Marshall & Burgess (2022) studied over 400 bills that were enacted, a third of which were passed by Republican-controlled administrations in the U.S. They found that these bills tended to frame climate change as an economic issue rather than a social or environmental one, opting to expand rather than restrict consumer and business choice. In summary, it is possible for legislation to move forward in the current political climate. And this should accelerate as the effects of climate change become more apparent, and solutions become more popular. Economic-based policy solutions such as subsidies, carbon taxes, and “cap and trade” are discussed below.

6.6 Economics

Environmental economists and political scientists have well-developed theoretical approaches for analyzing and addressing issues associated with climate change. Economists tend to view environmental problems such as climate change as economic problems at their core. Here we will give a simple overview of some of the factors economists consider when thinking about this topic.

6.6.1 Externalities

At its simplest, the problems caused by fossil fuels—not only climate change but air pollution and other health and environmental impacts—may be viewed as an “externality.” An externality is defined as an unintended consequence of an economic activity that affects a third party. More specifically, an externality occurs when the production or consumption of a good or service affects the welfare of individuals or entities who are not directly involved in the market transaction between the seller and buyer. Positive externalities are beneficial effects that spill over to others, such as the social benefits of education–an educated workforce provides benefits to the community beyond the individuals who receive the education, such as higher productivity and economic growth. Negative externalities

[12] https://www.epa.gov/newsreleases/epa-proposes-new-carbon-pollution-standards-fossil-fuel-fired-power-plants-tackle

Figure 6.2. [13]A comparison of air quality in Fanhe Town, China. The two images were taken ten days apart. Smog is an example of an externality—in this case induced by the use of coal to generate electricity. Residents experience costs that are direct (e.g., increased health care costs) and indirect (e.g., reduced productivity and quality of life). One way to think about externalities is: what would a person or group be willing to pay to not have to endure the externality, or what would they be willing to be paid *to* endure it? (Credit: Tomskyhaha).

are harmful effects that impose costs or harm onto others, such as pollution from a factory or traffic congestion (Figure 6.2). Externalities can lead to "market failures," where the price of a good or service does not accurately reflect the full costs or benefits of production and consumption. If not accounted for, negative externalities can be thought of as a subsidy as the industry in question is not paying for the consequences of its activities on others.

When discussing solutions to climate change, and in particular when deciding on which energy sources to use, we need to consider these externalities. Perhaps the most important one is the cost in life associated with the use of each energy source. Considering all factors (i.e., air pollution, accidents from installation and operation, and greenhouse gas emissions), the death rates associated with fossil fuels and biomass (e.g., wood, dung, and charcoal) far exceed that of other energy sources (Ritchie 2020). The death rates from fossil fuels are a hundred to a thousand times higher than that of nuclear and renewable energies (Figure 6.3).

From an economic perspective there is an optimal rate for the use of fossil fuels. While they cause climate change and other problems, the cessation of their usage (at least for now) would also have dire consequences. Since for the most part producers and users of fossil fuels currently don't pay for their externalities, existing market mechanisms under-price fossil fuel consumption; and as a result we use more than

[13] https://en.wikipedia.org/wiki/Smog#/media/File:Fanhe_Town_10_day_interval_contrast.png

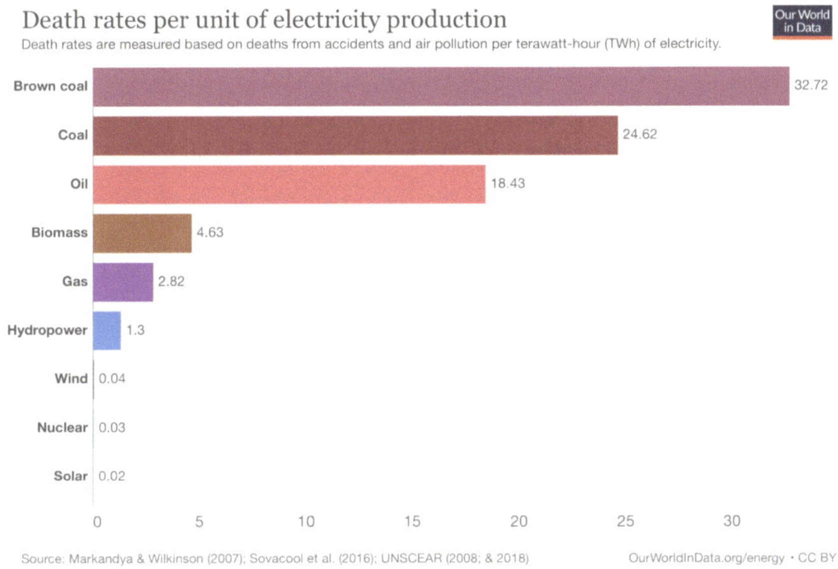

Figure 6.3. [14]The death rates for fossil fuels and biomass far exceed other energy sources largely because of the associated air pollution. It is worth noting that the calculated death rate of hydropower is primarily the result of a single incident—the Banqiao Dam Failure in China in 1975, which killed approximately 171,000 people. Otherwise it would have about the same death rate as wind, nuclear, and solar (Ritchie 2020).

the optimal amount. As discussed in Section 6.6.5, carbon pricing is a means to account for their externalities and enable market forces to determine their proper rate of use. However this is easier said than done. In general, solving environmental challenges caused by externalities are more difficult to address as the number and diversity of producers of an externality increases–and when the number and diversity of people affected increases as well. For a global issue like climate change, the challenges are formidable.

6.6.2 Supply and Demand

"Supply and demand" analysis is a powerful and deceptively complex theoretical tool for gaining insight into the production and consumption of any product (including fossil fuels and other energy sources). This analysis can inform us on how societies produce and consume a product, what causes changes in production and consumption over time, who benefits how much from its production and consumption, and how different strategies for reducing carbon emissions may affect its production and consumption. Prices are powerful incentives affecting production and consumption of any given product. And relative prices strongly determine the production and consumption of alternative products, including different kinds of energy.

[14] https://ourworldindata.org/safest-sources-of-energy

How much fossil fuels we use is a function of both what we produce and what we consume. We can't consume what we don't produce. And producers adjust production based upon what they think we want to consume. Production of fossil fuels depends on a variety of factors, including availability, economic viability and risk, technical challenges, regulatory constraints (e.g., environmental laws), and timescales. The last factor shows up in the "discount rate," a description of how much money is worth if invested today, versus waiting for it to be spent in the future. In the economics of climate change, burning fossil fuels today can have an immediate benefit, but result in greater harm in the future. As a result, a social cost to carbon needs to be added to offset the tendency to discount the present, but instead, to force a discount of the future.

To see how this works, take the example of someone who receives $100 today instead of waiting a year to receive the same amount. The money in hand can be invested, and if the available interest rate of return was 10%, then the money would grow to $110 after a year. To be equivalent to receiving $100 a year in the future, the present discounted value of that money today (assuming a discount rate of 10%) would be $90.91. The higher the discount rate, the less money is worth in the future compared to the present, meaning it should be spent now instead of saving and waiting to spend it. Economists will compare different policies by looking at their respective benefits minus costs over time, with the appropriate discount rate applied at each time step. If one policy has a higher total present value than a second policy, then the first policy is preferred. When evaluating the economics of climate policies, determining the discount rate to use requires estimating what the future return on investment is balanced with the cost of emissions reductions today weighed against the future damages from climate change. A higher return on investment in capital, education, and technology will mean a larger discount on climate damages in the future, meaning less of a need for emissions reductions now. A low return on investment means more needs to be done to reduce emissions now.

Since greenhouse gases persist in the atmosphere on timescales from decades to centuries, discounting should be done not just for those alive now. Intergenerational discounting requires an evaluation of the costs to future society and to people not yet born. Decisions on how to discount cannot be determined just from economics, but requires ethical considerations of how future generations are impacted and what they will want. Using a low discount rate of 1.4%, Stern (2006) estimated that damages from climate change could reduce global economic output by 5–20% by 2100, while mitigation today would cost only 1% of global GDP. While others criticized this model's use of a discount rate below that of observed market rates (e.g., Tol & Yohe 2006; Nordhaus 2008), Stern argues that the present generation has an ethical obligation to give equal weight to future generations, while the risks and uncertainties of climate change damages also calls for a low discount rate.

In global and local markets, prices of different energy sources are driven by both supply relationships (i.e., how much producers are willing to supply at any given price) and demand relationships (how much consumers wish to consume at any given price). Shifts in either supply or demand simultaneously affect (a) how much of different energy sources are produced; (b) how much of different energy sources are consumed; and (c) the prices at which different energy sources are bought and sold. At any given time and place, the energy from available different sources reflects the influences of many different factors that affect supply and demand, including but not limited to: population levels, technologies of energy production and consumption, available resources, energy transportation costs, economic structure (e.g., what the societies produce and how they produce it), and government regulations (e.g., subsidies and taxes). As any and all of these factors change, on widely differing time scales, energy consumption has changed, also on widely different timescales. For example, energy prices can vary daily (e.g., gasoline) to hourly (e.g., "peak" and "off-peak" electricity rates).

Economists think of prices as "signals" which encourage or discourage production and consumption. High prices encourage production and discourage consumption, while low prices do just the opposite. So economists tend to argue that if we want to reduce fossil fuel consumption, we need to make fossil fuels more expensive, both absolutely and relative to other energy sources. Paradoxically, efforts to reduce the supply of a particular energy source (e.g., by banning Arctic drilling to reduce oil production) tend to increase prices for that kind of energy by decreasing supply—which serves as an incentive for other producers of that same kind of energy to increase production. And high fuel prices increase profits of fossil fuel companies, which increases their political influence as well as attractiveness to investors. This is a significant challenge. The more we try to make it harder to produce oil or coal in one region, the more it encourages production in other regions. Likewise, taxes or subsidies which seek to make particular energy sources more or less costly to consume or produce disproportionately affect groups within a society depending on how dependent they are on consuming or producing those energy sources. For example, tax incentives to purchase EVs benefit only those for whom such a purchase is possible. While in theory the "winners" can compensate the "losers," in practice this is difficult to do in a way that is both fair and perceived as such.

Global fossil fuel consumption and climate change can't be blamed solely on—or fixed by only asking changes from–either producers or consumers. In principle, producers only produce when there is sufficient demand for their product. And consumers only consume because producers supply products that they want or need. If we use fossil fuels we can't put the blame for climate change solely on the countries and companies that supply the fossil fuels—especially if viable alternatives are available. However it is worth noting that producers influence the political and regulatory process to enhance supply (e.g., enabling resource extraction on government-owned land) as well as demand (e.g., stifle the development of mass transit so that people are reliant upon cars.) Consumers of course can also influence policy; e.g., advocating for neighborhoods that require less driving and are safer for cyclists and pedestrians.[15]

[15] https://en.wikipedia.org/wiki/New_Urbanism

Production and consumption respond to prices through different mechanisms on different time scales. For any given energy source, supply and demand relationships can shift much more dramatically in the long term than in the short term. As a result, it can be very difficult to change production and/or consumption of a particular energy source in the short term, but possible to make much more dramatic changes in the long term. Suppose, for example, we wish to reduce fossil fuel consumption by cars. If we tax gasoline at the pump, it will raise prices and in the short term (weeks to months), there will be relatively modest reductions in gasoline consumption. People may choose to cut back on voluntary driving or perhaps drive more slowly—but they will still need to use their cars for essential trips such as going to their jobs and shopping. In the medium term (months to years) people may opt to make larger changes such as telecommuting (i.e., working from home) or buy new kinds of cars that are more fuel efficient. If viable in their community, they may opt to carpool or take public transportation. In the longer term (five or more years) people may change where they live and work so that they don't need to travel as far in their daily lives. Rather than live in suburbs, they may choose to move back into cities or advocate for structural changes to their communities.

6.6.3 Comparing Energy Sources

In Chapter 7 we discuss each energy source in detail, at a level commensurate with its importance. Comparison is challenging because each has its strengths and weaknesses. But from an economic perspective the simplest—and perhaps most important—comparison is that of cost, which invariably dominates decision making in regards to investment and adoption.

Naturally, prices for different kinds of energy sources depend on the circumstances. And they change over time. Energy prices vary depending on supply and demand, on timescales of days (e.g., gasoline) to as short as hours (e.g., "peak" and "off-peak" electricity rates). Weather patterns can affect demand (e.g., heat waves and cold spells) as well as supply (e.g., from wind and solar). The cost of producing the energy; i.e., the raw materials for construction as well as fuel costs, also affect prices. Production costs can be impacted by international events such as economic sanctions and embargoes; e.g., the oil crises of the 1970s.[16] Changes in government policies such as taxes, subsidies, and regulations are also important. In the long term technological advancements, as well as "economies of scale" due to widespread adoption, can lower the cost of an energy type.

"Levelized cost of energy" (LCOE) is a common way in which energy costs are compared. LCOE refers to the estimated revenue required to build and operate a generator relative to the total energy produced over its lifetime. It is a useful metric because it allows for the comparison of different technologies of unequal cost, lifespans, and project size. It includes the costs of capital (i.e., the initial investment) and financing, operations and maintenance costs (including fuel), and the cost to shut down and decommission the facility once it is past its useful life. It also assumes

[16] The greatest decade for music. Not the greatest decade for everything else.

a rate of utilization for the energy produced; i.e., what percentage of the rated (i.e., "nameplate") capacity is actually generated and used. LCOE is usually given in units of cost per unit of energy; e.g., US$ per MWh (Megawatt-hour).

While LCOE is a useful means of estimating the overall cost per unit of energy, it doesn't tell the whole story. For example, even if an energy source has a relatively low LCOE over its lifetime, it is less likely to be adopted if it has high capital costs, which are a deterrent to investors who (naturally) wish to see a faster return on their investment. LCOE calculations also need to consider risk; i.e., what is the likelihood that the energy source will underdeliver or fail altogether. Examples of this include exploratory drilling for oil and gas as well as geothermal. There are other kinds of risk as well, such as cost risk (e.g. the project will cost more than expected to build or to operate) and regulatory risk (e.g. the government will change the rules about whether and how you can build and operate it.) How risk should be considered when calculating LCOE is debated (e.g., Aldersey-Williams & Rubert 2019). It is also worth mentioning that LCOE doesn't necessarily include the difference in value between "dispatchable" energy resources that can be scaled up quickly as demand requires (e.g., fossil fuels and geothermal) and intermittent resources that cannot be controlled (e.g., wind and solar).

Another important concept is the "levelized avoided cost of electricity" (LACE), which represents that power plant's value to the grid; i.e., it is a measure of the market value of the electricity it would generate. Power plants are considered economically attractive when their projected LACE (value) exceeds their projected LCOE (cost). Both LCOE and LACE are levelized over the expected electricity generation during the lifetime of the plant, resulting in values presented in dollars per MWh. These values vary by location, as resource availability, fuel costs, and other factors often differ by market. LCOE and LACE values also change over time as technology improves, tax credits and other taxes or subsidies expire, and fuel costs change (EIA 2018).

6.6.4 Subsidies

Comparisons of LCOE and LACE for different energy sources are further complicated by subsidies and externalities. In economics, a "subsidy" consists of financial assistance provided by the government to an individual or a business entity to support a particular activity or industry. The goal is to influence the market by reducing the cost of production or consumption of goods or services, thus making them more affordable and encouraging their use, potentially leading to increased economic activity and employment. However, subsidies can also have negative effects on the economy, such as distorting market prices, reducing incentives for efficiency, and creating inefficiencies in the allocation of resources. Examples of explicit subsidies are tax breaks and grants, while implicit subsidies are those that are provided through other means such as price controls or government regulations that reduce the cost of production.

In 2017, direct government subsidies of $634 billion was estimated to have been delivered to all forms of energy generation, with fossil fuels accounting for 70% of that total. Of the remaining subsidies, 20% were assigned to renewables, 6% for biofuels, and 3% for nuclear. Including unpriced externalities, the amount spent on fossil fuel subsidies outnumber the amount for renewables by 19-to-1 in 2017 (Taylor 2020). Fossil fuels are subsidized in various forms (Riedy & Diesendorf 2003). Explicit subsidies include financial assistance in the form of direct payments, tax breaks, and exemptions that reduce the cost of exploration, extraction, and production. Governments often invest in infrastructure, such as pipelines, refineries, and storage facilities. Trade policies can also be enacted to add tariffs to imports and negotiate the removal of tariffs for export. Governments also enable extraction of fossil fuels from public lands. Between 2005 and 2019, over 27% of U.S. oil, gas, and coal production came from federal lands and waters (ONRR 2020). However, the royalty rates for such extraction are often set below market value, resulting in a significant subsidy (Achakulwisut et al. 2021). Subsidies also exist via public funding of research and development by energy companies. There are also implicit subsidies of fossil fuels due to negative environmental and social externalities from their use that are not accounted for in their pricing. Fossil fuel companies are not held fully accountable for the environmental damage caused by their operations. Governments often cover the cost of environmental remediation, including cleaning up polluted water and air as well as restoring damaged land. Governments and individuals also pay for the associated costs from health problems caused by fossil fuels. The International Monetary Fund estimates that in 2020 global fossil fuel subsidies were $5.9 trillion, or 6.8 percent of global gross domestic product (GDP)[17] and could be as high as 8% in coming years (Parry, Black, & Vernon 2021). Overall, these subsidies make fossil fuels prices artificially low, discouraging investment in alternatives.

Subsidies are also a powerful tool for encouraging the adoption of new energy sources. Historically, emerging energy technologies were not competitive on cost with existing energy sources. It was only with government subsidies could new technologies like solar photovoltaic (PV) drop in cost over time to become commercially viable (Badcock & Lenzen 2010, Wen et al. 2021). Today, the LCOEs of nearly three-quarters of new utility PV installations in 2021 were cheaper than for fossil fuel generation (IRENA 2022). But when they first appeared commercially, photovoltaic cells had a cost of power that was three orders of magnitude greater than the wholesale price of electricity (Nemet 2014). After their initial introduction by Bell Labs in 1954, PV was used for extraterrestrial applications such as powering Earth-orbiting satellites and experiments left on the Moon during the Apollo missions, where the need for a compact energy source was price insensitive. Hart & Birson (2016) highlights the subsequent history of the terrestrial growth in PV installations. Widespread adoption of PV on the ground only started with the 1973 energy crisis, which spurred the U.S. government to fund research to improve solar cell efficiency and to reduce production costs. Around the time U.S. support for PV research and deployment slowed in the 1980s,[18] Japan

[17] https://www.imf.org/en/Topics/climate-change/energy-subsidies
[18] Not a bad decade for music, but not as good as people remember.

increased its own research funding—and followed with subsidies in the 1990s[19] that covered half the cost of PV installations. By the 2000s, Japanese subsidies had been reduced, but Germany then took the lead in supporting PV research, development, and deployment. It used "feed-in-tariffs" that guaranteed a fixed, high price for renewable energy to be paid to producers, with long-term contracts that would reduce their investment risk. After the Fukushima nuclear accident, similar feed-in-tariffs in Japan helped to increase PV adoption (Muhammad-Sukki et al. 2014). The rise in price of oil after the 2001 September 11 attacks prompted further changes in both U.S. federal and state policies that accelerated PV installations in the U.S. However, by the time of the 2008 global economic recession, cumulative PV capacity in the US (1.1 GW) still lagged behind Japan (2.1 GW), Spain (3.4 GW), and Germany (5.3 GW).

The PV story since then has been dominated by China, whose policies adopted after 2008 provided direct and indirect subsidies to its solar suppliers, and helped increase domestic demand. In one program started in 2009, systems connected to the electrical grid, such as building PV and large-scale utility PV were given total cost subsidies of 50%, while off-grid systems such as rural PV installations had total cost subsidies of 70% (Zhang & He 2013) An anti-dumping tariff enacted in China restricted US and Korean imports of polycrystalline silicon, the material used to create the ingots from which solar cell wafers are cut. These policies had an enormous impact on the Chinese PV industry. In 2006, China imported 95% of the polysilicon used to manufacture PV solar modules. In 2010, China, the United States, Germany, Japan and South Korea each had 15–30% of the polysilicon market. Five years later, China had increased its domestic supply by such a large amount that prices were driven down by 70%, undercutting producers in other countries. Today, China holds a commanding 80% share of the polysilicon market (IEA 2022).

The subsidization of PV development and deployment had a profound effect on its cost. By 2018, more than half of the new installations of utility-scale PV and onshore wind power had LCOEs that were cheaper than their fossil fuel analogs. In 2021, 70% of new solar PV and 96% of new onshore installations were cheaper than the cheapest fossil fuel-powered generators (IRENA 2022).

The positive effect of subsidies is also apparent for the adoption of other renewable energy sources. The Production Tax Credit in the U.S. provides a tax credit per kilowatt-hour for wind producers. Since it was enacted in 1992, wind energy capacity has grown steadily, but there are multiple plateaus in the cumulative installed capacity each time the tax credit expired; new wind installations would only resume after the law was renewed (Badcock & Lenzen 2010). Sendstad et al. (2022) found that a decrease in subsidies for renewable energies by the European Union during 2000–2017 decreased the investment rate by approximately 45% for solar photovoltaics and 16% for onshore wind. Finally, Norway's "right to charge"

[19] Truly an awful decade for music.

law—wherein tenants are granted the legal right to charge their EVs at their place of residence—is another example of legislation that can accelerate adoption.

From 2009 to 2019, the price of electricity from solar photovoltaic (PV) declined by 89% and onshore wind dropped by 70%. Over the same time period natural gas dropped by about 30%, whereas coal stayed flat and nuclear increased by 26% (IRENA 2020). As a result wind and solar are now as low or lower than natural gas, replacing it as the lowest-cost sources of electricity in most regions in the U.S., as determined by analyses of unsubsidized energy rates (EIA 2023; Lazard 2023; BlombergNEF 2022). As estimated by the BNEF study, in 2022 new-build onshore wind and solar projects are now around 40% lower than for new coal- and gas-fired power plants; and this gap continues to grow. These analyses therefore debunk claims that wind and solar won't work without subsidies. Renewable energy has benefited from government subsidies, as we have seen, like all other major forms of energy production. But they are now getting cheap enough that it is economically foolish to stick with fossil fuels. The conversation about renewables has changed from something we *need* to do despite their higher costs, to one of something we *want* to do regardless of climate change.

6.6.5 Carbon Pricing, "Cap and Trade," and Carbon Offsets

It is important to note that the EIA, Lazard, and BNEF analyses do not consider greenhouse gas emissions and the resultant consequences of climate change. This negative externality can be mitigated through policy measures collectively known as "carbon pricing." This includes carbon taxes and carbon fees that place a price on each ton of CO_2 or equivalent greenhouse gas emissions. The primary difference between carbon taxes and carbon fees is how the money is used. For a carbon tax—like other taxes—the fees are collected for government expenditures in general. For a carbon fee the funds are used specifically to offset the consequences of carbon emissions; e.g., to invest in renewable energy and other low-carbon technologies, fund public services that reduce emissions, or to reduce other taxes. Another option is for the revenue to be returned directly to citizens either to offset higher energy costs or assist with investing in renewable energy (e.g., residential solar or EVs).[20] Carbon pricing is seen as an effective policy tool because it provides a clear signal to the market that emissions have a cost. The goal is to create an economic incentive for individuals and businesses to reduce their carbon footprint and invest in cleaner technologies. Many countries, including Canada, Sweden, and Norway, have already implemented carbon pricing, and others are considering doing so.

The tax or fee is usually imposed at the point of production, import, or distribution of the fuels. It is then passed on to consumers, who pay more for the fuels or the products that use them. For example, burning a gallon of gasoline produces about 10 kg of CO_2, or 0.01 ton. A carbon price of $50 per metric ton, if paid by consumers, would therefore raise gasoline by about 50 cents per gallon (Marron et al. 2015). The price can be set at a fixed rate per ton of emissions, or it

[20] https://citizensclimatelobby.org/basics-carbon-fee-dividend/

can gradually increase over time to encourage further reductions. Ideally carbon taxes should be set at a value that matches the consequences of the emissions, not only from a climate change perspective but also for other health and economic impacts on society. Estimates of the "social cost of carbon" vary widely based upon the assumptions made, from values as low as \$30 ton^{-1} (Nordhaus et al. 2017) to over \$400 ton^{-1} (Ricke et al. 2018).

Another policy for reducing emissions is known as "cap and trade." The "cap" refers to an overall limit on annual emissions, which can get stricter over time. The "trade" part is a mechanism by which companies can transfer their rights to emit to other entities. Companies are granted "allowances" by the government—either for free or through an auction–that permit them to emit GHGs; e.g., one allowance may permit a company to emit one ton of CO_2. Companies that surpass their allowance are taxed or penalized, whereas those who are under their allowed limit can buy and sell their rights at market prices. The system thereby creates financial incentives to minimize emissions.

First implemented in 2005, the European Union's Emissions Trading System (EU ETS) is one of the first and currently the world's largest cap and trade market. Initially too many allowances were allocated, and permit prices dropped to nothing in 2007. However, reconfiguring the system and decreasing the cap has driven prices as high as 100€ in 2023. The U.S. currently lacks a national cap and trade policy, but individual states have taken action. Starting in 2012, the Regional Greenhouse Gas Initiative (RGGI) is a cooperative effort among twelve eastern states to reduce CO_2 emissions from power plants. Auctions for permits are held quarterly, with prices rising from \$2 in 2012 to about \$14 in 2022 (EIA 2022a). Likewise, California has a cap and trade program that—as part of a larger emissions reduction program—aims to make the state carbon neutral by 2045. Like the RGGI and EU systems, the cap decreases every year. Since 2012, most of the auctions have sold emissions credits at prices near the auction reserve price, the minimum price for allowances in each auction. However in 2022 the price jumped to \$29, well above the reserve price of \$19 (EIA 2022b).

For over thirty years cap and trade systems have been used successfully to reduce CO_2 emissions as well as other pollutants, such as SO_2 and NO_X (Schmalensee & Stavins 2017). As of 2023 there are 73 countries participating in carbon taxes and cap and trade systems, covering about a quarter of total GHG emissions.[21] Despite initial stumbles, these systems have demonstrated their viability for reducing GHG emissions through economic means.

Related to carbon pricing, "carbon offsets" can be purchased by corporations, organizations, or individuals to fund projects that either lower CO_2 emissions or sequester CO_2 from the atmosphere. Such projects include forestry (e.g., reforestation, or avoided conversion of existing forested lands), methane capture (e.g., from landfills or livestock), construction of renewable energy sources, or investment in more efficient energy sources (e.g., improved cookstoves). The idea is that the

[21] https://carbonpricingdashboard.worldbank.org/map_data

purchased offsets would compensate for emissions produced elsewhere; and can in some cases be purchased in connection with expected emissions; e.g., offsets can often be purchased with an airplane ticket. In general, offsets are sold to fund projects to reduce or avoid emissions in developing countries, and are purchased by governments or companies in developed countries to meet their emission reduction targets

To be effective, carbon offsets must absorb or prevent emissions equivalent to the activity in question. The purchased offset needs to result in a reduction that would otherwise not have occurred; e.g., it does no good to protect a tract of forest land if it wasn't going to be harvested in the first place—or if another tract is harvested instead (known as "leakage"). The offset should also be durable and effective in the short term; e.g., it can take a planted tree decades to grow to a size where it effectively sequesters CO_2. And trees are vulnerable to disease, drought, and of course fire. Finally, the effectiveness of the offset needs to be verified; e.g., a ProPublica report found that loggers cut down trees in Brazilian forests even though offsets were sold to protect them (Song 2019). Likewise, Calel et al. (2021) found that carbon offsets were often allocated to projects that would very likely have been built without assistance (i.e., these offsets did not have "additionality"). Worse, the net effect was to actually increase net CO_2 emissions, because the emissions reductions in the developing countries were in reality much smaller than the offset credits that were sold, while the developed countries continued to emit carbon.

There is considerable skepticism about the effectiveness of offsets as a carbon reduction tool. If done properly, they can be a means to reduce carbon emissions. But if not they can actually make the problem worse—as people will think the problem is being addressed while it is not. In the worst-case scenario, carbon offsets are merely used by corporations as a greenwashing tool—or by individuals wishing to assuage guilt—without having much benefit. Although Dargusch & Thomas (2012) argue that most offsets are ineffective but some can be beneficial. Monbiot (2006) compared the practice to purchasing indulgences from the Catholic church for the absolution of one's sins, and Kevin Anderson argues that "offsetting is worse than doing nothing" (Hudson 2012). A 2016 review of more than 5600 carbon offset projects revealed that less than 2% had a high likelihood of emissions reductions that were real, measurable, and were in addition to what would have occurred anyways (Cames et al. 2016).

Built into the carbon offsets structure is an incentive for double-counting, where the credit for emissions reduction is given to both the buyer and the seller. In this case, the host country selling the offset wants to make money from the sale, but also has an incentive to credit the offset toward its national goals for reducing emissions. Higher quality offsets that provide better proof that the offset is real and accounts for over crediting are more expensive, so entities in developed nations are incentivized *not* to buy them, but instead to opt for cheaper offsets and not ask any questions.

An agreement was made at COP26 in Glasgow in 2021 to make sure carbon credits were not counted twice (Zwick 2021). But even if the accounting for the carbon offset is done correctly, social injustice is still built into the framework.

Under the 2015 Paris Agreement, all countries have committed to reducing their carbon emissions to net zero by 2050. When a rich, developed country buys carbon offsets from a poorer, developing country, the rich country gets the credit while still emitting the carbon associated with that credit. Even though the poor country has proactively reduced carbon emissions (e.g., by reforesting or installing solar panels), it does not get the carbon credit for this work. Instead, its emissions are flat in the accounting of the transaction. Despite having pretended that its emission reduction has not occurred, the poor country must still meet its Paris Agreement emissions reductions target. If it was to go on the market to buy offsets somewhere else, it would be competing against richer nations that may have already bought up the cheaper offsets. In effect, the burden of emissions reductions would be shifted from rich to poor countries (Tripathi 2022). As developing countries recognize that they would rather keep carbon credits for themselves as opposed to selling them off, the cost of such offsets will become more expensive (World Bank 2023). If every country has to reduce its emissions to net zero by 2050, then there is no effective way this can be done by trading carbon offsets, which only shuffles emissions credits around. Instead, every country has to do the hard work to get to net zero themselves, since there will not be extra carbon offsets to sell (Romm 2023).

6.6.6 Stranded Assets

Finally, when discussing the economic factors of climate change we need to consider the value of existing fossil fuel deposits as well as the built infrastructure to extract, process, transport, and use them. It is now clear that the majority of the world's fossil fuel resources must stay in the ground. If we are to limit the temperature increase to 1.5°C, nearly 60% of oil and fossil methane gas, and 90% of coal must remain unextracted (Welsby et al. 2021). Different estimates (e.g., Rempel & Gupta 2021; Linquiti & Cogswell 2016) place the value of these fossil fuel resources in the tens to hundreds of *trillions* of dollars. The wide range of estimates is due to the assumptions made about the total amount of these resources available and their future value, particularly as the world transitions to renewable energy. If they go unextracted they will be "stranded assets," defined as resources that have suffered unexpected or premature devaluation. Needless to say the fossil fuel industry is reluctant to lose the value of these resources. Knowing their value helps to explain—as discussed in Chapter 13—why they have spent billions of dollars to stop or slow efforts to address climate change.

If fossil fuel infrastructure is abandoned to meet climate goals before the utilities have recovered their construction costs, they also become stranded assets. For instance, there is an estimated $486 billion of pipeline globally for delivering natural gas that is set to be built before 2030 (Langenbrunner et al. 2022,) while existing gas-fired power plants are already becoming more expensive to operate than new onshore wind and solar PV plants in Europe and the U.S. (Sims et al. 2021). Ultimately, the investment losses of these assets are borne by financial investors in the companies that hold these liabilities, such as pension funds in Western countries (Semieniuk et al. 2022).

To minimize the impact of the depreciating value of this infrastructure, there are calls to stop replacing aging gas mains and instead devote resources to electrifying heating in homes and buildings (Henchen & Kroh 2020). Individual municipalities and states in the U.S. are reducing subsidies for connecting new homes to gas lines (e.g., Hartford Courant 2022; Azhar 2022), requiring environmental reviews with a low discount rate for new large gas investments (CPUC 2022); and bans on new gas heaters and connections in construction (Cormany 2022; Swanson 2023). Even as U.S. gas utilities face fewer customers in the future, they will still need to recover their costs from an estimated $150–180 billion in infrastructure that they need to maintain reliability for their remaining customers (Brattle Group 2021). It will be up to policy makers and public utility commissions to determine who finances this transition and how to do it equitably, since without new policies and regulations it will be from the dwindling pool of customers—often low-income residents who cannot afford to electrify their homes who will see their utility bills increase (Davis & Hausman 2022).

6.7 Conclusions

In this chapter we've provided a framework for thinking about solutions to climate change from different perspectives—including sociological, political, and economic. While solutions need first and foremost be technically feasible, they must also address these other factors if they are to have a chance for widespread adoption. They need to adhere to perceived social norms and values. They need to be politically palatable, from the local to international levels. And they need to be economically viable. After all, in the end everyone is pro-money—regardless of their political stripe. In addressing solutions we also need to be aware that climate change is a social justice issue, and be aware of the impacts of different solutions on different groups.

In the next chapter we explore in more detail how we produce energy. Because the generation and use of energy ties into many of the physical principles we teach, it is an appropriate place for us to share our scientific knowledge. Since the use of fossil fuels accounts for 90% of climate change, the transition to carbon-free energy sources is the most important step.

References

Achakulwisut, P., Erickson, P., & Koplow, D. 2021, ERL, 16, 084023

Aldersey-Williams, J., & Rubert, T. 2019, EPo, 124, 169

Azhar, A. 2022, Maryland's Largest County Just Banned Gas Appliances in Most New Buildings —But Not Without Some Concessions. Inside Climate News. https://insideclimatenews.org/news/02122022/montgomery-county-maryland-gas-building-ban/

Badcock, J., & Lenzen, M. 2010, Subsidies for electricity-generating technologies: A review EPo, 38, 5038

Ballew, M. T., et al. 2022, ClCh, 173, 19

Barton-Henry, K., Wenz, L., & Levermann, A. 2021, NatSR, 11, 8571

BloombergNEF, 2022, Cost of New Renewables Temporarily Rises as Inflation Starts to Bite. https://about.bnef.com/blog/cost-of-new-renewables-temporarily-rises-as-inflation-starts-to-bite/

Bourdieu, P. 1985, TS, 14, 723

Brattle Group 2021, The Future of Gas Utilities Series Transitioning Gas Utilities to a Decarbonized Future Part 1 of 3.

Brulle, R. J., & Norgaard, K. M. 2019, EnPo, 28, 886

Coan, T. G., Boussalis, C., Cook, J., & Nanko, M. O. 2021, NatSR, 11, 22320

Calel, R., Colmer, J., Dechezleprêtre, A., & Glachant, M. 2021, Centre for Climate Change Economics and Policy Working paper 398/Grantham Research Institute on Climate Change and the Environment Working paper 371 (London: London School of Economics and Political Science)

Cames, M., et al. 2016, How Additional is the Clean Development Mechanism? Analysis of the application of current tools and proposed alternatives. CLIMA.B.3/SERI2013;/0026r. https://climate.ec.europa.eu/system/files/2017-04/clean_dev_mechanism_en.pdf.

Campbell, E., Kotcher, J., Maibach, E., Rosenthal, S., & Leiserowitz, A. 2022, Who is willing to participate in non-violent civil disobedience for the climate? Yale University and George Mason University (New Haven, CT: Yale Program on Climate Change Communication)

Chancel, L., & Piketty, T. "Carbon and inequality: From Kyoto to Paris Trends in the global inequality of carbon emissions (1998–2013) and prospects for an equitable adaptation fund World Inequality Lab." 2015,

Chancel, L. 2022, NatS, 5, 931

Chenoweth, E., & Stephan, M. J. 2011, Why Civil Resistance Works: The Strategic Logic of Nonviolent Conflict (New York: Columbia Univ. Press) http://www.jstor.org/stable/10.7312/chen15682

Clean Air Task Force 2022. https://cdn.catf.us/wp-content/uploads/2022/08/19102026/carbon-capture-provisions-ira.pdf

Climate Action Tracker 2022, The CAT Thermometer. November 2022. https://climateactiontracker.org/global/cat-thermometer/

Colgan, J. D., Green, J. F., & Hale, T. N. 2021, IntO, 75, 586

Cologna, V., Knutti, R., Oreskes, N., & Siegrist, M. 2021, ERL, 16, 024011

Corbett, J. B. 2021, Communicating the Climate Crisis: New Directions for Facing What Lies Ahead (Lanham, MD: Lexington Books)

Cormany, C. 2022, Why is California Banning Natural Gas Heaters? Efficiency First California. https://efficiencyfirstca.org/blog/whats-the-deal-with-california-and-natural-gas-in-buildings-it-started-with-cities-and-counties-requiring-new-construction-projects-to-be-all-electric-on-sept-22-2022-california-policymakers-vot/

CPUC 2022, CPUC Creates New Framework To Advance California's Transition Away From Natural Gas.https://www.cpuc.ca.gov/news-and-updates/all-news/cpuc-creates-new-framework-to-advance-california-transition-away-from-natural-gas

Dargusch, P., & Thomas, S. 2012, NatCC, 2, 470

Davis, L., & Hausman, C. 2022, JAEnvREco, 9, 1047

Donaghy, T., et al. 2007, Atmosphere of Pressure: Political Interference in Federal Climate Science (Cambridge, MA: UCS Publications) https://www.ucsusa.org/resources/atmosphere-pressure

EIA 2018, EIA uses two simplified metrics to show future power plants' relative economics. https://www.eia.gov/todayinenergy/detail.php?id=35552
EIA 2022a, RGGI carbon dioxide emissions price reached new quarterly record in June 2022. https://www.eia.gov/todayinenergy/detail.php?id=53759
EIA 2022b, Prices for California's emissions credits increase in early 2022 auction https://www.eia.gov/todayinenergy/detail.php?id=51918
EIA 2023. Levelized Costs of New Generation Resources in the Annual Energy Outlook 2022. https://www.eia.gov/outlooks/aeo/electricity_generation.php
Esposito, D., et al. 2021, Studies Agree 80 Percent Clean Electricity By 2030 Would Save Lives And Create Jobs At Minimal Cost.
Falkner, R. 2016, IntA, 92, 1107
Fielding, K. S., & Louis, W. R. 2020, Research Handbook on Communicating Climate Change (Cheltenham: Edward Elgar Publishing) 106
Friedlingstein, P., et al. 2022, ESSD, 14, 4811
Goldstein, N. J., Cialdini, R. B., & Griskevicius, V. 2008, JCoR, 35, 472
Hart, D., & Birson, K. 2016, Deployment of solar photovoltaic generation capacity in the United States. Office of Energy Policy and Systems Analysis, US Department of Energy.
Hartford Courant 2022, Connecticut utility regulators end state's natural gas expansion plan launched in 2013, citing rising costs, push for zero emissions. https://www.courant.com/2022/04/27/connecticut-utility-regulators-end-states-natural-gas-expansion-plan-launched-in-2013-citing-rising-costs-push-for-zero-emissions/
Henchen, M., & Kroh, K. 2020, A New Approach to America's Rapidly Aging Gas Infrastructure. RMI. https://rmi.org/a-new-approach-to-americas-rapidly-aging-gas-infrastructure/
Hudson, M. 2012, NatCC, 2, 307
IEA 2022, Solar PV Global Supply Chains—Analysis, https://www.iea.org/reports/solar-pv-global-supply-chains
IPCC 2018, Global Warming of 1.5°C. An IPCC Special Report on the impacts of global warming of 1.5°C above pre-industrial levels and related global greenhouse gas emission pathways, in the context of strengthening the global response to the threat of climate change, sustainable development, and efforts to eradicate poverty, ed. V. Masson-Delmotte, P. Zhai, H.-O. Pörtner, et al. (Cambridge: Cambridge Univ. Press) 3
IRENA 2020, Renewable Power Generation Costs in 2019. https://www.irena.org/publications/2020/Jun/Renewable-Power-Costs-in-2019
IRENA 2022, Renewable Power Generation Costs in 2021 (Abu Dhabi: International Renewable Energy Agency)
Islam, M. S., & Kieu, E. 2021, Clim, 9, 7
Juhola, S., et al. 2022, FrCl, 4,
Kahan, D. 2010, Natur, 463, 296
Karas, D. 2015, NewVisPublAff, 7, 9
Kelsey, E. 2020, Hope Matters: Why Changing the Way We Think is Critical to Solving the Environmental Crisis (Vancouver: Greystone Books)
Koepfler, J. A., Heimlich, J. E., & Yocco, V. S. 2010, ApEnEdC, 9, 233
Korten, T. 2015, In Florida, Officials Ban Term 'Climate Change'—Florida Center for Investigative Reporting. http://fcir.org/2015/03/08/in-florida-officials-ban-term-climate-change/

Langenbrunner, B., Joly, J., & Aitken, G. 2022, Pipe Dreams '22: Stranded Assets and Magical Thinking in the Proposed Global Gas Pipeline Buildout. https://globalenergymonitor.org/report/pipe-dreams-2022/
Lazard 2023, Levelized Cost of Energy Analysis—Version 16.0. https://www.lazard.com/research-insights/2023-levelized-cost-of-energyplus/
Lenton, T. M., et al. 2008, PNAS, 105, 1786
Lenton, T. M., et al. 2019, Natur, 575, 592
Levin, K., Cashore, B., Bernstein, S., & Auld, G. 2012, PoSci., 45, 123
Linquiti, P., & Cogswell, N. 2016, JEnvStudSci., 6, 662
Lynch, L. and Doinger, D. 2023, The EPA's Power Plant Carbon Rules Can Be Built to Last https://www.nrdc.org/resources/epas-power-plant-carbon-rules-can-be-built-last
Marlon, J. R., et al. 2019, FrC, 4,
Marron, D. B., Toder, E. J., & Austin, L. 2015, Taxing Carbon: What, Why, and How. https://doi.org/10.2139/ssrn.2625084
Marshall, R., & Burgess, M. G. 2022, ClCh, 171, 17
Meinshausen, M., et al. 2009, Natur, 458, 1158
Monbiot, G. 2006, Paying for our sins. The Guardian. https://www.theguardian.com/environment/2011/sep/16/carbon-offset-projects-carbon-emissions
Muhammad-Sukki, F. et al. 2014, Ren. Ener., 68, 636
Nemet, G. F. 2014, Energy Technology Innovation: Learning from Historical Successes and Failures, ed. A. Grubler, & C. Wilson (Cambridge: Cambridge Univ. Press) 206
Nordhaus, W. D. 2008, A Question of Balance: Weighing the Options on Global Warming Policies (New Haven, CT: Yale Univ. Press)
Nordhaus, W. D. 2017, PNAS, 114, 1518
Norton, P. D. 2008, Fighting Traffic: The Dawn of the Motor Age in the American City (Cambridge, MA: MIT Press) http://www.jstor.org/stable/j.ctt5hhckp
ONRR 2020, US Office of Natural Resources Revenue Calendar year production data (2005–2019) https://revenuedata.doi.gov
Otto, I. M., et al. 2020, PNAS, 117, 2354
Parry, I. W. H., Black, S., & Vernon, N. 2021, Still Not Getting Energy Prices Right: A Global and Country Update of Fossil Fuel Subsidies. IMF Working paper WP/21/236. https://www.elibrary.imf.org/view/journals/001/2021/236/article-A001-en.xml
Pielke, R. A. 2007, The Honest Broker: Making Sense of Science in Policy and Politics. Cambridge (Cambridge: Cambridge Univ. Press)
Rempel, A., & Gupta, J. 2021, WDev, 146, 105608
Ricke, K., Drouet, L., Caldeira, K., & Tavoni, M. 2018, NatCC, 8, 895
Riedy, C., & Diesendorf, M. 2003, EnPo, 31, 125
Ritchie, H., 2020, "What are the safest and cleanest sources of energy?" https://ourworldindata.org/safest-sources-of-energy
Rittel, H. W. J., & Webber, M. M. 1973, PoSci., 4, 155
Romm, J. 2023, Are carbon offsets unscalable, unjust, and unfixable—and a threat to the Paris Climate Agreement? University of Pennsylvania Center for Science, Sustainability, and the Media White paper. https://web.sas.upenn.edu/pcssm
Schmalensee, R., & Stavins, R. N. 2017, REEP, 11, 59
Semieniuk, G., et al. 2022, NatCC, 12, 532
Sendstad, L. H., Hagspiel, V., Mikkelsen, W. J., Ravndal, R., & Tveitstøl, M. 2022, EnPo, 160, 112675

Sims, J., et al. 2021, Put Gas on Standby. Carbon Tracker. https://carbontracker.org/wp-content/uploads/2021/10/Put-Gas-on-Standby-_Oct21.pdf

Song, L. 2019, An (Even More) Inconvenient Truth: Why Carbon Credits For Forest Preservation May Be Worse Than Nothing. ProPublica. https://features.propublica.org/brazil-carbon-offsets/inconvenient-truth-carbon-credits-dont-work-deforestation-redd-acre-cambodia/

Steel, D., DesRoches, C. T., & Mintz-Woo, K. 2022, PNAS, 119, e2210525119

Sterman, J. D., Siegel, L., & Rooney-Varga, J. N. 2018, ERL, 13, 015007

Stern, N. 2006, The Economics of Climate Change: the Stern Review. https://webarchive.nationalarchives.gov.uk/ukgwa/20100407172811/https:/www.hm-treasury.gov.uk/stern_review_report.htm

Stokes, L. C. 2020, Short Circuiting Policy: Interest Groups and the Battle Over Clean Energy and Climate Policy in the American States (New York: Oxford Univ. Press)

Swanson, K. 2023, California's Cities Lead the Way on Pollution-Free Homes and Buildings. Sierra Club. https://www.sierraclub.org/articles/2021/07/californias-cities-lead-way-pollution-free-homes-and-buildings

Taylor, M. 2020, Energy Subsidies: Evolution in the Global Energy Transformation to 2050 (Abu Dhabi: International Renewable Energy Agency)

Tol, R. S. J., & Yohe, G. W. 2006, WEco, 7, 233

Tripathi, B. 2022, Article 6: Will Corresponding Adjustments Tool Stop Double Counting? https://carboncopy.info/article-6-will-corresponding-adjustments-tool-stop-double-counting/

Welsby, D., Price, J., Pye, S., & Ekins, P. 2021, Natur, 597, 230

Wen, X., Ji, J., Song, Z., Li, Z., Xie, H., & Wang, J. 2021, ECM, 234, 113940

World Bank 2023, Corresponding Adjustment and Pricing of Mitigation Outcomes (Washington, DC: World Bank Working Paper) https://www.theclimatewarehouse.org/knowledge/papers/corresponding-adjustment-and-pricing-of-mitigation-outcomes

Zhang, S., & He, Y. 2013, RSER, 21, 393

Zwick, S. 2021, Article 6 and its Glasgow Rulebook: the Basics. https://www.ecosystemmarketplace.com/articles/article-6-and-its-glasgow-rulebook-the-basics/

Chapter 7

Energy Solutions to Climate Change

T A Rector and Ka Chun Yu

"There are no silver bullets, only silver buckshot."
—Bill McKibben (2006)[1]

This chapter focuses on how we produce energy, as the use of fossil fuels accounts for 90% of climate change. The good news is that most of the solutions we need are here now and are economically viable. The cost of renewable energies such as wind and solar have dropped below nuclear, coal, and natural gas. By switching to wind and solar, and electrifying as much of our energy usage as possible (e.g., with EVs and heat pumps) we can reduce our carbon emissions and our energy costs. Already these solutions are rapidly growing in popularity, but we need to accelerate the pace as what we do over the next decade is critical. Some solutions, such as sustainable aviation fuels and carbon capture, may in the future become technically and economically viable, but in the next decade or so are not likely to contribute meaningfully to keeping the temperature rise below 1.5°C or 2.0°C. That is not to say implementing these solutions will be easy—technically, socially, or politically. The key message here is that *it can be done,* and it's something we should be excited about.

7.1 Understanding the Problem

The focus of this chapter is on energy—how we create it, how we use it, and the solutions available. Fossil fuels—coal, oil and gas—are by far the largest contributor to global climate change, accounting for over 75% of global greenhouse gas (GHG) emissions and nearly 90% of all CO_2 emissions.[2] So our focus will be on how to transition away from fossil fuels to renewable and carbon-free energy sources.

[1] https://www.washingtonpost.com/archive/opinions/2006/05/27/welcome-to-the-climate-crisis-span-classbank-headhow-to-tell-whether-a-candidate-is-serious-about-combating-global-warmingspan/26b2ac5a-a4a3-46ff-b214-3fc07a3a5ab3/
[2] https://www.un.org/en/climatechange/science/causes-effects-climate-change

doi:10.1088/2514-3433/acfcb6ch7

The topic is intertwined with the physics and astronomy that we talk and teach about, so it is a natural fit for us to engage in this conversation. As a study case, this chapter will focus on the energy situation of the U.S., the largest cumulative GHG emitter, currently the second largest emitter per year (behind China), and among the largest GHG emitters on a per capita basis.

To find solutions we must of course first understand what's causing the problem. The answer is complex but also in a way simple. As mentioned above, approximately 90% of climate change is caused by the extraction and use of fossil fuels, i.e., coal, petroleum, and natural gas. The rest is mostly due to changes in land use; e.g., clearcutting of forests for agricultural use (Friedlingstein et al. 2022). Thus our focus needs to be drastically reducing our use of fossil fuels. To keep within a 1.5°C carbon budget, nearly 60% of known petroleum and natural gas reserves, and 90% of coal must remain unextracted (Welsby et al. 2021). Trout et al. (2022) find that achieving this goal requires not only a cessation of the development of new fossil fuel reserves, but also to prematurely decommission a significant portion of those already developed.

While fossil fuels are by far the primary cause of climate change, reducing their use is complicated by the fact that they are the dominant energy source across all sectors of usage. Figure 7.1 shows how energy was generated and used within the United States in 2021. For this year the total energy consumption was 97.3 "quads," where a quad is defined as 10^{15} BTU or 1.055×10^{18} J.

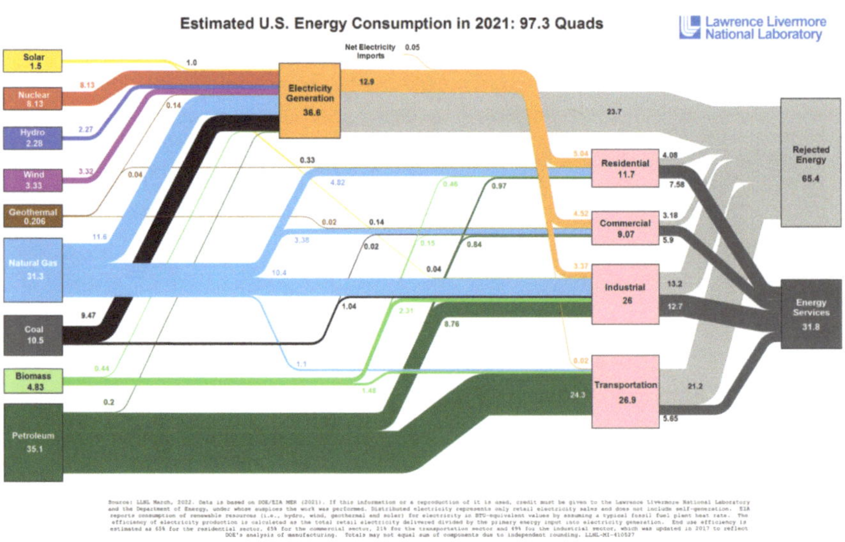

Figure 7.1. [3]The 2021 energy flow chart released by Lawrence Livermore National Laboratory (LLNL) details the sources of energy production, how Americans are using energy and how much waste exists.

[3] https://flowcharts.llnl.gov

The left side of the figure shows, in quads, the amount of energy generated from the eight most important sources. Fossil fuels (coal, petroleum, and natural gas) accounted for 79% of the total energy used. Energy sources that do not produce carbon emissions (solar, nuclear, hydro, wind, and geothermal) accounted for only 16%. Biomass fuel, which has lower carbon emissions, accounts for the remaining 5%. Of the total 97.3 quads generated, 36.6 quads (38%) were first converted into electricity. The energy generated is used by these four sectors (shown in pink on the diagram).

Residential: Consists of living quarters for private households. Common uses of energy associated with this sector include space heating, water heating, air conditioning, lighting, refrigeration, cooking, and other appliances. It varies somewhat by type of unit (e.g., single-family detached home, apartment, etc.) but overall half of the energy used is for HVAC (heating, ventilation, and air conditioning). The amount of energy usage for these can vary considerably with the seasons; i.e. more heating in the winter months and air conditioning in the summer. Water heating and refrigeration account for about a quarter of energy use, using roughly the same amount of energy regardless of the time of year or time of day. The remaining share of energy is for lighting, appliances for cooking and washing, as well as televisions and other electronics. The energy demand of these increases during evening hours when people are usually at home and active. How much energy a residence uses varies considerably, with the size of home, its geographic location and climate, and number of residents being the dominant factors. The energy efficiencies of the home (e.g., insulation) and its appliances are also important, with older homes and appliances being less efficient on average.

In 2021 residential accounted for 16% of all the energy used in the U.S.[4] Of this, 43% was electricity, 42% was natural gas, 8% was petroleum (e.g., heating oil, kerosene, and propane), and renewable energy, such as geothermal (e.g., ground source heat pump), solar, and wood biofuels, accounted for 7%.[5]

Commercial: Consists of service-providing facilities and equipment of businesses, federal, state, and local governments, as well as other private and public organizations, such as religious, social, or fraternal groups. Commercial buildings include a variety of building types; e.g., offices, hospitals, schools, police stations, places of worship, warehouses, hotels, and shopping malls. Common uses of energy associated with this sector include space heating, water heating, air conditioning, lighting, refrigeration, cooking, and running other equipment.

In 2021, commercial uses made up 12% of energy usage in the U.S. Of this, 61% was electricity, 32% was natural gas, and "District heating," where HVAC is supplied to multiple buildings from a central location, and fuel oil accounted for the remaining 7%.[6]

Industrial: Consists of all facilities and equipment used for producing, processing, or assembling goods. The industrial sector encompasses the following types of

[4] Note: How the relative percentages of these four categories are calculated vary depending on how electrical losses are included. Residential and commercial are often assigned a larger fraction because they currently use responsible for more energy usage.

[5] https://www.eia.gov/energyexplained/use-of-energy/homes.php

[6] https://www.eia.gov/energyexplained/use-of-energy/commercial-buildings.php

activities: manufacturing, agriculture, forestry, fishing and hunting, mining, oil and gas extraction, and construction. The combined energy use by six energy-intensive manufacturing subsectors–chemicals, petroleum and coal products, paper, primary metals, food, and nonmetallic minerals products–accounted for 87% of total industrial energy consumption (EIA 2018). Petroleum and coal products include gasoline, fuel oil, and plastics.[7] Two-thirds of the energy consumed is for heat used in industrial processes (e.g., manufacturing of glass and steel). A third is used as "feedstock"; i.e., raw materials used to produce chemicals, plastics, and other materials that are used in other sectors such as construction, transportation, and consumer goods.

In 2021, industrial accounted for 35% of all energy usage in the U.S. Of this, fossil fuels account for 78% of the energy used. Only 13% was electricity—although some industrial facilities also generate electricity for their own use. And the remaining 9% is from renewable energy.

Transportation: Consists of all vehicles whose primary purpose is transporting people and/or goods from one physical location to another. This includes automobiles, trucks, buses, motorcycles, trains, subways, aircraft, as well as ships, barges, and other waterborne craft. Vehicles whose primary purpose is not transportation (e.g., construction cranes and bulldozers, farming vehicles, and warehouse tractors and forklifts) are included elsewhere.

In 2021, transportation was responsible for 37% of energy usage in the U.S., essentially tied with industry for the highest energy use of any sector. Petroleum products accounted for about 90% of the total U.S. transportation sector energy use. Of the remaining 10%, biofuels contributed about 6% and natural gas accounted for about 4%. Electricity accounted for less than 1%.[8]

The flowchart also shows how the energy produced from each source is used. For example, of the 31.3 quads from natural gas, 11.6 quads (or 37%) are used to generate electricity whereas 4.82 quads (15%) went straight to residential. Ninety percent of coal and nearly all carbon-free emissions went to electricity generation. In contrast, 70% of all petroleum was used for transportation.

The right side of the flowchart shows how much of the energy generated actually gets used. Astonishingly, only 31.8 of the 97.3 quads (33%) is actually used for the desired purpose. The remaining two-thirds (labeled as "rejected energy") is in essence wasted. The two largest contributors to this inefficiency are electricity generation and transportation, which are only 35% and 21% efficient respectively. That's because coal, natural gas, nuclear, solar thermal and geothermal generate electricity via heat engines which, as we know from our introductory physics classes, are fundamentally limited in how efficiently they can convert heat into work. Internal-combustion engines (ICEs) are also heat engines. According to the second law of thermodynamics, in real-world conditions the maximum theoretical efficiency of a heat engine is around 40–50%. There are ways to improve upon these efficiencies; e.g., the vast majority of modern natural gas power plants in the U.S. use combined cycle technology—a combination of Brayton Cycle and Steam Cycle—which can have generation efficiencies between 45% and 57% (Storm 2020).

[7] https://www.eia.gov/energyexplained/oil-and-petroleum-products/refining-crude-oil-inputs-and-outputs.php
[8] https://www.eia.gov/energyexplained/use-of-energy/transportation.php

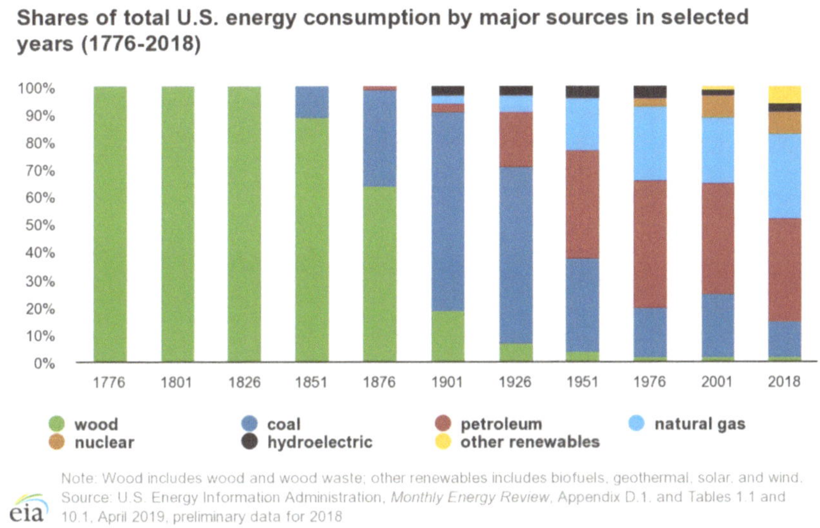

Figure 7.2. The fractions of each energy source used in the U.S. over time. For the first 100 years, wood from forest clearing was the primary source of energy, before being largely replaced by coal. By the mid 20th century the rapid rise of modern transportation caused a surge in the use of petroleum. Over the last twenty years fracking has led to the growth of natural gas, at the expense of coal. Hydroelectric and nuclear grew during the early and mid 20th century respectively, but haven't grown since then due to environmental and safety concerns. The small percentage of "other renewables," which includes wind and solar, is now rapidly growing.

As you can see from the flowchart, all sectors are currently heavily reliant on fossil fuels. Substantial reductions in carbon emissions will therefore require investing in a wide range of technologies.

The above numbers are a snapshot of how energy was used in the U.S. in 2021. Figure 7.2 shows how we've used energy over the years, from the founding of the U.S. in 1776 to the present. An important thing to note is that how we use energy has continuously changed over the years, and each new energy source was cleaner *and cheaper* than the previous. Figure 7.3 shows how CO_2 emissions in the U.S. have changed in recent decades by energy sector. Notably, emissions for the generation of electricity have dropped considerably because of the adoption of wind and solar.

7.2 Understanding Each Energy Source

Here we discuss each energy source in more detail, commensurate with its importance. Comparison is challenging because each has its strengths and weaknesses, which are discussed in detail below. The simplest—and perhaps most important—comparison is that of cost, which invariably dominates decision making in regards to investment and adoption. As discussed in the framework chapter, "Levelized cost of energy" (LCOE) is a common way in which energy costs are compared. LCOE refers to the estimated revenue required to build and operate a generator relative to the total energy produced over its lifetime. Another important concept is the "levelized avoided cost of electricity" (LACE), which represents that

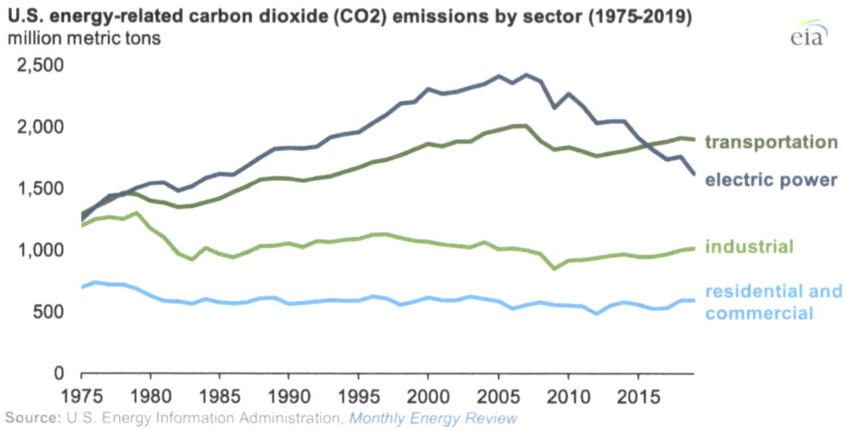

Figure 7.3. [9]U.S. energy-related CO_2 emissions by sector (1975–2019). Since 2005 the carbon emissions associated with electricity generation have dropped significantly because of transition from coal to natural gas as well as the rapid adoption of wind and solar. As a result the transportation sector has been the largest source of energy-related CO_2 emissions since surpassing the power sector in 2016. Other sectors have remained relatively flat, but in this graph some trends are not visible. For example, in 2019, the transportation sector's energy-related CO_2 emissions declined by only 0.7% as a decrease in motor gasoline-related CO_2 emissions (−0.7%) and diesel fuel-related CO_2 emissions (−1.1%) were offset by an increase in jet fuel-related CO_2 emissions (1.9%).

power plant's value to the grid; i.e., it is a measure of the market value of the electricity it would generate. LCOE and LACE are usually given in units of cost per unit of energy; e.g., US$ per MWh (megawatt-hour).

In the context of climate change we are primarily concerned with an energy source's greenhouse gas (GHG) emissions during its lifetime. An energy source may have GHG emissions associated with it even if it doesn't use fossil fuels, as energy is required for construction, operation, maintenance, and decommissioning. Assuming fossil fuels are used as the energy source for these, estimates of the "life cycle" greenhouse gas emissions for different energy sources (in g CO_2e/kWh) are given below (NREL 2021):

- Nuclear: 13
- Wind: 13
- Hydropower: 21
- Solar thermal: 28
- Lithium-ion battery storage: 33
- Hydrogen fuel cell storage: 38
- Solar PV: 43
- Natural gas: 486
- Oil: 840
- Coal: 1001

[9] https://www.eia.gov/todayinenergy/detail.php?id=43615

Clearly the carbon footprint of renewables, as well as for nuclear, are minimal compared to fossil fuels. And if we can fully decarbonize (i.e., fossil fuels aren't used for construction, etc.) then the emissions associated with renewables essentially go to zero.

In the subsections below, the technical details—as well as the strengths and weaknesses—of each of the eight energy sources shown on the LLNL energy flowchart are discussed in more detail.

7.2.1 Coal, Petroleum, and Natural Gas

Hydrocarbons are molecules that consist, as the name suggests, mainly of carbon and hydrogen atoms. The simplest, and most common, form is methane (CH_4). Others include ethane (C_2H_6), propane (C_3H_8), butane (C_4H_{10}), and octane (C_8H_{18}). Most people will recognize these as fuels.

Coal, petroleum, and natural gas are all hydrocarbons. The primary difference between them is the number of carbon atoms in their molecules. Natural gas is 70–90% methane, with the balance mostly ethane and nitrogen. Gasoline (petrol) hydrocarbons typically contain 5 to 12 carbon atoms, whereas diesel fuel contains hydrocarbons with 12 to 20. Coal consists of long-chain hydrocarbons, with more than a hundred carbon atoms. Natural gas, petroleum, and coal are found in gas, liquid, and solid form, respectively, at standard temperature and pressure because of the length of their hydrocarbons–longer hydrocarbons get entangled more easily, raising their boiling and/or melting temperature.

The hydrocarbons of fossil fuels store chemical potential energy that can be converted into thermal energy via combustion with oxygen. This thermal energy can of course be used for heating, but can also be used for locomotion or the generation of electricity via a heat engine (e.g., via the Rankine cycle). In the case of complete combustion, the primary byproducts are CO_2 and water. Incomplete combustion will also produce carbon or carbon monoxide (CO). For example, the balanced equation for the complete combustion of methane is:

$$CH_4 + 2O_2 \rightarrow CO_2 + 2H_2O$$

Coal, petroleum, and natural gas are non-renewable sources of energy; i.e., once the Earth's stores are used they will not be replaced on human timescales. They formed millions of years ago from dead ocean and land-based organic matter (e.g., plants, algae, and plankton) that were buried under sedimentary rock–hence the name "fossil fuels." Over the span of millions of years, the conditions of high pressure and temperature slowly transformed this organic matter into hydrocarbons. A common misconception is that they formed from the remains of dinosaurs.

It has taken hundreds of millions of years for the fossil fuels beneath the Earth's crust to accumulate. One way to think of the problem of climate change is that the Earth has naturally sequestered carbon over this timespan; and now we are releasing that carbon in a matter of only a few hundred years. The Earth's geologic record shows that such a rapid release of carbon, even when it occurs for natural reasons, can be catastrophic for existing life (e.g., Shen et al. 2011).

If the consumption of fossil fuels continues at the present rate, they will all be depleted within a few hundred more years; e.g., based on coal production rates in 2021, the recoverable coal reserves in the United States would last about 400 years.[10] Likewise, "peak oil" is a hypothetical point when maximum petroleum production is reached. It depends on the extractable supplies available (i.e., future production from existing fields as well as future discoveries) as well as demand. Estimates of the date of peak oil range from 2020 (i.e., we have already passed it) to 2050 or beyond. Improvements in extraction technologies are also relevant for estimating peak oil. For example, hydraulic fracturing, or "fracking," is the process of injecting high-pressure fluids into shale formations to shatter rock, facilitating the extraction of oil and natural gas. Along with other advances, such as horizontal drilling, fracking has caused oil production in the United States to rapidly increase from less than 6 million barrels per day (b/d) in 2007 to 12.2 million b/d in 2019 (EIA 2021), making the U.S. in 2022 the largest oil producer, and consumer, in the world.[11] It was once thought that fossil fuels would run out before climate change became a problem. Now it is now clear that the majority of known fossil fuel reserves cannot be used if we are to keep below a global average temperature increase of 1.5°C (Welsby et al. 2021) or 2°C (McGlade & Ekins 2015).

The dominance of fossil fuels across all sectors reflects its strengths as an energy source. It is a "dispatchable" source of energy—it can be easily turned on and off as demand requires. Its rate of energy production can also be precisely controlled, often simply by regulating the rate of combustion. In contrast, the available energy from wind and solar are obviously dependent on factors beyond our control. Fossil fuels are also energy dense. For example, petroleum has a specific energy of about 45 MJ kg^{-1}. For comparison, the best lithium-ion batteries currently have a specific energy of about 1 MJ kg^{-1}. It is therefore no surprise that petroleum currently makes up 90% of the energy used in the transportation sector.

Up until recently the costs of fossil fuels have also been considerably lower than most other energy sources, with competition occurring primarily between coal, petroleum, and natural gas. For example, from 1950 to 2008 coal-fired power plants provided nearly half of the electricity in the U.S. because the price of coal was significantly less than petroleum or natural gas. However, the jump in natural gas production from the rapid growth of fracking caused its prices to drop, bringing it below the cost of coal in 2016 in the U.S. (EIA 2016). Natural gas is now the most common fuel for electric power generation in the U.S., responsible for 41% of the total in 2023.[12] As coal is being replaced by gas, carbon emissions has gone down since burning gas results in less than half the carbon emissions of the equivalent coal.[13] In fact, the decline of coal in recent decades was primarily a result of the price

[10] https://www.eia.gov/energyexplained/coal/how-much-coal-is-left.php
[11] https://www.eia.gov/tools/faqs/faq.php?id=709&t=6
[12] https://www.eia.gov/outlooks/steo/data/browser/#/?v=22&f=A&s=&start=2013&end=2024&id=&linechart=NGEPGEN_US&ctype=linechart&maptype=0&chartindexed=1.
[13] https://www.eia.gov/todayinenergy/detail.php?id=48296.

drop of natural gas and not the adoption of renewables. In the U.S., coal consumption has decreased almost 50% from 2007 to 2019. An even more dramatic decrease occurred in Western Europe. However, China, India, Russia and some other countries have increased coal production and use during the same time period (Finkelman et al. 2021).

The biggest liability of fossil fuels, and the very reason for this book, is that their combustion produces carbon dioxide and water—both greenhouse gases. While the water vapor will precipitate out of the atmosphere within days, the CO_2 will remain for many hundreds of years. While in principle it is possible to capture and sequester CO_2, currently this is not feasible—as is discussed in detail below.

Fossil fuels have other liabilities as well. Although natural gas has been touted as being reliably dispatchable, compared to solar and wind that are not always available, the infrastructure that delivers the gas is vulnerable to extreme weather events that are becoming more common with climate change. Since the start of the fracking boom in the U.S., a network of pipelines has grown up to distribute just-in-time natural gas from wells. However, much of this infrastructure has not been weatherized for extreme cold. Abnormally low temperatures can cause water in gas streams to freeze at wellheads and compressor stations. When a power plant is knocked offline because it is not getting fuel due to frozen equipment, the failure can cascade, since electricity is needed to run the pumps and compressors that deliver gas to other power plants. Winter storm Uri caused widespread power loss in Texas in February 2021 because power outages caused gas production to drop, which led to further reductions in electricity generation (Cai, Douglas, & Ferman 2022). Winter storm Elliott in Christmas 2022 caused a quarter of the power normally supplied by the PJM Interconnection grid to mid-Atlantic states to go offline (Malik 2023). As measured in Megawatt-hours, natural gas plants were responsible for 63% of the outages, while wind and solar performed as expected given the conditions at that time (PJM 2023).

Another hazard of fossil fuels comes from their combustion. Unwanted emissions, particularly from coal, are generated in the form of sulfur oxides (SO_x), nitrogen oxides (NO_x), particulates, and heavy metals (e.g., mercury and lead). Note that the phrase "clean coal" originally referred to the use of technologies that mitigate these pollutants but do not necessarily reduce CO_2 emissions.

Health impacts from exposure to fossil-fuel pollutants can range from asthma and breathing difficulties, brain damage, heart problems, cancer, neurological disorders, as well as premature death (Munawer 2018). For example, exposure to fine-particulate air pollution has been associated with an increased likelihood of disease as well as shortened life expectancy (e.g., Pope et al. 2009). By 2060 this number could double or triple, and cost 1% of global GDP—around USD $2.6 trillion annually—as a result of sick days, medical bills, and reduced agricultural output (OECD 2016). Indoor air pollution is also an issue. For example, residential natural gas stoves expose occupants to NO_2, CO, and HCHO levels that exceed acute health-based standards and guidelines (Logue et al. 2014), with NO_2 levels exceeding those that are deemed safe for the outdoors (Seals & Krasner 2020). An estimated 10 million people per year die from pollution generated by fossil fuel combustion (Vohra et al. 2021).

The consequences of coal combustion is not limited to just air pollution nor are the impacts limited to people. For example, sulfur and nitrogen oxides, sulfates (SO_3^{2-}), and sulfuric acid (H_2SO_4) released from the burning of coal damage the environment in the form of acid rain (Burns et al. 2016). Likewise, half of the mercury released via human activities today comes from burning coal (Streets et al. 2011), part of which ends up in the ocean food chain, and is ingested by people and other mammals when they eat fish (Sackett et al. 2010). Coal combustion is also responsible for the release of other other heavy metals (e,g, lead, arsenic, cadmium, chromium, and antimony) into the air and deposited in coal ash residue ponds and landfills (Munawer 2018) that can further contaminate waterways and damage ecosystems by leaking or accidental release (Wang et al. 2022).

In addition to the harms from burning fossil fuels, there are numerous local health and environmental impacts associated with the extraction of fossil fuels. This includes air, water, and soil pollution as well as ecological degradation. The damage incurred will vary by location; e.g., mountaintop removal coal mining (Holzman 2011), upstream extraction of oil and gas (Johnston et al. 2019) or deep water oil and gas exploration (Cordes et al. 2016). Impacts also vary by method; e.g., fracking causes excessive water use (Kondash, Lauer, & Vengosh 2018), contamination of the local water table (Wollin et al. 2020), and the triggering of local earthquakes (Schultz et al. 2018). For these reasons, fracking is banned in many countries (e.g., Germany, France, and Australia) as well as some states in the U.S. (e.g., Vermont, Maryland, and New York).

Methane is not only a fossil fuel but also a potent greenhouse gas. Methane emissions from the U.S. oil and natural gas supply chain, per unit of natural gas consumed, produce a radiative forcing comparable to the CO_2 produced from the combustion of the natural gas itself (Alvarez et al. 2018). From a radiative forcing perspective, the use of methane is therefore no better than coal. This is important to point out because a common misconception is that natural gas is a "green" energy source—a myth perpetrated by industry. Abandoned wells are also a significant source of methane emissions (e.g., Townsend-Small & Hoschouer 2021). In the U.S., more than 3.2 million abandoned oil and gas wells together emitted 281 kilotons of methane in 2018 (EPA 2023). That's the greenhouse gas equivalent of consuming about 16 million barrels of crude oil, about as much as the United States uses in a day.

Working in the fossil fuel extraction industry is dangerous. Risks include fire and explosions, falls, being struck by or caught between heavy equipment, confinement in small spaces (in which suffocation or burial can occur), and exposure to hazardous chemicals—in the short and long term. For example, long-term exposure to coal mine dust causes a variety of pulmonary diseases, including coal workers' pneumoconiosis (CWP) and chronic obstructive pulmonary disease (COPD). Coal

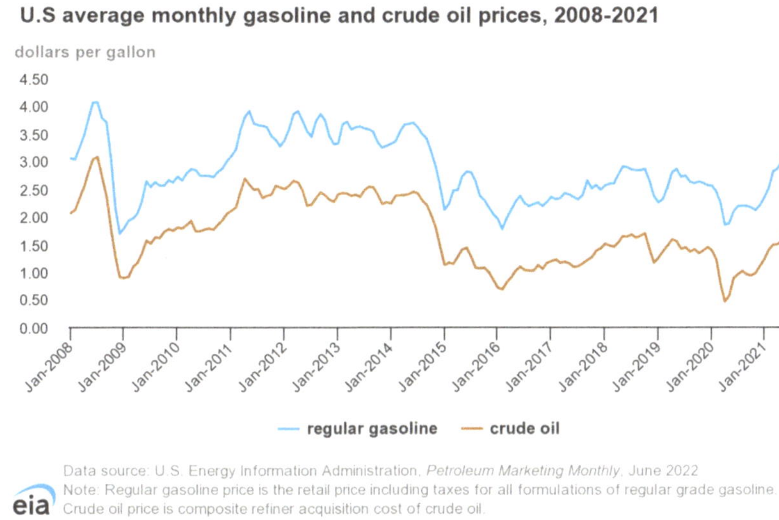

Figure 7.4. This plot shows how the price of gasoline (not adjusted for inflation) is tied to the cost of crude oil, which can fluctuate wildly based upon changes in supply and demand.[14] For example, crude oil prices reached record levels in 2008 as a result of high worldwide oil demand driven by significant growth in China, the Middle East, and Latin America. However, the 2008 financial crisis and Great Recession sent the price of a barrel of crude oil plummeting from $133 in June 2008 to $39 in February 2009.[15]

miners are also exposed to crystalline silica dust, which causes silicosis, COPD, and other diseases (CDC 2011).[16]

Another weakness of fossil fuels is that their prices fluctuate considerably based upon supply and demand. As shown in Figure 7.4, the price of gasoline depends primarily upon the cost of crude oil, which is refined into gasoline and other products. Prices can change rapidly if something disrupts crude oil supplies (e.g., the 1979 oil crisis in the wake of the Iranian Revolution) or refinery operations (e.g., in 2021 Hurricane Ida disrupted crude oil production in Louisiana). Similarly, the invasion of Ukraine by Russia in 2021 caused extreme volatility in the price of natural gas prices in Europe, which had been relying more on Russian exports since its domestic output has been decreasing over the last decade.[17]

As these examples show, while coal markets remain largely domestic for coal producing countries (Blondeel et al. 2021), economic competition in the oil and natural gas markets occurs on a global scale. Competition for access to these fossil fuels is a key component of geopolitics; e.g., the disputes and tensions over the Persian Gulf, the Caspian Sea, and the Arctic region. Known as the "resource curse," a country's natural resource wealth can have adverse effects on its economic, social, and political well-being. A country's wealth in mineral resources, particularly

[14] https://www.eia.gov/energyexplained/gasoline/price-fluctuations.php
[15] https://www.eia.gov/dnav/pet/hist/LeafHandler.ashx?n=PET&s=RWTC&f=M
[16] https://www.cdc.gov/niosh/docs/2011-172/pdfs/2011-172.pdf
[17] https://www.iea.org/reports/russian-supplies-to-global-energy-markets/gas-market-and-russian-supply-2

that of petroleum, tends to make authoritarian regimes more durable, to increase certain types of corruption, and to trigger violent conflict in low- and middle-income countries (Ross 2015). For example, Nigeria has exported over USD $300 billion in oil since 1958; however, the per capita income for Nigerians has not improved during that time (Bird 2004). For these reasons and others, renewable energy has many advantages over fossil fuels for international security and peace (Vakulchuk et al. 2020).

It is important to note that the market prices of fossil fuels, and their resultant CO_2, don't reflect their actual external cost to society, in the health and environmental damage that they do via their extraction and use. In addition to climate change they are a major source of airborne fine particulate matter ($PM_{2.5}$) that is responsible for killing more than 8 million people in 2018—about 1 in 5 deaths worldwide (Vohra et al. 2021). It's also important to note that these costs are not incurred equally, with most of the damage felt by the poor and people of color (Dennig et al. 2015), i.e., those who have the least resilience. For example, the region along the Mississippi River between Baton Rouge and New Orleans in Louisiana contains over two hundred chemical plants. It has been nicknamed "Cancer Alley"[18] because those who live near petrochemical plants and refineries in the area are at higher risk of cancer, with poorer and higher minority concentrated segments being more affected (James et al. 2012). Likewise, around 4% of world oil production is used as a feedstock to make plastics–and a similar amount is used as energy in the process. A third of current plastic production is used to make items of packaging, which are then rapidly discarded. The pollution caused by plastics has severe environmental impacts (Thompson et al. 2009).

Phasing out fossil fuels, and coal in particular, therefore yields substantial local environmental and health benefits that outweigh the costs due to decreasing the energy supply. It is worthwhile to phase out fossil fuels, even when only accounting for local effects and neglecting the global benefits from slowing climate change (Rauner et al. 2020). Considering their pervasive use in all energy sectors, the challenge will be to replace them as quickly as possible.

7.2.2 Biomass

Biomass is the original fuel. It has been in use since humans first began controlling fire more than 200,000 years ago (Barkai et al. 2017). Plants grow through the process of photosynthesis, which is in effect the opposite of hydrocarbon combustion —carbon dioxide and water (plus sunlight) are chemically reacted within plants to create glucose and oxygen:

$$6H_2O + 6CO_2 \rightarrow C_6H_{12}O_2 + O_2$$

These plants, collectively known as biomass, can be burned directly or processed to make biofuels. Wood is still the largest biomass energy resource today. Other sources include food crops (e.g., corn and sugar), perennial plants, waste from

[18] https://projects.propublica.org/louisiana-toxic-air/

agriculture or forestry, oil-rich algae, and the organic components of municipal and industrial wastes (e.g., the emissions from landfills, which contain methane). The most common biofuels are bioethanol and biodiesel, which are usually blended with gasoline and diesel fuel, respectively. An advantage of biofuels is that they can be made to "drop in"; i.e., they can be used in existing engines with little to no modification. Biomass can also be used for products that would otherwise be made from fossil fuels (e.g., bioplastics–many of which are biodegradable).

In principle the CO_2 released from the use of biomass and biofuels will be balanced by the CO_2 absorbed via photosynthesis during their growth. However, in practice this is not the case, since energy is required to harvest, process, and transport the fuel. It has been disputed whether or not biofuels actually produce more energy than is needed to make them (e.g., Pimentel & Patzek 2005, Hill et al. 2006).

Biomass was the largest source of energy in the U.S. and other developed countries, until the mid-1800s when it was largely replaced with coal. Biomass continues to be an important fuel in developing countries, especially for cooking and heating. The use of biomass fuels for transportation and for electricity generation is increasing in response to climate change. From 2008 to 2018, bioethanol production increased by 67%, and biodiesel production increased more than threefold. Despite this rapid growth, biofuels currently account for only 3.4% of total transportation fuels worldwide.[19] The global production of biofuels is dominated by the U.S. and Brazil–producing 69% of all biofuels in 2018—followed by Europe with 9% (REN21 2019). Production of bioethanol in the U.S. is almost exclusively from corn, whereas in Brazil it is from sugarcane (Jeswani et al. 2020). More than 98% of gasoline in the U.S. contains some ethanol. The most common blend of ethanol is E10 (10% ethanol, 90% gasoline). It is also available as E85 (which is also known as "flex fuel"), a high-level blend containing 51% to 83% ethanol.[20] When burned, pure biofuels generally produce less emissions of air pollution than their fossil-fuel derived counterparts. However, a downside is that the specific energy of ethanol (C_2H_5OH) is 29.7 MJ kg^{-1}, about 40% less than petroleum. This means that users will experience a reduction in their gas mileage.

The benefits of biomass are disputed, and depend strongly on how they are created and used. There are concerns that their wider deployment could lead to unintended environmental consequences. The biggest problem is how it impacts land use. Land dedicated to biofuel production can be taken away from other important uses, such as agriculture and timber, as well as for nature conservation and natural carbon sequestration.

Large-scale cultivation of dedicated biomass affects global food prices and water scarcity. For example, the price of corn in the U.S. has greatly increased since the 1970s,[21] in large part because of the rapid increase in production of corn-based ethanol since that time (Figure 7.5). In 2022, nearly 45% of total corn grown in the U.S. was used for ethanol production instead of food for human consumption or feed for animals (USDA 2023). The price of corn often fluctuates dramatically, sometimes

[19] https://www.iea.org/reports/renewables-2019
[20] https://afdc.energy.gov/fuels/ethanol.html
[21] The greatest decade for music. Not the greatest decade for everything else.

Figure 7.5. [22]Historical prices of corn in the U.S., 1959–2022, in USD per bushel. The price growth since the 1970s has been largely driven by the use of corn for bioethanol production. The dramatic variability in price has impacts on the costs of ethanol, as well as corn for food and feed.

doubling within the span of a year. These fluctuations impact ethanol costs in the same way that crude oil prices affect petroleum. For these and other reasons, it is better to use perennial plants for biomass, as they grow continuously, as opposed to seasonal plants like corn. Perennial bioenergy grasses and woody crops can be incorporated into agricultural production with reduced indirect land-use change, while increasing water quality benefits (Cacho et al. 2018). It is also better to grow biofuels on previously cleared land, such as under-utilized farmland.[23] Biofuels can also be extracted from aquatic plants, removing the need for land. For example, Liang et al. (2023) genetically modified duckweed plants to produce seven times more oil per acre than soybeans—currently the most commonly used biodiesel-producing plant.

Because of the uncertainty in their value and negative impacts, many see biomass fuel as a transition fuel, especially for difficult-to-decarbonize sectors like aviation, heavy transport, and manufacturing, until other renewable energy sources become viable (e.g., Reid et al. 2020). Where biofuels are harvested, grown, or used, it is clear that integrated policies for energy, land use, and water management are needed (Popp et al. 2014).

7.2.3 Solar Energy

Nearly all energy sources originate, directly or indirectly, from solar flux.[24] Above the Earth's atmosphere, the yearly average solar irradiation (also known as "insolation")

[22] https://www.ers.usda.gov/topics/crops/corn-and-other-feedgrains/feedgrains-sector-at-a-glance/.
[23] https://www.nrel.gov/research/re-biomass.html
[24] Nuclear and geothermal are notable exceptions.

is about 1360 $W{\cdot}m^{-2}$. This assumes the Sun is directly overhead at all locations; i.e., if the Earth was a circular disk of area πr^2 directly facing the Sun.[25] Assuming that the Earth is a sphere with a surface area of $4\pi r^2$, the average flux is therefore one quarter of this value, or 340 $W{\cdot}m^{-2}$. This is the value normally used for radiative forcing. Of course for solar power we are only interested in the time when the Sun is above the horizon. On average, 1.7×10^5 TW of solar radiation continuously strikes the Earth.[26] For comparison, global energy consumption for the year 2019 was 600 quads[27], which is an average rate of 20 TW. In the U.S. alone the annual incident solar energy is estimated at 400,000 TWh, which is one hundred times more than its electricity consumption for 2021.[28] Solar therefore represents an incredible resource of energy.

There are different types of solar energy technologies. The most common is photovoltaics (PVs), where solar insolation is converted directly into electricity via the photoelectric effect. The efficiencies of different PV technologies continue to improve. Though most commercial panels have efficiencies of 15–20%, researchers have developed PV cells with efficiencies approaching 50% (NREL 2023). Like charge-coupled devices (CCDs), solar PV efficiency increases as the PV panel's temperature decreases. Designs are being considered to combine PV with solar thermal, wherein the panels are cooled to improve their efficiency and the extracted heat can be used as well. Solar PV experiences an average degradation of output of less than 1% per year (Kim et al. 2021).

"Solar thermal" uses sunlight to warm materials to generate thermal energy. It consists of a range of designs; e.g., low-temperature collectors that are used to heat ventilation air or water for residential or commercial use. There is also concentrated solar-thermal power (CSP), where mirrors are used to focus sunlight onto a fluid (e.g., steam or molten salt) that is used by a heat engine to generate electricity. Solar thermal to electric has an efficiency similar to PV—about 20–30%. Overall, solar thermal represents a small portion of the solar energy market. In 2021, it consisted of less than 2% of the total solar power generated in the U.S. Because of the rapidly decreasing costs of PV, which solar thermal has not matched, it is expected that solar PV will continue to dominate the market.

From a climate change perspective, the biggest advantage of solar is that it produces no greenhouse gases during operation. The manufacture of PV panels does require energy; however, Wu et al. (2017) estimate that the "energy payback time" (i.e., the time required for it to produce as much energy as used during manufacture) is less than three years, a small fraction of their 25 to 30 year lifespans.

While painfully obvious, it is worth pointing out that sunlight is free. This stands in contrast to fossil fuels and biofuels, which cost money to produce and transport; and their prices fluctuate significantly. Because it consumes no fuel and has no moving parts, PV solar has the lowest operations and maintenance costs—less than \$15 kW $h^{-1}yr^{-1}$ (NREL 2020). They can also be placed on surfaces otherwise not in use, such as rooftops and building walls.

---

[25] Which, contrary to some opinions, is not true.

[26] https://doi.org/10.1038/454805a

[27] https://www.eia.gov/international/data/world/total-energy/total-energy-consumption

[28] https://www.eia.gov/energyexplained/electricity/use-of-electricity.php

When considering the unsubsidized LCOE, utility-scale solar PV is now effectively tied with wind as the lowest-cost source of electricity (Lazard 2023). The price of utility-scale solar PV has dropped an astonishing 90% over the last decade, from \$360 MW h^{-1} in 2009 to \$40 MW h^{-1} in 2019 (IRENA 2020), with prices predicted to drop an additional 34% by 2030 (BNEF 2022). In fact, solar is now so inexpensive in some cases it is financially preferable to decommission and replace coal-fired power plants before they reach the end of their operational lifetime.

Residential solar installations cost roughly three times as much per Watt as utility-scale solar (Ramasamy et al. 2022), but in many cases are still worthwhile. Residential installations have significant initial costs, but they are usually recovered within 3–10 years of operation. In markets with 'net metering" policies, utility companies will also compensate owners for excess electricity produced that is returned to the grid. As of 2017, thirty-eight states in the U.S. offer net metering.[29] Residential solar panels also increase the resale value of the home.

The biggest disadvantage with solar is the obvious one—it can only be used when solar insolation is available. It is not, by itself, a "dispatchable" source of energy. While seasonal and diurnal cycles are predictable, weather is less so. It must therefore be paired with other energy sources, or storage, to ensure enough power is available when needed. Solar insolation of course varies by region based upon latitude and regional climate (Figure 7.6). However, solar can be a viable option even in regions with less insolation; e.g, despite its far northern latitude, Alaska has insolation values that compare favorably to Germany, one of Europe's largest users of solar.[30]

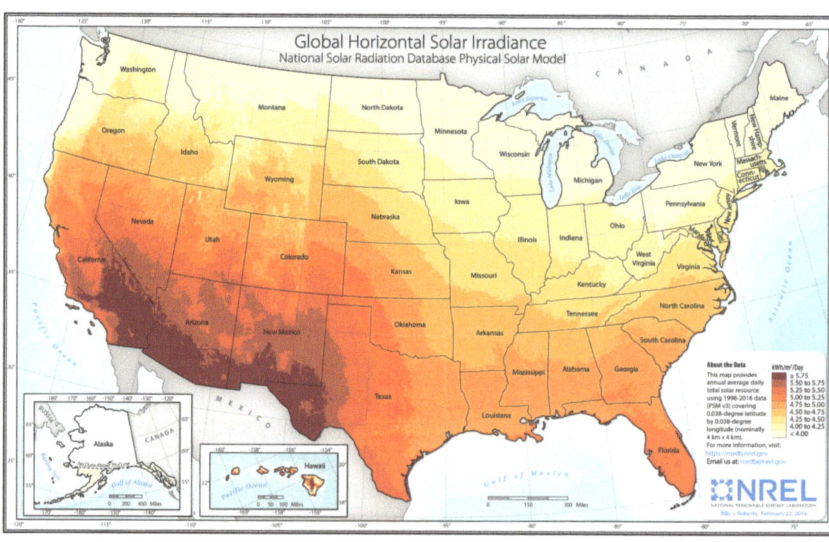

Figure 7.6. [31]Annual average daily total solar resource (in kW h m^{-2} day^{-1}) using 1998–2016 data.

[29] https://www.ncsl.org/research/energy/net-metering-policy-overview-and-state-legislative-updates.aspx
[30] https://www.energy.gov/sites/prod/files/2016/02/f29/Solar-Prospecting-AK-final.pdf
[31] https://www.nrel.gov/gis/solar-resource-maps.html

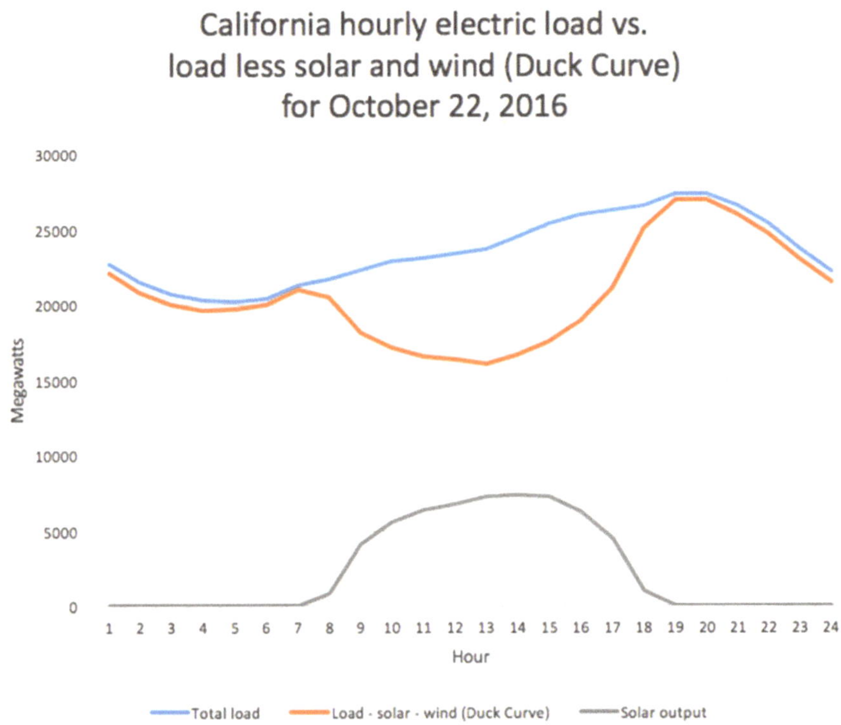

Figure 7.7. [32]Electricity demand (i.e., load) as a function of time of day, for the state of California on October 22,2016, a day when the wind output was low and steady throughout. The gray line along the bottom shows solar output for the day, which peaks around 1–2 pm local time. The blue line shows the total electricity load, which peaks in the evening around 7–8 pm. The orange line shows the difference in electricity demand (i.e., the blue line minus the gray line). It is known as the "duck curve," so named for its resemblance to the web-footed animal. Its shape is the result of the difference in time between peak solar output and the peak of total load. Data are from California ISO.[33]

Another issue with its output varies throughout the day; and in general its output isn't well matched to demand (Figure 7.7). Excluding variability due to weather (and solar eclipses), peak output occurs during the hours near solar noon. However, peak demand for electricity usually occurs in the evening.

Utility-scale solar installations require a lot of land. For example, Solar Star is one of the largest solar farms in the U.S. Capable of generating 579 MW, it has 1.7 million solar panels spread over 13 km^2 (3200 acres)—giving an energy density of about 45 MW km^{-2}. Developing land for energy production, solar or otherwise, can be in conflict with other land uses, such as the production of food and the conservation of ecosystems. PV panels have been linked to substantial impacts on species and ecosystems, the first and most obvious one being the degradation of

[32] https://en.wikipedia.org/wiki/Duck_curve#/media/File:Duck_Curve_CA-ISO_2016-10-22.agr.png
[33] http://www.caiso.com/market/Pages/ReportsBulletins/DailyRenewablesWatch.aspx

natural habitats but they may also lead to mortality of individuals and displacements of populations (Lafitte et al. 2022). Bird collision risk from solar panels is very low. There is likely to be more of a collision risk by infrastructure associated with solar PV developments, such as overhead power lines (Harrison et al. 2016). One option is to install solar on land that is otherwise deemed unusable, such as within urban environments, salt-affected or contaminated land, and water reservoirs (as "floatovoltaics"). In many cases use of such lands can be sufficient to meet energy needs (e.g., Hoffacker et al. 2017).

Another option is the co-location of solar installations on land used for agriculture, livestock, or other purposes. The solar panels cool vegetation underneath during the day due to shading, and keep them warmer at night. Depending on solar installation height and spacing, crops that are shade tolerant that grow to low height may become suitable for production, potentially opening new markets for agricultural producers. In some cases shade from solar infrastructure may reduce crop productivity, whereas in others it can help (e.g., Marrou et al. 2013). In a pasture landscape, elevated solar infrastructure could provide shade and cover for livestock while not having a substantial impact on productivity. However, agricultural activities involving large machinery may have limited options for co-location (Macknick, et al. 2013). Solar power can of course also be used to power the daily operations of farms and agricultural businesses, which can be particularly beneficial for those that are off grid.

Solar PV requires substantial amounts of mineral resources, in particular copper and silicon. They also represent a significant source of hazardous waste at their end-of-life disposal (Chowdhury et al. 2020). About 92% of solar panels are made from Si, followed by 5% CdTe (cadmium telluride), 2% CIGS (copper indium gallium selenide) and <1% a-Si:H (hydrogenated amorphous silicon). First generation solar panels, made primarily from Si, use thick wafers, requiring more minerals and incurring higher manufacturing costs. Second-generation thin-film solar cells, such as CdTe, CIGS, and a-Si:H, require less materials and are less expensive to manufacture. The solar cells can also be flexible, allowing for new uses. However, in CdTe solar cells, a toxic material (cadmium) is involved, and CdTe module disposal is an issue since these modules cause pollution during decommissioning (Ramanujam & Singh 2017).

Historically shade is a solar panel's greatest enemy, for occultation on even a portion of a PV panel could have dramatic effects. Intuition suggests that the power output of a panel will be reduced proportionally by the area that is shaded. However, this is not necessarily the case. Like rows in a CCD, PV cells in a solar panel are linked such that the current generated runs through all of the cells in the panel. Thus, when one or more cells are in shade, it reduces the current for the entire panel. For example, in earlier panel designs shading just one out of 36 cells in a small solar module can reduce total power output by as much as 75% (Masters 2004). Because of shading from foliage or other buildings, many commercial and residential locations were less suitable for solar PV. This is because of the inverter, which converts DC current to AC. However, newer inverter designs avoid this problem.

Solar energy production is reduced across regions that experience high levels of dust and/or anthropogenic particulate matter (PM) pollutants, including large areas of India, China, and the Arabian Peninsula. The degradation in output, up to 25%, is caused by roughly equal contributions from airborne PM and PM deposited on photovoltaic surfaces (Bergin et al. 2017). The NASA Mars Exploration Rovers and InSight lander can relate to the problem.

Because of their low cost and other advantages, solar power has finally taken off. In recent years global PV installations have skyrocketed from a cumulative of 39.2 GW in 2010 to 767.2 GW in 2020—an astonishing twenty-fold increase in ten years. However, even faster growth is necessary to achieve the goal of net zero emissions by 2050.[34] Only 2.8% of U.S. electricity was generated with solar technologies in 2021 (EIA 2022). However, the low cost of solar means that its use will grow rapidly; e.g., the DOE (2021) predicts it will account for 40% of the electricity market by 2035. Solar PV is now a mainstream source of electricity, and plays a vital role in the transition to renewables.

7.2.4 Wind Power

Wind power is simply the conversion of the kinetic energy of the wind into mechanical energy. Historically the mechanical energy was used for pumping water; however, in recent decades the primary use has been for electric power generation. Winds are driven by solar heating, meaning that the Sun is the true source of energy. The most common form are horizontal axis wind turbines (HAWTs), where the axis of rotation is parallel to the ground. There are also vertical axis wind turbines (VAWTs), where the axis is perpendicular to the ground. Recall from your introductory physics course that power is produced by the equation:

$$P = \vec{F} \cdot \vec{v}$$

where in this case F is the force on the turbine blades and v is the wind velocity. Most turbines produce energy from the force of lift, which is given by:

$$F = \frac{1}{2} C_l \rho A v^2$$

where A is the area swept by the turbine blades, ρ is the density of air, and C_l is the coefficient of lift, which depends on the shape of the turbine blades as well as their pitch. Combining these two equations you can see that the theoretical power of a turbine is a function of velocity *cubed*; and so a turbine's power output is, within a certain range, strongly sensitive to wind speed.

There are three wind speeds that are relevant to the operation of a turbine. The "cut-in speed," typically between 3–4 m s^{-1} (6–9 mph), is when the blades start rotating fast enough to generate power. As the wind speeds increase, more electricity is generated until it reaches a limit known as the "rated speed," typically 11–13 m s^{-1} (25–30 mph). This is the point that the turbine produces its maximum, or rated power. The rated energy is set by the amount of power the turbine's generator can handle. On modern turbines, when wind speeds surpass the rated speed the pitch of the blades is adjusted (i.e., "feathered") to reduce their angle of attack. This reduces the surface area on the blade exposed to the wind and therefore the force of lift. The power generated by the

[34] https://iea-pvps.org/wp-content/uploads/2022/01/IEA-PVPS-Trends-report-2021-4.pdf

turbine remains constant even if the wind speed increases beyond the rated speed. The "cut-out speed" is the point at which the turbine shuts down to prevent undue strain on the rotor, which usually occurs at 22–25 m s^{-1} (50–55 mph). Ideally then a turbine will reliably experience winds between the rated and cut-out speeds. Wind turbines are designed to different "classes" based on where they will be installed. Turbines are currently typically built to withstand speeds of 50 m s^{-1} (110 mph) and gusts up to 70 m s^{-1} (155 mph). Some are being designed in hurricane-prone regions and have special designs to withstand even higher wind speeds, which means they can withstand all but the strongest of storms (Worsnop et al. 2017).

There is a limit to how much kinetic energy can be extracted from the wind. Naturally, attempting to take 100% would bring the wind, and therefore the turbine, to a standstill. Known as Betz's Law (Betz 1966), the theoretical maximum efficiency, derived from conservation of mass and momentum, is about 59%. If this is indeed the limit is disputed; in practice turbines are able to extract 35–45% of the wind's kinetic energy.

Known as vertical wind shear, there is a wind speed gradient as a function of altitude because of the boundary layer on the surface of the Earth, as well as turbulence caused by topological features on land and heat fluctuations. It is therefore preferable to build taller turbines that can not only reach higher-speed winds but also have longer blades that sweep out a larger area. For utility-scale wind on land, tower heights typically are 80–160 m, with taller turbines in general having larger rated capacities (Lantz et al. 2019). On land, the size of a wind turbine is limited by the ability to move components from the point of manufacture to the installed location. Like solar, the availability of wind varies by region (Figures 7.8).

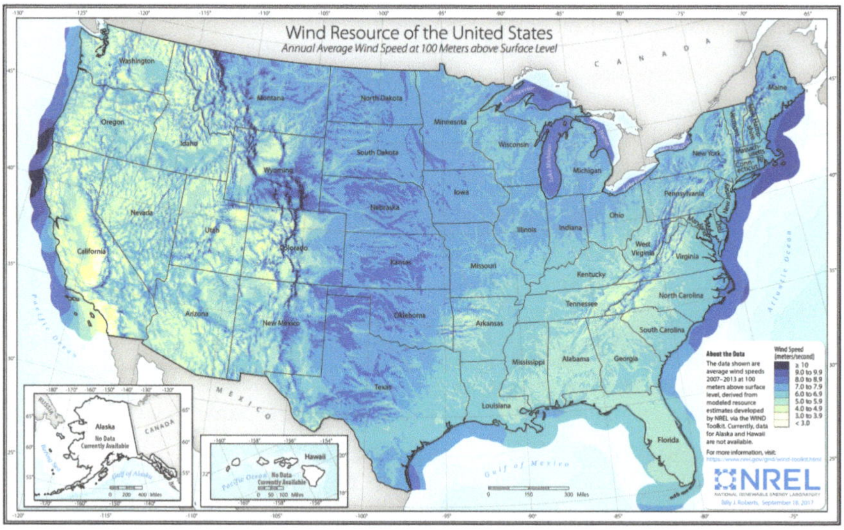

Figure 7.8. [35]Multiyear average wind speeds in the contiguous U.S. at 100 meters above surface level. Coastal regions and midwestern states show the highest average speeds.

[35] https://www.nrel.gov/gis/wind-resource-maps.html

"Offshore" turbines are placed over open water—usually in the ocean but also lakes. In general offshore wind speeds are higher and more consistent—leading to greater efficiency. Offshore turbines can also be larger because they don't face the logistical challenges of transportation by road. For example, the General Electric Haliade-X offshore turbine is 260 m tall and has blades over 100 m in length.[36] However, offshore wind faces more challenges. It is more difficult to transmit the electricity to market and maintenance costs are higher. It can also take longer to repair. If the depth of the water is greater than 60 m it is ineffective to affix them to the sea bottom. In these cases the turbines can be placed on floating platforms.

In many communities rooftop solar has become an important source of energy. However, small-scale and residential wind turbines have not had the same level of success for a few reasons. The wind is 30–40% slower in urban areas because tall buildings block the wind and make airflow turbulent (Micallef and van Bussel 2018). Vertical wind shear further reduces wind speeds near the ground, where such turbines are usually located. Wind speeds in a particular location tend to vary more than solar insolation, which means more variable power output. Residential turbines also require more infrastructure to install and more maintenance than solar. Residential-sized turbines tend to vibrate and make more noise than solar panels, particularly if they are attached to a house. Small-scale wind however can make sense in some locations, such in locations that are off grid or have unusually high wind resources.

The global capacity for wind power is estimated to be greater than 250 TW (Jacobson & Archer 2012; Miller & Kleidon 2016), which is more than ten times current global energy demand. In the U.S. alone the potential for wind is impressive. Most of the fixed-bottom capacity is along the Atlantic seaboard and in the Gulf of Mexico because of the steep shores on the west coast. The capacity potential for floating offshore wind is about twice that of fixed-bottom (Lopez et al. 2022). However, the technical potential for all offshore wind is less than half that of onshore wind power (Lopez et al. 2012). Veers et al. (2022) discusses challenges and opportunities for future wind turbine design.

In 2021, of the total 830 GW of wind capacity installed, 93% were on-shore systems, with the remaining 7% offshore wind farms[37] The potential for on- and offshore wind power in the contiguous U.S. could amount to more than 16 times current consumption (Lu et al. 2009). For comparison, the total technical potential capacity for solar in the U.S. is about nine times larger than that for wind (Lopez et al. 2012).

Like solar, the greatest benefits of wind power are that it consumes no fuel and produces no CO_2 during operation. A common misconception is that the energy required to make a wind turbine exceeds the amount it will generate. Even if we assume that fossil fuels are the energy source for the mining of materials as well as construction, the greenhouse gas payback time for the manufacture of wind turbines

[36] https://www.ge.com/renewableenergy/wind-energy/offshore-wind/haliade-x-offshore-turbine
[37] IEA (2022), Wind Electricity, IEA, Paris https://www.iea.org/reports/wind-electricity

is usually less than a year of operation (e.g., Dammeier et al. 2019), which is a small fraction of their overall lifetimes. Turbine blades can last more than 20 years, but sometimes they are replaced sooner with more efficient designs. Wind has operations and maintenance costs of about $11/MWh (Wiser et al. 2019), which is roughly double that of solar but half that of coal, nuclear, and geothermal.

Also like solar, the biggest downside to wind is that it is not dispatchable. It is vulnerable to changing wind speeds due to variable weather conditions. Wind doesn't have a diurnal variation like solar; generation typically peaks in the first half of the year, when there are more windy days (EIA 2022).

Wind power is not well suited for everywhere, because of available wind speeds as well as the large areas required. Turbines are typically spaced at distances equal to at least seven times their rotor diameter apart to minimize the wake effect. They are located relative to each other depending on the prevailing wind direction, as turbines placed within the wake of a neighboring wind turbine will produce less power. The wake of an offshore wind farm can extend for tens of kilometres.[38] As a result the energy density possible with wind is 3–5 MW km^{-2}. This is about ten times less than solar but assumes the land is only for wind generation. Because wind farms are usually located in remote locations, they require more infrastructure to bring electricity to market.

Wind farms can be co-installed on land used for other purposes, such as other forms of energy generation (solar or aquaculture offshore) or for ranching and agriculture—but not completely without impact. Wind turbines generate low-frequency noise (LFN) in the range of 20–200 Hz. LFN in general has been linked to an increased risk of epilepsy, cardiovascular effects, and coronary artery disease (Castelo Branco & Alves-Pereira 2004). In view of the adverse health impacts of LFN exposure, Chiu et al. (2021) argue for regulations on the requisite distances of wind turbines from residential communities. Some wish to be far from wind farms anyway because they consider them to be unattractive. However, others enjoy their appearance, and in some cases they attract tourists.[39]

Wind turbines also pose a risk to wildlife. Perhaps the most obvious threat is impact injuries to birds and bats. But turbine noise also impacts survival, social, and rearing mechanisms in certain species (Teff-Seker et al. 2022). Some mitigation of these effects are possible; e.g., painting one blade of a wind turbine rotor black resulted in 70% fewer bird collisions (May et al. 2020). It is useful to put bird deaths from wind power into context. Considering the effects of pollution from mining and operation, as well as the effects of climate change, Sovacool (2012) estimates that wind farms are responsible for roughly 0.27 avian fatalities per GWh of electricity, while nuclear power plants involve 0.6 fatalities per GWh and fossil-fueled power stations are responsible for about 9.4 fatalities per GWh. While every effort should be made to minimize deaths, it is clear that overall wind power is beneficial to bird and bat populations.

[38] https://maritime-spatial-planning.ec.europa.eu/practices/capacity-densities-european-offshore-wind-farms

[39] https://www.economist.com/britain/2023/01/05/britains-offshore-wind-farms-attract-tourists

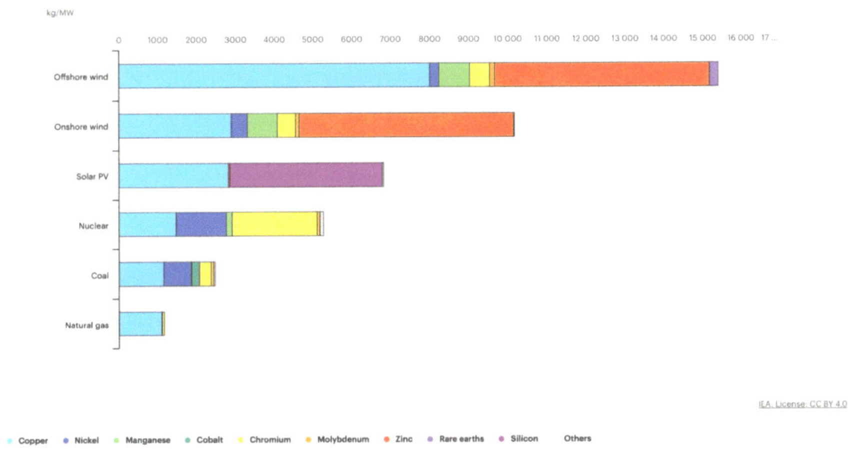

Figure 7.9. IEA, Minerals used in clean energy technologies compared to other power generation sources, IEA, Paris.

A concern is that wind power will adversely affect the climate by changing the Earth's overall wind patterns (e.g., Miller et al. 2011). Zhou et al. (2012) found that wind farms can cause the land near the turbines to be warmer than surrounding areas because the turbines redirect warmer air at higher altitudes down to the surface. However, studies found minimal impacts to the overall climate for the United States (Miller & Keith 2018) and Europe (Vautard et al. 2014).

Construction of offshore and onshore wind requires large amounts of minerals compared to other energy sources, in particular copper and zinc (Figure 7.9). This raises concerns about the mining operations necessary to provide these materials as well as their price volatility.[40] While most of the components of a wind turbine can be recycled, the end-of-life management and recycling of wind turbine blades has been frequently presented in the media as having an undue environmental impact. Due to the nature of the materials used in wind turbine blades, namely glass fiber reinforced thermoset polymer composite, wind turbine blades are technically difficult to re-process and convert into new valuable materials (Beauson et al. 2022). Pyrolytic recycling of decommissioned wind turbine blades could become a solution not only for their disposal but also for the recovery of high-value products (Yang et al. 2022). The blades represent only a small fraction of the weight of the turbine. Metals such as steel, aluminum, and copper represent 94% of the turbine weight (excluding the foundation) and are generally considered as materials that can be recycled (Schleisner 2000).

As dramatic as the images of turbine blades in landfills appear to be, they also need to be put into context. For comparison, the amount of solid waste from a megawatt-hour of coal electricity is 200 times the amount from a megawatt-hour of

[40] https://www.energymonitor.ai/tech/renewables/data-insight-the-cost-of-a-wind-turbine-has-increased-by-38-in-two-years/

wind electricity—and both of these pale in comparison to one's daily trash output. For perspective, if an individual U.S. resident gets all of their household electricity from wind energy, over 20 years their share of non-recyclable wind turbine blade waste will be 15 kg. That same mass of solid waste is produced by one person's share of a coal-fired power plant in 40 days, and it is just 7 days of municipal waste (Barnes 2020).

Like solar PV, wind power has grown dramatically in recent years. Global wind power capacity has jumped from a cumulative of 180.8 GW in 2010 to 735.7 GW in 2020—a four-fold increase in ten years. But even faster growth is necessary to achieve the goal of net zero emissions by 2050[41] Surpassing hydroelectric in 2019, wind power is now responsible for the largest fraction of renewable electricity generation in the U.S. In 2021, it also generated more than three times that of solar. Nonetheless, in that year it generated only 9.2% of total U.S. electricity.[42] Utility-scale wind and solar PV are currently tied as the lowest-cost sources of electricity (Lazard 2023); and wind energy costs are predicted to drop an additional 37% to 49% by 2050 (Wiser et al. 2021). As a result IRENA (2019) predicts a ten-fold increase in the global deployment of on- and offshore wind by that time. Wind power clearly plays a vital role in the transition to a low-carbon energy supply.

7.2.5 Nuclear Fission

A nuclear reactor uses the energy released from the fission of radioactive materials to produce electricity. Like a coal or natural gas power plant, a nuclear reactor is a heat engine. The energy released from fission is used to generate steam that turns a turbine (i.e., the Rankine cycle). Like all heat engines it has fundamental limitations on efficiency. Because they require fuel, nuclear reactors are not considered to be a renewable energy resource.

Uranium-235 (U-235) is the most commonly used fuel source because it is readily fissile. An example of one of the many fission reactions possible with U-235 is:

$$^{235}_{92}\mathrm{U} + \mathrm{n} \rightarrow {}^{144}_{56}\mathrm{Ba} + {}^{89}_{36}\mathrm{Kr} + 3\mathrm{n}$$

Energy is also released—in this example about 177 MeV per reaction—in the form of gamma rays as well as the kinetic energy of the fission products. The excess neutrons produced are then used to trigger more nuclear reactions, and to establish a self-sustaining chain reaction. The fissile material is placed within "fuel rods" that are placed within the nuclear reactor's "core," where the reactions occur.

The rate of fission reactions within the core is adjusted by controlling the number of neutrons that are able to induce further reactions. Nuclear reactors typically employ several methods of neutron control to adjust the reactor's power output. The fastest method for adjusting the rate of reactions in a reactor is with the use of "control rods" that absorb neutrons. When a control rod is placed among the fuel rods it absorbs more neutrons, thereby reducing the reaction rate and therefore the

[41] IEA (2022), Wind Electricity, IEA, Paris https://www.iea.org/reports/wind-electricity
[42] https://www.eia.gov/energyexplained/electricity/electricity-in-the-us.php

reactor's power output. Conversely, removing the control rod will result in an increase in the rate of fission events and an increase in power.

Nuclear fission is an incredibly dense source of energy; e.g., the energy released from the complete fission of 1 kg U-235 is equivalent to burning 2.7 million kg of coal.[43] And uranium is relatively abundant element in the Earth's crust—more than six hundred times more common that gold (Herring 2012). However, naturally occurring uranium must be processed before it can be used as fuel. U-235 has a half life of about 700 Myr, whereas U-238 has a half life of about 4.5 Gyr. As a result 99.3% of natural uranium is U-238 and only 0.7% is U-235, meaning that isotope separation must be used to increase the percentage of fissile U-235 over the relatively inert U-238. This process is known as uranium enrichment. Low-enriched, or "reactor grade," uranium is typically 3–5% U-235; whereas highly-enriched, "weapons grade," uranium is greater than 85% U-235. It is worth noting that U-238 can be transmuted via neutron capture into Plutonium-239 (Pu-239), which is commonly used in the production of nuclear weapons.

Research and development of thorium-based nuclear reactors is underway in several countries. Thorium-232 (Th-232) can be transmuted into U-233, which is usable as a fuel source. Thorium is three times as abundant as uranium in the Earth's crust. It consists nearly entirely of Th-232, so no enrichment is necessary. Mining of thorium is also safer and its use would in principle generate less nuclear waste.[44] While there are a few Th-232 research reactors, no commercial reactors yet exist.

From a climate change perspective, the greatest strength of nuclear power is that it is capable of producing large amounts of energy without generating any CO_2 during operation.[45] It requires a small land footprint—with an energy density of about 500 MW km^{-2} it is about ten times better than solar and a hundred times better than wind (assuming that the land is not also used for other purposes). Nuclear power is also a dispatchable source of energy; however, it typically requires hours for startup and cannot rapidly vary its output to adjust to changes in demand.

Nuclear power has a high "capacity factor," defined as the ratio of the actual energy output of a power plant to its theoretical (i.e., rated or "nameplate") output during a set period of time. They can run for more than a year without interruption. In 2021, U.S. nuclear power had a capacity factor of 93%. For comparison, solar PV and wind had capacity factors of 24% and 34% during the same time period.[46]

Nuclear power has been an important source of energy in the U.S. and globally. As of 2022 there are about 440 nuclear power reactors operating in over 30 countries, with a combined capacity of about 390 GW. In 2021 these provided 2653 TWh, about 10% of the world's electricity. About 60 power reactors are currently being constructed in

[43] https://www.euronuclear.org/glossary/coal-equivalent/
[44] https://www-pub.iaea.org/MTCD/Publications/PDF/TE_1450_web.pdf
[45] Since nuclear power requires fuel, there will be CO_2 production from its mining and processing.
[46] https://www.eia.gov/electricity/monthly/epm_table_grapher.php?t=table_6_07_b

17 countries. China, India, South Korea, and Russia are the biggest investors.[47] In 2021, nuclear power was responsible for 19% of the electricity generation in the U.S., which is half of natural gas but equivalent to coal and equivalent to all of renewables.[48] Nuclear reactors in principle can be placed nearly anywhere, but are usually located near sources of water for cooling purposes.

Despite its benefits, nuclear power has fallen out of favor in many countries, including the U.S. As of 2022 only two nuclear plants were under construction in the U.S.—the Vogtle 3 and 4 plants in Georgia. In 2021, only 41% of Americans felt the U.S. government should encourage the development of nuclear power, whereas 66% supported the development of wind and solar.[49] Unfairly or not, it is perceived by many as an unacceptably dangerous source of energy. A common misconception is that nuclear power plants release radioactivity as a normal consequence of operation, much in the same way as coal and natural gas power plants emit CO_2 and other pollutants. This perception may come from the steam clouds emitted from cooling towers. However, under normal operation the nuclear reactor itself releases no radioactivity or pollution. In fact, radiation doses from airborne particles emitted from a coal-fired plant can be greater than those from a nuclear plant (McBride et al. 1978).

That is not to say that nuclear reactors are completely safe. Much of their reputation is due to the risk of catastrophic failure. If the rate of nuclear reactions within the core is not properly controlled, excess heat can cause a dangerous buildup of steam pressure and/or hydrogen gas as well as an uncontrollable meltdown of the core itself. Major accidents that have resulted in the release of dangerous levels of radioactivity, as well as the loss of life, include Chelyabinsk (1957), Sellafield (1957), Chernobyl (1986), and Fukushima (2011)—the last two being catastrophic. These incidents and others, e.g., Three Mile Island (1979), have led to increased social and political resistance to nuclear power. Global nuclear energy capacity grew steadily in the U.S. until the Chernobyl disaster, but has since leveled off (Figure 7.10). Some countries have opted to ban nuclear power (e.g., Austria in 1978) or to accelerate its decommissioning (e.g., Germany after the Fukushima disaster). Whether or not this reaction to nuclear is justified is largely a matter of opinion, but it is worth noting that the number of deaths that can be attributed to nuclear power is low; e.g., Ritchie (2017) estimates that under 3000 deaths can be attributed to the Chernobyl and Fukushima accidents.

The operation of a nuclear power plant is not the only safety concern. They also produce nuclear waste that must be safely disposed of. Over time the irradiation of fuel rods within a reactor will reduce the amount of fissile elements (e.g., U-235) while increasing the amount of decay products, many of which are neutron absorbers that interfere with the rate of nuclear reactions. Eventually a fuel rod will no longer be useful for sustaining nuclear reactions. It must then either be reprocessed, to extract useful isotopes and remove unwanted decay products, or be stored as nuclear waste.

[47] https://world-nuclear.org/information-library/current-and-future-generation/plans-for-new-reactors-worldwide.aspx (Updated December 2023).
[48] https://www.eia.gov/energyexplained/electricity/electricity-in-the-us.php
[49] https://www.pewresearch.org/short-reads/2023/08/18/growing-share-of-americans-favor-more-nuclear-power/

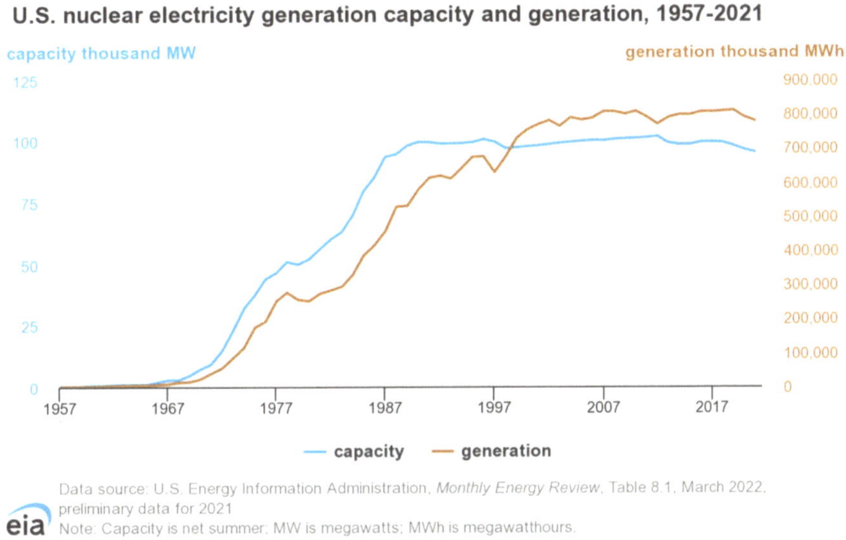

Figure 7.10. As of 2021, the United States had 93 operating commercial nuclear reactors at 54 nuclear power plants in 28 states. Nuclear capacity in the U.S. grew steadily during the 1970s and 80s, but has since stalled. No new capacity was built between 1996 and 2016. As a result, the average age of nuclear reactors in the U.S. is about 40 years old.[50]

For the last 60 years or so most nuclear waste has been stored near the reactor. However, the long-term storage of nuclear waste remains a largely unsolved problem. Decay products can remain radioactive for tens of thousands of years and continue to generate significant amounts of heat. It must somehow be stored for this length of time such that it doesn't leak into the ground and water table. In 2002 construction began on a storage site for high-level radioactive waste in the Yucca Mountain range in Nevada. However, safety concerns and public opposition led to its cancellation in 2011, leaving the U.S. without a long-term solution. Finland however has had success in opening a long-term repository for its nuclear waste (El-Showk 2022).

Nuclear power plants also represent a security risk. Nuclear reactors have been targeted during military conflict, and terrorists could target nuclear power plants in an attempt to release radioactivity. The United States 9/11 Commission found that nuclear power plants were potential targets originally considered for the September 11th, 2001 attacks. Nuclear power plants also represent a source of radioactive materials that could be used for nefarious purposes. According to the IAEA Incident and Trafficking Database, between 1993 and 2021 there were 320 incidents of theft or loss of nuclear materials connected with trafficking or malicious use.[51] Nuclear reactors used for the peaceful generation of electricity can also be used to produce Pu-239 for nuclear weapons. Notably North Korea and Iran have been internationally sanctioned because of their nuclear development programs.

[50] https://www.eia.gov/energyexplained/nuclear/us-nuclear-industry.php

[51] https://www.iaea.org/resources/databases/itdb

Producing the fuel for reactors is also problematic. Uranium mining has the same safety and environmental hazards as other forms of mining, with the complication that radioactive waste is also generated. Wind can blow radioactive dust from this waste into populated areas and can contaminate surface and groundwater. Uranium eventually decays to radon, a radioactive gas. Without proper ventilation, radon is a threat if inhaled by miners in underground mines. Considering the need for radiation protection for miners, long-term management of radioactive waste, and the lack of public support surrounding uranium mining in many countries, the industry's challenges are more complicated than in the case of other mined materials.

The economics of uranium mining is also risky. The price of uranium in recent years has shown the greatest volatility in history—with a peak of US $300 kg in 2007 and a low of US $41 kg in 2016.[52] Including expenses for land, exploration, drilling, production, and reclamation, the average cost to mine uranium in the United States is about $150 kg—well above usual market prices.[53] In 2021, Kazakhstan was responsible for 45% of global uranium mining production.[54] For such a large portion of a global resource originating in one country raises concerns about energy security, particularly so because of the corruption and political unrest in this former Soviet Socialist Republic.[55] The world's present measured resources of uranium are enough to last for about 90 years at current usage rates, although other sources (e.g., reprocessed weapons-grade uranium) are also available.[56]

Ultimately the biggest problem that nuclear power faces is that it is currently not financially competitive. The economics of new nuclear plants are heavily influenced by their capital cost, which accounts for at least 60% of their LCOE.[57] Recent projects have run far over budget and behind schedule; e.g., the required capital costs for two new nuclear reactors at Plant Vogtle—the first entirely new U.S. reactors in decades—has doubled from $14 billion to more than $30 billion; and construction is more than seven years behind schedule.[58] Cost overruns and delays have affected construction in other countries as well; e.g., Finland (Olkiluoto 3, factor of ~3×),[59] France (Flamanville, ~5×),[60] and the UK (Hinkley Point, ~1.6×).[61] South Korea and Japan have fared better, in large part because of a standardization of both design and manufacturing teams (Berthélemy & Escobar Rangel 2013). The decommissioning costs for a nuclear reactor can also be

[52] https://www.iaea.org/newscenter/news/uram-2018-ebb-and-flow-the-economics-of-uranium-mining
[53] https://www.epa.gov/sites/default/files/2015-05/documents/2013domesticuraniumreport.pdf
[54] https://world-nuclear.org/information-library/country-profiles/countries-g-n/kazakhstan.aspx
[55] https://www.world-nuclear-news.org/Articles/Uranium-sector-monitors-evolving-Kazakh-situation.
[56] https://world-nuclear.org/information-library/nuclear-fuel-cycle/uranium-resources/supply-of-uranium.aspx
[57] https://www.world-nuclear.org/our-association/publications/online-reports/nuclear-power-economics-and-project-structuring.aspx
[58] https://apnews.com/article/georgia-power-co-southern-climate-and-environment-business-3b1d6c65353c6a65b1ccfddede753ab7
[59] https://www.ft.com/content/6e90de68-cf5c-11e1-a1d2-00144feabdc0
[60] https://world-nuclear-news.org/Articles/Further-delay-to-Flamanville-EPR-start-up
[61] https://www.power-technology.com/news/hinkley-point-c-project-costs-rise-again/

expensive, reaching upwards of a billion dollars.[62] As a result, the LCOE cost of nuclear power is about five times that of wind and solar (Lazard 2023), which has stalled investment in new facilities. Excluding the capital costs, nuclear is at about the same price point as wind and solar, arguing that existing nuclear facilities should be kept operational but new capacity should not be pursued.

There is research underway to develop small modular reactors (SMRs) that could take advantage of simpler designs and economies of scale to have lower capital costs and shorter construction times. Newer designs also use "passive safety" features that do not require any active intervention to shut down the reactor in an emergency, reducing the risk of failure due to mechanical failure, loss of power at the reactor, or operator error. However, such systems remain largely undeployed. Licensing is potentially a challenge for SMRs, as design certification, construction and operation license costs are not necessarily less than for large reactors.[63] The U.S. Nuclear Regulatory Commission approved the NuStar SMR design for the first time in 2020,[64] but since that time the cost for power for that design increased by 50%,[65] resulting in a cancellation of its project in the U.S.[66] It is therefore not yet clear when this or other SMR projects will become commercially viable. SMRs are not forecast to hit the commercial market much before 2030,[67] during which time the costs of renewables will likely continue to decline. In the near future renewables with utility-scale battery backup may be cheaper than nuclear, and be just as dispatchable (Section 7.8). It therefore seems unlikely that nuclear power will be able to "catch up" and be competitive, at least not soon enough to play a major role in the rapid decarbonization necessary over the next decade.

7.2.6 Hydroelectric

A hydroelectric power plant uses moving water to turn turbines and generate electricity. The most common type of hydroelectric power plant is an "impoundment" facility, which uses a dam to store river water in a reservoir. Another design is a "diversion" dam, which redirects some, but not all, of a river through turbines. In both cases the kinetic and gravitational potential energy of the water in the river or dam is being converted into electrical energy. The power generation is controlled by the rate water passes through the turbines. A newer design is "pumped hydro," where water can be pumped back into a reservoir with an external energy source such as wind or solar. The energy from those sources is then stored as gravitational potential energy to be used when needed.

Hydroelectric is a renewable source of energy because it is driven by the water cycle. Solar insolation causes water from oceans and lakes to evaporate. This warm

[62] https://world-nuclear.org/information-library/nuclear-fuel-cycle/nuclear-wastes/decommissioning-nuclear-facilities.aspx
[63] https://www.world-nuclear.org/information-library/nuclear-fuel-cycle/nuclear-power-reactors/small-nuclear-power-reactors.aspx
[64] https://www.scientificamerican.com/article/first-u-s-small-nuclear-reactor-design-is-approved/
[65] https://ieefa.org/resources/eye-popping-new-cost-estimates-released-nuscale-small-modular-reactor
[66] https://www.reuters.com/business/energy/nuscale-ceo-defends-modular-nuclear-plants-after-project-cancellation-2023-11-14/
[67] https://www.energymonitor.ai/sectors/power/small-modular-reactors-smrs-what-is-taking-so-long/

water vapor rises by convection, where it cools and condenses into clouds. Eventually it will precipitate as rain or snow that will at some point move along rivers into oceans and lakes, and the cycle is completed. Hydroelectric extracts energy from the water as it returns to lakes and oceans. Like for wind, the Sun is the true source of energy.

Hydroelectric capacity was built in North America and Europe primarily between 1920 and 1970, when thousands of dams were constructed. Even larger hydropower dams have since been built along the Mekong River Basin, the Amazon River Basin, and the Congo River Basin (Moran et al. 2018). It has been the dominant source of renewable energy in the U.S., accounting for the far majority of renewable capacity until the rapid growth of wind and solar starting around 2010.[68]

Hydroelectric facilities come in a wide range of sizes. The seven largest power stations in the world of any kind are all hydroelectric.[69] The world's largest hydroelectric facility is the Three Gorges Dam in Hubei, China. It has a capacity of 22.5 GW, which is about three times bigger than the largest nuclear power plant. There are also "micro" hydroelectric plants, which are operated by individuals to produce enough electricity for a single home, farm, or ranch.

The capital costs for hydroelectric vary from \$1000–8000 kW^{-1}, which spans the range from wind and solar PV to the low end of nuclear. Operational and maintenance costs are quite low–typically 1–2% that of capital costs. Overall hydroelectric has a LCOE of \$20–270 MWh, which on the low end is competitive with wind and solar (IRENA 2012). Naturally, hydroelectric power can only be developed in locations with suitable resources (Figure 7.11). Hydropower also varies seasonally, typically peaking in the spring and early summer when the winter snowpack is melting. Many reservoirs must balance power output with competing water demand for irrigation, municipal, industrial, and other needs, as well as concerns with fish migration. As a result, hydroelectric facilities often do not run at full output. U.S. hydroelectric capacity factors average between 30% and 40% (EIA 2017).

From a climate change perspective the biggest benefit of hydroelectric is that it does not produce CO_2 during operation. It is a dispatchable energy source, capable of adjusting output within tens of seconds to minutes. Another benefit of hydropower is that it can provide a reliable source of electricity, especially in regions where other forms are not feasible. However, during drought conditions hydroelectric power can be less reliable, a concern in a climate changed world. Dams can also provide flood protection and water security, creating reservoirs that can be used for agriculture and recreation.

One of the main environmental concerns associated with hydropower is the impact of dam construction on rivers and aquatic ecosystems. The construction of dams can disrupt the natural flow of rivers, leading to changes in water temperature and the loss of fish and other aquatic species. In addition, dams can also disrupt the migration patterns of fish and other aquatic animals, leading to a decline in their

[68] https://www.eia.gov/energyexplained/electricity/electricity-in-the-us.php
[69] https://en.wikipedia.org/wiki/List_of_largest_power_stations

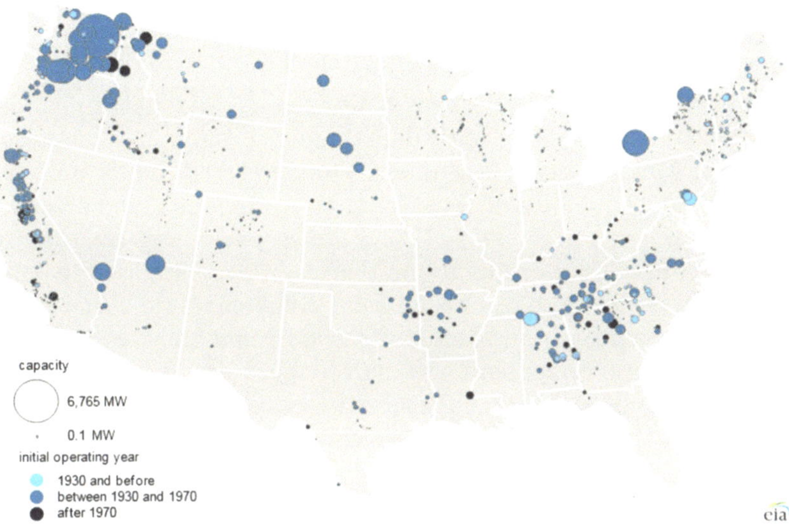

Figure 7.11. [70]Distribution of utility-scale hydroelectric plants in the Lower 48 states. The majority is located in Pacific-coast states; and most began operation before 1970. The average age of hydro facilities in the U.S. is greater than 65 years.

populations. Hydropower has been found to have negative impacts on many species, e.g., freshwater fish (Barbarossa et al. 2020) and macroinvertebrates (Trottier et al. 2022). The most vulnerable species are those with traits adaptive for habitats with fast flowing water, structural complexity, flood pulsing or those requiring connectivity across basins to complete their life cycle; e.g., salmon (Arantes et al. 2019). Dams can be built with "ladders" that facilitate the passage of fish upstream, but currently they do not work well. Overall less than 3% of fish successfully pass through fishways from the first dam to their spawning grounds (Brown et al. 2013). Another concern is the impact on the surrounding landscape. Dams and reservoirs can flood large areas of land, destroying wildlife habitats, and leading to the loss of biodiversity and the decline of endangered species. In some cases the costs of the environmental damages are greater than the purported benefits of emission reductions from hydroelectric generation (Ezcurra et al. 2019).

While hydropower doesn't produce greenhouse gases during operation, methane is produced by underwater microbes that digest the organic matter that piles up in the sediments trapped by dams. Collectively, reservoirs created by dams contribute 1.3% of the world's annual anthropogenic greenhouse gas emissions (Deemer et al. 2016). Dams can block the migration of sediment, which can lead to changes in water chemistry and the loss of important nutrients downstream. While hydropower has been found to degrade water quality, its effects are less than those caused by the use of fossil fuels (Alsaleh & Abdul-Rahim 2022).

[70] https://www.eia.gov/todayinenergy/detail.php?id=30312#tab1

Hydropower also impacts local communities. This can include degraded food systems, water quality, and agriculture, loss of livelihood, and the disruption of traditional ways of life. In some cases people are forced to relocate, and their homes and communities are destroyed. There is also the risk of catastrophic dam failure; e.g., the sudden breaching of a saddle dam on the Xe Nammoy hydroelectric-power reservoir in the Mekong basin (Latrubesse et al. 2020). Recently built hydropower dams are associated with reduced local economy, population, and greenness in areas near the dam sites (Fan et al. 2022). There are very few large hydropower projects under construction in developed nations. This is in part because the best sites have already been developed, but also because environmental and social problems have made them less desirable. In fact, more dams are being removed in North America and Europe than are being built. Several changes are needed to transform the hydropower sector to enable their benefits to exceed the costs and to ensure that dams contribute to sustainable energy systems (Moran et al. 2018).

7.2.7 Geothermal

Geothermal is a type of renewable energy that is generated by tapping into the heat within the Earth's crust. The process involves drilling a well into the Earth—typically a few hundred meters but can go several kilometers deep. Water is injected into the well, where it is heated by the Earth's internal heat. The hot water is then brought back to the surface and is used to generate electricity via a turbine. Geothermal electricity generation requires water or steam at high temperatures. "Open loop" geothermal power plants either use steam directly from the ground (known as "dry steam") or high-pressure hot water from the ground above 100°C that is allowed to expand rapidly to produce steam (known as "flash steam"). Newer "binary" plants are closed-loop, running hot water from the ground past a heat exchanger that heats a liquid with a lower boiling point; e.g., pentane. The gas is then used to turn a turbine. Dry and flash steam plants require heat of at least 200°C, whereas binary plants can generate electricity at temperatures as low as 100°C—opening up more geothermal resources for development. Geothermal energy can also be used for other purposes such as district heating, as well as for agricultural and industrial processes. For example, geothermal heat can be used to warm greenhouses and grow crops year-round. Geothermal energy is also valuable for recreational activities such as hot springs.

There is an abundance of geothermal resources that are not currently usable because the subterranean rock is not sufficiently permeable. Enhanced (or Engineered) Geothermal Systems (EGS) use fracking techniques to make cracks in the rock such that sufficient amounts of water can pass through. EGS shows the most potential to tap energy from high temperature (>200°C) rock at depths between 3–5 km (Moore & Simmons 2013). However, EGS is still in the research and development phase, and more research needs to be done to make it economically viable.

The U.S. leads the world in geothermal electricity generation. Nonetheless it still represents a very small fraction of total U.S. energy consumption. In 2021, there were geothermal power plants in seven states, which produced only 0.4% of total U.S. utility-scale electricity generation.[71] However, through technology improvements such as EGS, geothermal electricity generation capacity has the potential to increase to more than 60 gigawatts by 2050—providing 8.5% of all U.S. electricity generation (DOE 2023).

The main benefit of geothermal energy is that it is a renewable source of energy that produces no CO_2 during operation. It is a reliable source of power in that it is not susceptible to weather, like wind and solar. Nor is any fuel needed. Geothermal power plants are also compact. They use less land per gigawatt-hour than any other energy source except for nuclear.

Like the Sun, geothermal energy is practically unlimited. The current total heat flux from the Earth's interior to space is 44.2 ± 1.0 TW, about half of which is the result of radioactive decay (KamLAND Collaboration 2011). The total heat content of the Earth is about 10^{25} MJ, and that of the crust is roughly 10^{21} MJ. And the amount of heat within 10 km of the Earth's surface is estimated to contain 50,000 times more energy than all oil and gas resources worldwide (Shere 2013). Capturing even 2% of the thermal energy at depths of 3.5–10 km could provide two thousand times the annual energy use in the U.S. (Tester et al. 2006). The technical potential of geothermal resources currently suitable for electricity generation is 240 GW (Stefansson 2005). EGS could grow this number greatly.

Geothermal power also generates virtually no pollution during operation. Geothermal plants emit 97% less acid rain-causing sulfur compounds and about 99% less carbon dioxide than fossil fuel power plants of similar size.[72] Open-loop geothermal systems can emit pollutants such as hydrogen sulfide (H_2S), CO_2, and CH_4. H_2S, which has a distinctive "rotten egg" smell, reacts with air to produce sulfur dioxide (SO_2). However, most facilities use closed-loop water systems to minimize atmospheric emissions and contamination of the local water table.

Geothermal energy is also moderately inexpensive to produce. Its capital costs are much higher than wind and solar but comparable to hydro and less than nuclear. The LCOE of geothermal has remained relatively constant in recent decades at about \$70 MWh^{-1}, which is competitive with fossil fuels but not with wind and solar (Lazard 2023). However, if the cost of making wind and solar dispatchable is included—e.g., battery storage or pumped hydro—then geothermal is competitive. Since geothermal systems can also be used for heating and cooling, this can eliminate the need for separate systems and further reduce costs.

The biggest limitation of geothermal is that historically it has only been developed in areas with specific geologic conditions, such as volcanic regions or areas with hot springs (Figure 7.12). However, EGS has the potential to expand the amount of resources that are economically viable. Volcanic and seismic activity near the geothermal plant can affect power generation; e.g., in 2018 lava flows from the Kilauea volcano on the big island of Hawai'i led to the temporary shutdown of the Puna Geothermal Venture power plant.

[71] https://www.eia.gov/energyexplained/geothermal/use-of-geothermal-energy.php
[72] https://www.eia.gov/energyexplained/geothermal/geothermal-energy-and-the-environment.php

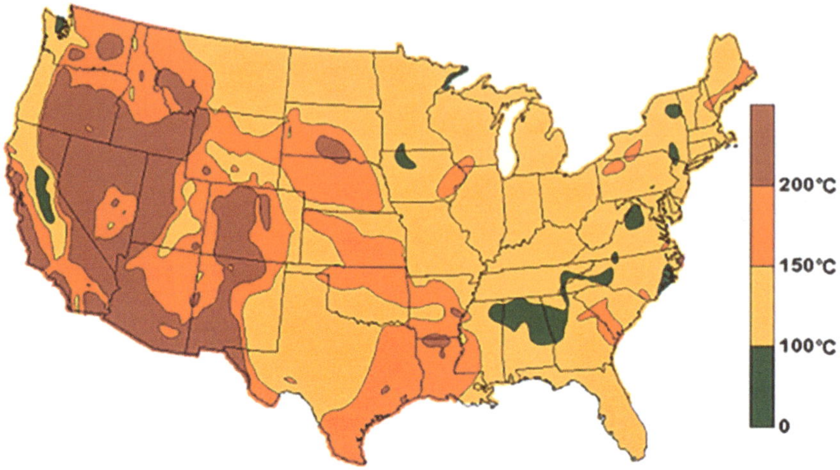

Figure 7.12. [73]Most of the geothermal power plants in the United States are in western states and Hawai'i, where geothermal energy resources are close to the Earth's surface. Newer plant designs will work at temperatures as low as 100°C, opening up a wider range of geothermal resources. Source: U.S. Department of Energy, Office of Energy Efficiency & Renewable Energy.

While geothermal energy can be relatively inexpensive to operate, the initial cost of installing a geothermal system is high. This includes the cost of finding and characterizing the resource, drilling deep wells, and building the necessary infrastructure. The amount of power that can be generated from a geothermal power plant depends on the temperature and flow rate of the geothermal resource. Power generation can fluctuate due to changes in these variables, making it difficult to predict the exact amount of power that will be generated. The test drilling stages of a project therefore present the most significant barrier to geothermal development as the developer is typically confronted with a combination of high risk and high capital expenditures for confirmation activities (Wall, Dobson, & Thomas 2017).

Geothermal systems require more maintenance than other renewable energy sources. This includes regular cleaning of the heat exchanger, replacing the fluid in the system, and ensuring the system is properly sealed. Since EGS technology is similar to natural gas fracking, the same concerns about contamination of groundwater exist for both technologies. Much like oil and gas drilling, some EGS projects have been associated with small earthquakes and problems with wastewater disposal, causing some projects to be abandoned.

Geothermal offers many benefits. It is a relatively clean and reliable renewable source of energy. It offers tremendous potential for expansion; however, even in the most optimistic scenarios for the near future it will likely remain a relatively small fraction of total electricity generation within the U.S. and globally.

[73] https://www.eia.gov/energyexplained/geothermal/where-geothermal-energy-is-found.php

7.2.8 Hydrogen

Hydrogen is plentiful. It is, after all, the most abundant element in the universe. However, on Earth it was, until recently, seldom found in its pure form. Electrolysis or other chemical processes (e.g., steam reforming) are needed to separate hydrogen from water, hydrocarbons, or other molecular hydrogen carriers. This of course requires energy. Hydrogen is often labeled based upon its origin: "Green hydrogen" is produced from the electrolysis of water with clean electricity from renewable energy sources, such as solar or wind. It is green in the sense that no CO_2 is produced in its generation or usage. Currently nearly all hydrogen is produced from fossil fuels because it requires less energy (e.g., the average bond strength in a CH_4 molecule is lower than for H_2O) and is cheaper. However, in this case CO_2 is a by-product. It is described as "blue hydrogen" if the produced CO_2 is captured and stored, and "gray hydrogen" if it is not. However, Howarth and Jacobson (2021) find that the greenhouse gas emissions associated from producing blue hydrogen exceed that of simply using the methane as the source of energy, effectively negating any benefits. Other colors are also used based upon the method of generation. At the end of 2021, almost 47% of global hydrogen production was from natural gas, 27% from coal, and 22% from oil—nearly none of which involved carbon capture. Only around 4% comes from electrolysis, and most of this is not from green energy sources (IRENA 2022).

Currently hydrogen is most commonly used in oil refineries, the chemical industry, and in steel production rather than as a source of energy. Refineries use hydrogen to lower the sulfur content of diesel fuel. Hydrogen is also used to produce ammonia (NH_3) for fertilizer. But it can also be combined with CO_2 to produce hydrocarbons and other synthetic fuels. Recently there has been more interest in using hydrogen as a fuel source. In particular because excess energy generated from renewables could be used to generate hydrogen, in effect using it as a means of storing this excess energy for later use.

Hydrogen can be stored as a gas or liquid; however, gas requires high-pressure storage and liquid requires cryogenic temperatures. The specific energy of hydrogen is 120 MJ kg^{-1}, about three times that of methane and gasoline. On a volumetric basis, however, the situation is the opposite. Methane is about nine times as dense as hydrogen, meaning that it has roughly three times the energy density of hydrogen. And liquid hydrogen has a density of 8 MJ L^{-1}, about a quarter that of gasoline.

Hydrogen can be used as a source of energy in two ways: combustion or fuel cell. Combustion is pretty obvious–when hydrogen is burned water and heat are produced. In principle no CO_2 or other pollutants are generated. However, it is worth noting that burning *anything* in air results in nitrogen oxides (NO_x) from chemical reactions between oxygen and nitrogen in the air, particularly at the temperatures at which hydrogen burns.

A hydrogen fuel cell doesn't use combustion. Instead, it is an electrochemical device that converts hydrogen and oxygen into electricity. The fuel cell is composed of two electrodes, called the anode and cathode, separated by an electrolyte membrane. Hydrogen is fed into the anode, while oxygen is available in the cathode.

The hydrogen ions (protons) are able to pass through the membrane and reach the cathode. The electrons, on the other hand, cannot pass through the membrane. Instead, they flow through an external circuit, creating electrical power. Water and heat are also produced. There are several types of hydrogen fuel cells, including proton exchange membrane (PEM) fuel cells, alkaline fuel cells (AFCs), and solid oxide fuel cells (SOFCs). However, PEMs are the most commonly used for transportation and stationary applications due to their high power density, quick start-up time, and relatively efficient operation. Hydrogen fuel cells have efficiencies of 40–60%, which is better than internal-combustion engines but not as good as battery-powered electric.

One of the main disadvantages of hydrogen power is its high cost of production and storage. Currently green hydrogen is two to three times more expensive to produce than fossil fuel equivalents; e.g., natural gas (IRENA 2022). However, there is now evidence that there may be large stores of naturally occurring H_2 underground, which is continually produced by water–rock reactions. Preliminary results indicate that there may be enough geologic H_2 to meet the growing global demand (Ellis & Gelman 2022).

Even if geologic H_2 greatly reduces the cost of production, the distribution of hydrogen is difficult because of its low volumetric energy density and the need for high-pressure containers. Hydrogen gas also causes steel to become brittle, which is a problem for equipment and pipelines. Because it is less dense than methane, hydrogen is less energy efficient to transport. It takes about three times as much energy to compress the same amount of heat energy if you supply it as hydrogen instead of natural gas. Currently most hydrogen production is therefore done very close to where it is used. The limited infrastructure and distribution network for hydrogen makes it difficult to access in many areas, limiting widespread adoption. Another challenge with hydrogen is its potential for leaks. H_2 is a very small molecule with low viscosity, and is therefore more prone to leakage than methane. Hydrogen is a safety risk because it is a highly flammable gas. It is also odorless and colorless, making it difficult to detect.

Hydrogen can in principle be used to replace fossil fuels in combustion systems, including internal-combustion engines. However, because of its low volumetric energy density, more of it is needed to generate the equivalent energy. And since heat engines are already inherently inefficient, it is usually more effective to use fuel cell technology.[74] Mixing hydrogen and methane is a possibility but the benefits are less than you might expect. For example, a 20% mixture of H_2 into natural gas by volume would result in only 7% less CO_2 emission to generate the same amount of energy.[75] Hydrogen burns at a temperature about 250°C higher than CH_4. Adding H_2 therefore increases the overall burn temperature, resulting in higher NO_x emissions.

While hydrogen can in principle be used to store excess energy generated by other means, there is considerable energy loss incurred by doing so. The "round trip"

[74] https://www.youtube.com/watch?v=1Ajq46qHp0c

[75] https://www.linkedin.com/pulse/hydrogen-replace-natural-gas-numbers-paul-martin/

efficiency of conversion of electricity into hydrogen and back is less than 30% (Escamilla et al. 2022); i.e., more than two thirds of the energy is lost, making this route unattractive for most purposes. Another problem is that, once produced, there is no way to differentiate green hydrogen from others—meaning it would be difficult to track and therefore incentivize green hydrogen production.

Hydrogen is a fuel source that is being considered for parts of the energy sector that cannot be easily electrified–including aviation, shipping, and heavy machinery. However, it also faces several significant challenges, including high costs, limited infrastructure, potential safety risks, and the potential for negative environmental impacts. These challenges must be addressed if hydrogen power is to become a viable and widespread alternative energy source. In fact, electrification has now superseded much of the originally envisaged hydrogen economy; and thus only a few, if important, roles remain for it in the transition to a global net-zero emissions economy (Whitehead et al. 2023), although the discovery of widespread geologic H_2 could change that perception.

7.2.9 Potential Energy Sources in the Future

The energy sources discussed above all have matured to the point where they are currently deployed to some degree and require relatively little, if any, innovation to become significant contributors to global energy generation. There are other energy sources under development that may also someday become relevant. It is worth noting that for the most part these technologies will not be ready to contribute to the rapid decarbonization that needs to happen over the next decade or two. But they may someday be contributors to a "net zero" energy mix.

7.2.9.1 Ocean Energy

"Ocean energy" refers to the generation of electricity from the movement of water. These devices can be placed on the surface of the ocean or submerged. Tidal energy production involves harnessing the kinetic energy of water moving from the tides, whether it be the rise and fall of water levels or the motion of water in and out of inlets. One example is the use of tidal turbines, which are similar to wind turbines but are designed to operate underwater. Wave energy technology uses wave energy converters (WECs) to capture the energy of ocean waves. There are several types of WECs, including point absorbers, terminators, and oscillating water columns.

Ocean energy offers several advantages. The total marine energy resource in the U.S. is equivalent to half of the electricity used in 2019 (Kilcher et al. 2021). Like wind and solar energy, it is a renewable energy source that produces no emissions during operation. Unlike wind and solar, ocean energy is highly consistent. Tides follow a predictable pattern, and the amount of energy that can be generated is directly proportional to the height of the tide. This makes it easier to forecast and plan for tidal energy production. Waves are created when wind blows across a water surface, so tidal energy is more reliant on weather patterns–but less so than wind turbines. In general ocean energy does not generate much noise, and they have a relatively low visual impact.

However, ocean energy production also faces a number of challenges. The cost of construction and maintenance is currently high. The ocean is a difficult environment in which to operate because of the strong forces on equipment and corrosion, making tidal energy production more expensive than other forms of renewable energy. Wave energy is widely available, as many of the world's largest cities are close to the ocean. However, there are a limited number of locations suitable for tidal energy production, as it is only possible in areas with strong tidal currents. The impact on marine life and coastal habitats also needs to be considered. Risks include the potential for turbines and WECs to cause disruption from emissions of underwater noise, generation of electromagnetic fields, changes in habitats, changes in oceanographic processes, and entanglement of large marine animals (Copping et al. 2020). The technology for ocean energy is still in its early stages, and further research and development is needed to make it safer, more efficient, and cost-effective.

7.2.9.2 Nuclear Fusion

Perhaps the holy grail of all energy generation is that of nuclear fusion. This is of course the energy source for the Sun, and stars in general. Nuclear fusion offers the promise of nearly limitless energy at potentially very low operational costs. Fusion fuel sources have an energy density more than four times higher than that used for fission. Advocates often point out that compared to nuclear fission, fusion has minimal to no risk of radioactivity, radioactive waste, or nuclear proliferation.

However, fusion on Earth is significantly different than the Sun. The dominant nuclear reaction in the Sun is the "proton–proton" chain reaction, the first step of which is:

$$p + p \rightarrow {}^{2}\mathrm{H} + \mathrm{e}^{+} + \nu_e$$

This reaction requires a minimum energy of $kT = 1.335$ keV, or a temperature of 15.5 million K, which along with quantum tunneling allows the Coulomb repulsion of the two protons to be overcome. Subsequent intermediate steps consume the deuterons to produce ^{3}He with the overall reaction consisting of:

$$4p \rightarrow {}^{4}\mathrm{He} + 26.1\ \mathrm{MeV}$$

Whereas solar fusion reactions actually fuse two nuclei together to produce a final, single nucleus, the terrestrial fusion channels typically produce two particles, making these "exchange" reactions as opposed to "true" fusion (Morse 2018):

$$\begin{array}{lll}
D + T & \rightarrow n + \alpha & + 26.1\,\mathrm{MeV} \\
D + {}^{3}\mathrm{He} & \rightarrow p + \alpha & + 18.3\,\mathrm{MeV} \\
D + D & \rightarrow p + T & + 4.03\ \mathrm{MeV}\ (50\%) \\
 & \rightarrow n + {}^{3}\mathrm{He} & + 3.27\ \mathrm{MeV}\ (50\%) \\
T + T & \rightarrow \alpha + 2n & + 11.3\ \mathrm{MeV} \\
{}^{3}\mathrm{He} + T & \rightarrow \alpha + p + n & + 12.1\ \mathrm{MeV}\ (51\%) \\
 & \rightarrow \alpha + D & + 9.5\ \mathrm{MeV}\ (43\%) \\
 & \rightarrow {}^{5}\mathrm{He} + p & + 9.5\ \mathrm{MeV}\ (6\%)
\end{array}$$

When there are two daughter particles, conservation of momentum means that the energy deposited is inversely proportional to the particle mass. For example, in the D–T (deuterium–tritium) reaction, which releases 17.6 MeV, 14.08 MeV is delivered to the neutron while only 3.52 MeV is left to the helium nucleus. In general neutrons receive 80% of the energy in terrestrial fusion reactions, in contrast to fusion in the Sun where no neutrons are produced. As a result, the energy of the neutrons has to be captured in a surrounding blanket that shields the rest of the reactor from the neutron and gamma ray flux. To generate power, energy captured by the blanket heats water circulating in pipes, which turns into the steam that drives turbines to produce electricity. If the blanket is made of Li-6, it can further act as a breeder layer to produce new tritium fuel via the exothermic reaction:

$$n + {}^{6}\mathrm{Li} \rightarrow T + {}^{4}\mathrm{He} + 4.78\ \mathrm{MeV}$$

Since each D–T reaction produces only one neutron which can go on to create a single triton, a breeder reactor will not be able to produce enough tritium to self-fuel itself as not all neutrons can be captured. A lithium blanket that contains other elements, like beryllium or lead that release two neutrons for each captured neutron, could be used as a neutron multiplier.

One of the bottlenecks for nuclear fusion is the availability of tritium. Tritium has a half-life of 12.32 years, so cannot be found in appreciable quantities in nature. For now, it must be produced in fission nuclear reactors. The Canada Deuterium Uranium (CANDU) reactors currently produce 0.5 kg yr^{-1}, and there is a global stockpile of ~25 kg (Clery 2022). A commercial fusion reactor would use up ~56 kg of tritium per gigawatt of electricity generated, so in order to be viable, D–T fusion reactors must breed tritium. Lithium breeding will be experimentally tested at the International Thermonuclear Experimental Reactor (ITER), which is currently under construction in southern France. It is expected to go online with the first plasma in 2025, and start D–T reactions not before 2035. There is still uncertainty about how much tritium can actually be bred operationally, with simulations showing that the excess tritium generated might only be 5–15% of the input tritium (Abdou et al. 2021). Whether there is any excess tritium at all will depend on what fraction of the original tritium is actually used up as fuel, in addition to the fraction of newly-formed tritium that can be recovered from the blanket, versus tritium that is lost through diffusion through the solid reactor walls. Simply put, if self-sustainability in tritium production is not feasible then neither is commercial fusion power.

The ITER fusion reactor has a goal to continuously generate 500 MW of power. It is an example of magnetic confinement fusion (MCF), where the plasma is constrained by magnetic fields–in this case, inside a toroidal "tokamak" (a word from the Russian acronym describing the original design). Whether a fusion power plant is successful is determined in part by three different parameters: the neutron production rate, the ratio of energy generated from the plasma to the energy put into the plasma (Q), and the "triple product" of plasma density, ion temperature, and confinement time, $nT\tau$–one of the variables that determines Q.

Because 80% of the energy released is lost via neutrons, the plasma heating has to be sustained by the 20% of the energy retained by the alpha particles. For the plasma to maintain its temperature in order to continue fusion reactions, Q must therefore be at least 5. For ITER, the additional external heating supplied to the plasma is expected to be 50 MW, so $Q = 10$. With the efficiencies expected for power generation versus the power required to sustain plasma ignition, the expected energy output roughly balances the input energy during operations: "ITER is about equivalent to a net (zero) power reactor when the plasma is burning … the minimum required for a convincing proof of principle."[76] At best, ITER will be able to demonstrate the net production of a few kilowatts of power (Jassby 2021).

In tokamaks, the triple product reached a peak in 1998 (Wurzel & Hu 2022). Although the length of the confinement times have grown to tens of seconds since then, the triple product has been decreasing even as the confinement times lengthen. This has led to some opinions that the future of terrestrial fusion is not in MCF but in inertial confinement fusion (ICF), where lasers are indirectly used to ignite fusion.[77] In this case, the laser target is a "hohlraum" (German for "cavity"), a gold chamber enclosing the D–T fuel that generates x-ray photons when irradiated. As the x-rays escape, the fuel pellet implodes via a rocket effect, increasing its density by 1000 times (Atzeni et al. 2022). In the past quarter century, ICF efficiencies have steadily progressed from $Q < 0.01$ to $Q \sim 1$ today.

In 2022 a milestone was achieved when, for the first time, the U.S. National Ignition Facility (NIF) produced more energy than the energy delivered to the plasma. NIF used 192 lasers to deliver 2.05 MJ of energy, resulting in 3.15 MJ of output (Clery 2022) from a millimeter-sized fuel pellet containing <0.002 g of tritium (Draggoo 2013). Because the NIF was designed for nuclear weapons research (Mecklin 2022) and not with the intent of generating power, its lasers were not optimized for fusion power research. With ~1% efficiency, the lasers consumed 322 MJ of energy for the shot (Tollefson & Gibney 2022). The fusion reactions generated more than 10 quadrillion Watts, if only for a few nanoseconds.

Some private fusion companies are focusing on D–^{3}He reactions, which do not produce neutrons. However, in order to achieve similar reaction rates as D–T fusion based on collisional cross-sections (Figure 7.13), the plasma temperature has to be raised by an order of magnitude, or up to ~1 billion K. ^{3}He is also a rare isotope on Earth, although it is 30,000 times more abundant on the Moon's surface (Fa & Jin 2007). Close (2007) points out that in the terrestrial reactor plasma environments where all of the different isotopes are intermingled, the deuterium in a D–^{3}He fuel mix will more likely interact with itself because of the higher D–D cross-section at the currently achievable plasma temperatures. The tritium that is produced has an even higher cross-section with deuterium, so a reactor that starts out with a D–^{3}He fuel mix effectively turns into a D–T reactor.

Because neutrons are an inescapable reality for terrestrial fusion for the time being, they negate the benefits that have been highlighted for fusion power.[78] Leakage of

[76] https://web.archive.org/web/20230413200303/https://www.jt60sa.org/b/FAQ/EE2.htm
[77] https://doi.org/10.37282/991819.22.30
[78] https://thebulletin.org/2017/04/fusion-reactors-not-what-theyre-cracked-up-to-be/

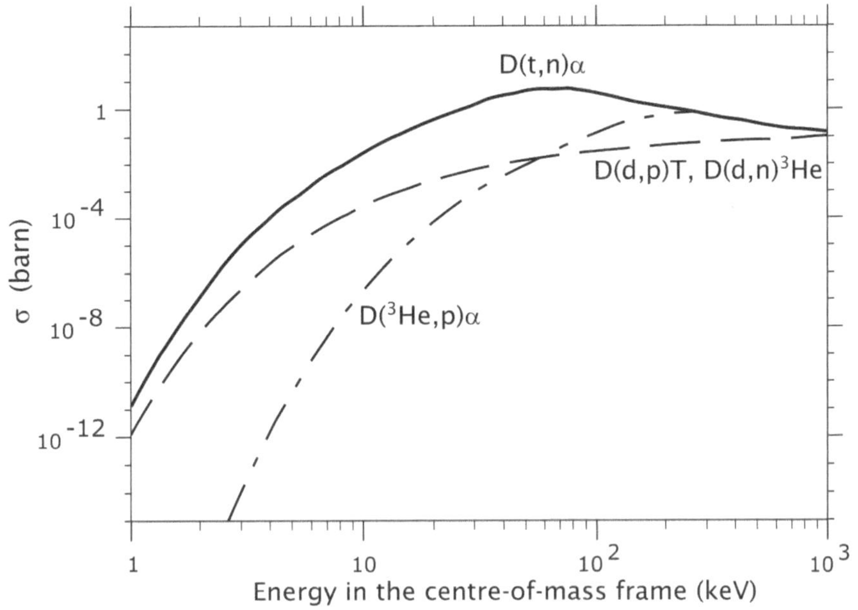

Figure 7.13. [79]Cross-section (1 barn = 10^{-24} cm^{-2}) as a function of energy of the colliding particles in a center-of-mass frame for the terrestrial fusion reactions mentioned in the text. In the reaction nomenclature, the first two letters represent the input particles, and the final two letters represent the output particles. Hence, "D(d,p) T" means that two deuterons interact to produce a proton and a triton. (From Bosch & Hale 1992).

neutrons has a deleterious effect on the steel enclosures, turning them brittle and sponge-like, filled with small cavities that weaken the metal. Irradiation of the surrounding infrastructure will lead to radioactive waste when these components have to be replaced. There is also a pathway to creating weapons-grade Pu-239 if U-238 is placed in the path of the energetic neutrons.

Fusion is still far from being a commercially viable energy source. ITER will not have D–T reactions until 2035 at the earliest and, as discussed earlier, will produce only a few kilowatts of energy at best. While the recent success at NIF has spurred additional focus on ICF (Clery 2023), NIF itself can only ignite one shot per day, far less than the 10 s^{-1} rate necessary for a commercial power plant. Progress towards a ICF power plant will require the development of new fabrication techniques to create fuel targets cheaply, and new technologies, including mechanisms to remove spent pellets and replace them with fresh fuel precisely inside the laser crosshairs multiple times per second, and a liquid metal waterfall that absorbs neutrons to breed tritium.[80] According to the 2023 report[81] of the Fusion Industry Association, there are 43 fusion companies that have obtained over $4.7 billion of private funding to date—indicating there's at least hopeful enthusiasm that alternatives to the

[79] https://doi.org/10.1088/0029-5515/32/4/I07
[80] https://doi.org/10.37282/991819.22.60
[81] https://www.fusionindustryassociation.org/wp-content/uploads/2023/07/FIA%E2%80%932023-FINAL.pdf

current MCF and ICF designs can be successful sooner than the 2030s. However, the announcements of new projects should be treated with caution since the commercial fusion startup industry has a history of over-promising for years without being able to show even the basic technical milestone of neutron production (Jassby 2019). There's a long-running joke that commercial fusion power is 20 to 30 to 50 years away, and always will be—which still appears to be true. We may therefore be best served to rely on the Sun—the only sustained fusion reactor to which we currently have access.

7.3 Carbon Capture

Carbon capture refers to a variety of different technologies used to capture, transport, store, and utilize CO_2. Although these processes include carbon capture and sequestration (CCS) and carbon capture and utilization (CCU) from power and industrial plants, we will use the term CCS to refer generally to any sort of carbon capture and storage from the emission source, whether the carbon ends up permanently sequestered or used in an industrial process. We distinguish these from carbon dioxide removal (CDR) or negative emissions technologies (NETs) which refer to methods that draw down and store CO_2 that has already been released into the environment.

CCS was first developed in the U.S. in the 1970s to remove CO_2 from natural gas in a processing plant before the gas could be sold. This CO_2 is then utilized for enhanced oil recovery (EOR): pressurized CO_2 is pumped underground where it mobilizes oil and gas that would otherwise not be extractable. It is still the most successful industrial use of CCS, and accounts for 80–90% of all captured carbon in the last 50 years.[82] Although part of the CO_2 used in EOR that would have been released is permanently sequestered in the rock formation, there is nonetheless a net carbon release from the burning of the newly liberated hydrocarbons.[83] CCS can be retrofitted to existing coal power plants to remove CO_2 from the post-combustion "flue emissions." CCS can also remove carbon from chemical manufacturing and from hard-to-decarbonize industries such as steel and cement production (Loria & Bright 2021); The first dedicated geological carbon sequestration project in the U.S. from CCS is at an Illinois ethanol plant, where CO_2 is emitted as part of the biomass fermentation process.

There are different kinds of CCS technologies, but the most technologically mature involves a solution of an amine chemical that is circulated through a "scrubber," where it comes into contact with flue gases, and binds chemically with the CO_2. The acidic solution is later heated by steam which strips the CO_2 from the amine. Water is condensed out of the CO_2-rich steam, and the concentrated CO_2 leaves as a compressed liquid. The amine solution is cooled and recycled back to the

[82] https://ieefa.org/resources/carbon-capture-crux-lessons-learned

[83] The exception is when the CO_2 used for EOR comes from biomass or from direct air capture, in which case, it is possible for there to be a net sequestration of carbon even when the carbon released from burning the newly liberated fossil fuel is included. https://www.iea.org/commentaries/can-co2-eor-really-provide-carbon-negative-oil

scrubber to absorb more CO_2. Because of the additional energy required to create the steam and to run the compressor, the amine process uses up part of the power generated by the plant, reducing its electricity output. For a coal power plant, this energy penalty is typically about 24%, which can increase the cost of electricity by 43% above the amount for a baseline plant without CCS (Rubin et al. 2015).

After the CO_2 is captured it can be stored in underground geological formations, such as depleted oil and gas reservoirs, deep saline aquifers, or unmineable coal seams. At a depth of 800 m, the hydrostatic pressure is 80 bar, which is above the minimum pressure necessary to keep CO_2 in a liquified state that can be trapped in the pores of sedimentary strata. However, even at 3 km below the surface, the specific gravity of CO_2 is still less than 1, meaning it is buoyant compared to any surrounding groundwater. The storage reservoir must not only be porous to the injected carbon, but it is also chosen to be under an impermeable caprock that the buoyant CO_2 plume cannot penetrate. This structural trapping along with other physical and chemical trapping mechanisms ensure that the CO_2 remains underground for thousands of years. The possibility of leakage via natural faults or defective wells means that such reservoirs also need to be monitored (IPCC 2005). Another option is mineral carbonation, by which CO_2 is permanently stored through a chemical reaction with naturally occurring minerals such as magnesium silicates (e.g., olivine) that can form stable carbonates with negligible risk of the CO_2 returning to the atmosphere. This process occurs naturally in the Earth's mantle and is one of the primary mechanisms by which carbon is cycled between the mantle, the atmosphere, and oceans. To accelerate mineral carbonation, captured CO_2 can be injected into a mineral-rich environment such as ground-up mine tailings or basalt formations. Carbon mineralization in basaltic rocks offers a global storage potential that exceeds anthropogenic emissions, although the process can be water intensive (Snæbjörnsdóttir et al. 2020).

A potential alternative to simply sequestering CO_2 is to use it to produce marketable compounds such as synthetic fuels that could be carbon neutral, if not carbon negative. This could in principle be a viable source of revenue. However, currently the costs associated with doing so are not economically viable (e.g., Redissi & Bouallou 2013); and the CO_2 necessary to meet the global demand for such products is minuscule compared to current levels of CO_2 emissions (Bui et al. 2018).

Although a small amount of U.S. federal funding for R&D appeared in the late 1990s,[84] CCS did not grow significantly until tax credits became available in 2008.[85] This led to a burst of investments in CCS in the U.S. (with a parallel wave of support elsewhere in the world following commitments made at the 2009 COP15 climate conference.) However, political support was weak, and most of the global funding promised for CCS was not delivered for a variety of reasons, including public and NGO opposition, long timelines, and high capital and operating costs (Lipponen

[84] Truly an awful decade for music.

[85] https://www.globalccsinstitute.com/resources/publications-reports-research/surveying-the-u-s-federal-ccs-policy-landscape-in-2021/

et al. 2017). Since 2009, CCS projects that have received government funding have also suffered a high rate of failure. Two thirds of the 149 CCS projects worldwide designed to geologically store carbon underground and were supposed to be in operation by 2020 were never completed or are on hold (Abdulla et al. 2021). While capture for EOR activities is still the most successful, with 70% of CCS proposed for gas processing plants still in operation, CCS for coal-fired power plants has a 90% failure rate. This is typically attributed to the declining price of natural gas which made coal power plants increasingly expensive to operate. Retrofitting such a plant with CCS made it even more complicated and less competitive to run. As a result, none of the coal CCS projects funded by the U.S. federal government in 2010–2017 are in operation (GAO 2022). By comparison, the same federal subsidy program also funded three industrial CCS projects with two completed and still running (while the third was withdrawn during the design stage).

The primary reason for the failure of such projects is that there has been little incentive to improve the technologies to sequester CO_2. Even if a technology is proven in the lab or at a pilot facility, it cannot be deployed at large, industrial scales unless it goes through a series of demonstration phases, where learnings from building and operating a plant at an earlier phase are used to improve the design, construction, and operation of the subsequent phases. It is expected that at least two cycles of development are needed before CCS can be rolled out at scale (Gibbins & Chalmers 2008). CCS has large capital costs, whereas historically, there has been little or no financial deterrence for releasing emissions into the atmosphere. This has resulted in a slow pace of CCS development and deployment, with the projected storage of CO_2 in coming decades an order of magnitude less than what is needed in mitigation scenarios meant to keep the global temperature rise to 1.5°C (Martin-Roberts et al. 2021).

Adopting CCS could be incentivized, for example, with a carbon tax or tax credit. In their empirical study of why 39 CCS projects in the U.S. failed or succeeded, Abdulla et al. (2021) showed the most important variables in determining whether a CCS project failed were tied to financial risk. Greater capital costs, using new untested technologies, and lacking a secure revenue stream for selling or sequestering the captured carbon were all risk factors that significantly increase the possibility of failure. A CCS project could be technologically feasible, but the market forces can make it impractical on a commercial scale. The Petra Nova coal-fired power plant opened in 2017 outside of Houston as the first commercial-scale CCS in the U.S., and delivered its captured CO_2 for EOR and sequestration in a nearby oil field. Although the plant was bedeviled by mechanical failures that resulted in frequent shutdowns, and its CO_2 capture rate was less than the promised 90%,[86] what caused its closure in May 2020 was the drop in oil prices after the onset of the COVID-19 pandemic, which made its EOR operations financially untenable.[87]

The CCS activities mentioned here reduce new carbon emissions from fossil fuel power plants, but do nothing for CO_2 already released into the environment.

[86] https://ieefa.org/resources/ill-fated-petra-nova-ccs-project-nrg-energy-throws-towel
[87] https://www.nrg.com/about/newsroom/2020/petra-nova-status-update.html

For that, we turn to negative emissions technologies, the most prominent of which might be the nascent industry of direct air capture (DAC), where CO_2 is pulled directly from the atmosphere. DAC is technically feasible, but is more challenging than CCS located at a power plant because the density of CO_2 in the atmosphere is less than 1% that of flue gases. Hence, large volumes of air must be processed, even before it is compressed and permanently stored, and this requires large, capital-intensive structures. A DAC plant that captures a million tons of CO_2 per year would require 46,000 m^2 of cross-sectional area for air flow (Herzog 2022). Electricity must be provided to power the fans that move the air over contactors where amine embedded in porous filters can extract the CO_2 from the inlet air source. A heat source is also necessary to release the CO_2 for storage, and regenerate the amine so that it can return to capturing carbon. The Swiss-based Climeworks is running a pilot DAC plant in Iceland that runs on geothermal energy, and is selling negative emissions for \$1300/ton CO_2. In his analysis, Herzog (2022) believes that the wide variety of factors that affect the cost of siting DAC plants means that their cost of net CO_2 removal will always be in a broad range, with \$600–1000 per ton possible by 2030. It is uncertain whether DAC can be scaled up economically to the billion ton per year removal rate needed to have any impact on climate change.

Another strategy is to sequester carbon from the oceans (National Academies of Sciences, Engineering, & Medicine 2022), which are already the Earth's main "carbon sink," absorbing 30 to 40% of anthropogenic emissions. The density of CO_2 in ocean water (primarily in the form of bicarbonate ions HCO_3^-) is 147 times greater than in the same volume of air. Thus removal of CO_2 from ocean water could be more efficient. One group of marine carbon dioxide removal (mCDR) approaches involve increasing "primary productivity," or growth of algae, plants, and bacteria at the surface of the oceans. These organisms would grow, absorb CO_2 from surface waters, and then sink to the sea bottom after they die, taking the carbon with them. Stimulating primary productivity could involve spreading iron-rich dust onto the oceans as a fertilizer (Martínez-Garcia et al. 2011), using pumps to artificially upwell nutrient-rich deep waters to the surface (Oschlies et al. 2010), or cultivating seaweed that is later sunk into the deep ocean (Silverman-Roati et al. 2022). Another approach proposes adding alkaline materials, such as lime or the mineral olivine, to the ocean, which increases its ability to absorb atmospheric CO_2 with the additional benefit of reducing ocean acidification (Renforth & Henderson 2017). Others propose to use electrochemical means to degas CO_2 or precipitate calcium carbonate ($CaCO_3$) from seawater (Eisaman 2020). There are also opportunities to preserve, restore, and sustainably manage coastal ecosystems, such as mangroves, seagrasses, and tidal marshes. Tidal inundation brings in sediment from outside the ecosystem boundaries, and carbon is incorporated into the coastal biomass and into the underlying soils. Because the soils accrete vertically with rising sea levels, coastal areas do not become saturated with carbon unlike their terrestrial counterparts. Despite having an area 1% that of terrestrial forests, they have comparable carbon capacity, since their storage rates per unit area are almost two orders of magnitude greater than even tropical rain forests (Mcleod et al. 2011).

One of the most visible, and popular, natural strategies for terrestrial carbon removal is the planting of trees and the management of forests to increase their capacity to absorb and store CO_2. This includes reforestation of land that had recently lost trees, as well as afforestation of land that did not have any trees in historic times. Some agricultural practices, such as conservation tillage, cover cropping, and the conversion of cropland to grassland or forest, can increase the amount of carbon stored in the soil. Bioenergy with carbon capture and storage (BECCS) is a carbon-negative energy source wherein biomass is used to generate energy, with the resultant CO_2 emissions captured and stored instead of being released into the atmosphere. One estimate of the maximum amount of global carbon storage that is possible after aggressively increasing a broad range of restoration, conservation, and land management practices is ~24 gigatons of CO_2 yr^{-1}(Griscom et al. 2017).

Unfortunately, methods of natural sequestration also pose risks. For example, pests, fires, storms, and land use changes can release the CO_2 stored in flora. Some of these risks are increasing as global temperatures rise, and they are not small liabilities. For instance, van der Velde et al. (2021) estimated that 715 million tons of CO_2 were released during the extensive wildfires in Australia during the 2019–2020 summer season, which was more than the entire country emits in a typical year from fossil fuels. van Oldenborgh et al. (2021) estimate that climate change made Australia's devastating fire season 30% more likely, illustrating the forest fires–CO_2 positive feedback loop and the risk of relying on them as a form of carbon sequestration. Fortunately in this case much of that excess CO_2 was absorbed by phytoplankton in the Southern Ocean downwind of Australia (Tang et al. 2021), although this too is not without problems, such as shifting the seasonal productivity of phytoplankton with consequences for the marine food web (Yamaguchi et al. 2022).

Natural storage also has constraints, such as competition from other ways that land is used (e.g., afforestation of grasslands currently used for grazing or other food production; Griscom et al. 2017); the reduced capacity for storage as natural carbon sinks slowly saturate over time; and the uncertainties about carbon fluxes between future natural terrestrial sinks and the atmosphere (Nolan et al. 2021). Because the carbon cycle consists of a complicated interplay between terrestrial, marine, and atmosphere reservoirs of carbon, future global cooling from the drawdown of atmospheric CO_2 can result in interactions between these different components. Thus, removing 1 gigaton of CO_2 from the atmosphere may not fully reduce atmospheric CO_2 by 1 gigaton, due to feedbacks that cause carbon to be released from land and ocean sinks (Keller et al. 2018).

Third party monitoring, reporting, and verification (MRV) is required to determine how much carbon has been removed, and whether it has been durably stored so that it cannot easily return to the atmosphere. Standards need to be defined for removal and storage, while certifications are created to ensure that the standards are met. As of 2022, there are 30 organizations that have developed 125 standards for CDR across 23 different CDR activities, and 27 different certifications (Arcusa & Sprenkle-Hyppolite 2022). Some CDR activities, such as afforestation, reforestation, and improving forest management, have multiple competing standards, while nine CDR approaches (including ocean fertilization and ocean liming) have no associated standards at all. Much work still needs to be done to reconcile the different standards, including ensuring that the amount of carbon sequestered matches independent verification methods. For example, the California Air Resources Board (CARB) reforestation standard uses satellite imaging, ground-truthed tree inventories, and simple modeling to determine the amount of carbon stored compared to a baseline with no human intervention. However, because the actual CO_2 fluxes between the atmosphere, forest, and soil are not directly measured, there can be dramatic inconsistencies between what the standard claims was stored versus an assessment that also quantifies the net exchange of carbon within the ecosystem (Marino et al. 2019). In contrast to the wide variety of MRV standards for terrestrial CDR, none exist for mCDR (Arcusa & Sprenkle-Hyppolite 2022). This shortfall is slowly being addressed, as workshops are held to discuss how to create new MRV frameworks for marine storage[88] as well as via U.S. government funding to help develop new MRV technologies.[89]

In general, most carbon credits sold for terrestrial natural sequestration are for biomass storage, where carbon is locked for ~100 yr. These credits are cheaper than credits for storing CO_2 in rocks and minerals, which can effectively lock carbon away for >1000 yr. Because the storage solutions with longer timescales are not as mature, they will require more investments to further their development to bring down their costs. Such "geosphere"-based storage will help fill in future storage gaps, as natural reservoirs saturate and become more expensive in coming decades (Joppa et al. 2021).

Carbon capture is just one part of the solutions portfolio necessary to limit warming to 1.5°C in all simulated global emission reduction pathways. Rapid reductions in carbon emissions still remain necessary. Because CDR at scale is unproven, it is risky to rely on its future deployment in order to achieve the goal of 1.5°C of warming, especially since there are considerable uncertainties about the amount of environmental and social disruptions such large-scale implementations could entail (IPCC 2022). Carbon removal therefore cannot be used as an excuse to avoid reducing our use of fossil fuels. Faster rates of decarbonization today means less, but not zero, reliance on future CDR. CDR is still expected to be needed in coming decades to mitigate hard-to-decarbonize industries, and to provide negative emissions in order to return warming to 1.5°C in case of overshoot. Of the current 2 gigatons CO_2 yr^{-1} removed globally by CDR, most is via conventional management of land (e.g., forestry measures) and only 0.1% is from new negative emission technologies (Smith et al. 2023). Increasing the conventional methods of CDR will

[88] https://www.us-ocb.org/marine-co2-removal-workshop/; https://www.youtube.com/playlist?list=PL2JK_uZ15iZCjCh89WsGJ0BH1dJLbCf2T

[89] https://arpa-e.energy.gov/news-and-media/press-releases/us-department-energy-announces-45-million-validate-marine-carbon

be constrained by other land usage, such as for food production and preserving biodiversity, so investments are needed to develop novel carbon removal technologies that can increase the global net removal of ~10 gigaton CO_2 yr^{-1} by 2050, and twice that rate by 2100 (National Academies of Sciences, Engineering, and Medicine 2019). In order for these novel technologies to scale up to large-scale deployments by 2050, policymakers have to balance incentives for their development, while continuing to pursue emission reductions given the range of uncertainties on when and at what scale these CDR technologies will be viable (EEA 2023).

7.4 Energy Storage

When one thinks of energy storage, batteries are usually what first comes to mind. The rapid growth in sales of electric vehicles (EVs) is increasing the demand for battery storage. Likewise, as non-dispatchable energy sources such as wind and solar become more widely adopted, it will be necessary to store excess generated energy that can be dispatched during peak periods of use. Batteries store electrical energy in chemical form. Several factors are relevant in battery design, such as energy storage density (by mass and by volume), discharge rates (i.e., power generated), time to recharge, cost and materials required for construction, degradation over the operational lifetime, and safety. Examples include lead-acid (Pb-acid), nickel–cadmium (Ni–Cad), and lithium-ion (Li-ion) batteries. Each has strengths and weaknesses; e.g., Pb-acid batteries have a lower energy density than Li-ion, but are more readily recyclable and are not flammable. Batteries also can have different ideal operational conditions; e.g., lithium iron phosphate batteries work best when regularly charged to 100%, whereas most Li-ion batteries can lose capacity over time if used in this manner. Ni–Cad batteries have a memory effect that requires complete charging and discharging before starting a new cycle, whereas Li-ion batteries can be damaged by a complete discharge.

Li-ion batteries are currently the most popular. They are used primarily in small-scale consumer products such as cellular phones and laptop computers, where energy density is a priority to minimize weight. They are also the most commonly used battery in EVs and grid-scale energy storage, currently making up more than 90% of both markets. They are projected to remain the most popular battery type for the foreseeable future because of the maturity of the technology. And growth of the EV market has driven down costs. The price of Li-ion cells in 2020, scaled by their energy capacity, has declined by more than a factor of thirty since their commercial introduction in 1991 (Ziegler & Trancik 2021); and their prices are expected to continue to drop another 25–70% by 2030 (Cole et al. 2021). The Li-ion batteries used for EVs are the same as those for grid-scale energy, meaning both sectors will benefit from price drops.

A marked strength of batteries is their fast response times for discharge, on the order of milliseconds. They are also efficient, typically having round-trip efficiencies of 70–80% for Pb-Acid and 85 to 95% for Li-ion (Kebede et al. 2022). They are ideal for uses on the timescales of hours, such as for use in EVs and to balance unpredictably variable energy production from wind and solar as well as changes in demand. Because of their high transfer efficiencies, batteries can be used in sophisticated ways. For example, bidirectional charging, also known as vehicle-to-grid (V2G) technology, refers to the ability of an EV to not only draw power from the grid to recharge its battery, but also to return power back to the grid when it is not being used. This technology allows electric vehicles to serve as a source of energy storage as well as provide grid services, such as frequency regulation and load balancing, which can help to improve the stability and reliability of the grid. EV batteries typically hold enough energy to provide back-up power to an average U.S. household for two days.

Batteries are not ideal for all applications. In general batteries lose charge over time, limiting their use for long-term storage. They also tend to lose their capacity over many cycles of charging and discharge. They also represent a significant safety risk. Li-ion batteries can be damaged if exposed to high temperature (>55°C) or if charged at low temperature (<0°C). They can also overheat, catch fire, or explode if they are damaged or charged improperly. Heat released during cell failure can damage nearby cells, releasing more heat in a chain reaction known as a thermal runaway. For these reasons Li-ion batteries have additional design considerations that increase their weight and cost. Because of the fire risk, transport of Li-ion batteries on aircraft is restricted. Other types of batteries have safety risks as well; e.g., during charging Pb-acid batteries produce flammable hydrogen and oxygen gases. They also contain sulfuric acid that can be dangerous if leaked.

There are other battery designs possible. For example, "flow batteries" consist of two tanks of electrolytes that are pumped into a membrane interface where they generate electricity. They operate in a similar way to fuel cells, and are therefore sometimes referred to as regenerative fuel cells. In contrast with conventional batteries, flow batteries store energy in the electrolyte solutions. They can therefore be recharged simply by replacing the electrolyte fluid. Flow batteries can be built in a wide range of capacities, from 25 kW to more than 100 MW, scaling simply by the amount of electrolyte used and the size of the membrane interface. Flow batteries can release energy continuously at a high rate of discharge for up to 10 hours (Chen et al. 2022), but typically at lower discharge rates than conventional batteries. Unlike most conventional batteries they are also well suited for longer-term, "seasonal" storage. Unfortunately they have round-trip energy efficiencies of 40–60% (e.g., Sánchez-Díez et al. 2021), lower than that of most conventional batteries. Flow batteries are starting to see use in large-scale projects. For example, construction was completed in 2022 on a 100 MW/400 MWh vanadium redox flow battery in China.[90]

From 2022 to 2030, the energy storage market in the U.S. is expected to grow to 3–5 times in size (DOE 2020). One of the major concerns with this rapid growth is the availability of the materials needed. For example, a single car lithium-ion battery pack (of a type known as NMC532) contains around 8 kg of lithium, 35 kg nickel, 20 kg manganese and 14 kg cobalt (Castelvecchi 2021). Of these, lithium receives the

[90] https://www.energy-storage.news/first-phase-of-800mwh-world-biggest-flow-battery-commissioned-

most attention. It is a relatively abundant element; however, currently mining capacity is struggling to keep up with rapidly growing demand, as three-quarters of all mined lithium is now used for batteries (USGS 2022)–leading to large price fluctuations. Another concern is that currently 90% of it is mined in only three countries–Australia, Chile, and China.[91] China also accounts for more than half of the refining capacity. Nickel is also in short supply–in large part because of surging demand for batteries as well as the disruption of nickel exports from Russia due to their invasion of Ukraine. Cobalt is also a major concern, as about three-quarters of global supply is mined in the Democratic Republic of the Congo. There are of course environmental concerns associated with any form of mining, and these are no exception. Lithium is extracted primarily by two methods. In the "lithium triangle" of Chile, Bolivia, and Argentina it is mostly extracted via solar evaporation of large brine pools in the Atacama Desert, which is a water intensive technique. In Australia, it is mainly extracted from the hard-rock mineral spodumene, which is energy intensive. Mining of cobalt in the Congo is causing severe health and environmental problems, especially in unofficial, "artisanal" mines (Banza Lubaba Nkulu et al. 2018). As a result some companies are moving towards batteries that don't use nickel or cobalt, such as lithium iron phosphate batteries.

New types of batteries that do not use any of these materials, such as dual-carbon batteries, offer the potential to be lower in cost, faster to recharge, easier to recycle, and safer (Tebyetekerwa et al. 2022). However, such batteries are still in development and at least many years from being ready to be brought to market.

Other energy storage options include:

Gravity: energy can be stored by moving mass to a higher gravitational potential. This could be done by moving weights from tall towers or inside old mine shafts. When excess energy is available, the masses are raised. When energy is needed, the objects are lowered—which turns an electrical generator. Gravity systems can also be located underwater, using the buoyant force. By far the most common form of gravity storage is "pumped hydro," where energy is stored by pumping water uphill to a reservoir, where it can be held until needed. Despite its relative obscurity, as of 2021 pumped hydro is still the most widely deployed grid-scale storage technology, accounting for over 90% of total global electricity storage.[92]

Flywheels: a flywheel is a rotating mechanical device that stores kinetic energy in its spinning motion. The energy can be extracted from the flywheel when needed by slowing down its rotation. Flywheels have great potential for rapid response, short duration, high cycle applications such as smoothing out the variable power output of wind and solar (Pullen 2022). Their designs are relatively complicated compared to batteries. Flywheels with magnetic bearings and in high vacuum can retain most of their energy; however, those using mechanical bearings face significant losses within hours because the flywheel continuously changes orientation due to the rotation of the Earth. Their operation at high angular velocities also represents a design challenge and a safety risk.

[91] https://www.weforum.org/agenda/2023/01/chart-countries-produce-lithium-world/

[92] https://www.iea.org/energy-system/electricity/grid-scale-storage

Hydrogen: hydrogen can be produced from excess energy and then stored in high-pressure tanks or as a liquid at low temperatures. When needed, it can be used to generate electricity through fuel cells. However, the round trip efficiency is less than 30% (Escamilla et al. 2022).

Thermal: energy can be stored in the form of heat using various methods, such as heating water or molten salt. The heat can then be converted into electricity using a turbine or used for heating or cooling buildings. This is often part of concentrated solar-thermal power (CSP), where the solar collector system heats an energy storage system during the day, and the heat from the storage system is used to produce electricity in the evening or during cloudy weather. The energy is either stored in the same fluid used to collect it, or transferred to another medium for storage. Solar thermal power plants can also be hybrid systems that use other fuels (usually natural gas) to supplement energy from the Sun during periods of low solar radiation.[93] A "sand battery" stores thermal energy using low-cost sand. They store five to 10 times less energy per unit volume but are 8–10 times cheaper than a Li-ion battery. The first commercial sand battery, an 8 MWh unit in Finland, was installed in 2023.[94] Phase Change Materials (PCMs) can store and release energy by transitioning from one phase to another. This takes advantage of the fact that the heat of fusion is generally much higher than the sensible heat.

Compressed air: energy can be stored by compressing air into a container, such as an underground cavern. When energy is needed, the compressed air is released and generates electricity by spinning turbines. There are currently only two large utility-scale compressed air energy storage (CAES) facilities in the world (Succar & Williams 2008). The 290 MW plant in Huntorf, Germany opened in 1978 to store baseload power generated during off-peak hours, and released when demand is higher. It is now increasingly used to balance power output from wind generation in northern Germany. Built following a similar design, a 110 MW plant in McIntosh, Alabama opened in 1991. For storage, both plants use cavities in salt domes that are solution mined to be tall and narrow, with less roof area for the air pressure to support. CAES can also be built to use saline and sandstone aquifers, and hard rock caverns, although any hard rock mining would substantially raise the cost of the project. Although there are many regions worldwide that have potential for CAES, the U.S. Department of Energy's Global Energy Storage Database[95] lists less than a dozen examples as of June 2023, with most pilot or demonstration plants, or announced but not yet built. So although there is much potential for CAES in the U.S. (DeVries et al. 2005) and elsewhere, there has been little financial support to develop it.

7.5 Efficiency

Needless to say, one of the most effective ways to reduce carbon emissions is simply to use less energy. By improving efficiency, it is possible to do so without reducing

[93] https://www.eia.gov/energyexplained/solar/solar-thermal-power-plants.php

[94] https://www.bbc.com/future/article/20221102-how-a-sand-battery-could-transform-clean-energy

[95] https://sandia.gov/ess-ssl/gesdb/public/projects.html

our quality of life. Following green building standards such as Energy Star[96] and Leadership in Energy and Environmental Design (LEED)[97] can reduce energy consumption and emissions and promote sustainable building practices. Net-zero buildings take this one step further—they are designed to produce as much energy as they consume, resulting in a zero net energy consumption. Design elements of efficient residential and commercial structures include:

- An energy-efficient building envelope that is an effective boundary between the conditioned interior of the building and the environment beyond it. The building envelope is designed to minimize heat loss and gain, which helps reduce the energy needed to heat and cool the building. This includes high-performance insulation, heat-resistant materials in the façade of the building, air sealing that reduces the amount of air that leaks in and out of your home (e.g., caulking and weather stripping), and high-performance windows.
- Efficient HVAC (heating, ventilation, and air conditioning) systems that reduce energy consumption, including ground-source heat pumps, air-source heat pumps, or solar thermal systems.
- "Dynamic glass" that changes its opacity to reduce or increase the amount of light and heat allowed to pass through as needed, reducing load on HVAC systems.
- Renewable energy systems such as solar PV, wind turbines, and geothermal heating and cooling systems that generate energy on-site.
- Energy-efficient LED lighting that reduces energy consumption and minimizes the need for frequent bulb replacements.
- Water conservation systems such as low-flow toilets and faucets, as well as rainwater harvesting systems.
- Building automation systems that optimize energy use by monitoring and controlling HVAC, and lighting systems.
- Natural ventilation systems that use natural airflow to heat and cool the building, reducing the need for mechanical HVAC.
- Sustainable materials such as bamboo, recycled content materials, and low-VOC (volatile organic compound) paints reduce the environmental impact of the building.
- "Green roofs" that incorporate vegetation and plant life, which help regulate temperature and provide natural insulation.
- Building design elements that naturally allow or block sunlight based upon the Sun's location in the sky, which of course varies daily and seasonally.
- Choosing a sustainable site and developing it in an eco-friendly manner, including the use of native plants and reducing stormwater runoff.

Retrofitting existing buildings, particularly with insulation, is one of the most cost-effective ways to make buildings more energy efficient. It is important to note that the poor usually live in less efficient homes, and therefore face higher energy costs.

[96] https://www.energystar.gov

[97] https://www.usgbc.org/leed

Efficiency retrofits are therefore especially beneficial to homes lived in by people with low incomes (Hills 2012).

Food waste is also a major source of inefficiency. Approximately one-third of all food produced for human consumption (1.3 billion tons of edible food) is lost and wasted across the entire supply chain every year. In low-income countries most of the loss is incurred during storage, poor infrastructure and transportation, inadequate market facilities, and a lack of refrigeration. In high income countries food is wasted because it doesn't meet aesthetic ideals, is lost during food manufacture, too much is produced, or leftovers are wasted (Gustavsson et al. 2011). Global food loss and waste are responsible for about 8% of total anthropogenic GHG emissions (Crippa et al. 2022).

7.6 Solutions by Sector

Since it is now possible to meet our electricity needs without fossil fuels, for all sectors the first step should be to electrify as much as possible. Below we discuss the impacts of doing so on the four energy-use sectors, and what would remain to be done if that were accomplished.

7.6.1 Residential

In 2021 in the U.S. 43% of residential energy usage was electricity, for a wide range of purposes. 42% was natural gas, used primarily for furnaces, water heaters, clothes dryers, and gas stoves. About 8% was petroleum (e.g., heating oil, kerosene, and propane) used for heating and cooking. And renewable energy accounted for 7%[5]. In most homes there are four appliances which may use natural gas: the furnace, water heater, kitchen stove, and clothes dryer. All of these are replaceable with electric. In fact already about 25% of homes in the U.S. use electricity for all their energy needs. This is especially true in southern states, where nearly half of all homes have only electric (EIA 2019).

Overall half of the energy used in residential is for HVAC (heating, ventilation, and air conditioning). So the greatest gains can be made by replacing furnaces with heat pumps. A "heat pump" is a machine that moves energy from a cold reservoir to a hot one using the Carnot cycle. Refrigerators are perhaps the most commonly known example of a heat pump, but they are also commonly used for residential and commercial heating, as well as in EVs. Heat pumps can also be used as water heaters. Air-source heat pumps extract heat from the outdoor air and transfer it into a building to provide heating. Ground-source (i.e., "geothermal") and water-source heat pumps use the stable temperature of the ground or a nearby body of water as a heat source. These are usually more efficient but more expensive. They also require drilling for vertical systems or tearing up landscape for horizontal systems, which makes them more difficult than air-source to retrofit on existing structures.

A heat pump is reversible, meaning that the same unit can be used for cooling as well as heating. To provide cooling, the heat pump simply reverses the flow of the refrigerant, absorbing heat from the indoor air and releasing it outside.

The efficiency of a heat pump is limited by the second law of thermodynamics. The "coefficient of performance" (COP) for a heat pump is the ratio of the energy moved to the energy to the work done to move it. The theoretical limit COP possible is:

$$\mathrm{COP}_{\mathrm{max}} = \frac{T_{\mathrm{H}}}{T_{\mathrm{H}} - T_{C}}$$

where T_{H} and T_{C} are the temperatures of the hot and cold reservoirs. Air-source heat pumps typically have COPs of 2.5 to 4.0, whereas ground- and water-source heat pumps range from 3.5 to 5.0 or higher. Heat pump water heaters typically have COP values of 2 or higher.[98] A heat pump with an efficiency of three is in effect 300% efficient because it leads to three times as much heating than if the energy were used to operate an electric or gas furnace. Heat pumps become less efficient at colder temperatures, but newer designs are still more than twice as efficient than furnaces at temperatures as low as −15°C (5°F) and can continue to operate down to -30°C (−22°F) or lower.[99] Below these temperatures supplemental heating may be necessary. A common misconception is that heat pumps somehow violate conservation of energy. However, a heat pump doesn't create energy, it *moves* it.

2021 was a record-high year for heat pump sales, as sales increased by more than 13% globally.[100] Heat pumps are especially popular in southern states of the U.S., where 30–40% of homes use them–as compared to about 10% of homes in the rest of the country.[101] This is because homes in the south require cooling in the summer and heating in the winter; and they are already predominantly all-electric.

The next step is to replace all natural gas appliances, e.g., clothes dryers and stoves, with electric ones. Many people[102] do not like traditional electric stoves for a variety of reasons: They are slower to warm up than gas stoves and are less energy efficient. They are also slower to cool down, making their surfaces dangerous for a longer time after use. However, "induction stoves" are a newer form of electric stove that address these problems.

An induction stove works by using electromagnetic induction to heat cookware directly. The stove surface is made of glass-ceramic that is not conductive to electricity. Underneath the surface a magnetic field induces electrical currents in the cookware. Because the heat is generated directly within the cookware—rather than through the surface of the stove—induction stoves are more energy-efficient than traditional electric or gas stoves. They also allow for more precise temperature control and faster heating and cooking times. Additionally, since the surface of the stove doesn't get hot, spills and splatters are less likely to burn onto the stove and are

[98] https://www.nrel.gov/docs/fy16osti/64904.pdf.
[99] https://www.energy.gov/articles/doe-announces-breakthrough-residential-cold-climate-heat-pump-technology
[100] https://www.iea.org/reports/heat-pumps
[101] https://www.eia.gov/consumption/residential/data/2020/state/pdf/State%20Space%20Heating.pdf
[102] Including the authors!

easier to clean up. However, it should be noted that the cookware must be made of iron or another ferromagnetic material; e.g., copper pans are not compatible.

Perhaps the biggest concern is whether or not a home has enough electrical capacity for all of its appliances and HVAC to be electric, especially if an EV charger is included. Most single-occupancy homes in the U.S. have either 100A or 200A service. *In principle* 100A could be insufficient, but only if all of them are running at full capacity at the same time—which is essentially never the case. And EV chargers can be set to operate at times when loads are usually low, such as in the middle of the night. So for the most part homes will not need to upgrade their existing electrical service. There are other options as well; e.g., smart circuit breakers can be installed that monitor and prioritize demand so that circuits aren't overloaded. They are also useful for integrating residential solar and battery storage.

By switching to heat pumps and electric appliances, all of the energy needs for most residential units can be met with electricity. Because of the efficiency of heat pumps and induction stoves, doing so is likely worthwhile even if that electricity comes from fossil fuels. Of course units can also benefit from adopting residential solar, wind, and geothermal where appropriate.

7.6.2 Commercial

In 2021 in the U.S. 61% of commercial energy consumption was electricity, used for a wide range of purposes including HVAC, lighting, refrigeration, and computers.[103] Natural gas accounted for 32% of usage, for HVAC, water heating, refrigeration, to cook, to dry clothes, and to provide outdoor lighting.[104] District heating and fuel oil accounted for the rest.

The solutions for commercial are largely the same as for residential, if on a larger scale. For larger buildings it is also worthwhile to install energy management systems that can track and manage energy use in commercial buildings. They can help identify areas of high energy consumption and opportunities for energy savings.

7.6.3 Industrial

Industry will be harder to decarbonize than residential and commercial. In 2021 in the U.S. fossil fuels account for 78% of the energy used in industrial processes, including those that use "process heating." Fossil fuels were used not only as fuels but also as feedstock. Only 13% of industrial energy consumption was electricity—although some industrial facilities also generate electricity for their own use. Gonk. And the remaining 9% is from renewable energy, most of which is biomass used as feedstock; e.g., for paper production.

Process heating is used in industrial applications such as metal smelting, refining, and chemical processing. This typically involves the use of furnaces or boilers to produce temperatures 400°C or higher, and is therefore highly energy intensive. Steel

[103] https://www.eia.gov/energyexplained/use-of-energy/commercial-buildings.php
[104] https://www.eia.gov/energyexplained/natural-gas/use-of-natural-gas.php

production is an example. It is responsible for approximately 8% of global CO_2 emissions. Coal currently meets around 75% of the energy demand for steel production.[105] However, hydrogen can also be used for process heating. Hydrogen has a high heating value and can be combusted to produce the necessary high temperatures. Known as "green steel," it is now possible to make steel with hydrogen, e.g., with the HYBRIT process (Åhman et al. 2018), that is cost competitive. We note that, while hydrogen is already in widespread use for many industrial processes, it is mostly derived from fossil fuels. Switching to green hydrogen for existing processes, as well as new ones, is therefore necessary.

Cement production is also a significant contributor to global greenhouse gas emissions, accounting for about 7% of total global emissions (Monteiro et al. 2017). It can be at least partially decarbonized by the use of clean energy for process heating as well as new materials and processes; e.g., carbonating concrete involves adding CO_2 to the concrete during production, which not only reduces emissions from cement production but also creates a stronger and longer-lasting material (Li et al. 2019).

Needless to say, the chemical and petrochemical manufacturing industries are also significant contributors to emissions, as they use fossil fuels as feedstocks as well as energy sources. It is therefore necessary to find alternate ways to make them—or to discontinue their use altogether.

7.6.4 Transportation

Transportation will also be a challenge to decarbonize. In 2021 petroleum products accounted for about 90% of the total U.S. transportation sector energy use. Gasoline, which is used primarily in cars, light trucks, and small airplanes, accounts for 53% of that. Distillate fuels (e.g., diesel), which are used mostly in heavy trucks, buses, trains and large boats, account for 23%. Jet fuel accounts for 11%. Of the remaining 10%, biofuels contributed about 6%. Natural gas accounted for about 4%, most of which was used in pipeline compressors that move the natural gas from well sites to end users. Electricity accounted for less than 1%[5]. Petroleum has long been the dominant fuel of choice for transportation because it is dispatchable, easy to store, and has a high energy density. Like industry, there are many distinct elements that need to be decarbonized—too many to be discussed here. Because of their importance in the discussion and debate over climate change solutions, particularly on the level of personal choice, we will focus on passenger vehicles and air travel.

7.6.4.1 Passenger Vehicles

The three most plausible ways to replace vehicles (i.e., buses, cars, trucks, and mopeds) that use petroleum in internal-combustion engines (ICEs) are vehicles that instead use hydrogen (H_2) combustion or vehicles that use electric motors powered either by batteries (EVs) or hydrogen fuel cells (FCEVs). Replacing petroleum

[105] https://www.iea.org/reports/iron-and-steel

has proven difficult because it has a specific energy of about 45 MJ kg^{-1}. For comparison, the best Li-ion batteries currently have a specific energy of about 1 MJ kg^{-1}. Electric cars therefore have roughly twice the mass of an equivalent ICE car because of the batteries. Hydrogen has a specific energy of 120 MJ kg^{-1}, which makes it ideal as a rocket propellant. However, in a passenger car it requires strong, high-pressure tanks that have a total mass of about 100 kg. The hydrogen itself is therefore only about 5% of the mass of the tank, giving an effective specific energy of roughly 6 MJ kg^{-1}.

EVs have been around since the beginning. At the turn of the 20th century, newly invented motor vehicles were powered either by steam, gasoline, or electric motors. The recent electrification of cities made EVs popular with urban residents—especially among women because they were easier to start and drive than the other two types of cars. However, the drastic drop in price of gasoline cars, as well as of gasoline itself, eventually doomed the EV.

However, in recent years EVs have become popular again. This is in part due to concerns over climate change, but also because they offer several advantages over ICEs. One of the biggest is that they are highly efficient. Only about 10% of energy from the grid is lost when stored in an EVs battery and later used to power the motors. During operation, "regenerative braking" converts the kinetic energy of the vehicle back into electric potential energy, typically with a 60–70% efficiency.[106] They also don't waste energy idling. In total EVs convert about 80% of the electrical energy from the grid to power at the wheels. For comparison, ICEs only convert about 12%–30% of the energy stored in gasoline to locomotion.[107] As a result, many EVs get 'miles per gallon equivalent" (MPGe) values over 100.

From a climate change perspective, the biggest advantage of EVs is that they don't produce any CO_2 emissions during operation. They also don't produce other forms of pollution, improving local air quality—which is particularly important in urban settings. It is true that most EVs are at least partially charged with electricity obtained with fossil fuels, meaning that there are "upstream" emissions associated with their operation. However, because of their efficiency, EVs still emit far less CO_2—less than one third of an equivalent ICE car if they are charged with the average mix of electricity generation in the U.S.[108] It is also true that EVs in general require more energy to manufacture than ICE cars, because of their batteries. However, because of the lower operational emissions, EVs quickly catch up to ICEs—achieving the "break even" point in about 1.5 years—and having about one third the total emissions of an ICE over the life cycle of a car (Woody et al. 2022). Note that all of these calculations assume that the energy for construction and operation of EVs are obtained from the current mix of electricity generation, which is heavily fossil-fuel dependent. Thus these numbers will favor EVs even more as the use of renewable energies continues to grow.

[106] The equivalent for an ICE car would be if the fuel tank refilled when you pressed the brakes!

[107] https://www.fueleconomy.gov/feg/evtech.shtml

[108] You can more accurately determine this number for make and model of car as well as location by using this calculator: https://www.fueleconomy.gov/feg/Find.do?action=bt2.

EVs have lower operating costs. For typical rates in 2022 the cost of energy to operate an EV is about one quarter that of an ICE vehicle, corresponding to a total projected fuel cost savings between $3,000 and $10,500 compared with gasoline vehicles over a 15 year timespan (Borlaug et al. 2020). EVs also have lower maintenance costs. A typical ICE vehicle has over 2000 moving parts in it, whereas most EVs have only about 20. The battery, motor, and associated electronics require little to no regular maintenance. They use fewer fluids, such as engine oil. And brake wear is significantly reduced due to regenerative braking.

The EV driving experience also offers several advantages. EVs are virtually silent, which means they produce less noise pollution and drivers are more aware of their surroundings. They also can offer better performance. An electric motor generates maximum torque at zero to low RPM, and delivers high torque and horsepower over a wide range of engine RPMs. In contrast, ICEs need to be "revved-up" to reach peak power and torque–and do so only over a relatively small range of RPMs. ICEs therefore require complex gearboxes so that the engine can operate in this small range of RPMs over a large range of car speeds. In general EVs do not have a gearbox and deliver power directly to the wheels, increasing power and efficiency. As a result even average EVs have 0–60 mph acceleration times that rival top-end ICE sports cars. Moreover, EVs have better weight distribution and lower center of gravity as the heavy batteries are usually placed in the center of the car or under the floor, leading to better handling and traction, which further improves stability and acceleration.

EVs are also safer. The motor in an ICE vehicle is usually in the front, where it can be pushed into the passengers during a head-on collision. EVs usually do not have the motor in front, eliminating this risk and giving more space for the car to "crumple" and absorb the energy of impact. Their lower center of gravity also greatly decreases their likelihood to roll. For these reasons, several EVs have earned top scores in NHTSA, IIHA, and Euro NCAP crash test ratings.

However, there are numerous downsides to EVs. One of the biggest concerns consumers have about EVs has been described as "range anxiety," which is the fear of getting stranded without a way to recharge. Many EVs have a limited range compared to traditional vehicles, which means they need to be recharged more frequently, and long trips may require more planning. And over time the battery in an EV may lose some of its capacity, which can reduce the car's range and performance.[109] The rate of degradation depends on factors such as how the car is driven (e.g., city vs. highway miles), how it is charged, and ambient temperature (Yang et al. 2018); however, a 10% decline over 100,000 miles is typical. Replacing the battery can be expensive, but usually not necessary during a car's lifetime.

Furthermore, EVs take longer to charge than it takes to fill up a gas tank. Currently there are three categories of EV chargers. Also known as "trickle charging," Level 1 chargers use a standard 120 volt AC household outlet and provide charging speeds of about 4–5 miles of range per hour. It can take more than

---

[109] It is worth noting that ICE engines also lose power and efficiency over time, due to deposits and other effects: https://www.youtube.com/watch?v=uj8hjAjI7p4.

24 hours to fully charge an EV with a Level 1 charger. Level 2 chargers use a 240 volt AC power supply and provide charging speeds of about 10–60 miles of range per hour, depending on the EV and the charger's output power. It usually takes between 4 to 8 hours to fully charge an EV with a Level 2 charger. Also known as DC Fast Charging or Supercharging, Level 3 chargers use direct current and provide charging speeds of 80–150 miles of range per hour, depending on the EV and the charger's output power. It can take less than an hour to charge an EV up to 80% with a Level 3 charger, making it ideal for long-distance travel. However, not all EVs can accept Level 3 charging, and not all Level 3 chargers are compatible with all EVs. Fast charging can also be hard on the battery, degrading capacity and performance over time. Level 1 and 2 chargers are typically found in households, workplaces, and some public charging stations. Level 3 are usually found at public charging stations in cities and along highways. Fast charging stations continue to improve speeds, and are getting close to ICE refueling times, but they're still not as widely available. As of 2022 there are currently about 150,000 gasoline stations in the U.S, each of which includes multiple pumps.[110] For comparison there were 50,000 publicly available EV chargers, and only 7,000 were Level 3.[111] However, the infrastructure is already essentially in place, as electricity is already available in pretty much any location where you would have EVs—all that is needed is the installation of chargers. It is also important to keep in mind that for most EV users the majority of charging occurs at home. About 50%–80% of all charging occurs at residences (Hardman et al. 2018), for which fast charging is not a priority. Another option is battery swapping, where the battery is physically removed and replaced with a fully charged battery in a matter of seconds or minutes. This solution works well for smaller vehicles such as mopeds, and is becoming popular in Asian markets (Aznar et al. 2021).

It is also worth noting that most EV batteries are currently Li-ion, which charge faster when mostly depleted. Most laptop and cellular phone owners have probably noticed with their devices– that it takes more time to charge the last 20% of a battery's capacity than the first 80%. For the most part EVs with Li-ion batteries are not charged to 100% for regular use, so cited charge times can be deceptively long.

The biggest hurdle for adoption for EVs may be their higher upfront costs. EVs are generally more expensive, costing as much as $10,000 more than an equivalent ICE—although government incentives and rebates may help offset the cost. Lower operational costs means that the higher price can be offset in 5–10 years, which is commensurate with the 8-year average length of car ownership.[112] Lack of residential charging availability is also a barrier to EV adoption (Funke et al. 2019). As technology and charging infrastructure continue to improve, EVs will become more accessible and practical for many people. About 2 billion EVs need to be on the road by

[110] https://www.convenience.org/Research/FactSheets/IndustryStoreCount
[111] https://afdc.energy.gov/stations/#/find/nearest?fuel=HY&ev_levels=dc_fast&country=US
[112] https://www.germaincars.com/average-length-of-car-ownership/.

2050 for the world to hit net zero,[113] but sales stood at just 6.6 million in 2022, and most carmakers are having difficulty keeping up with demand.

Another option is to use hydrogen (H_2) as a source of fuel, either in combustion or in a fuel cell. Unfortunately it is difficult to convert existing ICEs to H_2 combustion. The main challenge in converting a gasoline engine to run on hydrogen is the significant differences between the two fuels. Gasoline is stored and transported at ambient temperatures and pressures, while hydrogen is a gas that is typically stored and transported at high pressures and/or low temperatures—meaning that it is necessary to replace or modify the fuel injectors, fuel lines, and fuel storage tank. Gasoline and H_2 have similar adiabatic flame temperatures, but H_2 is about 57 times lighter than gasoline vapor and 14 times lighter than air, so the engine's ignition system also needs to be modified for proper ignition. The differences mean that existing engines cannot be easily modified, and new designs are necessary.

However, there's little incentive to do so.[114] Hydrogen combustion is only about 25% efficient, about the same as a gasoline engine. A fuel cell is more than twice that, so all current hydrogen production cars use fuel cell technology. FCEVs have EPA ratings of 60–70 MPGe, which is about double that of ICE cars but half as good as EVs. In a stroke of unit-conversion serendipity, one gallon of gas has roughly the same energy as 1 kg of H_2. For a 300 mile driving range, an FCEV will therefore need only about 5 kg of hydrogen.[115] That is a small amount of mass, but the tanks are heavy (again, about 100 kg for a passenger car) and the volume required is significant. H_2 tanks can store at up to 10,000 psi (700 bar), but even at this pressure it requires about six times as much volume as gasoline. FCEVs are also expensive—current models all cost over $50k, although this price can be reduced by tax breaks and other incentives.

Hydrogen offers fueling rates comparable to gasoline, but the biggest problem is that of cost and availability. Currently the cost of hydrogen is prohibitive. On a per-mile basis it is about twice as expensive as gasoline and eight times as expensive as electricity. It is worth noting that the benefits of H_2 are only realized if it is green hydrogen. Otherwise the CO_2 emissions from producing H_2 from CH_4, currently the most common method, are worse than just using hydrocarbons as a fuel in the first place (Howarth & Jacobson 2021). The cost of the hydrogen is 2–3 times higher if it is green H_2 (IRENA 2022).

Furthermore, it is hard for consumers to get H_2. As of 2022, there are only 90 publicly available hydrogen refueling stations in U.S., nearly all of which are in California.[116] Although the number of H_2 stations is expected to rapidly increase, it is not expected to catch up to EV charging stations because the cost of a H_2 fueling station is much higher. To install a Level 3 charging station is less than $50k.[117] The cost of a small commercial H_2 station that uses hydrogen delivered as a gas is about

[113] https://www.iea.org/reports/net-zero-by-2050
[114] Unless your car gotta go brap-brap.
[115] https://www.energy.gov/eere/fuelcells/articles/hydrogen-storage-fact-sheet
[116] https://www.nrel.gov/hydrogen/infrastructure-cdps-retail.html
[117] https://afdc.energy.gov/fuels/electricity_infrastructure.html

\$2M; and stations that make H_2 onsite from the electrolysis of water cost over \$3M (Melaina & Penev 2013). They also have a capacity of only 120–350 kg day^{-1}, which is enough to fully refuel only 25–70 vehicles per day. This is in large part because hydrogen is difficult to store. In contrast, EV stations are so inexpensive because the electrical infrastructure already exists everywhere—all you need to do is add chargers to the grid. For all of the reasons given above hydrogen is unlikely to play a major role in sustainable road transport (Plötz 2022).

Of course one of the solutions for transportation is simply to require less of it. This means building "walkable cities," where people can get around by foot, by bicycle, or by public transport. It can also happen by allowing people to telecommute to work.

7.6.4.2 Air Travel

Aviation is currently responsible for about 3% of total CO_2 emissions (Statista 2022) and as much as 5% of radiative forcing because of non-CO_2 impacts such as nitrogen oxides, water vapor, soot, and sulfate aerosols that are released at high altitude (Lee et al. 2021). In particular, the contrails produced by aircraft behave like high cirrus clouds that have a net warming effect (Avila et al. 2019).

Because its primary expense is fuel, the aviation industry has always been keen on reducing usage. The average fuel used by new aircraft on per passenger-km basis fell approximately 45% from 1968 to 2014, with a long-term trend of 1.1% improvement per year that is expected to continue (Kharina & Rutherford 2015). However, this trend is far too slow to meaningfully reduce emissions of the airline sector. In fact, aviation is expected to increase its carbon emissions dramatically. A surge in aviation demand is projected to result in 3.1 billion tons of GHG emissions by 2050, which is 4 times greater than the 2015 baseline (Doliente et al. 2020). The aviation sector will be one of the hardest to decarbonize. Because of its high specific energy and relative ease of storage, petroleum will be very difficult to replace with either electric or hydrogen-powered aircraft.

The specific energy of batteries rather than cost is the major constraint for battery-powered aviation (Viswanathan et al. 2022). Petroleum has a specific energy about fifty times higher than lithium-ion batteries. Furthermore, battery weight doesn't "burn off" during a flight as does jet fuel. Based upon current battery technologies, electric aircraft will be limited to distances under 500–1000 km. Due to the high cost of operation, the regional air travel market is currently very small. However, the much lower maintenance and operational costs of electric aircraft means that such markets will become viable; and several major airlines are planning to use hybrid and fully electric aircraft for regional flights within the next 5–10 years. However, about 95% of CO_2 emissions are from aircraft that fly longer distances (World Economic Forum 2020). Thus, for the foreseeable future, electric aircraft will not play a significant role in emissions reduction.

Hydrogen fuel cells could also be used to power electric aircraft, but face similar range and operational issues. Hydrogen combustion is an option that may be feasible for medium to long-haul flights (i.e., under 10,000 km). Compared to petroleum, hydrogen has a higher energy density by mass (which of course is what makes it ideal as a rocket propellant) but not by volume. It must be stored cryogenically, which means safe storage within an aircraft is difficult, as hydrogen takes up 4–5 times more volume than petroleum. Unlike petroleum, it cannot be stored within the wings, which greatly reduces the usable space within the fuselage. Hydrogen aircraft will also require longer refueling times as well as new airport infrastructure. While hydrogen combustion does not release CO_2, it produces water vapor that, at altitude, would result in radiative forcing. A hydrogen aircraft therefore would have a 50–75% reduction in emissions over current aircraft. Conversion of existing aircraft designs to hydrogen are underway, with certification as early as 2025.[118] And aircraft manufacturer Airbus has announced plans to develop hydrogen-powered commercial aircraft to be in operation by 2035.[119]

Sustainable aviation fuels (SAFs) are based on renewable hydrocarbon sources that are not based on fossil fuels. Depending on the type of SAF, carbon emissions are reduced by up to 80% (de Jong et al. 2017). Currently, most SAFs are biofuels derived from sources such as used cooking oil as well as agricultural and forest biomass. Technologies that will transform carbon dioxide into jet fuel are also being developed (e.g., Yao et al. 2020).

SAFs are called "drop-in fuels" because they have nearly the same chemical and physical characteristics of conventional jet fuel. An advantage is that they can be used in existing aircraft engines as well as use the same airport infrastructure. This is particularly important as aircraft can have operational lifetimes of up to 30 years, meaning that SAFs are the only way significant decreases of carbon emissions can occur within the aviation sector without decommissioning aircraft prematurely.

SAFs made up only about 0.1% of total fuel used in 2019. Cost of production is a major limiting factor. As of early 2020, SAFs cost more than twice that of traditional jet fuel. Large scale deployment of SAFs is a real challenge, as it requires large investments in new production facilities, strong reduction in production costs, and considerable investments in certification for usage in aircraft (Chiaramonti 2019). A vital aspect to biofuel production is ensuring sustainability criteria are met; e.g., by making sure that its generation doesn't create issues with food security or land usage/conservation (Cabrera & Sousa 2022).

It has been argued that aircraft are greener than automotives because they emit roughly half as much carbon per passenger per mile. But what's missed in this argument is how rapidly aircraft can produce carbon–they fly at over 500 miles per hour, after all. Aircraft facilitate trips that would otherwise be untenable; e.g., one is far less likely to drive from Chicago to Los Angeles for a one-day meeting. And that round-trip flight comes at great cost, as it produces as much carbon emissions as several months of driving. While not a viable solution for most of the U.S., travel by rail offers much lower emissions and should be used when possible. For example, a long-haul international flight is estimated to be about sixteen times worse than a Eurostar train on a per passenger per km traveled basis.[120]

Air travel is ultimately an activity of the wealthy. Globally, 1% of people caused half of aviation emissions in 2018, while 89% did not fly at all (Gössling & Humpe 2020).

[118] https://electrek.co/2023/03/02/universal-hydrogen-passenger-hydrogen-electric-plane-maiden-flight/
[119] https://www.airbus.com/en/innovation/zero-emission/hydrogen/zeroe
[120] https://assets.publishing.service.gov.uk/government/uploads/system/uploads/attachment_data/file/904215/2019-ghg-conversion-factors-methodology-v01-02.pdf

Even in developed countries, it is an elite minority of frequent flyers that cause most of the carbon emissions from aviation. For example, In the U.S., 12% of people took 66% of all flights (Hopkinson & Cairns 2020). For many people, air travel is the largest portion of their carbon footprint.

7.7 Working Together

Figures 7.14 and 7.15 show how U.S. electricity generation, in general and for renewables, has changed over the last 70 years. The last 10–15 years have been marked by the decline of coal and the rapid growth of wind and solar. These plots show how electricity generation and demand has changed over the years. However, demand also varies hourly, daily, and seasonally—and by geographic location. The increased use of intermittent wind and solar has raised concerns about the reliability of the electrical grid—i.e., that availability will always meet demand—as well as about rapidly changing energy costs as supply and demand vary. As we transition away from fossil fuels, can other energy sources provide the power we need at reliably low prices?

Fortunately the patterns of demand are generally predictable. Between 7 am and 10 pm on weekdays is considered to be "on peak" because it corresponds to higher demand. During winter and summer months demand is also higher because of the greater need for heating and air conditioning as compared to the spring and fall. The daily peak of demand usually occurs in the evening, but at what time

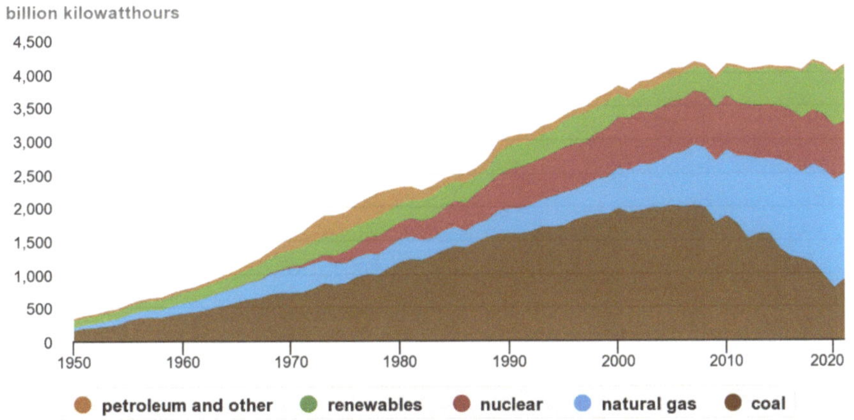

Figure 7.14. [121]U.S. electricity generation by major energy source, 1950–2021. Data source: U.S. Energy Information Administration, Monthly Energy Review, Table 7.2a, January 2022 and Elec. Power Monthly, February 2022. Includes generation from power plants with at least 1 megawatt electric generation capacity. Note the drop in coal since 2008 has largely been matched by growth in natural gas, with some wind and solar as well.

[121] https://www.eia.gov/energyexplained/electricity/electricity-in-the-us.php

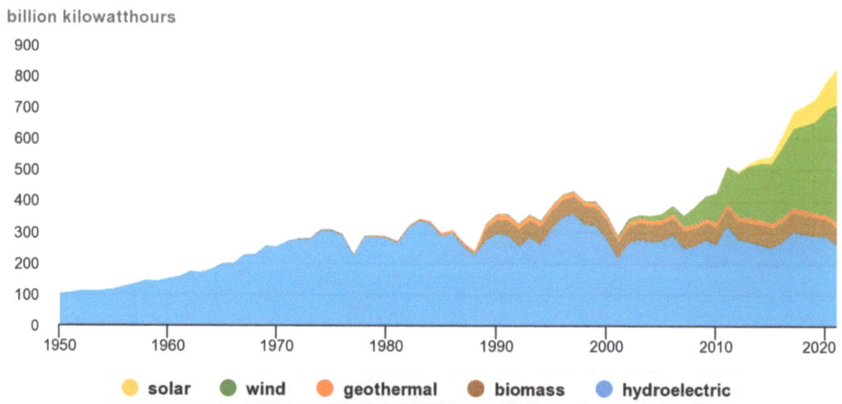

Figure 7.15. U.S. electricity generation from renewable energy sources, 1950–2021. Data source: U.S. Energy Information Administration, Monthly Energy Review, Table 7.2a, January 2022 and Elec. Power Monthly, February 2022. Includes generation from power plants with at least 1 megawatt electric generation capacity. Hydroelectric is conventional hydropower. Hydroelectric production has stayed relatively consistent since 1975. Note the rapid growth of wind and solar since around 2008.

exactly and by how much varies[122] considerably based upon the season and geographic region (EIA 2020). Supply of renewable energy also varies with the seasons. Hydropower and wind, which currently make up the majority of U.S. renewable energy, typically peak in the first half of the year, as there are typically more windy days in the spring and water levels are higher as the winter snowpack is melting.

Electrical supply must be matched to demand in real time, within tight tolerances. Historically, a "baseload" of power that meets the minimum daily requirement is provided by relatively constant sources, such as nuclear, natural gas, or coal-fired power plants. "Peaker plants" are power plants that run when there is a higher demand for electricity. Because they supply power intermittently, they operate at a lower capacity and therefore cost more to operate. That is why gas peaking power plants have roughly triple the LCOE of gas combined cycle in the Lazard (2023) analysis.

How quickly an energy source can be dispatched depends on its type. Grid-scale battery storage can respond within milliseconds to fluctuations in demand. Gravity and flywheel storage can respond within seconds. Hydroelectric can dispatch within tens of seconds to minutes. Gas turbine power usually requires 15–30 minutes to start up; and coal and nuclear require hours.

Wind and solar are complementary in that wind speeds tend to pick up as the Sun is setting due to the thermal changes. However, the increased reliance on wind and solar introduces the risk of being without energy when needed. The German word "dunkelflaute"[123] literally means "dark lull" and has come to be

[122] https://www.eia.gov/energyexplained/electricity/electricity-in-the-us.php.
[123] 'Boy those Germans have a word for everything!" https://www.youtube.com/watch?v=B01e7n4RzZc

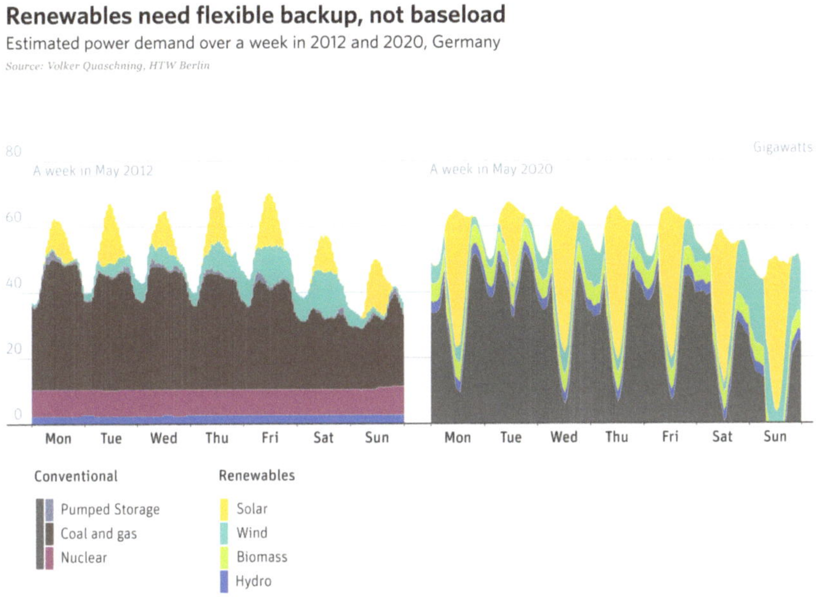

Figure 7.16. [124]Two examples of varying energy supply and demand in Germany throughout a week—May 2012 on the left and May 2020 on the right. The daily variation of demand is visible in both plots. How that energy is provided has changed dramatically in eight years. In 2012, a relatively constant baseload was largely provided by conventional sources, with wind and solar providing relatively small contributions. In 2020 solar is the biggest source of energy during the daytime. A dunkelflaute occurred on Tuesday on the right graph, where the total amount of wind and solar is less than half of normal. The ideal situation is when no conventional energy sources are required (e.g., Sunday on the right graph). Credit: energytransition.org.

used to describe the fear of having an insufficient supply of electricity when there is not enough wind or solar power due to weather conditions (e.g., Figure 7.16).

There are several strategies for how to deal with this problem. One is to overbuild wind and solar so that there's sufficient capacity for all but the worst dunkelflautes. There are criticisms that overbuilding can have decreasing benefits, as it can excessively drive down prices and lead to periods of unused capacity (Tong et al. 2021). Excess energy generated during peak production could be stored for use during peak demand; e.g., battery storage of excess solar from the daytime for evening use. However, storage is in general relatively expensive compared to the cost of energy production itself; and in the case of batteries it is not suitable for long-term, "seasonal" storage. Hydrogen can also be used to store excess energy, but its poor "round trip" efficiency makes this route unattractive in most situations.

---

[124] https://en.wikipedia.org/wiki/Dunkelflaute#/media/File:Renewables_need_flexible_backup_not_baseload.png

Another solution is to use a mix of renewables where possible. Hydroelectric and geothermal pair well with wind and solar because they are dispatchable, and reliably consistent, energy sources that can be used to provide baseload power. They are able to respond on the timescales of variation for wind and solar, especially when some gravity and/or battery storage is included.

If a renewable baseload energy source is not available, an option is to share resources over a large geographic region. For example, clouds typically range from tens to hundreds of kilometers across (Wood & Field 2011), meaning that solar power plants spread over distances larger than this scale can avoid being affected by the same meteorological conditions. In other words, electricity generated in sunny and windy regions can be distributed to areas where it is not. This is surprisingly efficient. Recall from your introductory physics classes that the power dissipated in an electrical circuit is given by the equation:

$$P = VI = I^2R$$

High-voltage power lines minimize current and therefore losses. In the U.S., electricity is transmitted on high-voltage alternating-current (HVAC) lines that run at 69 to 765 kV.[125] With HVAC lines, electricity can be transmitted up to thousands of km with only a few percent loss. Typically, greater losses occur during local distribution. Another promising option is using high-voltage direct-current (HVDC) lines. HVAC suffers from inductive losses due to the alternating current, as well as the "skin effect," the tendency of AC current to crowd along the outer surface of the cable. As a result HVDC is 30–50% more efficient than HVAC. The most notable advantage of HVDC transmission lines is their ability to transfer power underground and undersea over long distances.[126] Of course this requires a sizable and robust electrical grid, which is expensive but can be quite profitable; e.g., ElecLink is a 1,000 MW HDVC between the United Kingdom and France, passing through the Channel Tunnel that began operation in May 2022.

Our electrical grids not only need to be larger but *smarter*. The current strategy is primarily to have a "top down" approach, where a handful of large power plants supply power to a passive consumer. However, with "smart grids" energy generation and storage can be decentralized, with many smaller power producers that both provide and intake power. For example, EV batteries typically contain 40–80 kWh of energy. With vehicle-to-grid (V2G) technology an EV is sufficient to power a small home for 2–3 days. A decentralized grid is also more resilient to disruption, e.g., from extreme weather events.

Recent modeling has shown that at the utility level, having distributed energy resources (DER) in the form of home solar PV and battery storage not only smooths out the regional demand curve for electricity—by producing extra power during times of peak demand and absorbing extra energy if there is dip in demand—but also results in considerable cost savings for the entire grid (Vibrant Clean Energy 2020). VCE's *Weather-Informed energy Systems: for design, operations and markets*

[125] https://www.energy.gov/sites/prod/files/2015/12/f28/united-states-electricity-industry-primer.pdf

[126] https://cleanenergygrid.org/wp-content/uploads/2014/08/High-Voltage-Direct-Current-Transmission.pdf

Planning (WIS:dom*-P) optimization model accounts for different types of energy generation, storage, and transmission, accounting for local siting, weather, and climate change-related effects, at spatial scales as small as 3 km and temporal resolution down to 5 min dispatches. In scenarios where operators can shift supply around with distributed energy resources, there is less need to fire up power plants to meet peak demand, meaning peaker plants can be shut down early or not built at all. With a more level demand curve, power generators are run more consistently, with less ramp ups and ramp downs. When compared with the "business-as-usual" case in the U.S., where purely market forces are assumed to drive changes in the electrical grid, the total systems cost savings is $88 billion and 10 gigatons of CO_2 emission is avoided by 2050. When compared to a scenario where enough renewable energy sources have been added to reduce carbon emissions by 95% in 2050 compared to 1990 levels, a clean energy plus DER plan can avoid just as much emitted CO_2, but cumulatively save about $300 billion. The cost savings is much greater in this case, since the large utility-scale renewable deployment requires substantial capital cost investments, compared with the "business-as-usual" case.

In practice a combination of the above solutions will be implemented, each to varying degrees based upon region; e.g., Wärtsilä[127] gives examples of how this can be done in 145 countries and regions in the world. What is clear now is that it is possible for nearly every place on Earth to have access to reliable, and reliably cheaper, electricity using renewable and storage solutions without the use of fossil fuels.

7.8 Charting a Path to the Future

Driven largely by the rapid growth of wind and solar, there are now several possible scenarios for electricity generation in the U.S. to be carbon neutral as soon as 2035 (Denholm et al. 2022). And the good news is that the transition is starting to happen, if not necessarily fast enough. In 2019 renewables accounted for 72% of all new capacity additions worldwide (IRENA 2020). This is important because the widespread adoption of onshore wind and utility-scale solar PV are considered to be among the most impactful solutions for climate change.[128]

The growth of renewables has been in a large part due to their plummeting costs. In Figure 7.17, the inflation-adjusted costs for different types of fossil and non-fossil fuels energy sources are shown for the last 140 years (Way et al. 2022). Photovoltaic power has had a remarkable three orders of magnitude drop in price since its introduction in 1958. Wind power has a lower rate of exponential decrease, while battery costs have also been plunging steeply in the past quarter century. On the other hand, the costs for coal, oil, and gas were flat, but became more volatile in the last 40 years due to geopolitical shocks (e.g., the 1973 oil embargo, the 1979 Iranian Revolution) and as the most easily accessible reservoirs are depleted. Improvements in extraction technologies were needed before fossil fuel producers could access more

[127] https://www.wartsila.com/insights/article/mapped-renewable-energy-systems-across-145-countries-and-regions-100-renewable-energy/atlas-of-100-percent-renewable-energy#/.

[128] https://drawdown.org/solutions/table-of-solutions

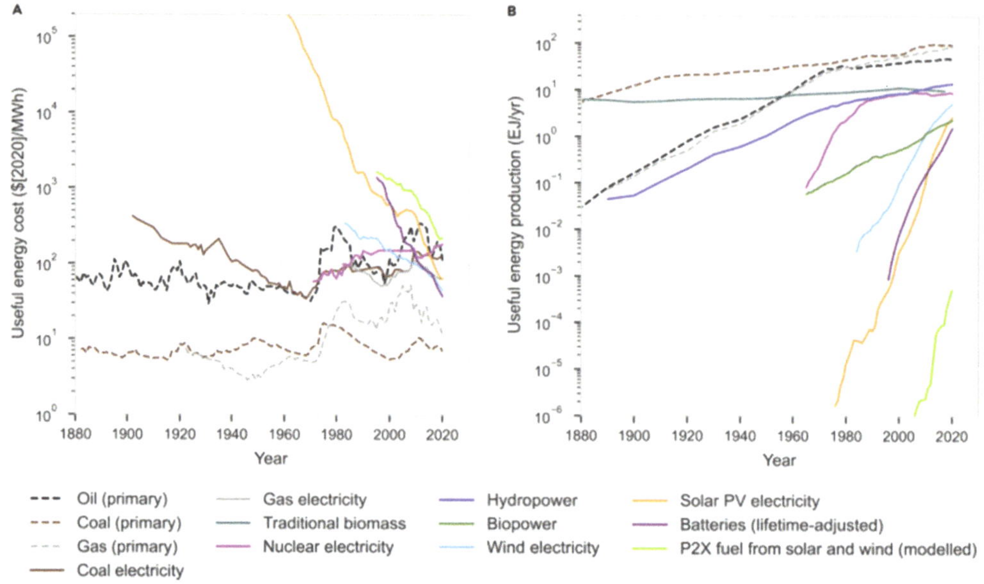

Figure 7.17. Left: Inflation-adjusted costs for "useful energy" which includes the effect of efficiency on the levelized cost of electricity (LCOE) for multiple types of energy technologies. P2X is "power-to-X" where X is hydrogen or ammonia created from renewable solar and wind. Right: The amount of energy produced over time for different types of energy technologies. (Credit: Way et al. 2022; Attribution 4.0 International CC BY 4.0).

difficult to reach deposits (e.g., the fracking boom starting in the early 2000s that caused natural gas prices to drop), but the end result is that inflation-adjusted fossil fuel prices are similar to what they were a century ago. Nuclear fission has seen its price for electricity increase since the introduction of commercial plants for the reasons mentioned in Section 7.2.5. Only PV, wind, and batteries show both exponential drops in prices while deployments have been exponentially increasing.

The exponential drops in price in renewables can be explained by Wright's Law, based on the idea that producers have a learning curve, where they gain experience and learn over time, increasing efficiencies in manufacturing and improving quality. Originally used to describe the industrial production of airplanes (Wright 1936), Wright's Law is generally described as the cost of the nth unit C_n, given the initial starting cost C_1 of the first unit, with the slope $\log_2 b$ defined so that $1 - b$ is the proportion reduction in cost with a doubling of production:

$$C_n = C_1 n^{\log_2 b}$$

Continuing R&D efforts to improve solar panel efficiencies, engineering improvements in the production of the source silicon ingots and wafers, reduction in costs as mining and processing scale up, and market competition are some of the factors that help decrease the costs of solar PV over time (Kavlak et al. 2018). This has resulted in a learning rate of $b = 20\%$ (LaFond et al. 2018), meaning that with a doubling in production of solar panels, the new unit cost is $1 - b = 80\%$ of the original price. Wright's Law also explains why nuclear power plants have not seen any similar price drops. Their complex nature requires enormous upfront capital costs, while the bespoke customized designs in the U.S. mean that there are no learnings from the building of a past plant that can be applied to a future plant. In contrast, solar PV modules have relatively simple, standardized designs that allow their costs to drop with mass market production (Malhotra & Schmidt 2020). Because renewables like solar PV and wind do not have fuel costs after they are deployed, their LCOEs scale with the per unit cost of the technologies, and as learning improves the manufacturing and deployment process over time, their energy costs continue to decline. If Wright's Law continues in future projections, then the more solar PV, wind, and batteries that are built and deployed, the cheaper those technologies become. Accelerating the replacement of fossil fuel energy generation with renewable power and batteries not only reduces the risk of catastrophic climate change, but results in substantially more savings than if a slower approach is taken. Way et al.'s (2022) analysis shows that electrolyzers may also be at the start of their learning curve that can reduce future costs for hydrogen production. Electrolyzer predictions have greater uncertainties than those for solar PV and wind, since their history of deployment is shorter than for the other two technologies.

Often it has been argued that addressing climate change (e.g., by transitioning to renewable energies) would be too damaging to the economy, or too expensive to make it worth it. Indeed, it has been estimated that it would cost $4.5 trillion—or a fifth of the annual U.S. GDP—to fully decarbonize the U.S. power grid, given the current state of technology (Wood Mackenzie 2019). To put that number in perspective, from 1960 to 1973 the entire Apollo program cost $25.8 billion (Dreier 2022), or about $250 billion in 2020 dollars when adjusted for inflation. However, that $4.5 trillion doesn't include costs of externalities. In the long term, it is clear that the economic consequences of *not* addressing climate change are much worse. Callahan & Mankin (2022). found that the top five emitters (the U.S., China, Russia, Brazil, and India) have collectively already caused US$6 trillion in income losses since 1990, comparable to 11% of annual global gross domestic product. And an analysis by the Deloitte Economics Institute found that inaction on climate change could cost the U.S. economy $14.5 trillion by 2070. And the U.S. economy could *gain* $3 trillion, and add nearly 1 million more jobs, if it rapidly decarbonizes over the next 50 years.[129] What's become clear is that addressing climate change is not a threat to our economy but an opportunity.

Of course, the actual cost of the energy transition would not be spent all at once, but would be paid out over decades. In order to meet a 1.5–2°C climate goal, it has been estimated that we would have to spend variously between 0.4% and 4% of global GDP (Stern 2015, IPCC 2022), with most climate economists settling in the range 2–3% (Nair 2021). Historian and author Yuval Noah Harari[130] points out

[129] https://www2.deloitte.com/us/en/pages/about-deloitte/articles/economic-cost-climate-change-turning-point.html

[130] https://www.sapienship.co/decision-makers/2-percent-more.

that the amount required is comparable to or less than the money spent on direct and indirect fossil fuel subsidies worldwide (as much as 8% of global GDP; Parry et al. 2021), food waste (3.5%), global military budgets (2.4%), and taxes evaded by wealthy individuals and corporations (0.5%). Therefore, the money is there, and it is economically feasible to pay for a climate transition. It is important not to be driven to despair that all is lost, and to instead convince politicians and policymakers that they only need to shift a small percentage of their countries' total spending to avoid climate catastrophe (Harari 2022).

What's also clear is that we need to do it *now*. Cars, appliances, and HVAC units typically have lifetimes of 10–15 years. Power plants and aircraft can last 30 years or longer. Delaying change means it will be necessary to replace them before their operational lifetimes expire. Some may be retired early because of the economic benefit, as is often done with commercial aircraft. The good news is that most of the solutions discussed above are available and are economically viable. In fact, most of them are such good ideas that we'd want to do even if climate change wasn't a problem. Some solutions, such as SAF fuels and novel CDR technologies, are decades away from being ready. But that doesn't mean we should wait—they can join the suite of solutions once they are.

7.9 Conclusions

In this chapter we've described how it is possible to transition most aspects of our energy usage to carbon-free sources, and how it can be done now or in the near future. There are also many other high-impact solutions that don't necessarily involve energy generation or consumption–such as education of women, adoption of plant-rich diets, and the use of refrigerants that aren't potent greenhouse gases—that also need to be done. Project Drawdown[131] offers a comprehensive breakdown of all the solutions available, and their impact. It's a wonderful resource to inspire people about what is possible.

Another valuable resource is the En-ROADS economic simulator.[132] It allows the user to explore the impact of roughly 30 policies—such as electrifying transport, pricing carbon, and improving agricultural practices—on hundreds of factors like energy prices, temperature, air quality, and sea level rise. It is a powerful tool for demonstrating that it is possible to keep the temperature increase below 1.5°C or 2.0°C while keeping energy costs down and growing the economy. Project Drawdown and En-ROADS make it clear that we need to implement many solutions. As has been said, "we can't Tesla our way out of this." Note that in this chapter we haven't talked about individual choices. As discussed in Chapter 13, climate change is often portrayed as a personal responsibility so as to shift the blame from the producers of fossil fuels to the consumers. But someone is not to blame if they live and work in a place that requires the use of fossil fuels; e.g., you can't fault people for driving to work if viable alternatives aren't available.

[131] https://www.drawdown.org

[132] https://www.climateinteractive.org/en-roads/

That's not to say that personal choices don't matter. Small actions can lead to big outcomes. They normalize concern about climate change and can be inspirational to others; e.g., people often don't consider purchasing an EV or residential solar until a friend, colleague, or neighbor did. Positive tipping points occur when small interventions trigger self-reinforcing feedbacks that accelerate systemic change (Lenton et al. 2022).[133] For example, in India annual sales of LED bulbs grew more than 130 times within 5 years, which was driven largely by government subsidies that lead to widespread adoption and substantial price reductions (Kamat et al. 2020).[134]

As talked about in Chapters 12 and 13, the barriers to achieving a carbon-free future are largely not technical—they're political and social. With the information from this chapter in hand, hopefully you are ready to help others understand what needs to be done and–importantly—that it *can* be done.

References

Abdou, M., et al. 2021, NuFu, 61, 013001
Abdulla, A., Hanna, R., Schell, K. R., Babacan, O., & Victor, D. G. 2021, ERL, 16, 014036
Åhman, M., et al. 2018, IMES/EESS Report, No. 109
Alsaleh, M., & Abdul-Rahim, A. 2022, EEnv, 33, 1304
Alvarez, R. A., et al. 2018, Sci, 361, 186
Arantes, C. C., Fitzgerald, D. B., Hoeinghaus, D. J., & Winemiller, K. O. 2019, COES, 37, 28
Arcusa, S., & Sprenkle-Hyppolite, S. 2022, CliPo, 22, 1319
Atzeni, S., Batani, D., Danson, et al. 2022, ENews, 53, 18
Avila, D., Sherry, L., & Thompson, T. 2019, TRIP, 2, 100033
Aznar, A., et al. 2021, Building Blocks of Electric Vehicle Deployment: A Guide for Developing Countries, Report NREL/TP-7A40-78776*, U.S. Department of Energy*
Banza Lubaba Nkulu, C., et al. 2018, NatS, 1, 495
Barbarossa, V., et al. 2020, PNAS, 117, 3648
Barkai, R., Rosell, J., Blasco, R., & Gopher, A. 2017, CA, 58, S314
Barnes, R. 2020, Adding Perspective to the Wind Turbine Waste Debate https://medium.com/climate-conscious/wind-turbine-end-of-life-waste-perspective-a561913dcbd9
Beauson, J., Laurent, A., Rudolph, D. P., & Pagh Jensen, J. 2022, RSER, 155, 111847
Bergin, M. H., Ghoroi, C., Dixit, D., Schauer, J. J., & Shindell, D. T. 2017, EnvSciTL, 4, 339
Berthélemy, M., & Escobar Rangel, L. 2013, EPo, 82, 118
Betz, A. 1966, Introduction to the Theory of Flow Machines, ed. D. G. Randall (Oxford: Pergamon)
Bird, F. 2004, In International Businesses and the Challenges of Poverty in the Developing World ed. F. Bird, & S. W. Herman (London: Palgrave Macmillan) https://doi.org/10.1057/9780230522503_3
Blondeel, M., Bradshaw, M. J., Bridge, G., & Kuzemko, C. 2021, GeCom, 15, e12580
BNEF 2022, New Energy Outlook 2022. https://about.bnef.com/new-energy-outlook/
Borlaug, B., Salisbury, S., Gerdes, M., & Muratori, M. 2020, Joule, 4, 1470
Bosch, H.-S., & Hale, G. M. 1992, NuFu, 32, 611

[133] https://doi.org/10.1017/sus.2021.30
[134] https://doi.org/10.1016/j.erss.2020.101488

Brown, J. J., et al. 2013, CL, 6, 280
Bui, M., et al. 2018, EES, 11, 1062
Burns, D. A., Aherne, J., Gay, D. A., & Lehmann, C. M. B. 2016, AtmEn, 146, 1
Cabrera, E., & Sousa, J. M. M. 2022, Energies, 15, 2440
Cacho, J. F., Negri, M. C., Zumpf, C. R., & Campbell, P. 2018, WIREsEEnv, 7, e275
Cai, M., Douglas, E., & Ferman, M. 2022, 'How Texas' power grid failed in 2021—and who's responsible for preventing a repeat." https://www.texastribune.org/2022/02/15/texas-power-grid-winter-storm-2021/
Callahan, C. W., & Mankin, J. S. 2022, ClCh, 172, 40
Castelo Branco, N. A., & Alves-Pereira, M. 2004, NH, 6, 3 https://pubmed.ncbi.nlm.nih.gov/15273020/
Castelvecchi, D. 2021, Natur, 596, 336
CDC 2011, Current Intelligence Bulletin 64: Coal Mine Dust Exposures and Associated Health Outcomes – A Review of Information Published Since 1995, Report, DHHS (NIOSH) Publication No. 2011-172 https://www.cdc.gov/niosh/docs/2011-172/
Chen, H., Xu, Y., Liu, C., He, F., & Hu, S. 2022, Storing Energy, ed. T. M. Letcher (Amsterdam: Elsevier) 2nd ed. 771
Chiaramonti, D. 2019, EProInnSET, 158, 1202
Chiu, C.-H., Lung, S.-C. C., Chen, N., Hwang, J.-S., & Tsou, M.-C. M. 2021, NatSR, 11, 17817
Chowdhury, M. D. S., et al. 2020, ESRv, 27, 100431
Clery, D. 2022, Sci, 378, 1154
Clery, D. 2023, Sci, 379, 625
Close, F. 2007, Fears over factoids, PhysicsWorld https://physicsworld.com/a/fears-over-factoids/
Cole, W., Will Frazier, A., & Augustine, C. 2021, Cost Projections for Utility- Scale Battery Storage: 2021 Update (Golden, CO: National Renewable Energy Laboratory) NREL/TP-6A20-79236. https://www.nrel.gov/docs/fy21osti/79236.pdf
Copping, A. E., et al. 2020, JMSE, 8, 879
Cordes, E. E., et al. 2016, FrES, 4
Crippa, M., et al. 2022, CO2 emissions of all world countries – JRC/IEA/PBL 2022 Report, (Publications Office of the European Union) doi:10.2760/07904, JRC130363.
Dammeier, L. C., et al. 2019, EnST, 53, 9289
de Jong, S., Antonissen, K., Hoefnagels, R., et al. 2017, BiBiof, 10, 64
Deemer, B. R., Harrison, J. A., Li, S., et al. 2016, BioSc, 66, 949
Denholm, P., Patrick, B., Wesley, C., et al. 2022, Examining Supply-Side Options to Achieve 100% Clean Electricity by 2035 (Golden, CO: National Renewable Energy Laboratory) NREL/TP- 6A40-81644. https://www.nrel.gov/docs/fy22osti/81644.pdf
Dennig, F., Budolfson, M. B., Fleurbaey, M., Siebert, A., & Socolow, R. H. 2015, PNAS, 112, 15827
DeVries, K. L., Mellegard, K. D., Callahan, G. D., & Goodman, W. M. 2005, Cavern Roof Stability For Natural Gas Storage In Bedded Salt, US DOE NETL Topical Report RSI-1829, DE-FG26-02NT41651, *United States Department of Energy National Energy Technology Laboratory*
DOE 2020, Energy Storage Grand Challenge: Energy Storage Market Report, NREL/TP-5400-78461
DOE 2021, Solar Futures Study Fact Sheet, https://www.energy.gov/sites/default/files/2021-09/Solar_Futures_Study_Fact_Sheet.pdf
DOE 2023, GeoVision. https://www.energy.gov/eere/geothermal/geovision (accessed 6.24.23)

Doliente, S. S., et al. 2020, FrER, 8,
Draggoo, V. 2013, HPhy, 104, 571
Dreier, C. 2022, SpPo, 60, 101476
EEA 2023, *Scientific advice for the determination of an EU-wide 2040 climate target and a greenhouse gas budget for 2030–2050*. Publications Office of the European Union. https://data.europa.eu/doi/10.2800/609405
EIA 2016, Natural gas expected to surpass coal in mix of fuel used for U.S. power generation in 2016. https://www.eia.gov/todayinenergy/detail.php?id=25392
EIA 2017, Hydroelectric generators are among the United States' oldest power plants, https://www.eia.gov/todayinenergy/detail.php?id=30312
EIA 2018, Manufacturing Energy Consumption Survey Consumption Results, https://www.eia.gov/consumption/manufacturing/
EIA 2019, One in four U.S. homes is all electric. https://www.eia.gov/todayinenergy/detail.php?id=39293
EIA 2020, Hourly electricity consumption varies throughout the day and across seasons. https://www.eia.gov/todayinenergy/detail.php?id=42915
EIA 2021, U.S. crude oil production fell by 8% in 2020, the largest annual decrease on record. https://www.eia.gov/todayinenergy/detail.php?id=47056
EIA 2022, In the first half of 2022, 24% of U.S. electricity generation came from renewable sources. https://www.eia.gov/todayinenergy/detail.php?id=53779
EPA 2023, Inventory of U.S. Greenhouse Gas Emissions and Sinks: 1990-2020. https://www.epa.gov/ghgemissions/inventory-us-greenhouse-gas-emissions-and-sinks-1990-2020
Eisaman, M. D. 2020, Joule, 4, 516
Ellis, G., & Gelman, S. E. 2022, Presented at the GSA Connects 2022 Meeting in Denver (Boulder, CO: GSA)
El-Showk, S. 2022, Sci, 375, 806
Escamilla, A., Sánchez, D., & García-Rodríguez, L. 2022, IJHE, 47, 17505
Ezcurra, E., et al. 2019, SciA, 5, eaau9875
Fa, W., & Jin, Y.-Q. 2007, Icar, 190, 15
Fan, P., et al. 2022, PNAS, 119, e2108038119
Finkelman, R. B., Wolfe, A., & Hendryx, M. S. 2021, EneG, 2, 99
Friedlingstein, P., et al. 2022, ESSD, 14, 4811
Funke, S. Á., Sprei, F., Gnann, T., & Plötz, P. 2019, TRD: TrEnv, 77, 224
GAO 2022, Carbon Capture and Storage: Actions Needed to Improve DOE Management of Demonstration Projects. https://www.gao.gov/products/gao-22-105111 (accessed 6.24.23)
Gibbins, J., & Chalmers, H. 2008, EPo, 36, 501
Gössling, S., & Humpe, A. 2020, GEC, 65, 102194
Griscom, B. W., et al. 2017, PNAS, 114, 11645
Gustavsson, J., Cederberg, C., Sonesson, U., Van Otterdijk, R., & Meybeck, A. 2011, Global food losses and food waste. https://www.madr.ro/docs/ind-alimentara/risipa_alimentara/presentation_food_waste.pdf
Harari, Y. N. 2022, The Surprisingly Low Price Tag on Preventing Climate Disaster. https://time.com/6132395/two-percent-climate-solution/
Hardman, S., et al. 2018, TRD: TrEnv, 62, 508

Harrison, C., Lloyd, H, & Field, C 2017, Evidence review of the impact of solar farms on birds, bats and general ecology, Technical Report, No. NEER012 (Natural England, Manchester Metropolitan University)
Herring, J. S. 2012, Uranium and Thorium Resources, ed. R. A. Meyers (New York: Encyclopedia of Sustainability Science and Technology. Springer)
Herzog, H. 2022, Direct Air Capture, in Greenhouse, ed. M. Bui, & N. Mac Dowell (London: Royal Society of Chemistry) 115
Hill, J., Nelson, E., Tilman, D., Polasky, S., & Tiffany, D. 2006, PNAS, 103, 11206
Hills, J. 2012, Getting the measure of fuel poverty: final report of the Fuel Poverty Review. CASEreport (72) (London: Centre for Analysis of Social Exclusion, London School of Economics and Political Science) http://eprints.lse.ac.uk/id/eprint/43153
Hoffacker, M. K., Allen, M. F., & Hernandez, R. R. 2017, EST, 51, 14472
Holzman, D. C. 2011, EHP, 119, a476
Hopkinson, L., & Cairns, S. 2020, Elite Status: global inequalities in flying. Report for Possible, March 2021. https://www.wearepossible.org/latest-news/elite-status-how-a-small-minority-around-the-world-take-an-unfair-share-of-flights
Howarth, R. W., & Jacobson, M. Z. 2021, ESE, 9, 1676
IPCC 2005, IPCC Special Report on Carbon Dioxide Capture and Storage. Prepared by Working Group III of the Intergovernmental Panel on Climate Change, ed. B. Metz, O. Davidson, H. C. de Coninck, M. Loos, & L. A. Meyer (Cambridge: Cambridge Univ. Press)
IRENA 2012, Renewable Energy Cost Analysis—Hydropower (Abu Dhabi: International Renewable Energy Agency) https://www.irena.org/publications/2012/Jun/Renewable-Energy-Cost-Analysis---Hydropower (accessed 6.24.23)
IRENA 2019, Future of wind: Deployment, investment, technology, grid integration and socio-economic aspects (A Global Energy Transformation paper) (Abu Dhabi: International Renewable Energy Agency) https://www.irena.org/-/media/files/irena/agency/publication/2019/oct/irena_future_of_wind_2019.pdf
IRENA 2020, Renewable Power Generation Costs in 2019 (Abu Dhabi: International Renewable Energy Agency) https://www.irena.org/publications/2020/Jun/Renewable-Power-Costs-in-2019
IPCC 2022, Global Warming of 1.5°C: IPCC Special Report on Impacts of Global Warming of 1.5°C above Pre-Industrial Levels in Context of Strengthening Response to Climate Change, Sustainable Development, and Efforts to Eradicate Poverty (Cambridge: Cambridge University Press) 93
IRENA 2022, Hydrogen (Abu Dhabi: International Renewable Energy Agency) https://www.irena.org/Energy-Transition/Technology/Hydrogen (accessed 6.24.23)
Jacobson, M. Z., & Archer, C. L. 2012, PNAS, 109, 15679
James, W., Jia, C., & Kedia, S. 2012, IJERPH, 9, 4365
Jassby, D. 2019, PhyS, 48, 13 https://vixra.org/pdf/1812.0382v1.pdf
Jassby, D. 2021, PhyS, 50, 5
Jeswani, H. K., Chilvers, A., & Azapagic, A. 2020, Proc. of the Royal Society A: Mathematical, Physical and Engineering Sciences Vol. 476, 20200351
Johnston, J. E., Lim, E., & Roh, H. 2019, ScTEn, 657, 187
Joppa, L., et al. 2021, Natur, 597, 629
Kavlak, G., McNerney, J., & Trancik, J. E. 2018, EPo, 123, 700
Kamat, A. S., Khosla, R., & Narayanamurti, V. 2020, ERSS, 66, 101488

KamLAND Collaboration 2011, Partial radiogenic heat model for Earth revealed by geoneutrino measurements NatGe, 4, 647
Kebede, A. A., Kalogiannis, T., Van Mierlo, J., & Berecibar, M. 2022, RSER, 159, 112213
Keller, D. P., et al. 2018, CCCR, 4, 250
Kharina, A., & Rutherford, D. 2015, Fuel Efficiency Trends for New Commercial Jet Aircraft: 1960 to 2014, International Council on Clean Transportation
Kilcher, L., Fogarty, M., & Lawson, M. 2021, Marine Energy in the United States: An Overview of Opportunities (Golden, CO: National Renewable Energy Laboratory (NREL/TP-5700-78773) https://www.nrel.gov/docs/fy21osti/78773.pdf
Kim, J., et al. 2021, Energies, 14, 4278
Kondash, A. J., Lauer, N. E., & Vengosh, A. 2018, SciA, 4, eaar5982
Lafitte, A., Sordello, R., de Crespin de Billy, V., et al. 2022, EnEv, 11, 36
LaFond, F., et al. 2018, TForSC, 128, 104
Lantz, E, et al. 2019, Increasing Wind Turbine Tower Heights: Opportunities and Challenges (Golden, CO: National Renewable Energy Laboratory (NREL/TP)) 5000 https://www.nrel.gov/docs/fy19osti/73629.pdf
Latrubesse, E. M., Park, E., Sieh, K., et al. 2020, Geomo, 362, 107221
Lazard 2023, Levelized Cost of Energy Analysis—Version 16.0. https://www.lazard.com/research-insights/2023-levelized-cost-of-energyplus/
Lee, D. S., et al. 2021, AE, 244, 117834
Lenton, T. M., Benson, S., Smith, T., et al. 2022, GS, 5, e1
Li, Z., He, Z., & Chen, X. 2019, Materials, 12, 3729
Liang, Y., et al. 2023, PBiJ, 21, 317
Lipponen, J., et al. 2017, 13th International Conf. on Greenhouse Gas Control Technologies, GHGT-13, 14–18 November 2016 *(Lausanne, Switzerland) Vol. 114, 7581*
Logue, J. M., Klepeis, N. E., Lobscheid, A. B., & Singer, B. C. 2014, EHP, 122, 43
Lopez, A., Green, R., Williams, T., et al. 2022, Offshore Wind Energy Technical Potential for the Contiguous United States NREL/PR-6A20-83650*, NREL*
Lopez, A., Roberts, B., Heimiller, D., Blair, N., & Porro, G. 2012, U.S. Renewable Energy Technical Potentials: A GIS-Based Analysis. Renewable Energy NREL/TP-6A20-51946
Loria, P., & Bright, M. B. H. 2021, EJ, 34, 106998
Lu, X., McElroy, M. B., & Kiviluoma, J. 2009, PNAS, 106, 10933
Macknick, J. et al. 2013, The 5 Cs of Agrivoltaic Success Factors in the United States: Lessons From the InSPIRE Research Study, (No. NREL/TP-6A20-83566) https://www.nrel.gov/docs/fy22osti/83566.pdf.
Malhotra, A., & Schmidt, T. S. 2020, Joule, 4, 2259
Malik, N. 2023, 'America's Biggest Power Source Wasn't Built for Extreme Weather." Bloomberg. https://www.bloomberg.com/graphics/2023-natural-gas-biggest-us-power-source-also-most-vulnerable/
Marino, B. D. V., Mincheva, M., & Doucett, A. 2019, PeerJ, 7, e7606
Marrou, H., Wery, J., Dufour, L., & Dupraz, C. 2013, EJAg, 44, 54
Martínez-Garcia, A., et al. 2011, Natur, 476, 312
Martin-Roberts, E., et al. 2021, OEart, 4, 1569
Masters, G. M. 2004, Renewable and Efficient Electric Power Systems (New York: Wiley)
May, R., et al. 2020, EcoEv, 10, 8927
McBride, J. P., Moore, R. E., Witherspoon, J. P., & Blanco, R. E. 1978, Sci, 202, 1045
McGlade, C., & Ekins, P. 2015, Natur, 517, 187

Mcleod, E., et al. 2011, FrEcoEn, 9, 552
Mecklin, J. 2022, The Energy Department's fusion breakthrough: It's not really about generating electricity. BuAtS. https://thebulletin.org/2022/12/the-energy-departments-fusion-breakthrough-its-not-really-about-generating-electricity/
Melaina, M., & Penev, M. 2013, Hydrogen Station Cost Estimates: Comparing Hydrogen Station Cost Calculator Results with other Recent Estimates, (No. NREL/TP--5400-56412, 1260510)
Micallef, D., & Van Bussel, G. 2018, Energies, 11, 2204
Miller, L. M., Gans, F., & Kleidon, A. 2011, ESD, 2, 1
Miller, L. M., & Keith, D. W. 2018, Joule, 2, 2618
Miller, L. M., & Kleidon, A. 2016, PNAS, 113, 13570
Monteiro, P. J. M., Miller, S. A., & Horvath, A. 2017, NatMa, 16, 698
Moore, J. N., & Simmons, S. F. 2013, Sci, 340, 933
Moran, E. F., Lopez, M. C., Moore, N., Müller, N., & Hyndman, D. W. 2018, PNAS, 115, 11891
Morse, E. 2018, Nuclear Fusion (Cham: Springer)
Munawer, M. E. 2018, JSM, 17, 87
Nair, S. 2021, 'Climate inaction costlier than net zero transition: Reuters poll." https://www.reuters.com/business/cop/climate-inaction-costlier-than-net-zero-transition-economists-2021-10-25/
National Academies of Sciences, Engineering, and Medicine 2019, Negative Emissions Technologies and Reliable Sequestration: A Research Agenda (Washington, DC: The National Academies Press)
National Academies of Sciences, Engineering, and Medicine 2022, A Research Strategy for Ocean-based Carbon Dioxide Removal and Sequestration (Washington, DC: The National Academies Press)
Nolan, C. J., Field, C. B., & Mach, K. J. 2021, NRvEE, 2, 436
NREL 2020, Model of Operation-and- Maintenance Costs for Photovoltaic Systems (Golden, CO: National Renewable Energy Laboratory (NREL/TP-5C00-74840)) https://www.nrel.gov/docs/fy20osti/74840.pdf
NREL 2021, Life Cycle Greenhouse Gas Emissions from Electricity Generation: Update. https://www.nrel.gov/docs/fy21osti/80580.pdf
NREL 2023, Best Research-Cell Efficiency Chart. https://www.nrel.gov/pv/cell-efficiency.html (accessed 6.24.23)
OECD 2016, The Economic Consequences of Outdoor Air Pollution (Paris: OECD Publishing)
Oschlies, A., Pahlow, M., Yool, A., & Matear, R. J. 2010, GeoRL, 37
Parry, I. W. H., Black, S., & Vernon, N. 2021, Still Not Getting Energy Prices Right: A Global and Country Update of Fossil Fuel Subsidies. IMF Working Paper WP/21/236. https://www.elibrary.imf.org/view/journals/001/2021/236/article-A001-en.xml (accessed 7.21.23)
Pimentel, D., & Patzek, T. W. 2005, NatRR, 14, 65
PJM 2023, Winter Storm Elliott Frequently Asked Questions https://www.pjm.com/-/media/markets-ops/winter-storm-elliott/faq-winter-storm-elliott.ashx
Plötz, P. 2022, NatE, 5, 8
Pope, C. A., Ezzati, M., & Dockery, D. W. 2009, NEJM, 360, 376
Popp, J., Lakner, Z., Harangi-Rákos, M., & Fári, M. 2014, RSER, 32, 559
Pullen, K. R. 2022, 11—Flywheel energy storage ed. T. M. Letcher Storing Energy (2nd ed.; Amsterdam: Elsevier) 207

Ramanujam, J., & Singh, U. P. 2017, EES, 10, 1306
Ramasamy, V., Zuboy, J., O'Shaughnessy, E., et al. 2022, U.S. Solar Photovoltaic System and Energy Storage Cost Benchmarks, With Minimum Sustainable Price Analysis: Q1 2022 (Golden, CO: National Renewable Energy Laboratory (NREL/TP-7A40)) 83586 https://www.nrel.gov/docs/fy22osti/83586.pdf
Rauner, S., et al. 2020, NatCC, 10, 308
Redissi, Y., & Bouallou, C. 2013, GHGT-11 Proc. of the 11th International Conf. on Greenhouse Gas Control Technologies*(18–22 November 2012) (Kyoto, Japan) Vol. 37, 6667*
Reid, W. V., Ali, M. K., & Field, C. B. 2020, GCBio, 26, 274
REN21 2019, Renewables 2019 Global Status Report (Paris: REN21 Secretariat). ISBN 978-3-9818911-7-1 https://www.ren21.net/wp-content/uploads/2019/05/gsr_2019_full_report_en.pdf
Renforth, P., & Henderson, G. 2017, RvGeo, 55, 636
Ritchie, H. 2017, What was the death toll from Chernobyl and Fukushima? Our World in Data. https://ourworldindata.org/what-was-the-death-toll-from-chernobyl-and-fukushima
Ross, M. L. 2015, Annu. Rev. Political Sci., 18, 239
Rubin, E. S., Davison, J. E., & Herzog, H. J. 2015, Int. J. Greenhouse Gas Control 40, 378
Sackett, D. K., Aday, D. D., Rice, J. A., Cope, W. G., & Buchwalter, D. 2010, Ecotoxicology, 19, 1601
Sánchez-Díez, E., et al. 2021, JPS, 481, 228804
Schleisner, L. 2000, RE, 20, 279
Schultz, R., Atkinson, G., Eaton, D. W., Gu, Y. J., & Kao, H. 2018, Sci, 359, 304
Seals, B. A., & Krasner, A. 2020, Health Effects From Gas Stove Pollution. https://rmi.org/insight/gas-stoves-pollution-health
Silverman-Roati, K., Webb, R. M., & Gerrard, M. B. 2022, Permitting Seaweed Cultivation for Carbon Sequestration in California: Barriers and Recommendations Available at: https://scholarship.law.columbia.edu/faculty_scholarship/3523
Shen, S., et al. 2011, Sci, 334, 1367
Shere, J. 2013, Renewable: The World-Changing Power of Alternative Energy (New York: St Martin's Press) 201
Smith, S. M., et al. 2023, The State of Carbon Dioxide Removal 1st ed. (OSF)
Snæbjörnsdóttir, S. Ó., et al. 2020, NRvEE, 1, 90
Sovacool, B. K. 2012, JIEnS, 9, 255 Vermont Law School Research Paper No. 04-13, Available at SSRN: https://ssrn.com/abstract=2198024
Stefansson, V. 2005, 'World Geothermal Assessment", Proceedings World Geothermal Congress 2005, Antalya, Turkey, 24–29 April 2005, https://www.geothermal-energy.org/pdf/IGAstandard/WGC/2005/0001.pdf
Statista 2022, Commercial airlines: worldwide fuel consumption 2023. https://www.statista.com/statistics/655057/fuel-consumption-of-airlines-worldwide/ (accessed 6.24.23)
Stern, N. 2015, PRSB, 282, 20150820
Storm, K. 2020, Industrial Construction Estimating Manual, ed. K. Storm (Houston, TX: Gulf Professional Publishing) 95
Streets, D. G., et al. 2011, EST, 45, 10485
Succar, S., & Williams, R. H. 2008, Compressed Air Energy Storage: Theory, Resources and Applications For Wind Power, Report, Princeton Environmental Institute
Tang, W., et al. 2021, Natur, 597, 370
Tebyetekerwa, M., Duignan, T. T., Xu, Z., & Song Zhao, X. 2022, AdEnM, 12, 2202450
Teff-Seker, Y., Berger-Tal, O., Lehnardt, Y., & Teschner, N. 2022, RSER, 168, 112801

Tester, J. W., et al. 2006, Impact of Enhanced Geothermal Systems (EGS) on the United States in the 21st Century (Cambridge, MA: Massachusetts Institute of Technology) https://www1.eere.energy.gov/geothermal/pdfs/future_geo_energy.pdf
Thompson, R. C., Moore, C. J., vom Saal, F. S., & Swan, S. H. 2009, RSPTB, 364, 2153
Tollefson, J., & Gibney, E. 2022, Natur, 612, 597
Tong, D., et al. 2021, NatCo, 12, 6146
Townsend-Small, A., & Hoschouer, J. 2021, ERL, 16, 054081
Trottier, G., Turgeon, K., Boisclair, D., Bulle, C., & Margni, M. 2022, PLoSO, 17, e0273089
Trout, K., et al. 2022, ERL, 17, 064010
USGS 2022, Mineral commodity summaries 2022, Report, U.S. Geological Survey 202
USDA ERS 2023 ERS—Feed Grains Sector at a Glance. https://www.ers.usda.gov/topics/crops/corn-and-other-feed-grains/feed-grains-sector-at-a-glance/ (accessed 6.24.23)
Vakulchuk, R., Overland, I., & Scholten, D. 2020, RSER, 122, 109547
van der Velde, I. R., et al. 2021, Natur, 597, 366
van Oldenborgh, G. J., et al. 2021, NHESS, 21, 941
Vautard, R., et al. 2014, NatCo, 5, 3196
Veers, P., et al. 2022, WESD, 1
Vibrant Clean Energy 2020, Why Local Solar For All Costs Less: A New Roadmap for the Lowest Cost Grid (Boulder, CO: Vibrant Clean Energy LLC)
Viswanathan, V., et al. 2022, Natur, 601, 519
Vohra, K., et al. 2021, ER, 195, 110754
Wall, A. M., Dobson, P. F., & Thomas, H. 2017, GRCT, 41
Wang, Z., et al. 2022, EST, 56, 14723
Wärtsilä, Atlas for 100% Renewable Energy https://www.wartsila.com/energy/towards-100-renewable-energy/atlas-of-100-percent-renewable-energy (accessed 6.24.23)
Way, R., Ives, M. C., Mealy, P., & Farmer, J. D. 2022, Joule, 6, 2057
Welsby, D., Price, J., Pye, S., & Ekins, P. 2021, Natur, 597, 230
Whitehead, J., Newman, P., Whitehead, J., & Lim, K. L. 2023, Sustain. Earth Rev., 6, 1
Wiser, R., Bolinger, M., & Lantz, E. 2019, REF, 30, 46
Wiser, R., et al. 2021, NatEn, 6, 555
Wollin, K.-M., et al. 2020, ArTox, 94, 967
Wood, R., & Field, P. R. 2011, JCli, 24, 4800
Wood Mackenzie, W. 2019, Deep decarbonisation: the multi-trillion dollar question. https://www.woodmac.com/news/feature/deep-decarbonisation-the-multi-trillion-dollar-question/ (accessed 6.24.23)
Woody, M., et al. 2022, ERL, 17, 089501
World Economic Forum 2020, Clean Skies for Tomorrow: Sustainable Aviation Fuels as a Pathway to Net-Zero Aviation https://www3.weforum.org/docs/WEF_Clean_Skies_Tomorrow_SAF_Analytics_2020.pdf
Worsnop, R. P., Lundquist, J. K., Bryan, G. H., Damiani, R., & Musial, W. 2017, GeoRL, 44, 6413
Wright, T. P. 1936, JAS, 3, 122
Wu, P., Ma, X., Ji, J., & Ma, Y. 2017, Energy Procedia, 8th International Conf. on Applied Energy, ICAE2016, 8–11 October 2016 *(Beijing, China) Vol. 105, 68*
Wurzel, S. E., & Hsu, S. C. 2022, PhPl, 29, 062103
Yamaguchi, R., et al. 2022, NatCC, 12, 469

Yang, F., Xie, Y., Deng, Y., & Yuan, C. 2018, NatCo, 9, 2429
Yang, W., Kim, K.-H., & Lee, J. 2022, JCP, 376, 134292
Yao, B., et al. 2020, NatCo, 11, 6395
Zhou, L., et al. 2012, NatCC, 2, 539
Ziegler, M. S., & Trancik, J. E. 2021, EES, 14, 1635

Part IV

Formal and Informal Education

Chapter 8

Climate Change in Astro 101

T A Rector

> "We all travel the Milky Way together, trees and men."
>
> —John Muir

An introductory physics or astronomy class is an excellent opportunity to teach climate change. Many of the topics regularly covered in these classes can be used to lay a solid conceptual foundation upon which a deeper understanding of climate change can be built. With this foundation in place we can teach climate change in a way that addresses students' concerns and fears, cognizant of common misconceptions. To do so we need to discuss the consequences of climate change, as well as describe solutions. And we should be willing to talk about why climate change is controversial. Students need to understand the efforts to spread misinformation; and we should develop their skills to identify it. This chapter describes how to achieve these goals.

8.1 Introduction

Every semester I teach climate change, in some capacity, in all of my classes. I usually find it easy to connect it to the content of the course. As an assignment I have my students ask a question that they have about climate change. Here are some representative responses from a recent semester:

> "My question is how much can a few degrees really affect the world?"
>
> "Will the long term effects of renewable energy be able to undo the negative short term effects of the production of those energy sources? For example, will solar panels be able to undo the negative environmental effects of mining for the materials used to create the panels? Or is it a losing battle?"
>
> "I sometimes wonder whether there may be any agenda in the science community. As in, maybe funding must preserve the research and therefore we should make a big deal of the issue at hand, regardless of

doi:10.1088/2514-3433/acfcb6ch8

its actual importance. Not sure if that makes sense, but I also consider about foreign engagement, such as China that doesn't reduce its carbon emissions, and whether they would rather themselves be more powerful with oil and gas than everyone else."

"This seems like such an insurmountable issue, is it even possible to bring the whole world together to solve this before it's too late?"

These questions reflect common concerns among students. Some—but not most—don't understand the severity of the problem ("What's the big deal about a few degrees?") while many fear that the problem is so big and "insurmountable" that already "it's too late." We also see influence from common climate denial talking points; e.g., that scientists are fabricating or exaggerating the problem, other countries (especially China) are to blame, or that the solutions are as bad or worse than the actual problem.

First off, I challenge you to set a goal of incorporating climate change into every course that you teach, regardless of the topic or level. Fortunately that's easy to do in an introductory physics or astronomy class, because the underlying physics of climate science is interwoven into many of the topics we normally cover. There are many points at which you can start a discussion about climate change. In this chapter the focus will be on incorporating climate change into your introductory astronomy classes; and there will also be discussion of introductory physics courses as well.

The focus in this chapter is on science content, but it's important to emphasize that you need to talk about more than the science alone. As scientists we naturally would prefer to focus on just that—the science of what is causing climate change, what are the consequences, how do we know this, and how confident we are. As discussed in more detail in Chapter 12, much of our teaching is implicitly based on the "information deficit model"—a model that assumes apathy or skepticism about climate change is due to lack of knowledge about the problem. In other words, if we explain the science well enough people will be naturally motivated to take action. However, there is now a wealth of research that indicates that this isn't enough (e.g., Sturgis & Allum 2004, Suldovsky 2017). And, as the student questions above demonstrate, they have misconceptions that need to be addressed.

So you need to be ready to talk about those misconceptions. And you need to talk about solutions. Most students do not need to be convinced that climate change is happening, or that it is a serious problem. What they need to be convinced of is that the problem is surmountable. If you don't talk about solutions, they may incorrectly infer that there's nothing we can do.

Climate change is unlike other topics we normally teach. First of all, it is an emotionally charged topic. For the most part students are interested in climate change, and welcome the opportunity to talk about it. But many are reluctant because they are scared for their future. Hickman et al. (2021) surveyed youth aged 16–25 and found that 59% were very or extremely worried about climate change, and 84% were at least moderately worried. And more than 50% reported at least one of the following emotions: sadness, anxiety, anger, powerlessness, helplessness, and guilt. While we are scientists—not therapists—we need to be aware that these emotions are in the room and will affect the conversation. Climate change is also a socially and politically charged topic, meaning that many people have established opinions on climate change based upon their peer groups, values, and political identity. Finally, as discussed in detail in Chapter 13, misinformation about climate change is widespread—fueled by an extensive disinformation campaign intended to confuse and delay action. Chapter 12 discusses ways to talk about climate change in ways that can help people overcome these barriers.

8.2 Pre-teaching

Whether it be for an astronomy or physics class, it helps to "pre-teach" related topics before the section on climate change to lay a solid conceptual foundation upon which their understanding can be built. It also helps to see these topics outside of the context of climate change, so that students understand that climate change is a natural consequence of these physical principles rather than a specialized field of study that may be biased or wrong. It is of course not necessary (or desired!) to cover all of these topics, but here is a relevant list:

Big numbers: To most people the words "million," "billion," and "trillion" lack an appropriate scale. All are interpreted as being "really big," with their names suggesting that billion and trillion are two and three times larger than a million, respectively. Of course a billion (in the U.S. at least) is a thousand times larger than a million. And a trillion is a thousand times larger than a billion. These numbers are hard to fathom, even for professional scientists. So it is important to put these numbers into perspective; e.g., a million seconds is about 11.5 days, whereas a billion seconds is over 31 years. You can ask your students to reflect upon what they were doing a million seconds ago, and a billion seconds ago. Naturally, for the latter many of your students had not yet been born.

Timescales: Physical processes in astronomy tend to occur on timescales of millions to billions of years, so we naturally want students to develop a feel for their relative lengths. In particular we want them to understand that the Earth's climate changes naturally on timescales of thousands to millions of years. We also want them to know that the Earth's climate has only been as it is now for the last 10,000 years—a period known as the Holocene. This is a fantastically short time period compared to the overall history of the Earth but still much longer than the 100 years or so during which we have rapidly increased CO_2 levels in the Earth's climate system by using fossil fuels.

Physical scales: Most of us set as an educational goal for students to understand the relative scales of size and distance for astronomical objects. A standard 30 cm (1 ft) globe of the Earth is a useful starting point. On this scale, the Moon would roughly be the size of a tennis ball and about 10 meters (30 ft) away. The Sun would be about 3 km (2 mi) away and 35 m (110 ft) tall—about the height of an eight-story building. Mars will therefore be between one and five miles away. This is a useful starting point for explaining the difficulty of human spaceflight to Mars.

The Sun: A common misconception about climate change is that it can be attributed to variability in the brightness of the Sun. The Sun's brightness does vary with its 11-year cycle, but that variability is less than 0.05% of its nominal value of 1361 W m^{-2}. The Sun also varies on longer timescales. Since the 1960s, the overall trend is that the Sun's brightness has decreased at the same time that the Earth's climate has warmed the fastest (Foukal et al. 2006). So we know that solar variability is not the cause of climate change.

The electromagnetic spectrum: Developing a deeper grasp of the greenhouse effect requires understanding infrared and optical light, and the different opacities for greenhouse gases in these bands. If you have access to an infrared camera, a useful demonstration is to show how glass is transparent to visible light but opaque to the infrared, and that a black trash bag is just the opposite. You can also describe a hot car interior on a sunny day, a phenomenon most people have experienced. If you teach spectroscopy, absorption lines are a natural transition to talking about how atoms and molecules absorb specific ranges of light.

Reflectivity (albedo): Most people can relate to the fact that a dark surface is less reflective, and therefore warms up more. The temperature of a terrestrial planet depends on its distance from the Sun, the greenhouse effect from its atmosphere (if any), and its albedo. Understanding albedo is also important for understanding the ice-albedo positive feedback loop from Arctic sea ice melt, which is causing the Arctic to warm at about three times the rate as the rest of the Earth (Ballinger et al. 2021)

The Earth's atmosphere: The Earth's atmosphere is about 100 km (60 mi) thick, which is incredibly thin compared to the radius of Earth—6,378 km (3,963 mi) at the equator. For scale, on a standard 30 cm (1 ft) globe of the Earth the atmosphere would be about the thickness of a sheet of paper. Our atmosphere of course creates the conditions necessary for life—not only providing the air we breathe but the atmospheric pressure necessary for liquid water on the surface. It is a major component of the Earth's overall climate system, enabling the movement of energy and material from land, ocean, and ice. Extending to an altitude of about 10 km, the lowest layer of the atmosphere—the troposphere—is where nearly all weather events occur. The troposphere contains about 75–80% of the mass of the atmosphere and nearly all of the water vapor. In general in the troposphere temperature and pressure decrease with altitude. Above it lies the stratosphere.

Greenhouse effect: Greenhouse gases (GHGs), such as water vapor, carbon dioxide (CO_2), and methane (CH_4), are relatively transparent to visible light but mostly opaque to the mid-infrared. Roughly half of the Sun's visible light is absorbed by the Earth's surface, where it is re-radiated in the infrared. Because of their high opacity, GHGs slow the flow of energy back to space, causing the surface of the planet to be warmer. The greenhouse effect of course is why Venus is so hot—much hotter than Mercury even though it is further away from the Sun. It is also why the temperature of Venus doesn't vary much day or night, unlike objects without an atmosphere (e.g., Mercury and the Moon). Overall the greenhouse effect on Earth is a good thing. Without it, Earth's surface would be on average about

33°C (59°F) cooler. But Venus is a good demonstration of how it is possible to have too much of a good thing.

Geologic histories of Venus, Earth, and Mars: Why did Venus turn into a raging furnace, even though it is similar in size (and therefore internal geologic activity) as the Earth? It is widely understood that Venus' dense CO_2 atmosphere is due to the loss of its ocean to space, but when this occurred is still unclear (Bullock et al. 2014). Why did Mars turn into a dry, barren world? One-third of the surface of Mars is thought to have been covered in water and ice (Clifford & Parker 2001) before the red planet lost its magnetosphere (e.g., Kass & Young 1995). And how did Earth's climate manage to avoid either of these fates? The geologic histories of all three planets give important perspectives on the "habitability" of a planet. An important lesson is that—regardless of its size, distance from its star, or other physical conditions—there's no guarantee a planet will become habitable, or remain so. Because of its importance, Chapter 2 is devoted to the geologic climate history of the Earth.

The carbon cycle: Unlike Mars and Venus, the Earth has remained remarkably habitable for life over the last four billion years. The carbon cycle is the process wherein CO_2 released into the atmosphere by volcanic activity is transported into the ocean via rain, where it forms carbonate rock and is eventually subducted via plate tectonics. The cycle is regulated by a "carbon thermostat" on the timescales of thousands to millions of years. Understanding this natural process, and the timescales on which it occurs, is useful for understanding the differences and impacts of anthropogenic CO_2 emissions.

Feedback loops: Understanding the carbon thermostat requires a basic apprehension of feedback loops. This is also useful to know when discussing other feedback loops in climate change, e.g., the ice-albedo positive feedback loop. Understanding feedback loops is also helpful for recognizing that many phenomena do not have a simple "cause and effect" relationship. For example, while there is clearly a close connection between global temperatures and atmospheric CO_2 levels in the Earth's geologic history, global warming often preceded the increase in CO_2 concentrations; e.g., during the last deglaciation by 200 to 1000 years (Shakun et al. 2012). If CO_2 causes warming, how could that be? Shouldn't it be just the opposite? In reality several feedback loops are at play. In particular, a warming event (e.g., by Milankovitch cycles) would warm the oceans and cause them to release more dissolved CO_2—which would set up a positive feedback loop. John Cook has an excellent analogy: "Do sparks cause fire? Or does fire cause sparks?" The answer of course is yes to both. Likewise, CO_2 in the atmosphere causes warming, but warming also causes CO_2 to be released into the atmosphere. Note that when talking about feedback loops it is more intuitive to describe negative feedback as "self regulating" (like a thermostat) and positive feedback as "self reinforcing" or a "vicious cycle." People will naturally think that a positive feedback loop is a good thing—it is positive after all—but of course it is usually just the opposite.

Seasons, Orbits, and Milankovitch cycles: Over the last two million years the Earth's climate has oscillated between cooler "ice ages" and warmer interglacial periods. These are the result of Milankovitch cycles, which combine variations in the

Earth's axial tilt, axial precession, and orbital ellipticity. Understanding that the seasons are caused by the tilt of the Earth's axis is a primary goal of most introductory astronomy classes. It's a natural step to extend that knowledge to include that a larger tilt causes more extreme seasonal variability. Likewise, understanding orbital eccentricity and precession are important.

Volcanic Eruptions: A common misconception is that volcanic activity is primarily to blame for climate change; i.e., that CO_2 emissions from volcanoes far exceeds that from human activity. But the reality is just the opposite. For example, the amount of CO_2 emitted by human activity in 2019 was 50–100 times larger than the annual emission rate associated with volcanic activity (e.g., Fischer et al. 2019). Even major volcanic eruptions such as Agung (1963), El Chichon (1982), and Pinatubo (1991) had no noticeable impact on atmospheric CO_2 levels. Whether intentionally or by accident, these numbers are often confused and volcanoes are claimed to be greater emitters—where it is just the opposite. Volcanic eruptions can, however, have dramatic short-term impacts on the climate by ejecting sulfate aerosols into the stratosphere that block sunlight and cool the Earth. For example, after the Mt. Pinatubo eruption in 1991 temperature dropped 0.5°C (Parker et al. 1996), but only for a few years before the aerosols were naturally removed from the atmosphere.

Mass extinction events: Whether human caused or not, the geologic history of the Earth also shows that rapid changes in its climate have led to mass extinction. For example, the mass extinction event at the end of the Permian period, about 252 million years ago, is thought to have been triggered by the massive release of CO_2 from Siberian Traps volcanism (Burgess et al. 2014).

Exoplanets: There are two common misconceptions regarding planets around other star systems. The first is that some of them are suitable for us to live on. This comes from our use of the phrase "habitable zone," which of course merely means a planet is the right distance from its star such that *in principle* its surface temperature could be within the right range for liquid water to occur on its surface. It doesn't mean that it actually does, much less that the world is actually habitable for the flora and fauna of Earth. As mentioned before, the planet would also require a suitably dense atmosphere. The phrase "Earth like" also suggests this, even though it is used only to suggest that it is similar *in size*. From that perspective Venus is "Earth like." Press releases about such planets also often have realistic-looking graphic artistic renderings of what these planets might look like, illustrations that are often very similar to Earth's appearance (e.g., with swirling white clouds). These illustrations are so detailed that many people don't realize that they aren't actual images, suggesting that we know much more about these worlds than we really do.

Second, these planets are too far away. The closest exoplanet to us, Proxima Centauri b, is over four light years away. In comparison, Mars at its closest to us is 3 light minutes; and currently we cannot send a single human this distance. It is useful for students to compare 3 minutes to 4 years to get a sense of this scale.

Astrobiology: This topic can be (and often is) an entire course on its own. Understanding the origin and evolution of life on Earth can give a profound appreciation of our uniqueness. In particular, the last term of the Drake equation—the lifetime of advanced civilizations—is an opportunity to reflect on humanity's very brief tenure on Earth, and to speculate on what challenges we face for surviving far into the future.

Radiometric dating and isotopes: We often teach about isotopes in the context of radiometric dating of Earth and moon rocks, as well as meteorites. Understanding isotopes can naturally lead to a discussion about how we know that CO_2 being added to the atmosphere is from the use of fossil fuels. Radiometric dating also plays an important role in discovering the widespread contamination of the environment from the use of leaded fuels (Patterson 1965).

Science media literacy: A worthwhile goal is for students to be better consumers of information about science as presented in the media. Stories online are often written as "click bait" to entice people to read them so that the website earns ad revenue. Astronomy is a popular topic that often invites such articles that are either incorrect or at least stretch the truth. There are of course also websites that spread misinformation and disinformation about climate change. So we'd like our students to be less vulnerable to them. Science media literacy is about developing those skills, and is discussed in Chapter 13.

Below is a list of topics in physics that are relevant to climate change:

Conservation of energy: This is of course one of the most fundamental principles in physics and a guiding force in understanding the "energy budget" of our planet. It is also important in understanding how we generate energy; e.g., fossil fuels represent chemical potential energy that originated from photosynthesis in plants millions of years ago; wind and hydropower represent the conversion of kinetic energy to electric energy; "pumped hydro" is the storage of energy as gravitational potential; and "regenerative braking" in electric vehicles is the conversion of kinetic into electric potential energy. This is also an opportunity to talk about "free energy" and "perpetual motion" devices that have long been proposed by pseudoscientists but are also now falsely touted as solutions to climate change.

Power and work: Power can be defined as the rate of doing work, or the rate at which energy is used or created. And the Work–Energy Principle states there is a relationship between the change in kinetic energy of an object and the work done on it. These are relevant, e.g., in the conversion of kinetic energy of the wind into electric potential energy, where the power generated is a function of wind velocity cubed.

Thermodynamics: A heat engine does work by converting "disordered" thermal energy into "ordered" kinetic energy. Internal-combustion cars as well as coal, natural gas, and nuclear power plants are examples of heat engines. Of course the second law of thermodynamics asserts fundamental limits on the efficiency of heat engines. The Carnot cycle describes the movement of energy from a cold to a hot reservoir, which is done by heat pumps.

Thermal capacity and expansion: The oceans have absorbed about 90% of the excess heat added to the climate system because of the high heat capacity of water. About a third of ocean level rise right now is being driven by the thermal expansion of water as it warms.[1] The rest is the addition of water to the oceans from melting land ice, from glaciers (like those in Alaska or the Himalayas) as well as Greenland and Antarctica.[2]

Stefan–Boltzmann law: The energy radiated by a blackbody is proportional to the temperature to the fourth power, and is therefore very sensitive to variations in temperature. It is then natural to wonder if variations in the Sun's temperature have changed its luminosity; and if this could explain climate change. The luminosity of the Sun—measured precisely from spacecraft since 1978—varies less than 0.1% during its 11-year cycle and is therefore not a factor in climate change (Foukal et al. 2006).

8.3 Climate Change in Astro 101

Like many universities we offer two introductory astronomy courses—one on the solar system and the other on everything beyond (i.e., stars, galaxies, and cosmology). For both classes I do a three-lecture section (about four hours in total) on climate change. In our "solar system" astronomy class this occurs after we've talked about terrestrial planets and atmospheres; and in our "stars, galaxies, and cosmology" class it occurs after exoplanets. The lectures are broken into three parts of roughly equal length: on the causes, the consequences, and the solutions.

Each section is intended to address common questions, concerns, and misconceptions. Common misconceptions about climate change include:

- It's cold out today, so how can there be global warming?
- It's natural, not human caused.
- Climate change is a problem for the future.
- It's being caused by the Sun.
- It's not that bad (and might actually be good).
- There isn't consensus among scientists.
- Scientists are fabricating or exaggerating the problem (e.g., "Climategate").
- Other countries (e.g., China) are to blame.
- Dealing with climate change will harm the economy.
- Natural gas is a "green" energy source.
- Solutions such as wind and solar are worse for the environment than fossil fuels.

People also often confuse climate change with other atmospheric problems, such as acid rain and the hole in the ozone layer.

The first misconception on the list shows a misunderstanding about the difference between climate and weather. The primary difference is that of location and timescale. Weather is the daily and hourly variation in conditions in a particular place, whereas climate is the long-term pattern of weather over the span of decades. The average temperature of Seattle in January is a question of climate. If you'll need

[1] https://phys.org/news/2022-02-sea-one-foot-decades-frequent.html

[2] https://sealevel.nasa.gov/faq/11/how-much-rise-should-we-expect-from-greenland-and-antarctica/

a raincoat tomorrow is a question of weather. Marshall Shepherd has a nice analogy: "weather is your mood and climate is your personality."

The other misconceptions are addressed in the three sections. The goal of the "causes of climate change" section is for students to understand the difference between natural and human-caused (anthropogenic) climate change, the sources and sinks of GHGs (collectively known as the "carbon budget"), and that 90% of climate change is therefore the result of the use of fossil fuels. In the "consequences" section the goal is for students to understand why climate change is such a serious problem, how it is impacting them now, and what could come in the future. The "solutions" section is to help them understand that we already have the solutions we need to solve climate change, what needs to be done, and why it has been so difficult to do so. More detail on the goals for each of these sections is discussed below.

8.3.1 History of Climate Science

A common misconception is that climate change was the invention of politicians—and in particular Al Gore, the former U.S. vice president. This is largely because of the 2006 film *An Inconvenient Truth*, which documents his efforts to educate people about global warming. The implication is that climate change was essentially fabricated for political purposes.

However, the study of climate science is much older than many imagine. Students are often surprised to learn that the first evidence for the greenhouse effect was discovered over 200 years ago, when in 1820 Joseph Fourier calculated that the Earth is warmer than it should be if only solar flux is considered. He speculated it could be related to the Earth's atmosphere, but had no evidence. That evidence came a few decades later when Eunice Foote (1856) measured that water vapor and CO_2 held the Sun's heat longer than other gases. She also recognized that these gases must play a role in the Earth's temperature, and that if the quantities of these gases varied in the Earth's atmosphere it would result in changing temperature. Further contributions by other scientists, such as John Tyndall, Svante Arrhenius, and T. C. Chamberlain established the foundation for the greenhouse effect and its role in the climate before the end of the 19th century. From pencil to supercomputer, Anderson et al. (2016) provide a review of the history of climate science calculations.

Of course since then the scientific evidence has continued to grow; and is now well established. Cook et al. (2013) quantified the scientific consensus on climate change by examining approximately 12,000 peer-reviewed scientific articles from 1991–2011. Their findings, and that of similar studies, showed that over 97% of these papers endorsed the view that climate change is real and the result of anthropogenic emissions. An analysis of the 2% of papers that reject anthropogenic global warming found that they contain a number of methodological flaws and common mistakes (Benestad et al. 2016). Lynas et al. (2021) found that greater than 99% of papers published since 2012 endorse anthropogenic global warming. As discussed in Chapter 12, understanding that there is an overwhelming scientific consensus is particularly important. It has been described as a "gateway belief" in that people are unlikely to accept other elements of climate change unless they first understand that scientists overwhelmingly agree that climate change is real and anthropogenic.

It's also worth noting that Al Gore was not the first major U.S. politician to talk about climate change. For example, in February 1965 the newly inaugurated U.S. president Lyndon Baines Johnson talked about the "steady increase in carbon dioxide from the burning of fossil fuels" in a special address to congress about environmental issues.[3]

8.3.2 The State of the Art

Our understanding of climate change is constantly improving. While the fundamentals of what causes climate change are firmly established, there is still much to learn—particularly in the areas of understanding the consequences for different temperature anomalies.

The Intergovernmental Panel on Climate Change (IPCC) is the United Nations body for assessing the science related to climate change. It was created in 1988 to provide regular scientific assessment reports on the impacts and risks of climate change, as well as options for addressing it through adaptation and mitigation. The IPCC prepares comprehensive Assessment Reports about the state of scientific, technical, and socio-economic knowledge on climate change, its impacts and future risks, and options for reducing the rate at which climate change is taking place. It also produces Special Reports on topics agreed to by its member governments, as well as Methodology Reports that provide guidelines for the preparation of GHG inventories. The IPCC is organized into three Working Groups that deal with specific aspects of climate change and a Task Force that develops methodologies to estimate national GHG emissions. It is important to note that the IPCC does not carry out its own research. Instead, hundreds of scientists volunteer their time and expertise as authors to assess the available scientific literature.

As of 2023 the IPCC has produced six Assessment Reports. The most recent assessment cycle (2015–2023) consists of eight reports in total. This includes three Special Reports (Global Warming of 1.5°C, Climate Change and Land, and The Ocean and Cryosphere in a Changing Climate) and a methodology report. These are followed by three reports, each of which is written by a separate Working Group: The Physical Science Basis (Working Group 1), Impacts, Adaptation, and Vulnerability (Working Group 2), and Mitigation of Climate Change (Working Group 3). The final report—the Synthesis Report—brings together the findings of all of these. In total 782 scientists are working across all three working groups, with hundreds more contributing authors. More than 66,000 peer-reviewed studies in total were analyzed by the working groups.[4]

IPCC reports also make predictions. The biggest uncertainty about the future is, arguably, how humanity will respond. In the Fifth Assessment Report (AR5), five

[3] https://www.presidency.ucsb.edu/documents/special-message-the-congress-conservation-and-restoration-natural-beauty

[4] https://www.ipcc.ch/site/assets/uploads/2022/04/AR6_Factsheet_April_2022.pdf

Representative Concentration Pathways (RCPs) were used to simulate future climate change. The RCPs were chosen to represent a broad range of possible climate outcomes, based on a literature review. The RCPs are defined by the amount of GHGs released in the future and therefore the total radiative forcing attained by 2100; e.g., RCP2.6 assumes 2.6 $W{\cdot}m^{-2}$ by that year. There are now a total of seven RCPs: from RCP1.9, which would limit the temperature anomaly to below 1.5°C, up to RCP8.5, which is likely a worst-case scenario. Chapter 3 describes in detail how climate models work.

The IPCC reports do not predict the probability of each scenario. There is disagreement on which scenario is most likely. RCP1.9 and RCP 2.6 are increasingly unlikely based upon current emission levels without the significant use of negative emission technologies in the future. Schwalm et al. (2020) argue that RCP8.5 is most consistent with historical CO_2 emissions and best matches out to 2050 under current policies. However Hausfather & Peters (2020) argue that RCP4.5 or RCP6.0 better match fossil fuel emissions relative to International Energy Agency scenarios. It has also been argued that existing fossil fuel reserves make RCP8.5 unrealistic (Ritchie & Dowlatabadi 2017). To be clear, even RCP4.5 is catastrophic. The IPCC Sixth Assessment Report has switched from RCPs to use Shared Socio-Economic Pathways (SSPs), scenarios of projected socioeconomic global changes up to 2100. They are used to derive GHG emissions scenarios with different climate policies. There's a greater focus on lower degrees of warming because of these scenarios. Levels of warming like 1.5°C and 2°C can be assessed more rigorously than in AR5.

Historically, the IPCC reports have underestimated the certainty that climate change is anthropogenic (Wigley & Santer 2013; Cook et al. 2013) as well as its severity (Brysse et al. 2013 and references therein). This is likely the result of the IPCC reports being written by consensus, where only the most certain results will be included. It may also reflect the political climate in which the IPCC reports are generated, where there is a reluctance to be perceived as being "alarmist." Another criticism is that the length of time between reports is too long, meaning that the reports don't necessarily reflect the latest understanding of the science.

The U.S. also releases its own reports on climate change. The Global Change Research Act of 1990 mandates that the U.S. Global Change Research Program (USGCRP) conduct a National Climate Assessment (NCA) every four years to evaluate scientific findings and uncertainties related to global change, analyze the effects of global change, and analyze the current and projected trends in global change, both human-induced and natural. As of the writing of this book, the most recent report is the Fourth National Climate Assessment (NCA4). Volume I (USGCRP 2017) consists of an assessment of the science of climate change, with a focus on the United States. Volume II (USGCRP 2018) assesses its impacts across the United States, now and throughout this century. A particularly valuable aspect of this report is that it contains chapters by region, making it easier to learn the local impacts as well as actions to address it. Development of the Fifth National Climate Assessment (NCA5) is currently underway, with anticipated delivery in 2023.

Other regions of the world also provide resources. For example, the Copernicus Climate Change Service (C3S)[5] provides authoritative information about the past, present, and future climate in Europe and the rest of the world.

8.3.3 Natural Climate Change

Perhaps the most common misconception about current climate change is that it is natural; i.e., it is not the result of human activity but part of natural cycles that we cannot (or should not) control. It's true that the climate changes for natural reasons; e.g., during the last three million years the Earth's transitions between ice ages and interglacial periods can be explained by Milankovitch cycles (e.g., Spiegel et al. 2010). It's also true that there have been abrupt periods of climate change that have occurred for natural reasons, e.g. during the Younger Dryas event about fifteen thousand years ago. That the Earth's climate has changed naturally, and even rapidly, of course doesn't prove that current climate change is natural any more than the fact that people regularly die of natural causes proves that someone wasn't murdered. It is helpful to understand natural climate change to gain a perspective on anthropogenic drivers.

When teaching natural climate change you may wish to explain what a "proxy measure" is, and how they are used to determine temperature prior to 1880—a time before which there were sufficient global measurements with accurate thermometers. A simple example can be found on the beach: the location of debris serves as a proxy measure of high tide. There are many different methods for the proxy measure of temperature, from tree rings to lake sediment. In the context of climate change perhaps the most important method uses ice cores from Greenland and Antarctica. These samples have been used to determine the atmospheric content of CO_2 and temperature over the last 800,000 years. This time scale is particularly relevant because modern humans, *Homo sapiens*, appeared between 200,000 and 400,000 years ago (Schlebusch et al. 2017).

Over the last 800,000 years eight transitions between ice ages and interglacial periods have occurred, during which there is a strong correlation between CO_2 and temperature. During this timespan atmospheric CO_2 levels vary between roughly 180 and 280 parts per million (ppm) on timescales of thousands of years. This natural variation stands in clear contrast to its spike to above 420 ppm (as of 2022) in the last century. The current epoch, known as the Holocene, is an interglacial period that started about 11,700 years ago (e.g., Walker et al. 2009). The start of the Holocene coincides with the rise of agriculture and civilization. A key learning goal is to understand that the Earth's climate, as we experience it now, has only existed for this very short period of time. There were times when the Earth was much colder; e.g., the Neoproterozoic "snowball Earth" about 600 million years ago (Ma) when atmospheric CO_2 levels were only about 100 ppm. There were also times when it was much hotter; e.g., during the Cambrian Explosion (540 Ma) when CO_2 levels were around 7000 ppm (Cummins et al. 2014).

The purpose of this content is to address a common misconception that we don't need to be worried about climate change because the Earth has been hotter before.

[5] https://climate.copernicus.eu

While it's true that the Earth has been hotter than now, *we weren't here then*, and neither were the plants and animals upon which we rely to survive. We also need to worry about how quickly we are changing the climate. The geologic history of the Earth is clear: rapid climate change leads to major mass extinction events (e.g., Song et al. 2021). In the last 500 million years there are five known such mass extinction events. We are now experiencing the sixth such event, wherein species are currently going extinct at 100–1000 times the normal rate because of climate change and other human activity (e.g, Pimm et al. 2014). Chapter 2 gives a detailed look at the history of climate on Earth, and the wide variety of proxy measures that allow us to understand what past climate was like.

The key points of this section are:

- The Earth's climate changes for natural reasons on timescales of thousands to millions of years.
- There were times when Earth was much colder and much hotter than now, but we weren't here!
- Over the last 3 million years, transitions between ice ages and interglacial periods have been caused by Milankovitch cycles.
- The climate as we experience it now has only been like this for the last 11,700 years, a miniscule fraction of the overall history of the Earth.

8.3.4 Human-caused Climate Change

The goal of this section is to help students understand what is causing anthropogenic climate change, and how we are certain it is not from natural causes. There is a wealth of evidence that shows that current climate change is anthropogenic; e.g., the increase of CO_2 (Boden et al. 1999) and the decline of O_2 (Keeling & Manning 2014) in the Earth's atmosphere matches the use of fossil fuels and other human activity; and the decline in the ^{13}C and ^{14}C isotope percentages in atmospheric CO_2 show that the added amounts must be from fossil fuels (e.g., Graven et al. 2020).

It is important for people to understand that the current state of the Earth's climate isn't natural. The last time there was this much CO_2 in the Earth's atmosphere, humans didn't exist. Humanity's fossil-fuel use, if unabated, risks taking us by 2050 to values of CO_2 not seen since the early Eocene 50 million years ago (Foster et al. 2017).

The National Network for Ocean and Climate Change Interpretation (NNOCCI)[6] program has introduced a useful framing of "Regular" vs. "Rampant" CO_2. "Regular" CO_2 is the result of natural processes on Earth, such as volcanoes and life. "Rampant" CO_2 is the result of unnatural processes, such as burning fossil fuels. It is these rampant emissions that are driving climate change.

Figure 8.1 illustrates the effects of these rampant emissions. Over the last 800,000 years atmospheric CO_2 levels have varied slowly and naturally between 200 and 300 ppm. Over the last century or so we have spiked CO_2 levels to over 420 ppm.

[6] https://nnocci.org

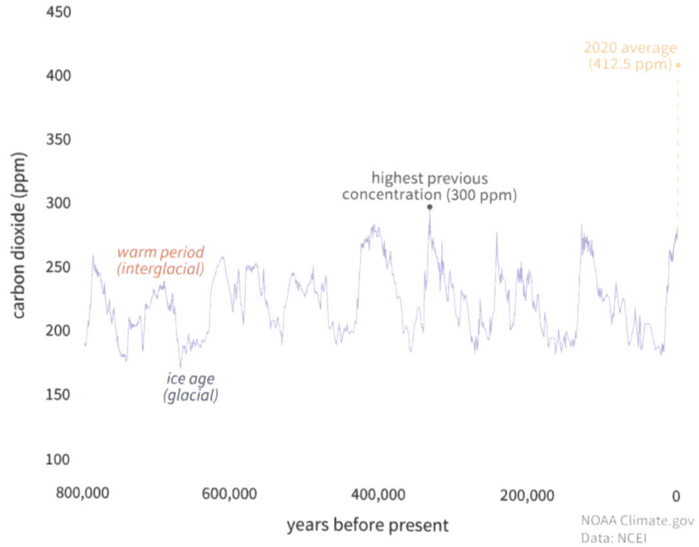

Figure 8.1. [7]Global atmospheric CO_2 concentrations in parts per million (ppm) for the past 800,000 years. The peaks and valleys track ice ages (low CO_2) and warmer interglacials (higher CO_2). During these cycles, CO_2 was never higher than 300 ppm. On the geologic time scale, the increase (orange dashed line) looks virtually instantaneous. Graph by NOAA Climate.gov based on data from Lüthi, et al., 2008, via NOAA NCEI Paleoclimatology Program.

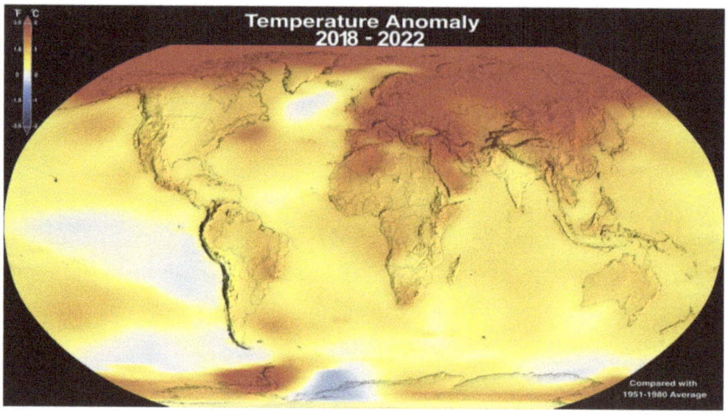

Figure 8.2. [8]This color-coded map in Robinson projection shows Earth's 5-year global temperature anomalies from 2018–2022, according to an analysis by NASA's Goddard Institute for Space Studies. Normal temperatures are calculated over the 30 year baseline period 1951–1980. Normal temperatures are shown in white. Higher than normal temperatures are shown in red and lower than normal temperatures are shown in blue. Credit: NASA's Scientific Visualization Studio.

Not all parts of the Earth are warming at the same rate, as shown in Figure 8.2. As mentioned before, the Arctic is warming at about three times the rest of the Earth largely because of the ice-albedo positive feedback loop (Ballinger et al. 2021).

[7] https://www.climate.gov/media/12987
[8] https://doi.org/10.25923/13qm-2576

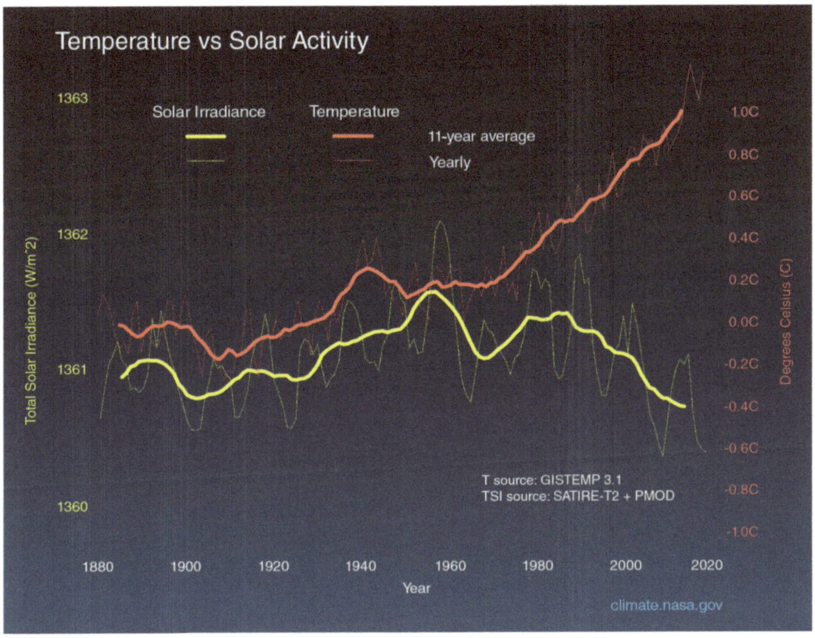

Figure 8.3. [9]The above graph compares global surface temperature changes (red line) and the Sun's energy received by Earth (yellow line) in watts (units of energy) per square meter since 1880. The lighter/thinner lines show the yearly levels while the heavier/thicker lines show the 11-year average trends. Eleven-year averages are used to reduce the year-to-year natural noise in the data, making the underlying trends more obvious. The two lines diverge after 1950, making it clear that variations in solar irradiance are not responsible for climate change. (Credit: NASA).

As the Earth absorbs energy from the Sun, it must eventually emit an equal amount of energy back out to space. The difference between incoming and outgoing energy is known as a planet's "radiative forcing" and is also known as its "energy budget." It is analogous to the balance of your bank account. The temperature of the surface of the Earth depends on these radiative forcings. If the amount of energy entering the Earth's climate system exceeds the amount leaving, the temperature will go up. Primary factors in the Earth's energy budget are, the Sun, the Earth's albedo, the GHGs in the atmosphere, aerosols, and internal variability in the Earth's climate system (e.g., El Niño).

The Sun is of course the source of energy that drives the Earth's climate system. So it is natural to suspect that variations in solar flux are responsible for climate change—and that is a common misconception. Figure 8.3 shows that the Sun's brightness does vary slightly—changing less than 0.1% over the last 150 years. However, during the timespan of 1960 to the present the Sun has dimmed slightly at the same time that the Earth's temperature has rapidly climbed. Thus it is clear that solar variability is not an explanation.

Clouds play an important role in the Earth's energy budget in a complex manner. Low-altitude clouds (at altitudes around one mile or less) tend to have a negative

[9] https://climate.nasa.gov/faq/14/is-the-sun-causing-global-warming/

radiative forcing (i.e., they work to cool the Earth) because they reflect sunlight and therefore increase the Earth's albedo. However high-altitude clouds have the net opposite effect. High-altitude cirrus tends to be thin, allowing solar flux to pass through but absorb infrared radiation. High clouds are also much colder, meaning they radiate less energy to space. The overall effect is that they contribute to warming the Earth.

Aerosols also play an important factor in the Earth's energy budget. To many people the word "aerosol" conjures images of spray cans or COVID-19, however aerosols are defined generally as solid or liquid particles suspended in the atmosphere—and can come from many different sources. Aerosols are less than one micron in size, and as such can remain airborne for long periods of time. Fog and dust are examples of naturally occurring aerosols. Another natural source are volcanic eruptions that inject sulfate aerosols into the stratosphere, where they are spread globally by strong winds. The net effect is to cool the Earth for a few years (e.g., Robock 2000). Anthropogenic aerosols include particulate air pollutants from the burning of fossil fuels, particularly diesel and coal. As discussed in the consequences chapter, the inhalation of particulate air pollutants results in the premature deaths of millions of people per year (Cohen et al. 2017).

Most aerosols reduce how much of the Sun's energy reaches the Earth's surface, either by absorption or scattering back out into space. A notable exception is black carbon, i.e., "soot." It is the dominant absorber of visible solar radiation in the atmosphere and is second only to CO_2 as a driver of climate change (Ramanathan & Carmichael 2008). However, the net effect of aerosols has been to cool the Earth. It is estimated that, without aerosols, the Earth would overall be 0.5°C to 1.1°C warmer than it is now (Samset et al. 2018). There have been concerns that reducing aerosol pollution would therefore cause global warming to worsen, particularly in heavily polluted areas. Nonetheless the health benefits from the reduction of air pollution would have an overall positive effect (Shindell & Smith 2019).

Over its geologic history, many of the naturally occurring changes in the Earth's climate have been driven (or "forced") by external factors, such as the Milankovich cycles. However there is internal variability that is the result of interactions between different elements of the Earth's climate system, such as ocean-atmosphere interactions. There are many internal variations that affect different regions of the Earth, and on different timescales. Perhaps the most famous example is the "El Niño Southern Oscillation" (ENSO) of ocean temperatures in the south Pacific ocean. The name "El Niño" (Spanish for "The Boy") originated from the discovery of the phenomenon by South American fisherman, who noticed that in some years the ocean was significantly warmer. The phenomena would start in December around the Christmas holiday, the time of the Christ child. In "El Niño" years, surface waters will be more than 0.5°C warmer than normal, whereas in "La Niña" ("The Girl") years it is just the opposite. Figure 8.4 shows how temperatures vary worldwide during El Niño and La Niña. The three most recent extreme El Niño events occurred in 1982–83, 1997–98, and 2015–16, but less extreme ones have happened in between (Ren et al. 2018). While ENSO oscillations are most pronounced in the southern Pacific ocean, their effects can be felt worldwide. It is

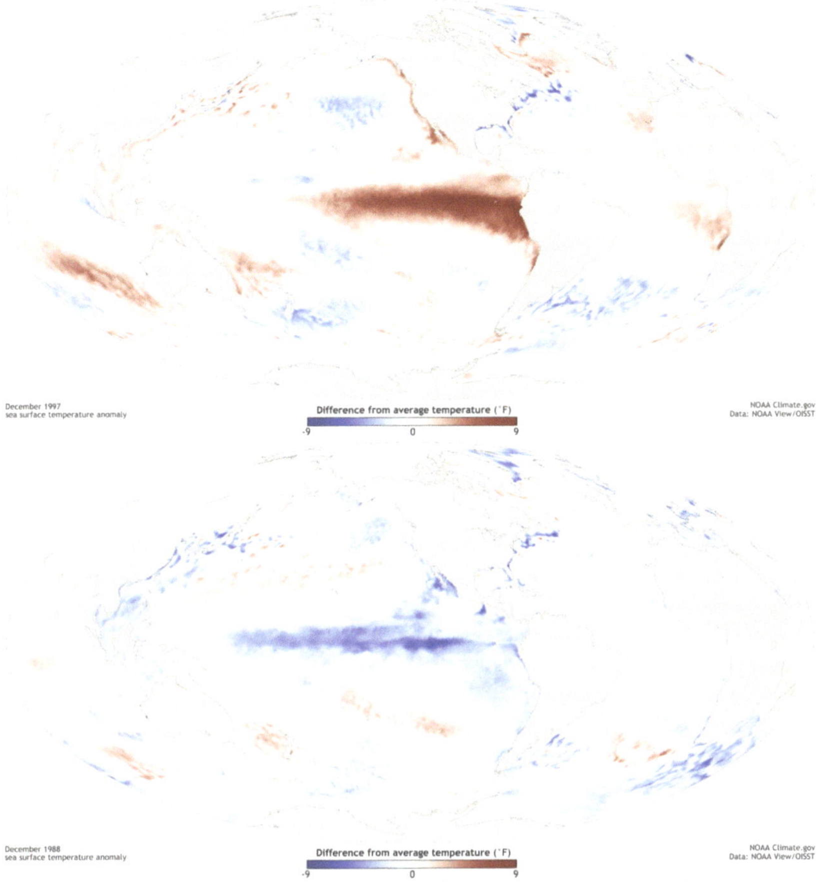

Figure 8.4. [10,11]Temperature changes in the ocean during the El Niño year of 1997 (top) and the La Niña year of 1988 (bottom). Credit: NOAA

important to note that internal variations like ENSO are separate from climate change. It is not yet clear how climate change is affecting these variations, but it is an active area of research (e.g., McPhaden et al. 2020; Geng et al. 2023).

8.3.5 Carbon Budget

The changes in temperature to the troposphere and stratosphere (Figure 8.5) are one of the many pieces of evidence that confirm that the source of global warming is the increase in GHGs in the atmosphere. When considering solutions, it is therefore important to think about the sources and "sinks" of these gases. This is collectively known as the carbon budget. While it is focused on CO_2, other GHGs are also

[10] https://www.climate.gov/sites/default/files/iconic_ENSO_elNino_lrg.jpg
[11] https://www.climate.gov/sites/default/files/iconic_ENSO_laNina_lrg.jpg

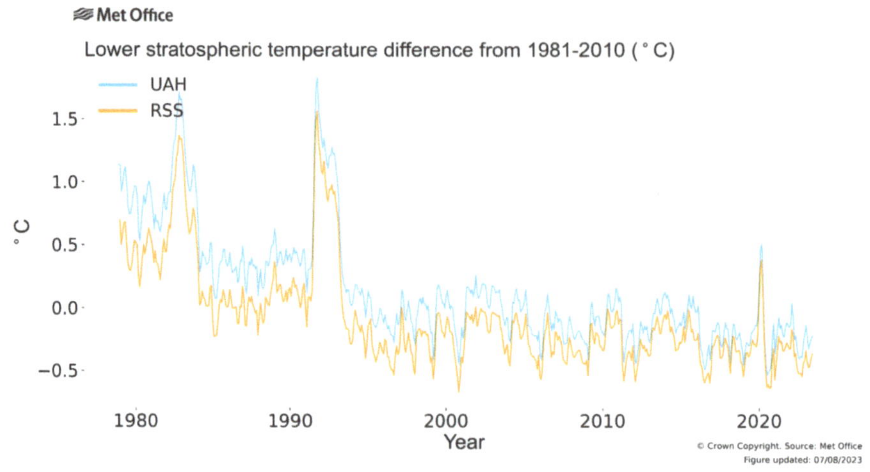

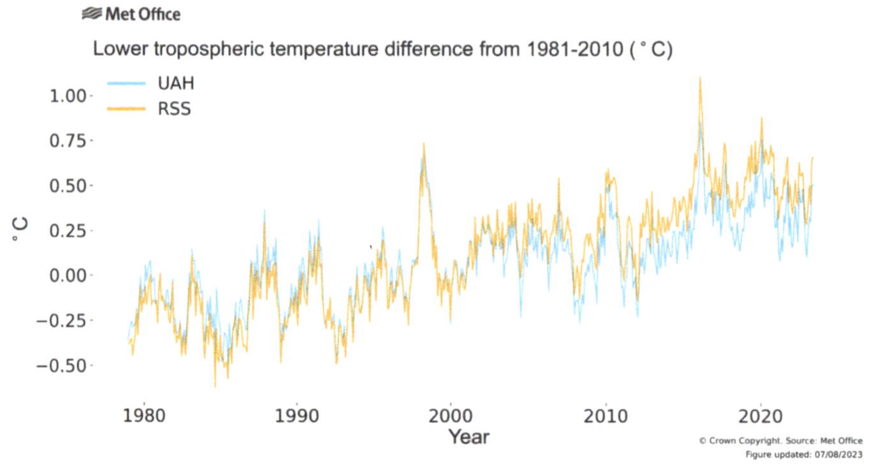

Figure 8.5. [12]These plots show the temperature record for the lower stratosphere (top) and lower troposphere (bottom) from 1981 to 2023. The troposphere extends from the Earth's surface to an altitude of 10 km; and the stratosphere extends over 10–50 km. Notice that the stratosphere is cooling as the troposphere warms, which is consistent with warming from GHGs and inconsistent with solar warming. The temperature spikes in the stratosphere are due to the El Chichon (1982) and Pinatubo (1991) volcanic eruptions. Ash emitted from these eruptions settled into the stratosphere and absorbed sunlight. The temperature spike in the troposphere in 1998 is the result of an unusually strong El Niño.

considered. Their effects are often included as a "carbon equivalent"—see Hofman et al. (2006) for a detailed explanation.

A common misconception is that CO_2 can't be that dangerous because it is a very small fraction of the overall composition of the atmosphere. However, chemical compounds can still have a significant impact even in small amounts; e.g., according

[12] https://climate.metoffice.cloud/upper_air.html.

to the U.S. Environmental Protection Agency (EPA), arsenic in drinking water is considered dangerous if it is more than 10 parts per *billion*.[13] Many people also confuse climate change with other atmospheric environmental problems, e.g., the ozone hole and air pollution (Dunlap 1998). This conflation is understandable since CO and CO_2 are also often conflated.

It is important to consider for how long the gas will naturally remain within the atmosphere. Because of the double covalent bonds between the C and O atoms, CO_2 is not highly susceptible to chemical reactions and ultraviolet photolysis. Once added to the atmosphere, it remains for 300 to 1,000 years.[14]

Of particular importance is methane (CH_4), which is a potent GHG. Methane is responsible for around 30% of the current rise in global temperature (IEA 2022). While more potent than CO_2, It lasts on average only about a decade in the atmosphere. It is slowly destroyed via reactions with the hydroxyl radical (OH) and ultraviolet photolysis. Thus a timescale needs to be specified when comparing the two gases. For example, over a 20-year period, CH_4 traps 84 times more heat per mass unit than CO_2, and has 105 times the effect when accounting for aerosol interactions (Shindell et al. 2009).

There are other important GHGs. Nitrous oxide (N_2O) has a global warming potential (GWP) 273 times that of CO_2 for a 100-year timescale, whose emissions are rapidly growing (Tian et al. 2020). Fluorinated gases include chlorofluorocarbons (CFCs), hydrofluorocarbons (HFCs), hydrochlorofluorocarbons (HCFCs), perfluorocarbons (PFCs), and sulfur hexafluoride (SF_6). They are powerful GHGs that are emitted from a variety of household, commercial, and industrial applications and processes; e.g., HFCs are often used in air conditioning units. The GWPs for these gases can be thousands or tens of thousands times worse than CO_2. In 2020, N_2O and fluorinated gases accounted for 7% and 3% of greenhouse emissions, respectively, in the U.S.[15] The phase out of CFCs through an international treaty, the 1989 Montreal Protocol, is also a climate success story. The reduction of global warming from CFCs has delayed the first ice-free summer in the Arctic by 15 years, with each ton of CFCs avoided responsible for preventing 7 km^2 of ice loss compared to a scenario with no elimination of fluorinated gases (England & Plvani 2023).

Water vapor also plays an essential role in the Earth's greenhouse effect—in fact, it is responsible for more than half of the overall greenhouse effect. However, water vapor behaves differently than other GHGs in that the amount of it in the atmosphere is controlled primarily by air temperature. If there's too much vapor it simply precipitates out as snow or rain. Other GHGs, primarily CO_2 and CH_4, are therefore necessary to sustain the presence of water vapor in the atmosphere. For that reason, it is usually considered to be a feedback agent, rather than a forcing. Anthropogenic emissions of water vapor (e.g., through irrigation or power plant cooling) have a negligible impact on the global climate because water vapor persists in the atmosphere for, on average, around nine days before precipitating (Myhre et al. 2013). Because of

[13] https://www.epa.gov/dwreginfo/chemical-contaminant-rules
[14] https://climate.nasa.gov/news/2915/the-atmosphere-getting-a-handle-on-carbon-dioxide/
[15] https://www.epa.gov/ghgemissions/overview-greenhouse-gases

this strong temperature dependence, the amount of water vapor in the atmosphere varies dramatically with latitude. A typical column of air extending from the surface to the stratosphere in polar regions may contain only a few kilograms of water vapor per square meter, while a similar column of air in the tropics may contain up to 70 kg.

It's important to note that CO_2 is a small but relatively heavy molecule—its atomic mass is 44 amu, only 12 amu (27%) of which comes from the carbon atom. For comparison CH_4 is only 16 amu. Liquid and solid forms of fossil fuels (i.e., petroleum and coal) consist of chains of hydrocarbons that can get quite long; e.g., 5–12 carbon atoms for gasoline (petrol) and several hundred carbon atoms for coal. In gasoline there are roughly two hydrogens per carbon atom, such that 87% of the mass is from carbon atoms. As counter-intuitive as it may be, the burning of one gallon (6.3 lb or 2.9 kg) of gasoline will generate about 20 lb (9 kg) of CO_2. That added mass of course comes from oxygen in the atmosphere. Unfortunately there has been a lack of consistency when reporting the mass associated with carbon emissions. Often carbon emissions are reported in units of C only, neglecting the oxygen atoms. To convert mass measurements in units of C to CO_2 you therefore need to multiply by a factor of 3.7.

When discussing carbon emissions, large numbers—in units of "gigatonnes" (Gt) and "teragrams" (Tg) are often used. Adding 2.1 GtC (i.e., 7.8 $GtCO_2$) to the atmosphere increases the atmospheric concentration by 1 ppm (part per million). For perspective, in 2019 an estimated 36.7 $GtCO_2$ were emitted by human activity (Friedlingstein et al. 2020). Figure 8.6 shows the average annual global carbon

Figure 8.6. [16]This graphic shows the average emissions of CO_2 for 2015, in billions of tons. About 90% of emissions are related to the use of fossil fuels, whereas most of the rest is due to changes in land use; e.g., the clearing of forest land for agricultural use. Roughly half of the emitted CO_2 remains in the atmosphere, whereas a quarter each is absorbed by land and oceans. This absorption has therefore slowed the buildup of CO_2 in our atmosphere, but there are consequences; e.g., ocean acidification. Credit: Global Carbon Project (Image: CC BY-SA) (Data source: Friedlingstein et al. 2022).

[16] http://www.globalcarbonatlas.org/en/content/global-carbon-budget.

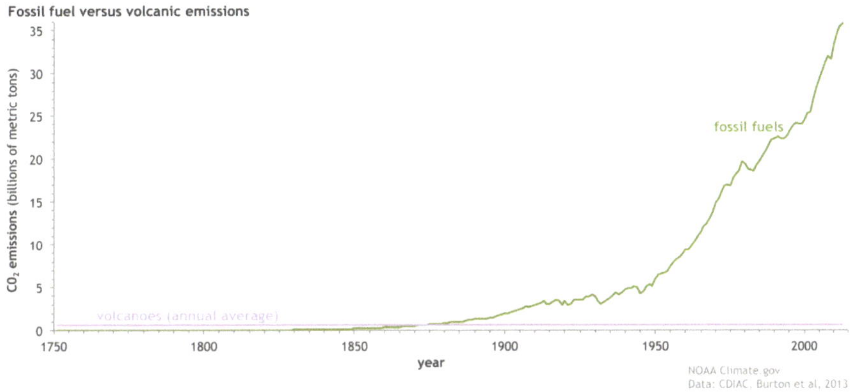

Figure 8.7. [17]Fossil fuel emissions of CO_2 in comparison to volcanic activity during 1750–2013. Since the start of the 20th century fossil fuel emissions have greatly exceeded volcanoes.

budget for the years 2011–2020, and Figure 8.7 shows how fossil fuel emissions have grown over time. Discussing the mass of CO_2 can be helpful for students to understand why the numbers are so large, but it can also lead to the misconception that the mass of CO_2 is the actual source of the problem.

Other natural sources of CO_2 include: forest fires, thawing permafrost, as well as organic respiration and decomposition. The signature of organic respiration and decomposition can be seen in the yearly atmospheric CO_2 levels (Figure 8.8). The growth and photosynthesis of vegetation in spring and summer months absorbs CO_2, causing global levels to drop. In fall and winter months, the cessation of photosynthesis and decay of vegetation causes CO_2 levels to rise. This oscillation is dominated by seasonal cycles in the Northern Hemisphere, which contains 68% of the world's land mass. This annual cycle has been described as the inhalation and exhalation of the Earth's lungs. However it doesn't contribute to the overall increase in CO_2 levels.

The Earth's biomass stores a tremendous amount of CO_2. For example, about 50 GtC are stored in forest biomass in the U.S. alone.[18] Much of this CO_2 is at risk of being released during a fire. The massive forest fires in Australia in 2019–20 released an estimated 715 $MtCO_2$ (van der Velde 2021), and in 2019 the forest fires in British Columbia, Canada, emitted an estimated 190 million tonnes of GHGs, nearly triple B.C.'s annual carbon footprint from all other sources.[19]

Where does the carbon go? Our oceans are by far the largest carbon sinks, storing about 37,000 GtC dissolved inorganic carbon—which is more than 40 times than the atmosphere (Friedlingstein et al. 2022). Soil is also a carbon sink because it consists of dead plant life (which stores carbon) as well as microscopic life forms (e.g., bacteria, fungi, and algae). Soils store about 1700 GtC. But over half of agriculture soils have been degraded or eroded due to intensive farming practices

[17] https://www.climate.gov/media/7289

[18] https://cfpub.epa.gov/roe/indicator.cfm?i=86

[19] https://www.cbc.ca/news/canada/british-columbia/it-s-alarming-wildfire-emissions-grow-to-triple-b-c-s-annual-carbon-footprint-1.4259306

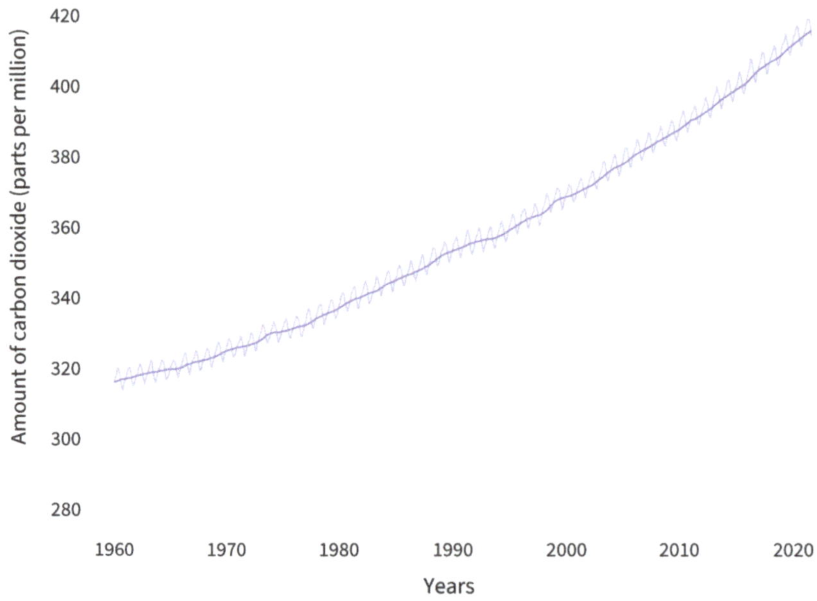

Figure 8.8. [20]The modern record of atmospheric CO_2 levels obtained from Mauna Loa Observatory in Hawai'i. The seasonal cycle of highs and lows (small peaks and valleys) is driven by organic respiration and decomposition from summertime growth and winter decay of Northern Hemisphere vegetation. Notice that the Agung (1963), El Chichon (1982), and Pinatubo (1991) volcanic eruptions had no effect on atmospheric CO_2 levels.

such as the use of synthetic fertilizers that can kill microbial life. This causes the soil to be less able to store carbon (Kopittke et al. 2019). Permafrost, as the name implies, is a layer of Earth that remains permanently frozen throughout the year. It occurs primarily in arctic and subarctic regions. Permafrost is important because it traps plant and animal material. It stores about 1400 GtC—more than subterranean gas, oil, and coal reserves and about twice that found in the atmosphere (Schuur et al. 2015). Forest fires (e.g., Sidik 2020) and the melting of permafrost (e.g., McCalley et al. 2014) are positive feedback loops (i.e., "vicious cycles") that can trigger further CO_2 and methane emissions.

What is clear is that the increase in atmospheric CO_2 levels is overwhelmingly the result of human activity (i.e., "anthropogenic emissions"), and in particular our use of fossil fuels. The plot below shows how atmospheric CO_2 levels have increased from pre-industrial levels of 280 ppm to current levels greater than 410 ppm. This increase matches anthropogenic emissions over the same time period (Figure 8.9).

[20] https://www.climate.gov/media/13611

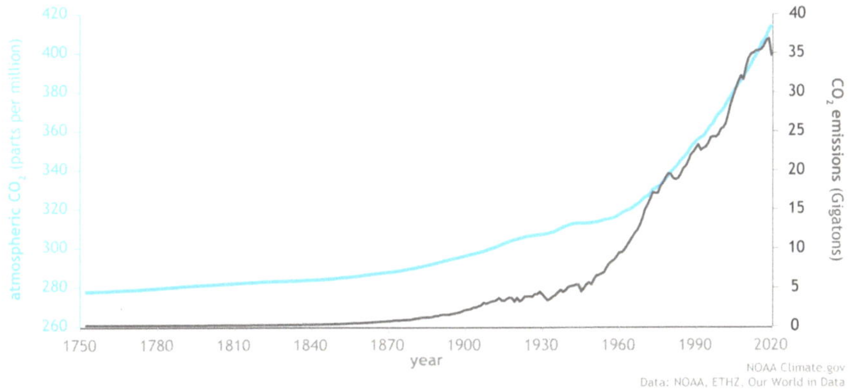

Figure 8.9. [21]The amount of carbon dioxide in the atmosphere (blue line) has increased along with human emissions (gray line) since the start of the Industrial Revolution in 1750. Emissions rose slowly to about 5 billion tons per year in the mid-20th century before skyrocketing to more than 35 billion tons per year by the end of the century. NOAA Climate.gov graph, adapted from original by Dr Howard Diamond (NOAA ARL). Atmospheric CO_2 data from NOAA and ETHZ. CO_2 emissions data from Our World in Data and the Global Carbon Project.

In summary, scientists have ruled out any natural phenomena as the explanation for the increase in atmospheric CO_2 levels. It is also clear that the measured increases in temperature on Earth are a result of these increases. Indeed, these temperature increases are exactly what one would expect. As part of its Sixth Assessment Report, the IPCC (2021) has stated that, "It is unequivocal that human influence has warmed the atmosphere, ocean and land. Widespread and rapid changes in the atmosphere, ocean, cryosphere and biosphere have occurred."

Who then is responsible? Is China to blame? It's true that in 2006 China overtook the U.S. as the world's largest producer of CO_2, and emitted roughly twice that of the U.S. in 2022 (IEA 2023). At the same time that China's emissions rates have skyrocketed, the U.S. has declined—as of 2021 emitting 20% less CO_2 than at our peak in 2007.[22] However on a per capita basis, Americans still have one of the highest rates of CO_2 emissions in the world. The average American emits about 15 tons of CO_2 per year, which is twice that of China, and seven times that of India.[23] The U.S. has produced more cumulative emissions than any other country—comparable to the 28 countries of the European Union and twice that of China (Hickel 2020).

How much can we continue to add to the atmosphere? As of 2022 the temperature anomaly has already reached about 1.1°C above the 20th century average (IPCC 2023). Assuming emissions continue at 2022 levels, the remaining carbon budget for

[21] https://www.climate.gov/media/12990
[22] https://www.eia.gov/todayinenergy/detail.php?id=52380
[23] https://ourworldindata.org/explorers/co2

a 50% likelihood to limit global warming to 1.5°C, 1.7°C, and 2.0°C from the beginning of 2023 are equivalent to 9, 18, and 30 years (Friedlingstein et al. 2022).

8.3.6 Consequences

This is of course why we care. The consequences of climate change are far reaching, and Chapters 4 and 5 discuss many of them. It is impossible to cover them all, nor should you try. In choosing what to discuss, focus on consequences that are happening now where your audience lives, as talking about local impacts is more motivational (van der Linden 2015). And focus on consequences that impact people's livelihoods, their health, and their culture. Fortunately there are resources available to explain this for many parts of the world. For example, in the U.S., the National Climate Assessment report (USGCRP 2018) assesses the impacts, risks, and adaptation by region. For example, when talking about climate change in Alaska I often focus on ocean acidification, and how it threatens several major species important to our state—such as king and tanner crab as well as salmon. While most people know about global warming, few know about ocean acidification. The fishing industry is the largest employer in Alaska, and is therefore a major economic driver of the state. Fishing is also important for recreation and subsistence, and is part of our culture and identity. It is therefore a topic that most Alaskans care about. However, localized events can have global impacts. As the Canadian wildfires of spring and summer 2023 have shown to a majority of the residents of the U.S., even communities that are thousands of kilometers away from the nearest fire may have its air quality negatively affected by smoke that has been blown across a continent.

In general most places are being affected by warmer temperatures. Severe climate events (such as drought and flooding) are becoming more common and extreme; e.g., floods that would historically occur on average every 100 years now occur every 10 years. Severe weather events (e.g., blizzards, tornados, hurricanes, and fires) are also becoming more common and more extreme. If one of these is common in the region of concern you may wish to offer more detail why. For example, for tropical cyclones (i.e., hurricanes and typhoons) it is helpful for people to understand that such storms build over oceans in the summer and fall, deriving their energy from the warm water. Warmer water from climate change therefore means stronger storms. Warmer air can also hold more water, so tropical cyclones are now stronger *and* wetter, meaning more flooding when they make landfall. Hurricanes Katrina in 2005, Sandy in 2012, Harvey in 2017, and Ian in 2022 all resulted in heavy flooding, resulting in hundreds to thousands of fatalities as well as heavy economic losses.

Often there are multiple contributing factors to a situation. For example, in the western U.S.—and many other parts of the world—forest fires are now a widespread concern. Higher temperatures mean less snowpack, leading to drought conditions. Higher temperatures also mean drier conditions, so once healthy forests are now dangerously dry. Fires that were once small and more easily contained are now turning into "mega fires" that rapidly run out of control. Poor forest management is often blamed for forest fires—and that may be in some cases a contributing factor. It is important to acknowledge when it is, but that doesn't change the fact that climate change is making forest fires more likely, and more dangerous. In general we need to take a different mindset about natural disasters. Don't ask, "did climate change cause this?" Instead ask, "did climate change make it worse?"

As discussed in detail in Chapter 5, a major consequence of climate change is that it is adversely impacting the Earth's capacity to provide nutrition for its inhabitants—a worsening problem as the population continues to grow. In other words, there will be less food when we need more. Without adaptation, by 2050 climate change will reduce global agriculture yields up to 30%[24] while food demand goes up 50–100% (Valin et al. 2014) as the population of the Earth approaches 10 billion people (United Nations 2022).

8.3.7 Solutions

As mentioned before, it is critical that you talk about solutions. Students need to know that there is a path forward. Chapter 6 offers a framework on just that. On one hand the solution is fairly simple. About 90% of climate change is the result of the extraction and use of fossil fuels; i.e., coal, petroleum, and natural gas. So addressing climate change is as straightforward as not using them. The challenge is that fossil fuels are used for just about everything. Chapter 7 goes into depth about every renewable and carbon-free energy source available to us, and how we can use them to become carbon neutral. The good news is that most of these solutions are now available and are economically viable. In fact, most of them are such good ideas that we'd want to adopt them even if climate change wasn't an issue. In particular, the air pollution caused by the use of fossil fuels is the world's leading killer—responsible for 1 in 5 deaths (Vohra et al. 2021).

"Geoengineering" is particularly well suited to discussion in a physics or astronomy class. It is defined as the deliberate large-scale manipulation of an environmental process that affects the Earth's climate, in an attempt to counteract the effects of global warming. This includes the injection of aerosols into the stratosphere, likely by airplane or balloon, to block sunlight from reaching the surface. In principle it could cool the Earth but wouldn't undo other effects of elevated CO_2 levels, such as ocean acidification. Potential side effects include that the aerosols could cause health problems, affect the ozone layer, and decrease photosynthesis in plants (leading to lower crop yields). Aerosols would need to be continuously supplied to the atmosphere, and temperatures would spike if they were stopped. Other geoengineering techniques include carbon sequestration, albedo modification (e.g., marine cloud brightening), and thinning of stratospheric cirrus.

There are also many other high-impact solutions that don't necessarily involve energy generation or consumption—such as education of women, reforestation, adoption of plant-rich diets, and the use of refrigerants that aren't potent GHGs. Project Drawdown[25] offers a comprehensive breakdown of all the solutions available, and their impact. Every semester I give my students an assignment to

[24] https://gca.org/reports/adapt-now-a-global-call-for-leadership-on-climate-resilience/
[25] https://www.drawdown.org

use Project Drawdown to identify the most impactful solutions as well as find a solution that interests them. The hope is that they may be inspired to develop a deeper understanding of that solution, and possibly advocate for it or pursue it professionally. Again, many people are wondering what it is they can do. Here is a way to find solutions that resonate with their interests and values.

Another valuable teaching resource is the En-ROADS global climate simulator.[26] It is an economic simulator that allows the user to explore the impact of roughly 30 policies—such as electrifying transport, pricing carbon, and improving agricultural practices—on hundreds of factors like energy prices, temperature, air quality, and sea level rise. Every semester I challenge my students to use the En-ROADS simulator to create a scenario that limits the temperature increase to below 1.5°C while keeping energy costs down and growing the economy. It's not hard. The key lesson is that there's no one single solution. But when we use the solutions available to us it's possible.

You should be careful when talking about individual choices. As discussed in Chapter 12, the use of fossil fuels is often portrayed as a personal responsibility so as to shift the blame from the producers to the consumer. But someone is not to blame if they live and work in a place that requires the use of fossil fuels; e.g., you can't fault people for driving to work if viable alternatives aren't available. And you can't fault them if they can't afford an EV (yet!) That's not to say that personal choices don't matter. Small actions can lead to big outcomes. They normalize concern about climate change and can be inspirational to others; e.g., people often don't consider purchasing an EV or residential solar until a friend, colleague, or neighbor did. To paraphrase Gandhi (as others have), "Be the change you wish to see in the world."

8.3.8 Teach the Controversy

If the solutions to climate change are ready—and if the problem is as bad as the science indicates—why aren't we doing more? And why did it become so politicized? Needless to say, fossil fuels are a big business. Combined, the top ten fossil fuel companies generated $2.24 trillion in revenue in 2022. If the fossil-fuel industry was its own country it would be among the top ten biggest GDPs in the world.[27] As discussed in Chapter 13, the sad reality is that fossil-fuel and related industries have worked for decades to stop or slow the transition to renewable energies, even though they have known since the 1980s about the danger. Much of the climate disinformation and politicization is the result of these efforts.

Many scientists are reluctant to engage in discussion about climate change because of this politicization. But it's important to keep in mind that we didn't politicize it—the fossil-fuel industry did. And it's ok to share with students the well-documented evidence of their efforts to do so; e.g., funding by fossil fuel corporations of organizations and think-tanks that spread information that is

[26] https://www.climateinteractive.org/en-roads/
[27] https://time.com/6254086/fossil-fuel-2022-revenue-vs-country-gdp/

skeptical or in denial about climate change and its solutions. You can also share examples of other industries that have followed similar strategies regarding products that were found to be dangerous through the scientific process; e.g., tobacco, leaded gasoline, and asbestos. We wish for students to understand that scientific results are often politicized when they are a threat to an industry—what's happening now is nothing new.

You can also share examples of how through science we've learned that something we do is dangerous, and that led to systematic changes that solved the problem. The above examples illustrate that—while it took many decades to overcome resistance from industry—leaded gasoline and asbestos are mostly banned, and rates of smoking have greatly decreased. There are other examples where change occurred more rapidly because there wasn't as much resistance from industry; e.g., DDT and its effects on bird populations, and CFCs and the ozone layer. These success stories are discussed in more detail in Chapter 12.

8.4 Conclusions

Regardless of the class, the outcomes that I wish for my students to understand include:

- The difference between natural and anthropogenic climate change
- That 90% of anthropogenic climate change is from the use of fossil fuels
- How climate change is a threat to them personally
- The worst consequences can still be avoided if we act quickly
- Solutions are available now that will make our lives better and grow the economy
- There are ways for them to work on parts of the solution

Ultimately students should feel concerned about the problem but hopeful that there will be a world worth living in *after* we solve climate change. Of all the goals we may have for our students perhaps that is the one that is most important. Chapter 12 goes into detail about how we can talk with students (and the public) to overcome despair and increase engagement.

You are welcome to use my Climate Change 101 slideset. It consists of five parts, each about 30 minutes long:

1. Natural climate change.
2. Anthropogenic climate change.
3. Consequences.
4. Solutions.
5. Why the controversy?

You may modify them as you wish. Note that many of these slides, particularly the graphs and information about solutions, need regular updating.

References

Anderson, T. R., Hawkins, E., & Jones, P. D. 2016, Endeavour, 40, 178

Ballinger, T. J. 2021, NOAA Arctic Report Card 2021: Surface Air Temperature

Benestad, R. E., et al. 2016, ThApC, 126, 699

Boden, T. A., Marland, G., & Andres, R. J. 1999, Global, Regional, and National Fossil-Fuel CO_2 Emissions (1751—2014), Report (V. 2017), *U.S. Department of Energy*

Brysse, K., Oreskes, N., O'Reilly, J., & Oppenheimer, M. 2013, GEC, 23, 327

Bullock, M.A., Harder, J.W., Simon-Miller, A.A., & Mackwell, S.J. 2014, Comparative Climatology of Terrestrial Planets (Tucson, AZ: Univ. Arizona Press) muse.jhu.edu/book/28817.

Burgess, S. D., Bowring, S., & Shen, S. 2014, PNAS, 111, 3316

Clifford, S. M., & Parker, T. J. 2001, Icar, 154, 40

Cohen, A. J., Brauer, M., Burnett, R., et al. 2017, Lancet, 389, 1907

Cook, J., et al. 2013, ERL, 8, 024024

Cummins, E. P., Selfridge, A. C., Sporn, P. H., Sznajder, J. I., & Taylor, C. T. 2014, Cell. Mol. Life Sci., 71, 831

Dunlap, R. E. 1998, IntSociol, 13, 473

England, M. R., & Polvani, L. M. 2023, PNAS, 120, e2211432120

Fischer, T. P., et al. 2019, NatSR, 9, 18716

Foote, E. AmJSA, 22, 382 1856,

Foster, G. L., Royer, D. L., & Lunt, D. J. 2017, NatCo, 8, 14845

Foukal, P., Fröhlich, C., Spruit, H., & Wigley, T. M. L. 2006, Natur, 443, 161

Friedlingstein, P., et al. 2020, ESSD, 12, 3269

Friedlingstein, P., et al. 2022, ESSD, 14, 4811

Geng, T., et al. 2023, Natur, 619, 774

Graven, H., Keeling, R. F., & Rogelj, J. 2020, GBioC, 34, e2019GB006170

Hausfather, Z., & Peters, G. P. 2020, Natur, 577, 618

Hickel, J. 2020, LaPH, 4, e399

Hickman, C., et al. 2021, LaPH, 5, e863

Hofmann, D. J., et al. 2006, TellB, 58, 614

IEA 2022, Global Methane Tracker. https://www.iea.org/reports/global-methane-tracker-2022 (accessed 6.23.23).

IEA 2023, CO_2 Emissions in 2022—Analysis. https://www.iea.org/reports/co2-emissions-in-2022 (accessed 6.23.23).

IPCC 2021, Summary for Policymakers. In: Climate Change 2021: The Physical Science Basis. Contribution of Working Group I to the Sixth Assessment Report of the Intergovernmental Panel on Climate Change (eds.) Masson-Delmotte, V. et al.

IPCC 2023, Summary for Policymakers. In: Climate Change 2023: Synthesis Report. Contribution of Working Groups I, II and III to the Sixth Assessment Report of the Intergovernmental Panel on Climate Change [Core Writing Team, H. Lee and J. Romero (eds.)]. IPCC, Geneva, Switzerland, pp. 1–34, doi: 10.59327/IPCC/AR6-9789291691647.001

Kass, D. M., & Yung, Y. L. 1995, Sci., 268, 697

Keeling, R. F., & Manning, A. C. 2014, ed. H. D. Holland, & K. K. Turekian Treatise on Geochemistry (2nd edn.; Oxford: Elsevier) 385

Kopittke, P. M., Menzies, N. W., Wang, P., McKenna, B. A., & Lombi, E. 2019, EnI, 132, 105078
Lynas, M., Houlton, B. Z., & Perry, S. 2021, ERL, 16, 114005
McCalley, C. K., et al. 2014, Natur, 514, 478
McPhaden, M. J., Santoso, A., & Cai, W. 2020, El Niño Southern Oscillation in a Changing Climate (Washington, DC: American Geophysical Union) 1
Myhre, G., et al. 2013, Climate Change 2013: The Physical Science Basis. Contribution of Working Group I to the Fifth Assessment Report of the Intergovernmental Panel on Climate Change, ed. T. F. Stocker, et al. (Cambridge: Cambridge Univ. Press) https://www.climatechange2013.org/images/report/WG1AR5_Chapter08_FINAL.pdf
Parker, D. E., Wilson, H., Jones, P. D., Christy, J. R., & Folland, C. K. 1996, IJCli, 16, 487
Patterson, C. C. 1965, AEnH, 11, 344
Pimm, S. L., et al. 2014, Sci., 344, 1246752
Ramanathan, V., & Carmichael, G. 2008, NatGe, 1, 221
Ren, H.-L., Lu, B., Wan, J., Tian, B., & Zhang, P. 2018, JMetR, 32, 923
Ritchie, J., & Dowlatabadi, H. 2017, EE, 65, 16
Robock, A. 2000, RvGeo, 38, 191
Samset, B. H., et al. 2018, GeoRL, 45, 1020
Schlebusch, C. M., et al. 2017, Sci., 358, 652
Schuur, E. A. G., et al. 2015, Natur, 520, 171
Schwalm, C. R., Glendon, S., & Duffy, P. B. 2020, PNAS, 117, 19656
Shakun, J. D., et al. 2012, Natur, 484, 49
Shindell, D. T., & Smith, C. J. 2019, Natur, 573, 408
Shindell, D. T., et al. 2009, Sci., 326, 716
Sidik, S. M. 2020, Feedback Loops of Fire Activity and Climate Change in Canada. EOS. http://eos.org/articles/feedback-loops-of-fire-activity-and-climate-change-in-canada (accessed 6.23.23).
Song, H., et al. 2021, NatCo, 12, 4694
Spiegel, D. S., Raymond, S. N., Dressing, C. D., Scharf, C. A., & Mitchell, J. L. 2010, ApJ, 721, 1308
Sturgis, P., & Allum, N. 2004, Public Underst Sci., 13, 55
Suldovsky, B. 2017, The Information Deficit Model and Climate Change Communication Oxford Research Encyclopedia of Climate Science (Oxford: Oxford Univ. Press)
Tian, H., et al. 2020, Natur, 586, 248
United Nations, Department of Economic and Social Affairs, Population Division 2022, World Population Prospects 2022: Ten Key Messages. http://population.un.org/wpp/publications/
USGCRP 2017, Climate Science Special Report: Fourth National Climate Assessment, Volume I, ed. D. J. Wuebbles, et al. (Washington, DC: U.S. Global Change Research Program) 470
USGCRP 2018, Impacts, Risks, and Adaptation in the United States: Fourth National Climate Assessment, Volume II, ed. D. R. Reidmiller, et al. (Washington, DC: U.S. Global Change Research Program)
Valin, H., et al. 2014, Agric. Econ., 45, 51
van der Linden, S., Maibach, E., & Leiserowitz, A. 2015, Perspect Psychol Sci., 10, 758
van der Velde, I. R., et al. 2021, Natur, 597, 366
Vohra, K., et al. 2021, ER, 195, 110754
Walker, M., et al. 2009, JQS, 24, 3
Wigley, T. M. L., & Santer, B. D. 2013, ClDy, 40, 1087

Chapter 9

Teaching with Inspiration, Not (Only) Fear

Jeffrey Bennett

> "Humanity is now standing at a crossroads. We must now decide which path we want to take. How do we want the future living conditions for all living species to be like?"
>
> —Greta Thunberg

Many of today's students will live beyond the year 2100, yet future impacts of global warming are often portrayed only as a choice between bleak and bleaker. This is counterproductive for teaching, since it risks scaring our students and alienating their parents, particularly those who may not yet take the threats seriously. In this chapter, I will discuss how you can teach about global warming in a scientifically honest way while providing students with inspiration, not fear. Part of the key to this lies in focusing on the fact that, if we deal with this issue successfully over the next few decades, our students could someday live in a "post-global warming world" in which the threat of global warming will have largely become a thing of the past.

9.1 Introduction

Consider this simple fact: Most of today's students entering college were born after the year 2000, and elementary and middle school students after the year 2010. These students therefore have a very good chance of still being alive beyond the year 2100, particularly if medical science can continue to increase life expectancies. Now ask yourself: when was the last time you heard the media talk about the impacts of global warming beyond about the year 2050, let alone 2100? You'll quickly realize that the media time horizon is quite different from the time horizon that matters to our students, and the latter is the one that matters if we want to provide meaningful and engaging climate education.

This fact also leads to an important corollary. If we focus only on the threats posed by global warming, the future can look very bleak. Therefore, if we want to teach our students successfully, we must find ways to inspire them about what the

doi:10.1088/2514-3433/acfcb6ch9

future could be like, since inspiration is far more likely to create positive outcomes than fear.

More specifically, I've found it useful in the classroom (and for the public) to focus on how to build what we might call a "post-global warming world," meaning a world in which global warming can be discussed as a past problem that was solved, rather than as an ongoing threat. This kind of world may seem far-fetched in light of current political realities, but in principle it is achievable. Indeed, it really requires only two "simple" steps: (1) stop greenhouse emissions quickly, so that we can stop making the problem worse; and (2) develop technology to reduce the atmospheric carbon dioxide (CO_2) level back down to a more reasonable level,[1] so that we could actually restore the climate to a better state. If we can successfully achieve both steps, it's quite possible that many or most of today's students could live to see a post-global warming world that will be far more prosperous, peaceful, and just than our world today. That kind of vision is not only inspirational, but can transcend the kinds of political differences that hinder action today. In this chapter, I'll discuss some of the ways that we, as astronomers, can put this vision to work in our classrooms and with the public.

9.2 The Role of Astronomers

In the past, most astronomers tended to shy away from teaching about a topic like global warming or climate change.[2] However, as the recently formed group Astronomers for Planet Earth emphasizes, it is important to put aside that shyness and use our platform as astronomers to help educate both students and the public. Chapters 6, 8, and 12 address these reasons more in depth, but here are a few key points:

- While we may think of climate science as being very different from astronomy, the differences are not so important to the public, since all practitioners of science are considered "scientists" first, with disciplines being only secondary.
- Climate science has deep connections to astronomy (and astrobiology). A few examples:
    - The underlying physics (e.g., fluid dynamics, radiative transfer) used to understand Earth's climate is the same as that used to understand planets, stars, and the interstellar and intergalactic medium.
    - We use other planets, such as Venus and Mars, in building scientific understanding of the greenhouse effect and of planetary atmospheres in general.

[1] For example, climate scientist James Hansen, currently at Columbia University, has calculated that a good target would be 350 ppm.

[2] I tend to use the two terms interchangeably, but note the subtle distinction that *global warming* generally refers to the rise in global average temperature, while *climate change* refers to the resulting changes in local and regional climates.

- The modeling used to understand global warming and its consequences is essentially the same as that used to study other planetary atmospheres and to explore the potential atmospheres of exoplanets.
- Astronomical understanding of the Sun and stars is necessary for understanding Earth's climate, since the Sun is the primary source of energy for Earth's surface, atmosphere, and oceans.
- An understanding of Earth's climate regulation processes is crucial to astrobiology, because it helps us learn the requirements for habitability on other worlds.
- The question of how well we can deal with the problems of climate change is very important to the search for extraterrestrial intelligence (SETI), because other civilizations must inevitably confront similar threats if they reach our level of technological development. Therefore, the long-term survivability of civilizations depends on whether they overcome such threats or are permanently inhibited by them.

- The great public interest in astronomy gives us (as astronomers) credibility and a platform with the public. Note, for example, that we rarely encounter political polarization when discussing exotic topics like black holes. Therefore, if we stick to science and speak respectfully to audiences of all political persuasions, we have an opportunity to break through the polarization that often accompanies the topic of global warming.
- Perhaps most important, especially for those teaching non-major courses at the college level, it is quite likely that your astronomy or astrobiology course will be the only course in which your students will ever have the opportunity to learn about the science of global warming. It is therefore critical that we take the opportunity to build upon the many connections between astronomy and climate science so that our students will understand the key scientific ideas.

9.3 Three General Principles for Climate Communication

I will spend most of this chapter providing a brief overview of strategies I've found effective in using astronomy to teach about climate change. First, however, I'll list three general principles that are important always to keep in mind:

1. Always be respectful to your audience, maintaining the "radical civility" described in Chapter 12. Moreover, in the context of teaching, be sure to remember that your "audience" includes not only your students, but their parents and perhaps their other family members and friends as well. After all, they will surely discuss what you teach them with others, so make sure you've presented it in a way that will alleviate rather than exacerbate the kinds of polarization that often occurs with this topic.
2. Be sure to focus not only on the basic science and consequences of global warming, but also on solutions. Of course, as you do this, you must also take care to stick to a scientifically justifiable viewpoint on solutions. For example, emphasize how various energy technologies actually work and

data concerning their viability. Note: while it's fine to talk about policy options, and even to state your own opinions, be careful not to make your opinions sound like scientific facts, since that would be counterproductive to building confidence in science for your students and their families and friends who come from different political viewpoints.

3. As noted in the title and introduction to this paper, always teach with inspiration, not (only) fear. This is a bit subtle, since global warming *is* scary. But you should continually remind students that it is also solvable if we, as a society, act wisely and in a way that is consistent with our scientific understanding. If you do this, you can inspire students to help our society take the action that will create the kind of world in which we'd all like to live.

9.4 Discussing the Science: "Global Warming 1-2-3"

I'll now turn to a few suggestions for presenting the science, consequences, and solutions to your classes or in special presentations to groups of students or the public. There are, of course, many ways to present these topics. As astronomers, however, I think it's useful if we begin by emphasizing the astronomical ties. To do that, I generally start with "a tale of two planets" in which I show Venus and Earth to scale (Figure 9.1). It is immediately obvious that Venus and Earth are essentially the same size, and I explain that other data show that they also are quite similar in most other ways (e.g., composition, mass, density), with one major exception: their great difference in global average surface temperatures. I then ask my audience (of any age) to "think like a scientist," in which case they should wonder: Why would two planets that are generally so similar differ in this one major way?

Figure 9.1. Earth and Venus are shown to scale. You can see that they are both about the same size, and other data show they are also quite similar in most other ways, with one major exception: Earth has a pleasant global average surface temperature, while Venus is hot enough to melt lead. If you think "like a scientist," this should make you wonder: Why would two otherwise similar planets have such different surface temperatures? Photo credits: NASA.

Figure 9.2. This photo shows the locations of the Sun, Venus, and Earth in the Voyage Scale Model Solar System in Washington, DC, which shows both sizes and distances on a 1 to 10 billion scale. The building to the left is the National Air and Space Museum. Note: If you would like to put a similar scale model solar system in your community, it is now possible through the Voyage National Program; see voyagesolarsystem.org or email me (jeff@bigkidscience.com) for more information. (Photo by the author.)

Most students will guess that the explanation lies with Venus's closer distance to the Sun, but it is easy to show that this cannot be the full story. The definitive way to show it is through calculation, but in this case it is easy to demonstrate it qualitatively by making or showing a scale model of the inner solar system. Figure 9.2 shows such a model. You can then ask students to imagine a bonfire at the location of the model Sun. While we'd expect it to be warmer at the location of the model Venus than at the location of the model Earth, it's fairly clear that it's not going to be enough to account for the huge difference in surface temperature. If you have time, you can further explain that the situation becomes even more mysterious when you recognize that Venus is completely covered in clouds and that clouds reflect sunlight. This cloud cover means that Venus actually gets less sunlight on its surface than Earth, despite being closer to the Sun, which might make you expect it to be colder than Earth. So what is making Venus so hot?

Depending on the level of your class and how much time you have available, you could continue to have students hypothesize and investigate their hypotheses, but at this point I usually get straight to the answer: Venus is hot because of the greenhouse effect. You should now explain the physics of the greenhouse effect, along with the atmospheric differences between Venus and Earth, at a level of depth appropriate to your audience. Moreover, you should explain that the greenhouse effect also makes Earth much warmer than it would be otherwise, and is in fact the only reason our

Figure 9.3. The conclusion of the tale of two planets. The explanation for Venus's high surface temperature is the extreme greenhouse effect caused by its vast amount of atmospheric CO_2.

planet is not completely frozen. (Earth's "no greenhouse" temperature, based solely on distance from the Sun and reflectivity, is –16°C.) In that sense, the naturally occurring greenhouse effect is a very good thing for life on Earth and us, while Venus provides absolute proof that it is possible to have too much of this good thing (Figure 9.3).

Continuing the theme of "think like a scientist," I next encourage the audience to ask how scientists can be so confident that they truly understand the greenhouse effect. This leads to a discussion of the two key lines of evidence for greenhouse warming: (1) laboratory measurements of how various gases interact with different wavelengths of light; and (2) real-world validation of those measurements through the fact that models can reproduce actual planetary temperatures only when the greenhouse effect (as measured in the laboratory) is included in the models. Again, the level of detail to which you should go in presenting the evidence will depend on the level of your audience and the time available. Similar considerations apply to how detailed you get in the discussion of other greenhouse gases, and why CO_2 drives the greenhouse effect on Earth, even though water vapor is the dominant greenhouse gas in Earth's atmosphere.[3]

Having established the importance of CO_2 to Earth's greenhouse warming, the next step is showing the measured rise in the CO_2 concentration.[4] The data are straightforward, so the "think like a scientist" follow-up question concerns how we know that the rise is caused by us, rather than by any natural process. In brief, there

[3] For a brief, middle school level explanation of how water vapor can be the dominant greenhouse gas even while CO_2 is the driver, see: https://grade8science.com/7-1-1-what-is-the-greenhouse-effect/i-was-wondering-4/.
[4] The latest CO_2 data are available at: https://gml.noaa.gov/ccgg/trends

are three major lines of evidence, which you can again discuss as appropriate to your audience:

- The amount of CO_2 released into the atmosphere by human activity has risen in lockstep with the CO_2 concentration (though only a little less than half of the released CO_2 stays in the atmosphere, with the rest absorbed into the oceans and the ground) and can easily account for the rise in concentration since the dawn of the industrial age. In contrast, the total amount of CO_2 released by natural sources such as volcanoes and the oceans is only about 1% of what would be needed to explain the observed CO_2 rise.
- The burning of fossil fuels consumes oxygen at the same time that it releases CO_2, which means that if the rising CO_2 concentration comes from fossil fuels, there should be a corresponding decrease in the concentration of oxygen in the atmosphere and oceans. The expected decrease is very small, because oxygen makes up about 21% of Earth's atmosphere in comparison to only about 0.04% for CO_2. Nevertheless, a small decrease in oxygen concentration has been measured[5] and matches the expectation (Stolper et al. 2016).
- A third line of evidence comes from carbon isotope ratios and the fact that fossil fuels represent the remains of living organisms that died hundreds of millions of years ago. Because C^{14} has a relatively short half-life (5700 years), fossil fuels don't contain any (because any that was once present would have decayed long ago). Therefore, if fossil fuels are the source of the rising CO_2 concentration, then the relative abundance of atmospheric C^{14} (compared to C^{12}) should be falling as the total CO_2 rises—and this is just what has been observed. In addition, the fraction of C^{13} found in living organisms is slightly lower than it is in non-biological sources, and fossil fuels have this same lower fraction. So again, if fossil fuels are the source of the rising CO_2 concentration, then the atmospheric C^{13} to C^{12} ratio should be dropping in tandem with the rise in CO_2, and data confirm that it is (Rubino et al. 2013).

Together, these lines of evidence leave no reasonable doubt that human activity is indeed driving the strengthening of the greenhouse effect.

We can now summarize the basic science of global warming with what I like to call "global warming 1-2-3," to emphasize that it is as easy to understand as 1-2-3:

"Global Warming 1-2-3":

1. **Fact: It is scientifically indisputable that CO_2 and other greenhouse gases trap heat and therefore make Earth (or any other world) warmer than it would be otherwise, and that more greenhouse gas means higher temperatures.**
2. **Fact: Human activity, especially the use of fossil fuels, has been adding significantly more CO_2 and other heat-trapping greenhouse gases to Earth's atmosphere.**

[5] The latest O_2 data are available at: https://scrippso2.ucsd.edu/.

3. Logical conclusion: Given that more greenhouse gas means warmer temperatures and we are adding more greenhouse gases to Earth's atmosphere, it is inevitable that global warming must occur as a result.

You can then show the data confirming the warming (Figure 9.4) as evidence that the logical conclusion is indeed occurring as expected.

I also recommend showing the 800,000-year ice core record (Figure 9.5), which you can use to point out several key ideas:

- First, notice that this record shows that Earth's climate has substantial and natural swings, with the dips representing ice ages and the peaks the warmer, interglacial periods. Notice also the close correlation between the CO_2 concentration and temperature, which again confirms the basic logic of the "1-2-3" science.
- During the 800,000-year period shown, the CO_2 concentration never exceeded about 280 parts per million (ppm) until the dawn of the industrial age. We have since pushed the level to more than 420 ppm, representing a more than 50% increase, and the increase is continuing at a rapid rate.
- The recent temperature rise of about 1°C (see Figure 9.4) looks much smaller in comparison. However, this is largely an artifact of the way the graph is scaled (with scaling to overlay the CO_2 and temperature data, and a highly compressed time scale). In addition, climate models indicate that we should expect some lag time between the CO_2 rise and the temperature rise (due to complexities of the climate system, especially the ocean circulation), suggesting that the warming will continue for at least a decade or two even if we were to immediately stop adding greenhouse gases to the atmosphere.
- The data show that past increases of 50% in the CO_2 concentration have gone with temperature increases of as much as 10°C or more. However, because the temperature was already near the 800,000-year maximum when the

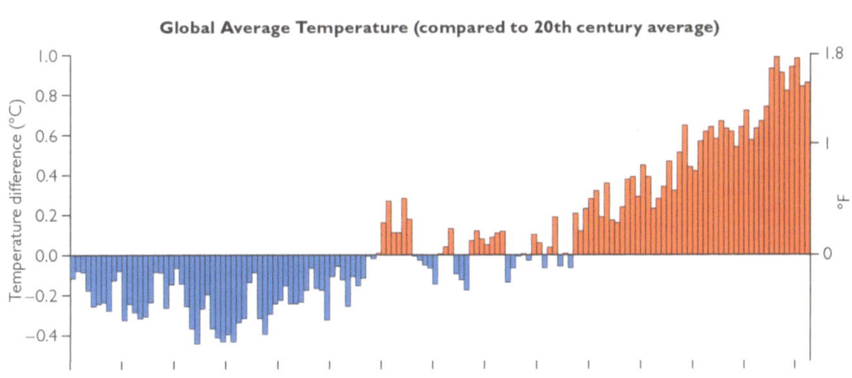

Figure 9.4. This graph shows the clear rise in global average temperature since the late 1800s. The 0°C line represents the average temperature for the 20th century. The uncertainties are not shown, but range (with 95% confidence) from about 0.1°C for the late 1800s to less than about 0.03°C today. Data from the National Climate Data Center.

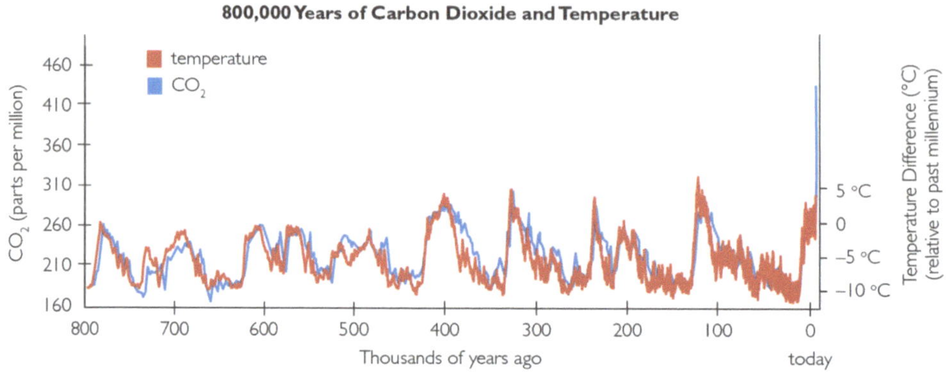

Figure 9.5. This graph shows ice core data for both Earth's global average temperature (red) and the CO_2 concentration (blue) over the past 800,000 years. Notice the close correlation. Note: the zero line for temperature represents the average from the last millennium (A.D. 1000 to 2000). Adapted from the 4th National Climate Assessment, using data from the European Project for Ice Coring in Antarctica.

current global warming began, we cannot be sure that the response to the current CO_2 increase would be the same.[6] This is a major reason why we need sophisticated climate models to predict how much our planet will warm in the coming decades.

Keep in mind that this brief outline of the "1-2-3" science is only the beginning of what you might do with your students, again depending on the level of your course and the time you can devote to this topic. For example, you could go into more detail on how temperature measurements are made, how the ice core record is obtained and analyzed, the reasons behind the natural climate cycles, the ways in which climate modeling adds to our understanding, and much more.

9.5 Discussing the Consequences

Like skeptical members of the public, students might wonder why a relatively small increase in global average temperature, such as the 1°C to date or projections of 2°C to 5°C by the end of the century should be of great concern. It is therefore important to spend at least some time discussing the various consequences of global warming. There are many ways to list or categorize these consequences, but I've found it useful to group them into five major categories: local and regional climate change, storms and extreme weather, melting of sea ice, sea level rise, and ocean acidification. I assume that readers of this paper will be familiar with all of these, so I will just briefly review some key points worth emphasizing in your courses.

[6] This response is what is often called the "climate sensitivity," meaning how sensitive the climate is to changes in the greenhouse gas concentration. When you occasionally hear a legitimate scientist claiming that global warming won't be as bad as models suggest, they are in essence arguing that the climate sensitivity is being overestimated. However, if that were the case, it would be difficult to understand why current models seem to work so well.

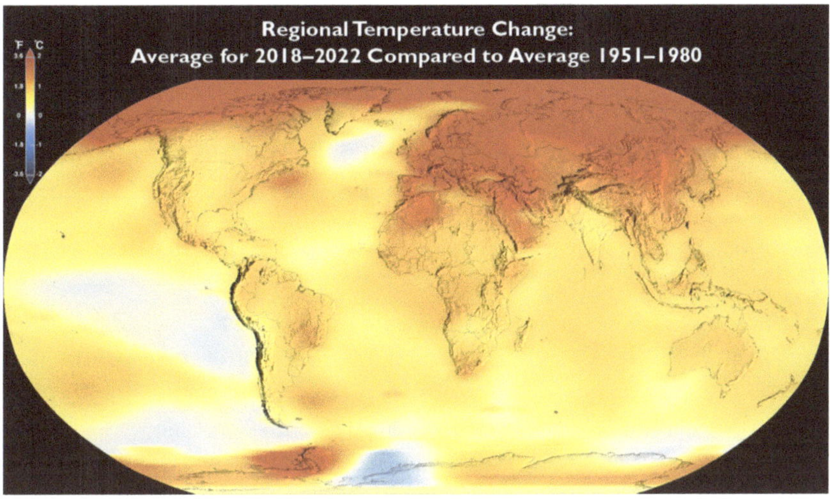

Figure 9.6. This map shows how regional average temperatures for the period 2018–2022 compared to their averages from 1951–1980. It is a still frame from the video posted at svs.gsfc.nasa.gov/5060. Source: NASA Scientific Visualization Studio. you can generate similar maps for any time period with the NASA GISS Surface Temperature Analysis mapping tool.

The first category is local and regional climate change. The point to emphasize is that the temperature rise will not be uniform everywhere, nor will other climate related changes in such things as precipitation be uniform. Figure 9.6 shows a still frame from a video that is worth sharing with your students (or any other audience), as it makes the temperature changes strikingly clear. Students will also notice that the strongest warming is occurring in Arctic regions.

The second category is an expected increase in the frequency and/or severity of storms and extreme weather. Here, it is worth noting that the underlying science is easy to understand: Global warming means more heat and energy being trapped in the lower atmosphere and oceans, and heat and energy are the drivers of weather events. It's also worth noting that "extreme" applies to all types of weather. For example, in some cases winter blizzards can become more severe, which may be somewhat counterintuitive to your students. Similarly, while overall global precipitation should increase (because warmer temperatures mean more evaporation from the oceans and therefore more moisture in the atmosphere), some places may experience more extreme dry spells and drought.

The third category is the melting of sea ice, meaning the ice floating in the Arctic Ocean. This melting does not have a significant effect on sea level because the ice is already floating, but it has other worrisome effects. Two of the most important are that the melting reduces the local salinity (because the ice is essentially fresh water), which can have repercussions throughout the ocean food chain and possibly to ocean currents, and that the reduction in ice coverage means more absorption of solar energy (because ocean water has a much lower reflectivity than ice), which can further amplify the effects of the warming caused by greenhouse gases.

The fourth category is an increase in sea level, which is traceable to two factors. First, there has been a rise of more than 20 centimeters in the past century due largely to thermal expansion as the oceans warm; this has already led to more flooding along coastlines, particularly when the rise is magnified by a storm surge. Second is the effect of glacial melting, with the runoff water entering the ocean. Depending on the future melting rate, current models predict that the overall sea level rise by 2100 (including the 20 centimeters to date) could be anywhere from about 25 centimeters to about 2.5 meters (8 feet), with the "intermediate" models suggesting a rise close to 1 meter (Sweet et al. 2022).

The final category in this organizational scheme is ocean acidification, in which the CO_2 that is absorbed into the oceans reacts chemically to make the water more acidic. Ocean acidification gets much less media coverage than other consequences of global warming, but it could be just as serious in terms of its effects on humans. For example, the combination of ocean warming (as global temperatures rise) and ocean acidification is already causing significant damages to coral reefs and other ocean ecosystems. Because billions of people depend on the ocean for much of their food supply, these damages could potentially lead to increasing hunger or starvation.

In addition to these five categories of consequence, there is also the possibility of reaching various "tipping points" that would amplify some or all of them. Examples of such tipping points include release of CO_2 and methane from thawing Arctic tundra, shifting ocean currents, cascading ecological changes, and rapid disintegration of certain glaciers. These are worth at least noting, while also pointing out the associated risks, which the latest IPCC report summarizes as follows: *Low-likelihood outcomes, such as ice sheet collapse, abrupt ocean circulation changes, some compound extreme events and warming substantially larger than the assessed very likely range of future warming cannot be ruled out and are part of risk assessment* (IPCC 2021).

Again, you should adjust the depth to which you explore the consequences based on the particulars of your course. But also remember: the consequences are scary, so if you are going to teach with inspiration, not (only) fear, keep reminding your students that we still have time to avoid the worst consequences, and possibly to undo some of the damage already done. And that brings us to the topic of solutions.

9.6 Discussing the Solutions

On a purely scientific level, the solution to the problem of global warming is simple. Because we know that the burning of fossil fuels is the primary cause of global warming, the obvious solution is to stop using them. Of course, this is much more difficult on a practical level, because our civilization depends on the energy we get from fossil fuels, so we can't realistically stop using them unless and until we have replacement energy sources. The good news is that we already have technologies that could in principle allow us to meet all our energy needs without continued use of fossil fuels. Chapter 7 goes into more detail on these energy sources.

In particular, three general categories of existing technology could in principle allow us to get just as much energy as we have today without any further increases in greenhouse gases, and quite possibly at lower cost than we pay for energy today:

- Energy efficiency, which allows us to meet the same energy needs with less total energy, thereby reducing the need for fossil fuels or any other energy source.
- Renewable energy sources, including wind, solar, geothermal, hydroelectric, and biofuels (depending on how they are produced). While none of these sources are perfect—for example, solar panels contain toxic chemicals, wind turbines can kill birds and bats, dams for hydroelectric power can damage river ecosystems, and crops grown for biofuels use land that could otherwise be used to grow food—they have the advantage of producing energy without increasing the concentration of CO_2 or other greenhouse gases in our atmosphere.
- Nuclear energy, which also does not release greenhouse gases. Note that while nuclear energy has some well-known downsides, such as the problem of disposing of radioactive waste, it also has upsides. For example, nuclear power can provide much steadier and more reliable power than renewables, and when done "right" (as opposed to, for example, Chernobyl), nuclear technology has a long track record of success. France, for example, gets more than 70% of its electricity from its more than 50 nuclear power plants. Moreover, newer nuclear power plant designs, including those known as "small modular reactors," should be much safer and lower in cost than past designs. Note: as you are undoubtedly aware, many environmental activists are opposed to nuclear power, but much of the rationale they use is not much more scientific than the claims made by "skeptics" of global warming. As scientists, I therefore believe it is very important for us to treat nuclear power in a scientifically honest way, which means clearly discussing its benefits along with any potential dangers. As a further note, while I'm no expert on nuclear power, I am beginning to wonder if, especially with the newer technologies, nuclear power might prove to be "greener" than even solar and wind power. The reason is its energy density: it takes far less total material—and therefore far less mining with its environmentally destructive consequences—to obtain energy from nuclear power plants than from solar or wind installations.

The future holds even greater promise, as many new energy technologies are in various stages of development. But for the purpose of teaching with inspiration, let's focus for the moment on nuclear fusion. As astronomers, we are perfectly positioned to explain the enormous potential for generating vast amounts of clean, safe energy through fusion, the energy source of the stars. While fusion has been the "energy source of the future" for a long time now, there are good reasons to think that we will finally achieve commercially viable fusion power within the next few decades—which means well within the lifetimes of today's students. Even if we don't achieve it, there are other energy sources that may offer similar potential, such as solar power from space (huge solar collectors in high orbit, beaming energy down to Earth), advanced biofuels, and more. One way or another, if we can avert a climate catastrophe, it seems highly likely that, not long after mid-century, our energy problems will be solved in a way that no longer adds any greenhouse gas to the atmosphere. That brings us to the most inspirational part of this story.

9.7 Building a Post-global Warming World

The climate future is often portrayed in the media as a choice between "bleak and bleaker," but it is possible to envision something far more promising. Consider, for example, a few things that we might do with the vast potential of fusion power:

- If we hope to restore the climate to a better state, we will need to bring the CO_2 level back down, which will likely require technology for active CO_2 removal. This technology will almost certainly be energy intensive, which means that putting it into place will require a significant increase in global energy consumption. While this will certainly be challenging, fusion power would make it very possible.
- Global poverty is closely tied to energy availability. With fusion, we can imagine having enough energy to raise living standards all around the world, perhaps eliminating poverty once and for all.
- Fresh water availability is a major problem around the world. But with fusion, we could simply desalinate seawater.
- Fusion might allow us to obtain resources from the Moon or asteroids at a low enough cost that we could imagine an end to the environmental damage of mining on Earth, perhaps even turning much of our planet into protected national parks.

This is the vision of a post-global warming world that I believe we must share with our students and the public. It is certainly true that global warming poses an existential threat to the prosperity we enjoy today, but we have it in our power not only to end this threat, but to build a far better world than the one we live in today. This is a vision that can inspire our students—and perhaps their parents and grandparents as well—to work together to build the world as we'd all like it to be.

9.8 Resources

This chapter is necessarily brief, but for those of you teaching particular courses or audiences, I have helped to develop several resources that you may find to be of use:

- For high school, college, and the general public: *A Global Warming Primer*, posted freely at globalwarmingprimer.com (and also available in book form), uses a Q&A format to go into much more depth on the science, consequences, and solutions than I have provided in this article.
- For college astronomy and astrobiology courses: you'll find sections discussing global warming in my textbooks for astronomy (*The Cosmic Perspective*) and astrobiology (*Life in the Universe*).

- For middle/high school: I have posted a free, digital textbook for Earth and Space Science at grade8science.com; Chapter 7 focuses on climate change and can be used for as much as about a month-long unit on the topic.
- For younger kids: I've written a children's book, *The Wizard Who Saved the World*, that follows a young child as he learns about global warming and how he can help us solve the problem and build the kind of promising future discussed in this article. More info at bigkidscience.com/wizard. You can also freely view a video reading of this book from the International Space Station at https://storytimefromspace.com/stories/the-wizard-who-saved-the-world/.
- I also encourage you to download the free app *Totality by Big Kid Science* which, in addition to providing interactive maps of upcoming total solar eclipses, includes a section on global warming; more info and download links at bigkidscience.com/eclipse.
- Scale model solar systems provide a great way to help people get a "cosmic perspective" on our planet and also to understand many ideas of global warming. Available now to communities around the world. Info at voyagesolarsystem.org or by emailing me (jeff@bigkidscience.com).

References

IPCC 2021, Summary for Policymakers. Climate Change 2021: The Physical Science Basis. Contribution of Working Group I to the Sixth Assessment Report of the Intergovernmental Panel on Climate Change ed. V. Masson-Delmotte, et al. (Cambridge: Cambridge Univ. Press) 3

Rubino, D. M., et al. 2013, JGRD, 118, 8482

Stolper, D. A., Bender, M. L., Dreyfus, G. B., Yan, Y., & Higgins, J. A. 2016, Sci., 353, 1427

Sweet, W. V., et al. 2022, Global and Regional Sea Level Rise Scenarios for the United States: Updated Mean Projections and Extreme Water Level Probabilities Along U.S. Coastlines. NOAA Technical Report NOS 01, *National Oceanic and Atmospheric Administration* https://oceanservice.noaa.gov/hazards/sealevelrise/noaa-nos-techrpt01-global-regional-SLR-scenarios-US.pdf

Chapter 10

Global Warming: A Case Study in Science

David J Helfand

> "Civilization is in a race between education and catastrophe. Let us learn the truth and spread it as far and wide as our circumstances allow. For the truth is the greatest weapon we have."
>
> —H. G. Wells

Global warming provides an excellent opportunity to illustrate generally applicable scientific reasoning skills in the context of a topic of great interest to most college students. This chapter provides an illustration of this opportunity as realized in a course required of all first-year students at Columbia University as part of the institution's famed Core Curriculum. Included are the philosophy that informs the subject's presentation (including the distinction between measured facts and feedback uncertainties and between complex computer models and physical laws), a complete list of the scientific habits of mind we illustrate and apply in this course, reading samples, and in-class seminar activities that allow students to construct knowledge of the key physical processes involved in this problem, as well as to develop the ability to communicate effectively about the science of global warming.

10.1 Introduction

Forty-five years ago, at 6:00 pm on a Monday evening in early September, I walked into a lecture hall at Columbia University to teach my first class. I had never taught anyone anything. I wasn't a TA in graduate school, and my six-week stint one summer while in college substituting for a seventh- and eighth-grade math teacher in my hometown was one of the most dispiriting experiences of my life. Given this complete lack of preparation for one of my principal roles at the University, I had at least spent the previous month thinking about what I wanted to accomplish in teaching a Physics for Poets course to a class of largely non-traditional (e.g., older and employed) students.

doi:10.1088/2514-3433/acfcb6ch10

While I hoped to make atomic physics—and perhaps even projectile motion—interesting, it was clear to me that the content of the course was largely irrelevant to these students' lives. But the modes of thought that have provided us with insights into the atomic world and the rest of the material Universe—what John Dewey (1910) called "scientific habits of mind"—could indeed be of use to my students as they made decisions in their financial, medical and political lives. I thus structured the course around a set of topics designed to illustrate the logical and quantitative reasoning skills characteristic of scientific thinking and strove to provide everyday examples of their use beyond the arcane world of physics. The course was a success, and I maintain its approach to this day.

Global warming is an ideal topic around which to structure a lecture, a unit, or an entire course in which a primary goal is to inculcate scientific habits of mind in one's students. First, it is inherently transdisciplinary, including crucial topics from astronomy, physics, chemistry, earth science, and biology and can thus serve to illustrate subtle differences in the disciplinary thinking that defines each of these fields of inquiry. Second, it provides many opportunities to incorporate basic scientific skills: reading graphs, performing back-of-the-envelope calculations, engaging in both deductive and inductive reasoning, dealing with uncertainty through statistics and probability, illustrating feedback loops, and linking observations with models. Third, it contains many connections to economics, politics, psychology, sociology and other fields students may be studying simultaneously. Finally, and perhaps most important of all, it is a topic in which today's students should be, and indeed generally are, deeply interested and engaged.

In the 1980s, before the first IPCC report, I read an article on paleoclimate which I found fascinating. This spurred my interest in the subject of climate, an interest easily indulged at Columbia because of our world-leading research center at the Lamont-Doherty Earth Observatory. In 1975, my late colleague Prof. Wally Broecker had written a paper in *Science* entitled "Climatic Change: Are We on the Brink of Pronounced Global Warming?" (Broecker 1975) in which he stated:

> *"The exponential rise in the atmospheric carbon dioxide content will tend to become a significant factor, and by early in the next century will have driven the mean planetary temperature beyond the limits experienced during the last* 1000 *years."*

In 1975 when this paper was published, the mean global temperature was −0.03 °C from the mid-twentieth century average. Beginning in 1977, however, we have experienced 46 straight years above that average, with the last five years averaging +0.9 °C above the mean, significantly higher than anything seen for many thousands of years. Wally clearly had it right.

At the same time I was developing my interest in climate change, I also began my quest to change Columbia's famed Core Curriculum. When I arrived at Columbia in the 70s, I was delighted to see our undergraduates were provided with a truly common experience in their first two years: every student was reading the same book in the same week and discussing it in twenty-student seminars. I was simultaneously appalled, however, that this curriculum, described in the course catalog as the institution's "intellectual coat of arms," consisted of seven humanities courses, zero math courses and zero science courses; this didn't strike me as an adequate coat of arms for the twenty-first century graduate. In 1982, I chaired a committee whose report, entitled "Science in a Liberal Curriculum," called for adding science to the Core.

Twenty-two years later (academia moves on geologic timescales), I succeeded. All 1150 Columbia College first-year students now take a course called "Frontiers of Science." Each year, we choose four topics at the boundaries of our scientific knowledge designed to engage all students, regardless of their planned majors, in the scientific enterprise and, as important, to teach them the scientific habits of mind essential for life as an informed citizen in our technology-saturated world. We switch out topics every two to three years as faculty go on leave or as new topics of interest arise. The one topic that has remained in place in each of the 19 incarnations of the course is Climate Change.

In this chapter, I will discuss the approach we take in teaching this subject to the very broad range of first-year students enrolled in this required, general education course. I describe the way lectures are structured, the reading materials we select, and the habits of mind covered in the course. I will also provide some of the seminar activities we use to deepen students' understanding of the topic and how we provide, within the context of a science course, links to the many other subjects relevant to this complex global issue.

10.2 Facts, Physics, Feedback, Foreshadowing, and Fictions: Being Explicit About What We Know and What We Don't Know

It is crucially important that first-year university students abandon their notion that science consists of a bunch of facts that are tedious to memorize, plus some esoteric mathematics they will never be able to master. Rather, science is a process for creating falsifiable models of the natural world. Its conclusions are contingent, and progress is most often made by showing that our current model is wrong rather than "proving" it correct. Nonetheless, there are some things we do know (to within some specified precision, at least) and there exist models so extensively tested that we can use them with confidence. To be explicit about some of these subtleties, I entitle my lecture series "Global Warming: What We Know and What We Don't Know," and I begin my first lecture by explaining that every slide will be clearly labeled as a "fact," as arising from "physics," as involving "feedback" (the principal source of uncertainty in our predictions about the future), as an observation that may, or may not, indicate "foreshadowing" of that future, or as a "fiction" that I do my best to demolish using the preceding four tools.

A "fact" I define as a measurement of the material world, done with the best available instruments and techniques, always with an assigned uncertainty, vetted by skeptical peer review and, preferably, arrived at independently by multiple investigators. We do not conclude the global temperature is rising because one scientist uses an alcohol thermometer to record the temperature at his location over a decade or two, but by collating data from several independent networks of over 9000 sensors, spread over the land and oceans, carried out by several different international teams and checked for consistency with satellite observations. It is therefore a "fact" that the global temperature was 1.0° C warmer in 2020 than the mid-20th century average. While it is always appropriate to scour one's data for biases and to question the details of how measurements are made, there comes a time when the level of certainty is such that it makes sense to call this observation of nature a "fact." And it is pointless to argue about facts.

"Physics" is a model of the material world. It is by far the most successful predictive model humankind has ever created, but it is important to recognize that it is a model, not the material world itself. Some facets of the model are so well-tested, it is safe to accept them as highly accurate representations of reality: the gravitational perturbations manifest in Milankovitch cycles and the absorption properties of greenhouse gas molecules are beyond reasonable questioning. General Circulation Models are not.

"Feedback" requires the explanation that it is not the red ink on a returned student essay, but a specific logical construct applicable in situations where the effect influences the cause: in positive feedback, the effect amplifies the cause and in negative feedback the effect suppresses the cause. Care must also be taken to make clear that in most situations, "positive" feedback is bad, as it leads to a runaway situation, and "negative" feedback is generally good as it describes a stable situation. It is also important to point out that in the climate system, both types of feedback occur, sometimes with the same initiating action, but that rarely is the feedback loop perfect (e.g., in a negative loop where the effect suppresses the cause, it almost never does so exactly—higher temperatures lead to greater ocean evaporation, which leads to more cloud cover, which leads to more sunlight reflected and less solar energy absorbed which leads to lower temperatures, but clearly this has not balanced out perfectly over the past few decades).

"Foreshadowing" refers to observations that roughly follow (or sometimes exceed) past predictions and suggest the path Earth's climate is on. Melting Arctic sea ice is one example. In the 1990s, predictions were that the Arctic ice would largely be gone in late summer by the year 2100. In the ensuing three decades, the retreat of multi-year ice has been so dramatic that a largely ice-free Arctic is now expected within 25 years. This is partly both a result, and a cause, of the more rapid heating at high northern latitudes—less ice means more absorbed sunlight, which means warmer waters and more ice melt.

Finally, "fictions" are abundant in discussions of global warming, and while it is not worth spending too much time debunking them, I find it worthwhile to provide a few cases, as these illustrate yet further examples of how to use scientific habits of mind in the search for models that comport with reality. The fiction that a brighter Sun is responsible for global warming can be refuted by presenting the sunspot record over the past century and the direct measurements of solar luminosity from satellite measurements over the past fifty years; solar flux at the top of the atmosphere typically varies by about 0.1% from the maximum of the solar sunspot cycle to the minimum. The most recent cycle from 2010–2021 saw the lowest solar activity in the last century, indicating less solar energy input, the opposite of the denialist claims.

However passionate one is about the subject of global warming, and however committed to the scientific view of the world, it is important to highlight both what we know, and what we don't know, about this subject *as scientists.* That means acknowledging that all measurements have uncertainties, that our models should be treated with varying degrees of confidence, and that the solution to the problem of global warming requires using expertise far from that which we possess. We are taught that in science we must be humble before nature, as it is the arbiter we must trust. Our exposition of the global warming issue is most effective when that humility is manifest in our presentations to our students.

10.3 Scientific Habits of Mind

As noted in the Introduction, the global warming issue is an excellent topic with which to illustrate the scientific habits of mind we hope our students take away from general education courses. In our Columbia class, we explicitly list learning objectives for each week divided into subject content and habits of mind. The latter include 24 specific skills grouped in nine categories: basic frequentist statistics, basic probability, graphical representation of data, standards for scientific claims, the scientific process, a sense of scale (length, time, mass, etc.), calculating with units, basic algebraic manipulation of equations and calculations derived therefrom (including scientific notation and back-of-the-envelope calculations), and feedback loops. The list of habits is included here:

1. *Demonstrate an understanding of mean, standard deviation, and standard error. [Statistics]*
2. *Estimate mean and standard deviation from a distribution (histogram). [Statistics]*
3. *Calculate confidence intervals and determine statistical significance (mean ± 2SE confidence intervals) from a bar graph. [Statistics]*
4. *Determine statistical significance from p-values. [Statistics and Scientific Claims]*
5. *Distinguish between random and systematic errors and describe their respective implications for precision and accuracy. [Statistics and Scientific Claims]*
6. *Describe the role of falsifiability in science. [Scientific Claims]*
7. *Distinguish between observations and experiments, identify strengths and weaknesses of observational and experimental studies. [Scientific Claims]*
8. *Distinguish between correlation and causation. Identify potential explanations for an observed correlation, and evaluate whether a causal relationship is well-supported. [Scientific Claims]*
9. *Determine the purpose of controls in an experiment, assess what makes a good control group or control task. [Scientific Claims]*

10. *Read and interpret graphical information in histograms, bar graphs, scatter plots using axis labels, numerical information, units, error bars, legends, and figure captions. [Graphs]*
11. *Evaluate hypotheses in the context of presented evidence, propose experiments to test hypotheses or to extend previous studies. [Scientific Claims]*
12. *Use units and unit conversions in calculations; utilize dimensional analysis. [Calculating with Units]*
13. *Manipulate powers of ten. [Calculating with Units]*
14. *Develop a sense of scale for the topics we study. [Sense of Scale]*
15. *Utilize logarithmic scales, read values off a log scale, add values to a log scale. Identify when a log scale is useful, and when it cannot be used. [Sense of Scale]*
16. *Distinguish between scientific hypotheses and scientific theories. [Scientific Process]*
17. *Consider the role of models and thought experiments in understanding physical processes; utilize models and thought experiments; identify their limitations. [Scientific Process]*
18. *Use familiar and unfamiliar equations to calculate quantities of interest, paying special attention to algebraic manipulation, unit conversions, and powers of ten. [Calculations]*
19. *Identify when to use back-of-the-envelope estimation; make estimates using back-of-the-envelope calculations; identify and state assumptions; check answer plausibility. [Calculations]*
20. *Assign probabilities to events; calculate probabilities for multiple events. [Probability]*
21. *Calculate probabilities of events occurring at least once, given N opportunities. Approximate probabilities of rare events. [Probability]*
22. *Consider the role of quantitative and qualitative models in understanding physical processes. [Scientific Process]*
23. *Identify feedback loops and draw feedback loops. Determine whether a feedback loop is positive or negative. [Feedback]*

In each case we have developed adaptive learning exercises that allow students with strong backgrounds to quickly review material while providing complete pedagogical lessons for those unfamiliar with these skills; we have also prepared written tutorials on each subject that are also provided to the students.

The Climate unit is almost always the final unit of the term. While it introduces feedback loops for the first time and offers extensive practice in graph reading (including recognizing the slightly perverse geological practice of plotting time axes backwards), we are able to use all of the habits developed earlier in the semester, as well as reinforce concepts and practices covered in other units (e.g., they learn how fMRI machines use blood flow as a "proxy" for neural activity, so can recognize $^{18}O/^{16}O$ isotope ratios as proxies for temperature, and can do a back-of-the-envelope calculation to show that the O_2 decline in the atmosphere roughly equals the amount of O_2 appearing in the newly accumulating CO_2). I illustrate the application of these myriad scientific habits of mind to the climate problem in Chapter 10 of my book "A Survival Guide to the Misinformation Age" (Helfand 2016).

10.4 Readings: Connecting to the Students' World

Each week the students have assigned readings. Some weeks all 20 students in a seminar will have the same reading, while on other occasions subgroups are assigned different readings and must come prepared to present their articles to the rest of the class. In the climate unit we use papers from the scientific literature, the IPCC Summary for Policymakers report, and news articles and essays that focus on social justice issues around climate. This past year we included the following:

Wally Broecker's landmark paper entitled "Climatic Change: Are We on the Brink of Pronounced Global Warming?" (Broecker 1975)

A *New York Times* article by Christopher Flavelle entitled "A Climate Plan for Texas Focuses on Minorities. Not Everyone Likes It" discussing how flooding mitigation plans for Houston shifted from protecting wealthy neighborhoods to focusing on those disadvantaged communities most at risk.

A *Vogue* magazine article by Leah Thomas entitled "Why Every Environmentalist Should be Anti-Racist" discussing climate social justice issues.

An *NPR* story by Hansi Lo Wang on how "Climate Change Complicates Counting Some Native Alaskan Villages for the Census"

A lengthy *New York Times Magazine* article by Abraham Lustgarten entitled "The Great Climate Migration Has Begun" documenting the plight of climate refugees from various areas of the world.

The IPCC Synthesis report issued in 2022 *Summary for Policymakers*. See below for the activity we run based on this reading.

10.5 Energy Balance and Feedback Loops

I have developed an in-class exercise (usually done in groups of 3–5) based on the widely available energy balance diagram for Earth (see Figure 10.1). The first section of the exercise is to drive home the basic equation that determines a planet's temperature—Energy In = Energy Out—and to recognize the primary components of the processes that are important for this balance on Earth: reflection and absorption of incoming energy, evapotranspiration, thermal heat transfer from the ground/ocean to the atmosphere, and the insulating effect of greenhouse gasses.

The questions include the following:

Using the attached diagram (Figure 10.1) *that illustrates the energy balance of the Earth, consider the following questions.*

1. *Does this diagram represent balance between incoming and outgoing energy? What three numbers show this?*
2. *How much energy is emitted by the atmosphere plus the clouds?*
3. *What five components add up to account for the energy emitted by the atmosphere plus the clouds ($=165 + 30 = 195\ W\ m^{-2}$)?*

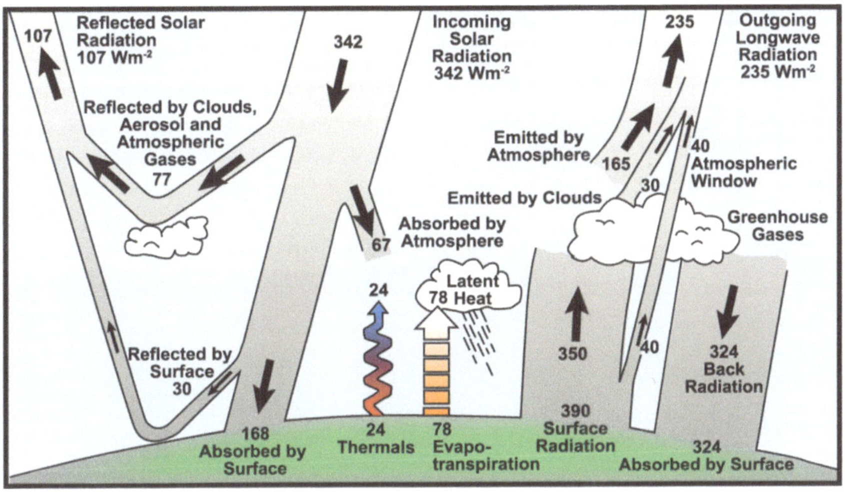

Figure 10.1. Estimate of the Earth's annual and global mean energy balance. Over the long term, the amount of incoming solar radiation absorbed by the Earth and atmosphere is balanced by the Earth and atmosphere releasing the same amount of outgoing longwave radiation. About half of the incoming solar radiation is absorbed by the Earth's surface. This energy is transferred to the atmosphere by warming the air in contact with the surface (thermals), by evapotranspiration and by longwave radiation that is absorbed by clouds and greenhouse gases. The atmosphere in turn radiates longwave energy back to Earth as well as out to space. Source: Kiehl and Trenberth (1997). (IPCC AR4 Climate Change 2007: The Physical Science Basis (WG1).

4. *What three numbers combine to make up the outgoing long-wave radiation of 235 $W\,m^{-2}$?*

   The second part of the exercise, usually done in the second or third week, is to demonstrate feedback effects and how the interlocking character of these in the climate system makes predictions difficult. This section is considerably more challenging and few, if any, students get to the correct answers without considerable guidance. Indeed, when I presented this exercise to a roomful of faculty at a conference on Astronomy teaching some years ago, most of them struggled as well. But I find the process of working on these problems leads the students to gain an appreciation for the complexity of the climate system; they come to appreciate how simple, well-determined processes can quickly lead to a cascade of complications that make precise, definitive predictions problematic.

5. *Cloud formation is aided by cosmic ray ionization of air molecules in the atmosphere. If energetic particles from the Sun increase because of a particularly active solar cycle with lots of huge sunspots (thus an increased cosmic ray flux), average cloud cover might increase by 5%. If this happens, and assuming all changes are proportionate (*i.e., *if one quantity changes, the change is distributed among the various components that make up that quantity proportionally),*
   a) *How many extra $W\,m^{-2}$ will be reflected by the cloud cover?*
   b) *How much will the total reflected energy increase? (careful!)*

c) *Given that the latent heat energy is what forms clouds, will there be a negative or positive feedback on total cloud cover? Explain.*

d) *How much (in %) will this feedback reduce or increase the cloud cover?*

6. *If a little extra snow one year makes the ground more reflective and surface reflectance grows by 10%,*

   a) *by how much (in %) will this increase or decrease the cloud cover?*

   b) *by how many* $W\ m^{-2}$ *will the total reflected energy increase?*

7. *If the atmospheric window through which 40* $W\ m^{-2}$ *from the surface now escapes was narrowed 10% by an increase in anthropogenic greenhouse gasses (and thus is turned into back radiation), by how much (in %) would this increase or decrease cloud cover?*

10.6 IPCC Reports: Climate Change Conference

Another activity we use takes students directly to the IPCC reports. The students are instructed to prepare as follows:

For this activity, you will become a climate change expert in one of the four sections of the Summary Report for PolicyMakers (SPM) that covers important takeaways from the IPCC Sixth Assessment Report: SPM A (Current State of the Climate pp. 4–11), SPM B (Possible Climate Futures, pp. 12–23), SPM C (Risk Assessment and Regional Adaptation, pp. 23–27), or SPM D (Limiting Future Climate Change, pp. 27–31)(IPCC 2021). *During the seminar you will discuss your assigned SPM section with other students in your group, and you will collectively decide what you think is the most important information to share with top government officials and policymakers (in this case, your classmates!) at our semi-annual Frontiers of Science Climate Change Conference. Your goal is to succinctly provide the state of our knowledge on various aspects of climate change, based on the IPCC AR6, along with considerations for adaptation and mitigation.*

As you read your assigned SPM section, ponder the questions listed below. Be prepared to discuss them with your group, so that you can develop a compelling summary of key findings for your classmates. Be sure to also consider and interpret all of the figures in your assigned SPM section.

a) *What are the most important takeaways from your SPM section? If you had to summarize general results for your family and friends, how would you do it?*

b) *What did you find most surprising or interesting?*

c) *If you had to choose one important figure or table from your SPM section to share with family and friends, what would it be?*

d) *What information would you like to learn more about? In other words, was there anything that was not addressed in your assigned section of the summary report that you think is important and should be addressed in a future report?*

The presentations are generally of high quality and have the benefit of peer-to-peer instruction that is particularly effective on an issue of such innate importance and interest to the students.

10.7 Conclusions

Global warming and its associated effects will have a profound impact on the lives of today's college students. Because of the inherent interest it holds for this population, it provides an excellent opportunity for pairing relevant content knowledge with the quantitative reasoning skills essential for graduates' future success. While the subject has large socioeconomic, psychological, and ethical dimensions, I believe it is important to cleanly delineate its scientific components from these other aspects, and to present clearly what we know and what we don't know, making the unassailable facts distinct from the uncertain predictions. It is also important to emphasize that this is a soluble problem within the capabilities of our current knowledge in science and engineering: the solar energy reaching Earth's surface is more than 8000 times greater than current total world energy consumption. It is time to get moving on the solution.

References

Broecker, W. 1975, Sci., 189, 460

Dewey, J. 1910, Sci., 31, 121

Helfand, D. J. 2016, A Survival Guide to the Misinformation Age (New York: Columbia Univ. Press)

IPCC AR4 Climate Change 2007, The Physical Basis (Geneva: IPCC)

IPCC AR6 Synthesis Report 2021, Summary for Policymakers (Geneva: IPCC)

Kiehl, J.T., & Trenberth, K.E. 1997, BAMS, 78, 197

Chapter 11

Addressing Climate Change with Informal Science Education

Ka Chun Yu

> "If museums can create experiences that consistently and genuinely support both individual and societal well-being … Then museums need not fear for their future."
>
> —John Falk[1]

Informal science education that occurs out-of-school is not a lesser form of learning, but is instead the predominant way that adults learn science. The types of informal education that are perhaps most familiar to the reader are traditional exhibits or films experienced passively by a visitor to a museum or science center. But much more active engagements about climate change are becoming more common, including forums and dialogues, community science projects to gather local environmental data, and programs targeting youth. Many of these get the public to think about how to increase community resilience to the hazards of extreme weather and other phenomena, empowering them to be change makers who can take action.

11.1 Introduction

Although there is a considerable movement to improve climate change education in the formal classroom (Hoffman & Barstow 2007, Niepold et al. 2018), informal education efforts have a large role to play as well. Evidence suggests that most science learning in adults actually takes place outside of the classroom, occurs throughout life, and from a wide variety of sources and contexts (Falk et al. 2007). Only 5% of the waking hours in a typical American's lifetime takes place in school, with the time spent learning science an even smaller fraction of that total (Falk & Dierking 2010). And although Trends in International Mathematics and Science

[1] https://rowman.com/ISBN/9781538149218/The-Value-of-Museums-Enhancing-Societal-Well-Being

doi:10.1088/2514-3433/acfcb6ch11

Study (TIMSS) and Programme for International Student Assessment (PISA) test scores show U.S. students lagging their top peers internationally in science and math (OECD 2019, US DOE 2021), science interest in U.S. adults later rebounds (NSB 2018). Science museums and science centers play an important role in the public's knowledge of and interest in science (Falk et al. 2016), with two-thirds of Americans visiting them at least once a year, a fraction that is greater than those in other countries (Falk et al. 2018, NSB 2018).

Given the importance of informal learning experiences, I will review some of the ways that climate change education has appeared in museum and other informal settings. This includes traditional museum exhibits and via visualization technologies like spherical displays and planetariums. However, in the second half of this chapter, I will also examine how cutting-edge informal science education now goes far beyond the traditional museum or science center experience. Instead, effective ways of increasing public climate change literacy targets at a community level, instead of towards the individual. Instead of exhibits, educators use deliberative public forums, community (citizen) science projects, and engagement strategies outside the walls of traditional informal science institutions. These can lead not just to a greater understanding of climate science, but such activities can have significant, concrete impacts on the ability of communities to prepare for, mitigate, and adapt to climate change.

11.2 Museum Exhibits and Projection Technologies

In recent years, museums worldwide have brought to the public more climate change-related exhibits. These include new permanent exhibits at larger natural history museums in the U.S., such as the American Museum of Natural History, the California Academy of Sciences, and the Smithsonian Museum of Natural History. Smaller institutions dedicated to climate change have also appeared throughout the world, including the Jockey Club museum in Hong Kong, the Museu do Amanhã in Rio de Janeiro, the Climate Museum in New York, and the Klimahuset in Oslo (Newell 2020). These exhibits cover the science behind global warming, including impacts happening at global and local scales. Others go beyond the basic facts. *Altered State: Climate Change in California* at the California Academy of Sciences has interactives allowing visitors to see the amount of carbon generated by different everyday activities, as well as suggestions for reducing their food carbon footprint. This type of approach helped the public understand that they could be part of the solution in dealing with climate change (RKA 2010).

Specialized projection technologies, not generally found in formal educational settings, have appeared in many museums and science centers with "wow" factors that draw in and engage audiences. Perhaps the most powerful tool for climate change education are the digital globes, exemplified by the Science On a Sphere (SOS). Invented at the U.S. National Oceanic and Atmospheric Administration (NOAA), SOS consists of a 1.7-meter sphere illuminated with global datasets (Figure 11.1; McDougall et al. 2007). Researchers and content creators have generated visualizations about Earth's atmosphere, hydrosphere, cryosphere,

Figure 11.1. A Science On a Sphere presentation at the Denver Museum of Nature & Science. Image credit: Eddie Goldstein.

lithosphere, and biosphere. The datasets can be as varied as near real-time satellite imagery of Earth's clouds, animations of plate tectonics through geologic time, tracks of migrating animals, and local sky brightness due to light pollution. From visitor evaluation studies, we know that SOS captivates and intrigues audiences with realistic-looking visualizations, while giving them appreciation of the interconnectedness and dynamism of Earth systems (Apley 2004, Goldman et al. 2010). Recognizing that museum visitors may not gain much insight viewing unfacilitated geospatial datasets with minimal interpretation, many have advocated for the use of story to enhance understanding (e.g., Niepold et al. 2007). Since 2006, content creators have produced narrated movies formatted specifically for spherical displays, with more than 80 now available (NOAA 2021). Assessments show such narratives can lead to significantly greater learning in visitors compared to those in control groups (Schollaert Uz et al. 2014, Beaulieu et al. 2015). A community of practice has developed among the 170+ institutions that have SOS, which has collectively developed and shared hundreds of datasets and films.[2] Because NOAA has nurtured the use of this technology through regular professional development workshops and trainings,[3] SOS remains a popular and powerful tool in museums and science centers for climate change and Earth science education.

Another venue designed to draw in audiences using spectacular visuals are planetariums equipped with "fulldome" video projection systems that can show

[2] sos.noaa.gov/catalog/datasets

[3] sos.noaa.gov/education/education-forum

filmed or computer-generated content. Astronomy and space-related themes remain the focus of the vast majority of programming in the more than 1800 digital dome theaters worldwide (Loch Ness Productions 2022). However, in recent years, there has been a small but growing number of fulldome films covering climate change. A number of planetariums have created in-house productions to explain the basic climate science and climate change impacts to their local audiences, including *Our Living Climate* (2009) from the Melbourne Planetarium, *Green Planet* (dir. T. W. Kraupe 2010) from the Hamburg Planetarium, *The Cosmic Climate Cookbook* (dirs. J. Hoffman & J. Bartel, 2018) from the Science Museum of Virginia, and *Climate Change in Our Backyard* (dir. T. Metzinger, 2021) from the University of Colorado Boulder. Independent film productions that are licensed to planetariums also have also sprung up, such as *Earth's Climate* (dir. T. Matsopoulos, 2021). Three California Academy of Sciences productions (dir. R. Wyatt)—*Fragile Planet* (2008), *Expedition Reef* (2018), and *Living Worlds* (2021)—are about various Earth systems but make explicit references to climate change. *Worlds of Ice* (dir. P. Baylaucq, 2022) examines the role of ice in Earth's past and present, climate change, and human culture through an Inuit lens. A number of recent shows highlight the constellation of Earth-observing satellites that are documenting climate and other global change. *Atlas of a Changing Earth* (dir. T. Lucas, 2021) combines satellite-based remote sensing of Earth systems with sophisticated computer simulations, to show the complex interactions between different Earth systems and human drivers that result in climate impacts on our civilization (Lucas 2021). *We Are Guardians* (dir. M. Crow, 2022) describes how satellite observations show the many ways that humans are impacting ecosystems and biodiversity. Finally, *Preserving Our Precious Planet* (dir. T. Matsopoulos, 2022) details the observations from a fleet of European Space Agency (ESA) satellites that can help decision makers on Earth.

Finally, giant screen films like IMAX are popular in science centers because of their ability to immerse audiences, allowing them to feel that they "are there" in the scenes on screen (Flagg 1999). While there are many shows that delve into natural history, conservation, and prehistoric life, the number of documentary films with climate change as an explicit theme has been smaller but have been growing in the past decade. *To The Arctic* (dir. G. MacGillivray, 2012) focuses on polar bears living on ice floes in the Arctic, and examines the impact of global warming on their habitat. Other films appear to tiptoe around the topic. In *The Earth Wins* (dir. J. Grayson, 2013), aerial footage paired with rock music is used to tell the story of the growing human impact on our planet. However, only in the educator's guide is there a direct connection made between increasing wildfires and climate change (Morrison & Beeson 2013). In recent years, films have become more explicit about the problem. Not surprisingly, they often use the dramatic changes in the cryosphere to drive home their points. *Antarctica* (dir. F. Devas, 2020) shows how climate-related transformations to sea ice in the eponymous continent affects the rest of the planet. *The Arctic: Our Last Great Wilderness* (dir. F. Schulz, 2021) highlights the threat that fossil fuel development poses to Alaska's Arctic National Wildlife Refuge. *The Last Glaciers* (dir. C. Leeson, 2022) is blunt about climate impacts on glaciers and the peoples reliant on their meltwater. National Geographic's *Extreme*

Weather (dir. S. Casey, 2016) has dramatic footage of tornadoes, wildfires, and calving glaciers. However, it has also come under criticism about not making more explicit the connection between the phenomena suggested by its title with human-caused climate change (O'Driscoll 2016).

11.3 Climate Change as Controversy

This last film is not the only instance when an informal science educational product has come under fire for not dealing explicitly enough with climate change. The Science Museum in London has been accused of watering down its *Atmosphere* climate exhibit in order not to offend donors from the oil and gas industries (Nesbit 2015). Although the Smithsonian Museum of Natural History's *Deep Time* exhibit received praise for using the history of CO_2 over geologic time to bring the audience's attention to global warming (Kaplan 2019), it has also been criticized for being too restrained in linking the burning of fossil fuels to climate change (Svoboda 2019). Other museums have been open about not wanting to cause controversy with the public or with donors (Kuchment 2014).

The delicate line that museums and science centers walk is due, in part, to the status of many as non-profit organizations that cannot afford to offend private donors or corporate and government sponsors. In addition, delivering more information about a scientific topic to those already skeptical about the science is unlikely to change minds. Although majorities of Americans are worried about climate change, roughly a fifth are skeptical or outright hostile to the science (Leiserowitz et al. 2021). Within this minority, the deficit model of science communication—the assumption that people are simply misinformed and only need more facts—is ineffective (Mooney 2010, NAS 2018). Worse, delivering more information can backfire: these individuals dig in their heels even more (Hart & Nisbet 2012). Even for those receptive to the message, learning more about climate change does not necessarily motivate an individual to change their behavior (Allen & Crowley 2017, Spitzer & Fraser 2020).

Instead of the amount of content knowledge they have, personal identity and the values associated with that identity is a more important predictor of a person's beliefs around controversial scientific issues (Kahan 2010, 2012). The more knowledgeable a climate skeptic is, the more dismissive they may be about climate science (Kahan et al. 2012). Hence, people are predisposed to reject information that does not conform to their personal viewpoints. The Big Bang and evolution are other topics where increased knowledge about scientific facts does not lead to an acceptance of those facts because of religious or other cultural reasons (Kahan 2016).

11.4 Local, Hopeful Messages for Community Education

So what are some of the preferred ways of communication that do not turn off an audience because of their identity? An alternative is crafting climate change stories that highlight issues at the local level, where climate impacts are visible and individual action is possible. Stories about the challenges of a changing environment are more effective when their focus is as narrow as possible. The larger scale of the problem cannot be ignored when explaining the science, but most people will care less about what will occur in the future versus the present, and less about events happening on the other side of the globe versus something occurring in their neighborhood. For those living in the U.S., the National Climate Assessment[4] shows the expected or current effects of climate change on a region or state, while specific state and city-level studies can provide details at a more granular level. Educators can use local instances of drought, wildfires, heat waves, and other extreme weather events that have or are happening to point out the type of environmental change that will become more common as Earth continues to heat up.

Stories should also give hope and not focus only on doom-and-gloom scenarios (Kristof 2009), something all too easy to do when discussing the impacts from climate change. The general public wants to know more than just the basic facts; they want to learn about what actions they can take to address climate change (Koepfler et al. 2010). Depressing messages can turn off individuals, leading them to apathy and hopelessness if they feel they are powerless against an enormous global problem (O'Neill & Nicholson-Cole 2009). Negative messages can even backfire and increase skepticism about climate change (Feinberg & Willer 2011). A story that invokes fear can lead to an irrational, emotional response, while a story that triggers a mix of hope and worry can motivate people to think about their choices and act (Ojala 2012; Smith & Leiserowitz 2013). To convey hope, narratives can showcase projects where groups are making a difference in their own small ways to reverse environmental damage (Hes & du Plessis 2015), or to highlight the large-scale efforts needed to counter and reverse climate change (Hawken 2017). These examples show that these problems are not intractable; that scientists, engineers, and many others are creating plans to deal with the impacts from climate change; and that everyone can play a role.

Climate change mitigation requires action at all levels of society, whereas most informal climate change education has focused on what the individual can do. Instead of engaging only at the individual level, educational focus should also be directed to collective action (Salazar 2015; Spitzer & Fraser 2020; Bey et al. 2020). Since identity is an important determinant of whether behavior change is possible, Allen & Crowley (2017) argue that making people identify with their community and believe their community can respond effectively will empower them to act. Furthermore, since extreme weather is an immediate threat from climate change, education should focus on addressing such risks. Compared to their counterparts, communities that are educated about natural hazards tend to be better at risk perception, are better prepared against environmental hazards, and recover from disasters faster (Muttarak & Lutz 2014).

These recommendations for climate change education are put into practice in *Climate Solutions*, a traditional museum exhibit which opened at the Wild Center in Tupper Lake, NY in July 2022. In a space only 3000 ft^2 (280 m^2) in size, the exhibit

[4] nca2023.globalchange.gov

designers' goals were to build awareness that climate solutions exist today, and to empower visitors to find their own agency in addressing climate change in their lives and spheres of influence. In addition to informational panels about the need to change how we generate energy, grow food, and support Earth's ecosystems, the exhibit is also devoted to personal stories from community members who are tackling the climate crisis in their own personal ways. These include students who started a non-profit composting program at their high school, an architect who designs sustainable buildings, and an indigenous leader who encourages the use of traditional ecological knowledge to rethink our relationship with nature. The positive messages in the exhibit have given visitors both a sense of hope and urgency for action, while encouraging many to commit to climate action (Kera Collective 2022).

11.5 Digital Planetariums for Community Dialogues

Many of the best practices described above—centering dialogues on problems relevant to a community, creating narratives with solutions-oriented, hopeful messages, and focusing on community education—gave inspiration to the creation of the *Worldviews Network*. This collaborative web of interdisciplinary scientists, artists, and educators was funded by NOAA from 2010–2014, and turned the planetarium, a space traditionally used only for astronomy education, into a venue for facilitating dialogues around local issues at each partner institution (Yu et al. 2012, 2013). The Worldviews Network created more than a dozen "bioregional community dialogues,"[5] each with audiences viewing a presentation in the planetarium, followed by a discussion often involving professionals with expertise on the topic at hand (Figure 11.2).

The presentations were built around the themes of "Seeing," "Knowing," and "Doing." "Seeing" involved experiencing stories built around geospatial datasets in

Figure 11.2. Using a digital planetarium to engage audiences about Earth systems at the Denver Museum of Nature & Science (left), and a post-dome discussion session at the California Academy of Sciences (right). Image credits: Denver Museum of Nature & Science and Ka Chun Yu.

[5] www.worldviews.net

immersive planetariums. "Knowing" meant getting a deeper understanding of Earth's interconnected systems. "Doing" was via interaction among the participants (including the invited outside experts) to help audience members remain engaged in the problems identified in the story. The Worldviews Network model was able to create a diverse set of dialogue topics at a wide variety of institutions, from large natural history museums to science centers to small non-profit educational organizations. These events allowed participants to understand Earth systems, and helped them visualize phenomena at different scales. However, the success of the discussions were often dependent on the expert voices present including that of a dialogue facilitator (Sickler & Hayde 2016).

11.6 Community Education through Deliberative Forums

The community dialogue model of increasing community environmental literacy with the help of expert scientific voices has appeared in many other contexts. The guiding principle behind community involvement is that in a democracy, citizens should have input in policy debates. For climate policy to succeed, it has to have buy-in from all sectors of society (Kasemir et al. 2003), and one way to engage those groups is by including them in deliberative forums. One of the earliest examples of a community forum around climate change occurred at the Arizona Science Center in May 2009. In the one-day workshop, the general public met with scientists and local experts to learn about the impact of climate change and what steps the Phoenix, AZ community could take to increase resilience against the threat of increasingly severe heat waves and drought (Dahlman et al. 2009). At roughly the same time, the Nurture Nature Center in Easton, PA used deliberative forums to engage local communities about flooding in Pennsylvania's Lehigh Valley. A broad cross-section of the public shared their opinions in a first round of discussions after learning about the science behind flooding in their communities from scientists and other experts. In later engagements, community decision-makers learned about the public's concerns from earlier forums, including the need for more mitigation efforts and early warning systems (Apley & Goldman 2010). In discussions targeted towards professional audiences, the Science Museum of Minnesota used an energy retrofit of its building to highlight energy reduction solutions. It convened energy symposia attended by architects and engineers, who could learn more about renewable energy and how to achieve energy efficiencies in both new and old buildings (Hamilton & Christian Ronning 2020).

On a global scale, the *World Wide Views* is an initiative run by the Danish Board of Technology Foundation that brings together citizens for daylong deliberations about complex science issues that impact society. The results of these deliberative forums are used to inform policy makers in venues like the U.N.[6] The 2009 forums on global warming engaged 4000 people in 38 countries on six continents, with the final report sent to policymakers at the 2009 COP15 (United Nations Climate Change Conference) in Copenhagen. The 2015 forums on climate and energy

[6] wwviews.org

brought together 10,000 citizens in 76 countries. The reports from these forums show the enthusiasm of an informed and engaged citizenry, and add an often-neglected voice to international negotiations that usually involve only decision-makers, representatives from special interest groups, and scientists (Bedsted et al. 2015).

Since 2003, Boston's Museum of Science has hosted over 100 public engagement of science forums on a wide variety of topics, including nanotechnology, genetically modified organisms, and autonomous cars. In a recent project funded by NOAA, MOS partnered with the Arizona State University's Consortium for Science, Policy, & Outcomes to create a six-hour-long forum on community resilience planning.[7] Participants are guided through one of four different topics (drought, extreme heat, extreme precipitation, or sea level rise). They adopt fictional communities, but use data and visualizations from real cities. Participants also consider the perspectives of different stakeholders in the community, from business owners to farmers to outdoor enthusiasts. Informed by their own personal values, but also empathizing with the values and interests of the stakeholders, the participants deliberate among themselves to construct a resilience plan for their community. Evaluations showed that those engaging in these role-playing exercises had increased knowledge, awareness, and engagement in the topic, and became more interested in undertaking actual resilience planning themselves (Todd et al. 2019).

11.7 Other Community Education Strategies

Other educational strategies include taking educational efforts outside of the confines of an informal science institution. Four museums in different major urban areas (Philadelphia, Pittsburgh, New York City, and Washington, DC) participated in the *Climate and Urban Systems Partnership* (CUSP). Building on the idea of community education, the founders of CUSP argue that knowledge of and response to climate change should be embedded in the fabric of an urban community for residents to discover (Snyder et al. 2014). CUSP developed education kits that went to city festivals, libraries, and recreational centers. They created new resources to help community organizations deliver climate educational programming to those in their networks. These additional channels of outreach broadened the impact of the participating museums, by targeting audiences who normally would not consider visiting those institutions (Knutson 2018).

The *Throwing Shade* campaign at the Science Museum of Virginia was a community science effort where volunteers, equipped with heat sensors and GPS units, drove through Richmond, VA (Figure 11.3). Community-based organizations that were already familiar with the landscape of the neighborhoods they were associated with worked with scientists and policymakers to create routes that would go through the greatest variation in land use and tree cover. Traversing such routes would give maximum possible measured temperature variations. Volunteer community (citizen) scientists, many from marginalized neighborhoods, drove to provide the hyperlocal data needed to create a detailed extreme heat map for

[7] www.mos.org/pes-forum-archive/noaa-forum

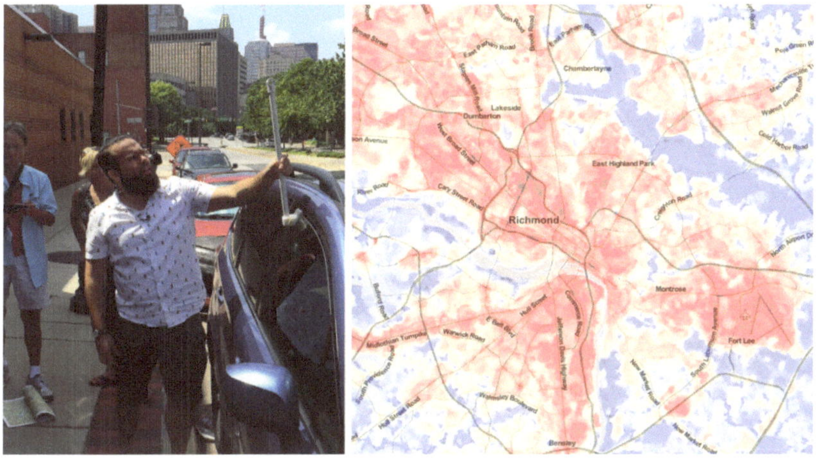

Figure 11.3. Jeremy Hoffman, chief scientist at the Science Museum of Virginia, shows volunteer community scientists how to use a thermal sensor probe attached to a car to measure the heat island effect in Richmond, VA (left). A temperature map of Richmond on the afternoon of 13 July 2017. Image credits: Jeremy Hoffman and Jackson Voelkel.

Richmond. Combining this data with other environmental and socioeconomic variables, researchers generated a heat vulnerability map to indicate where residents were most sensitive to extreme heat events. The data gathering and final map turned the global issue of climate change into a local problem to be addressed with local solutions. One answer came in a six-week-long follow-up pilot program that engaged teens from an at-risk neighborhood. The youth learned about the increasing hazards from a warming world, and explored the heat vulnerability map along with the scientific and socioeconomic data that fed into it. Using the map as a guide, the teens decided to plant fruit trees at a high school campus, a solution that maximized benefits to local youth by creating shade, improving air quality, and providing a source of fresh food (Hoffman 2020).

11.8 Empowering Youth as Agents of Change

Youth are also prominent in our final two example programs. Middle and high school students in Tucson, AZ made their schools more resilient to extreme precipitation through another NOAA-funded project. Tucson is one of the driest metropolitan areas in the U.S., averaging less than 30 cm (12 inches) of rainfall annually. When the precipitation comes during the summer monsoons, water hitting impermeable city surfaces flood the streets after storm drains reach their capacity. Grassroots efforts have sprung up to alter the urban infrastructure so that rainwater is redirected to native plants in designated rain basins, instead of being allowed to run off into the streets (Lancaster 2008). Following in the same spirit, the Watershed Management Group and Arizona Project WET created the *Recharge the Rain* program. Teachers and students learn about climate science and climate change

Figure 11.4. Drachman Montessori K-8 Magnet School in Tucson, before (left) and after (center) the installation of rain basins on the east side of the building, and 17 months later (right). Image credits: Watershed Management Group and Arizona Project WET.

impacts in their community. They then embark on a project to make their school more resilient to extreme precipitation and flooding events by designing and building rain basins on school grounds to harvest rainwater, including planting native trees, shrubs, and grasses (Figure 11.4). With the help of program staff, students and teachers have created 12 water harvesting systems, with the capacity to reduce runoff by 873,000 liters (192,000 gallons) of rainwater per storm (Wilkening & Murrieta-Saldivar 2020; Wilkening 2021).

At the other temperature extreme of Alaska, the traditional lifestyles of Alaskan Natives are under threat from global warming. For instance, in the category of food security, thawing ice and tundra result in conditions that are more dangerous for hunters venturing out to look for game, while melting runways can impede supply planes from landing. As a response to these challenges, the Chugach School District's *Environmental Literacy through Alaskan Climate Stewards* (ELACS) program won grant funding from NOAA. One of the project goals was to help Alaskan Native youth learn about climate change, document the environmental change happening in their communities, and use this new knowledge to develop resilience and action plans for their communities (Sotero 2021). Students in different schools undertook different community science efforts. Those in Chenega Bay took measurements of plankton in water to determine how the changing climate is altering the timing of salmon runs. Youth in Whittier documented landslides occurring because of melting glaciers that have unsettled the soil. To preserve a record of indigenous knowledge, students made video interviews with tribal elders, to learn from them firsthand about how the climate and environment have changed in their lifetimes (Figure 11.5).

Figure 11.5. Frames from a video interview with an Alaskan Native elder in Tatitlek, AK. Video credit: M. Vlasoff.

11.9 Conclusions

This review shows how thinking about informal science education for climate change has evolved from traditional museum exhibits to programs that actively engage communities. The latter types of experiences include public forums and dialogues about climate change issues, and citizen scientists understanding more about climate impacts by gathering environmental data in their communities. The just described programs from Virginia, Arizona, and Alaska also show how youth can be an important target audience for climate education programs. These adolescent participants went beyond merely learning about climate change to documenting and responding to the climate consequences in their hometowns. Such engagement strategies helped to empower them to act, leading to one necessary outcome of climate literacy: preparing communities for the numerous types of hazards they face from climate change (Haynes & Tanner 2015; Trott 2020). The youth of today are inheriting a world where the burdens of climate and other environmental change will be increasingly visible around them. Instead of being victims helpless against the vicissitudes of a warming world, empowered young people become agents of change. The change makers and leaders of tomorrow will come from their ranks, and well-designed informal science education programs can help prepare them today.

References

Allen, L. B., & Crowley, K. 2017, Int. J. Global Warming, 12, 299

Apley, A. 2004, Science On a Sphere Front-end Evaluation (Portsmouth: RMC Research)

Apley, A., & Goldman, E. 2010, Engaging Citizens in Science Dialogue: An Evaluation of the Nurture Nature Foundation's Flood Forum Project (Portsmouth: RMC Research)

Beaulieu, S. E., Emery, M., Brickley, A., et al. 2015, J. Geosci. Edu, 63, 332

Bedsted, B., Mathieu, Y., & Leyrit, C. 2015, World Wide Views on Climate and Energy (Copenhagen: Danish Board Tech. Foundation)

Bey, G., McDougall, C., & Schoedinger, S. 2020, Report on the NOAA Office of Education Environmental Literacy Program Community Resilience Education Theory of Change (Washington DC: NOAA)

Dahlman, L., et al. 2009, Summary Report: Community Conversations on Climate and Water (Washington DC: NOAA, ASTC))
Falk, J. H., et al. 2016, ScEd, 100, 849
Falk, J. H., & Dierking, L. D. 2010, AmSci., 98, 486
Falk, J. H., Storksdieck, M., & Dierking, L. D. 2007, Pub. Understanding Sci., 16, 455
Falk, J. H., Pattison, S., Meier, D., Bibas, D., & Livingston, K. 2018, J. Res. Sci. Teach., 55, 422
Feinberg, M., & Willer, R. 2011, Psych. Sci., 22, 34
Flagg, B. 1999, Informal Learn. Rev., https://www.informalscience.org/sites/default/files/GSTA_Flagg%20paper.pdf
Goldman, K. H., Kessler, C., & Danter, E. 2010, Science On a Sphere: Cross-Site Summative Evaluation (Edgewater: ILI)
Hamilton, P., & Christian Ronning, E. 2020, J. Mus. Edu, 45, 16
Hart, P. S., & Nisbet, E. C. 2012, Commun. Res., 39, 701
ed. Hawken P. (ed) 2017, Drawdown (New York: Penguin)
Haynes, K., & Tanner, T. M. 2015, Child. Geographies, 13, 357
Hes, D., & du Plessis, C. 2015, Designing for Hope (Abingdon: Routledge)
Hoffman, J. S. 2020, J. Mus. Edu, 45, 28
Hoffman, M., & Barstow, D. 2007, Revolutionizing Earth System Science Education for the 21st Century (Cambridge: TERC)
Kahan, D. 2010, Natur, 463, 296
Kahan, D. 2012, NatNws, 488, 255
Kahan, D. M. 2016, J. Risk Res., 20, 995
Kahan, D. M., et al. 2012, NatCC, 2, 732
Kaplan, S. 2019, Washington Post, https://www.washingtonpost.com/national/health-science/the-smithsonians-renewed-fossil-hall-sends-a-forceful-message-about-climate-change/2019/05/25/bc896212-78d2-11e9-b3f5-5673edf2d127_story.html
Kasemir, B., Jaeger, C. C., & Jäger, J. 2003, Public Participation in Sustainability Science: A Handbook, ed. B. Kasemir, J. Jäger, C. C. Jaeger, & M. T. Gardner, 3
Kera Collective 2022, Climate Solutions Summative Evaluation (Washington DC and New York: Kera Collective)
Knutson, K. 2018, Routledge Handbook of Museums, Media and Communication, ed. K. Drotner, V. Dziekan, R. Parry, & K. C. Schröder (Abingdon: Routledge) 101
Koepfler, J. A., Heimlich, J. E., & Yocco, V. S. 2010, App. Env. Edu. Commun., 9, 233
Kristof, N. 2009, Outside Magazine, https://www.outsideonline.com/outdoor-adventure/nicholas-kristofs-advice-saving-world/
Kuchment, A. 2014, Dallas Morning News, https://www.dallasnews.com/news/news/2014/06/14/museums-tiptoe-aroundclimate-change
Lancaster, B. 2008, Rainwater Harvesting for Drylands and Beyond Vol 2 (Tucson: Rainsource)
Leiserowitz, A., Roser-Renouf, C., Marlon, J., & Maibach, E. 2021, Curr. Opinion Behav. Sci., 42, 97
Loch Ness Productions 2022, Fulldome Theater Compendium, https://www.lochnessproductions.com/lfco/lfco.html
Lucas, T. 2021, private communication
McDougall, C., McLaughlin, J., Bendel, W., & Himes, D. 2007, Proceedings of the 5th International Symposium on Digital Earth, June 2007, (San Francisco, CA)
Mooney, C. C. 2010, Do Scientists Understand the Public? (Cambridge: AAAS)

Morrison, J., & Beeson, P. 2013, The Earth Wins: An Educator's Guide, https://paperzz.com/doc/7916698/the-earth-wins-educator-s-guide---australian-literacy-edu
Muttarak, R., & Lutz, W. 2014, Ecology Soc., 19, 42
NAS 2018, The Science of Science Communication III: Inspiring Novel Collaborations and Building Capacity (Washington DC: NAP)
NOAA 2021, SOS Data Catalog, https://sos.noaa.gov/catalog/datasets/
NSB 2018 2018, Science and Engineering Indicators (Alexandria: NSF)
Nesbit 2015, US News & World Report, https://www.usnews.com/news/blogs/at-the-edge/2015/06/01/science-museum-exhibit-funded-by-shell-under-fire
Newell, J. 2020, Mus. Manage. Curatorship, 35, 599
Niepold, F., Herring, D., & McConville, D. 2007, Proceedings of the 5th International Symposium on Digital Earth, June 2007 (San Francisco, CA)
Niepold, F., Poppleton, K., & Kretser, J. 2018, Green Schools Catalyst Quarterly, 5, 14
O'Driscoll, B. 2016, Pittsburgh City Paper Blogh, https://www.pghcitypaper.com/Blogh/archives/2016/10/11/science-centers-imax-film-on-climate-change-is-silent-on-its-human-causes
O'Neill, S., & Nicholson-Cole, S. 2009, Sci. Commun., 30, 355
OECD 2019, PISA 2018 Results: Combined Executive Summaries, Volume I, II & III (Paris: OECD)
Ojala, M. 2012, Env. Edu. Res., 18, 625
RK&A 2010, Summative Evaluation: Altered State: Climate Change in California Exhibition (Alexandria: Randi Korn & Associates)
RK&A 2018, Summative Evaluation: Climate Change and Resiliency prepared for Science Museum of Virginia (Alexandria: Randi Korn & Associates)
Schollaert, Uz S., et al. 2014, J. Geosci. Edu, 62, 485
Sickler, J., & Hayde, D. 2016, J. Mus. Edu, 41, 66
Sotero, S. 2021, Private Communication
Salazar, J. F. 2015, Climate Change and Museum Futures, ed. F. Cameron, & B. Neilson (New York: Routledge) 90
Snyder, S., et al. 2014, Future Earth: Advancing Civic Understanding of the Anthropoceneed G Roehrig and P Hamilton (New York: Wiley) 103
Spitzer, W., & Fraser, J. 2020, J. Mus. Edu, 45, 5
Svoboda, M. 2019, Yale Climate Connections, https://yaleclimateconnections.org/2019/07/new-koch-funded-fossil-exhibit-at-the-smithsonian-is-curiously-quiet-on-fossil-fuels/
Todd, K., et al. 2019, Science Center Public Forums: Summative Evaluation Report 2019-03 (Boston, MA: Museum of Science)
Trott, C. D. 2020, Env. Edu. Res., 26, 532
US DOE 2021, TIMSS 2019 US Highlights Web Report NCES 2021-021 (Washington DC: DOE, IES, NCES)
Wilkening, B., & Murrieta-Saldivar, J. 2020, AGU Fall Meeting 2020 abstract #ED030-0001
Wilkening, B. 2021, Private Communication
Yu, K. C., Hamilton, H., Connolly, R., McConville, D., & Gardiner, N. 2012, ASTC Dimensions, 2012, 42
Yu, K. C., et al. 2013, Proceedings of the 8th International Symposium on Digital Earth, August 2013, Kuching (Sarawak, Malaysia)

Part V

Communication

Chapter 12

Communication and Climate Change

T A Rector and P Banchero

> "Fight for the things that you care about, but do it in a way that will lead others to join you."[1]
>
> —Justice Ruth Bader Ginsburg

Scientists like to "geek out" on the details. But communication about climate change needs to be more than just about the facts. Our communication has to connect with people's belief systems, cultural values, and social structures. It has to help them overcome psychological, emotional, and social barriers. In addition, it has to rise above misinformation—as there are individuals and organizations working to confuse and sow doubt about the causes and consequences of climate change. In other words, *how* you say something—and *where* and *when* you say it—are just as important as *what* you say. Being right doesn't always win the argument. For climate change, we want people to understand the causes and consequences of the problem, as well as the solutions. But we also want them to feel hopeful that it's not too late. And we want them to become active participants in the process. After all, understanding the problem of climate change isn't terribly helpful unless it inspires you to do something about it. In this chapter we'll talk about the basics of communication so that your interactions with students and the public—in your role as an astronomer—are effective. We'll also talk about interpersonal communications; i.e., how you can interact with individuals in ways that are fulfilling and productive.

12.1 Effective Communication

Simply put, communication is the conveyance of information. Humans are far from the only animals that do it, but how we do it is sophisticated and subtle—via verbal and nonverbal cues. Much of what we share is abstract, including emotions

[1] https://www.radcliffe.harvard.edu/news-and-ideas/ruth-bader-ginsburg-tells-young-women-fight-for-the-things-you-care-about

doi:10.1088/2514-3433/acfcb6ch12

or concepts like hope and freedom. Even telling a simple story—such as when your family adopted a puppy—requires making choices about what you say. What pictures do you show them? Do you mention the chew marks on the leather sofa? Do you play up the delighted expressions on your children's faces when you brought the puppy home? Do you emphasize why you adopted a mutt over buying from a dog breeder? There are other aspects that might affect your storytelling. Maybe it's loud in the park where you are talking. Perhaps your friend is distracted by her cell phone. If you were telling this story to a stranger, it's possible you'd use different words than you would with a friend who is accustomed to your informal speech and sense of humor. Communication, therefore, depends on the nature of your message, the audience, and the environment (i.e., "taking the temperature of the situation.")

To be effective communicators about climate change, astronomers need to know how messages are constructed, and we need to keep in mind the challenges of communicating about climate change as other forces seek to command attention with their own messages. Much of our communication about climate change is implicitly based on the "information deficit model," which assumes apathy or skepticism about climate change is due to lack of knowledge about the problem. We hope that, if we explain the science well enough, people will be naturally motivated to take action. However, there is now a wealth of research that indicates that this isn't sufficient (e.g., Sturgis & Allum 2004, Suldovsky 2017). Our communication also has to resonate with our audience's beliefs and values. It has to help them overcome psychological, emotional, and social barriers, such as fear and guilt. As you communicate about climate change, it's important to think about: What is your goal? How do you want people to respond to your information? How do you want them to feel? How do you want them to act? And what barriers need to be overcome for them to do so? All of these are part of "strategic communication," which is defined as the purposeful use of communication by an individual or organization to convey its message. Strategic science communication is a robust field of study that has many insights to offer; e.g., see Besley & Dudo (2022) for a useful guide of the field.

One of the first rules of effective communication is to "know thy audience." That's especially true for climate change, as there's a wide range of opinions and levels of engagement. When having a conversation about climate change, with an individual or a group, consider the following:

What is your relationship to your audience? Are you with friends or family? Is it an audience member at a public talk or planetarium? A student in your class? Or maybe a stranger on social media? A member of the public at a talk may recognize you as an authority figure whereas someone on social media may not know (or care) that you are a scientist. You may be perceived as "one of them," or perhaps belonging to the "other" group—with whom they may have had little experience but may already possess preconceived notions about. This may make people wary but it's also an opportunity to connect and build upon shared values.

Do you have a short- or long-term relationship? If you have a long-term relationship, then the focus should be to engage in a way that maintains and strengthens that relationship so that they will be willing to talk with you about climate change again. If it is a short-term relationship, think about how you can engage in ways that improve their attitudes toward scientists, climate science, and climate change solutions. It is unlikely that a single interaction will radically change a person's viewpoint, but a positive experience can soften resistance in the wary and strengthen engagement in the concerned.

What is the power dynamic? Are you in a position of authority? For example, a student may not feel comfortable speaking freely about climate change if they feel their views aren't shared by the professor or other students. This can make it difficult for students to get answers to their questions. For this reason you can often get better dialog when students can ask questions anonymously, or if you are a guest lecturer in another instructor's class. There also exists a power dynamic between family members—parents in particular may be challenging to talk with because they were the original authority figures in your life.

What are their values? People will respond more positively if you talk about what they care about, and put climate change in terms that resonate with their interests and values. If you're talking with an individual, what do you know about them? If not much you can start with questions to learn more. If you're talking with an organized group, what is their purpose? They may have a mission or values statement that can help you.

Is it the right time? If it is a public speaking engagement, have you been asked to talk about climate change? If not, do the organizers know you plan to? If it is a private conversation, are they interested in hearing about it at that moment? Naturally if someone brings up the topic it is an opportunity to share with them what you know. Otherwise think carefully about how you will be received. For example, it may be helpful for people to understand how an extreme weather event was made worse by climate change. But if they are stressed by the event shortly after it happened they may not be ready to hear about it. It may also come off as "I told you so." For some people it may never be the right time to talk about it, and you need to be respectful of that.

Do you think a person is acting in good faith? As we'll discuss, there is a difference between a skeptic–someone who still has doubts about climate change but can be swayed—and a denier—someone who is determined to play the role of the antagonist. Unless you have reason to believe otherwise, assume that a person's questions are legitimate and respond as best you can. If not, you may be best served to end the conversation gracefully.

Are you alone or around others? People often act differently when in a group. They may act aggressively or defensively to "save face" with their peers. If it is someone that you would like to talk with, find a way to talk privately. A common scenario is to be confronted by a doubter during a public talk. Your response should therefore be for the benefit of everyone. Keep in mind that your audience will contain a mix of perceptions and attitudes, but that most people are concerned about climate change. They are also there to hear you speak, not the doubter. The body language of the audience will help you understand how the conversation is going.

12.2 Demographics of Climate Change

Especially when talking with larger groups, it helps to understand the overall distribution of climate change attitudes. Maibach et al. (2009) discuss a framework for understanding these attitudes within the U.S. They identify "Six Americas": six unique audiences within the American public that each respond to the issue in distinct ways. The "Alarmed" represent the highest level of concern. They are convinced of the seriousness of climate change and are likely already taking action to address it. Next are the "Concerned," who are also convinced that global warming is a serious problem, but have not yet engaged the issue personally. Three other groups—the "Cautious," the "Disengaged," and the "Doubtful" represent decreasing stages of understanding and acceptance of the problem. Finally, the "Dismissive" are certain climate change is either not happening, not anthropogenic, or not a problem. They tend to actively oppose efforts to reduce greenhouse gas emissions. Similar audience segmentation analyses have been done for other countries; e.g., Australia (Richardson et al. 2022) and India (Leiserowitz et al. 2023a).

Results from the Six Americas survey from 2012 to 2022 reveal several trends (Figure 12.1). First, concern about climate change overall has steadily grown during this

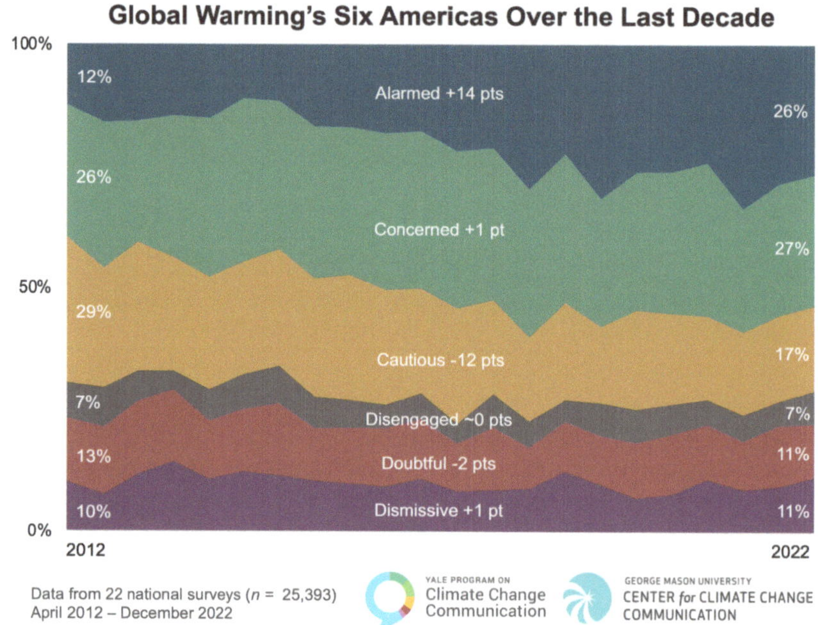

Figure 12.1. [2]The evolution of the "Six Americas" categories over 2012–2022. The Alarmed segment has more than doubled in size, growing from 12% of the U.S. population in 2012 to 26% in 2022. The Alarmed segment is now similar in size to the Concerned (27%). Conversely, the Cautious segment has decreased in size from 29% in 2012 to 17%. The Disengaged (7%), Doubtful (11%), and Dismissive (11%) segments have remained relatively similar in size over the last decade.

[2] https://climatecommunication.yale.edu/publications/global-warmings-six-americas-december-2022/

time period. Half of the U.S. population are now in the Alarmed and Concerned categories. The Cautious group has steadily declined—presumably increased exposure to information about climate change as well as personal experience with extreme weather and climate events has moved many people from Cautious into the Concerned and Alarmed categories. Second, the number of people in the Disengaged, Doubtful, and Dismissive categories has stayed largely unchanged, suggesting that individuals in these groups are less interested, able, or willing to change their minds (Leiserowitz et al. 2023b). Whereas the Doubtful are largely uninterested, the Dismissive are quite engaged on the topic. In fact, they are second only to the Alarmed in the percentage that say that they pay "a lot" of attention to global warming (Roser-Renouf et al. 2015).

Attitudes toward climate change vary according to the aspect highlighted by the survey questions. While 72% of Americans think that global warming is happening, only 57% believe it is caused mostly by human activity, and while 71% think that global warming will harm future generations, only 64% think that it will harm people in the U.S.; and only 47% think that it will harm them personally. Over 75% support funding for research into renewable energy, or tax rebates for energy-efficient vehicles. But only slightly more than half thought that local and national political officials should do more to address it (Yale Climate Opinion Maps 2021; see Howe et al. 2015 for data and methodology).

People's views are correlated with different characteristics. The strongest correlations are between political ideology and education level, and are also influenced by race, gender, and age (Figure 12.2). Climate change, like many other environmental issues, is

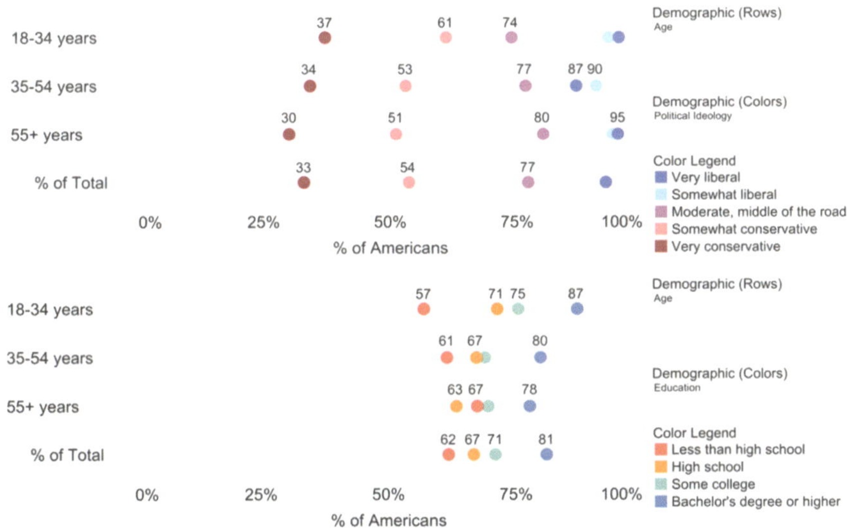

Figure 12.2. Strong trends in acceptance that climate change is happening can be seen as functions of age, political ideology, and education level (YPCC and Mason 4C 2022[3] and Ballew et al. 2019). The bottom row of each plot shows the trends regardless of age. (Plots created with the CCAM Explorer[4] for the timespan 2018–2022.)

[3] https://doi.org/10.17605/OSF.IO/JW79P

[4] https://climatecommunication.yale.edu/visualizations-data/americans-climate-views/

often portrayed as a concern primarily of wealthy white liberals. However, Pearson et al. (2018) found the anticipated levels of concern among minorities and low-income Americans was highly underestimated. Ballew et al. (2023) found that overall people of color are more concerned than whites, women are more concerned than men, and younger people are more concerned than older generations. For comparison 68% of young (Gen Z and Millennial) women of color fall into the Concerned or Alarmed categories, whereas only 44% of older (e.g, Baby Boomer) white men do.

There are also geographic differences; e.g., counties along the U.S.–Mexico border are more likely to think global warming will harm them personally. This may reflect the higher percentage of Hispanics living in the area who may have already been directly impacted by climate change. You can explore regional variations with the Yale Climate Opinion Maps as well as generate custom climate opinion factsheets for states, counties, or congressional districts within the U.S.

Religious beliefs also play a part, and in complex ways. For example, there is considerable variance among different Christian groups. Many Protestant denominations and the Roman Catholic Church have made explicit declarations of support for global climate action. Southern Baptists and other evangelical Protestants, in general, have done the opposite (Zaleha & Szasz 2015). There is an inverse correlation between concern and the proportion who identify as "born-again" or evangelical, where less than 20% of the Alarmed and more than half of the Doubtful and Dismissive identify as such (Leiserowitz, Maibach, & Roser-Renouf 2009). This may be an induced correlation, as evangelicals are more likely to identify as conservative. Believers in Christian end-times theology are less likely to support policies designed to curb global warming (Barker & Bearce 2013).

While there is a strong connection between climate skepticism and political ideology in the U.S. (Hornsey, Harris, & Fielding 2018) and Australia (Fielding et al. 2012), this correlation is much weaker in most other countries, such as those in the European Union (McCright, Dunlap, & Marquart-Pyatt 2016). The U.S. Republican Party has been an anomaly in denying anthropogenic climate change, even among conservative political parties worldwide (Båtstrand 2015).

Personal experience also plays an important role. Those who have recently experienced hot weather are more inclined to believe in climate change, whereas those recently exposed to cold weather are just the opposite (Kaufmann et al. 2017). This creates a challenge for persuading climate skeptics if their recent weather experiences seem inconsistent with global warming. Leiserowitz & Smith (2017) find that Americans increasingly associate extreme weather events (e.g., heat waves and hurricanes) with climate change, and experiencing one of these events increases concern (Bergquist et al. 2019).

Views about climate change also change over time, and not always toward greater concern (Figure 12.3). These changes reflect variations in the overall political mood of the country, as well as recent events. For example, Americans' belief that global warming was already creating problems peaked at 61% in 2007 after Hurricane Katrina but then dipped to 49% in 2011 before returning to 2007 levels in 2017.[5]

[5] https://news.gallup.com/poll/206030/global-warming-concern-three-decade-high.aspx

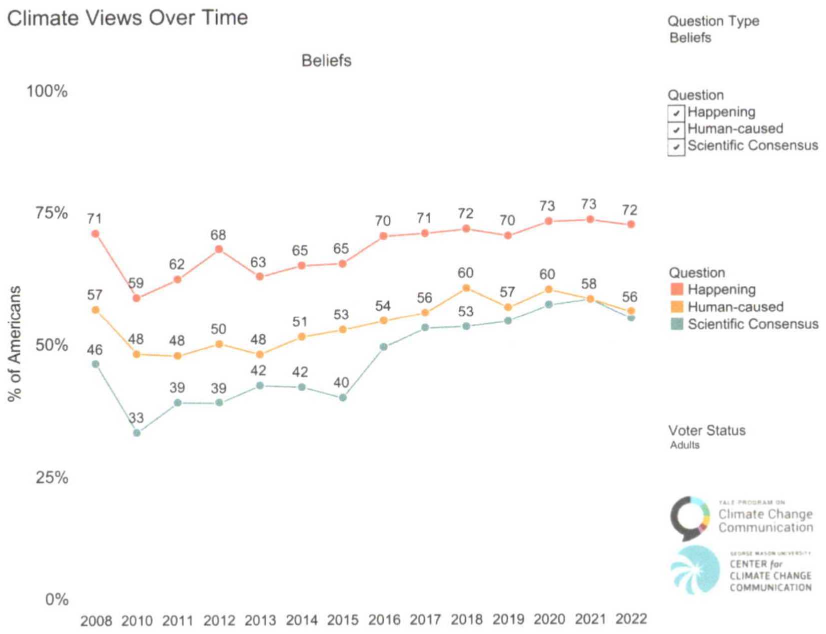

Figure 12.3. [6]Attitudes towards climate change as a function of time over 2008–2022[7]. A gap exists between acceptance that climate change is happening (shown in red on plot), that it is caused by human activity (orange), and that there is a widespread consensus among scientists that this is indeed the case (green) (YPCC and Mason 4C 2022 and Ballew et al. 2019). (Plot created with the CCAM Explorer.[8])

Since then attitudes have remained fairly stable at that level, perhaps reflecting an increased exposure to information and to extreme climate events.

While public acceptance and concern about global warming increased overall from 2012 to 2022, they have increased faster among young adults (aged 18–34). Youth have also surpassed older Americans on several measures of policy support, including funding more research into renewables. There is also evidence for increased political polarization. For example, during this same time interval support for renewable energy research has grown for all political ideologies except conservative Republicans (Marlon et al. 2022).

12.3 Barriers of Climate Change

The majority of people—in the U.S. and beyond—now believe climate change is a problem that needs to be solved. But there are still barriers to overcome before this will lead to widespread action. One set of barriers is simply how we think and feel about the problem.

---

[6] https://climatecommunication.yale.edu/visualizations-data/americans-climate-views/

[7] https://doi.org/10.17605/OSF.IO/JW79P

[8] https://climatecommunication.yale.edu/visualizations-data/americans-climate-views/

12.3.1 Psychological Barriers

Evidence from behavioral and brain sciences suggests that the human judgment system is not well equipped to identify climate change—a complex, large-scale and unintentionally caused phenomenon—as a physical threat or moral imperative (Markowitz & Shariff 2012). Klöckner (2020) and others describe several elements as to why climate change is difficult for us to deal with psychologically:

1. Humans are well suited to solve problems that have a clear cause and effect. Climate change is a poorly defined problem in the sense that there are multiple causes as well as multiple consequences.
2. Climate change is occurring on timescales longer than the human psyche is accustomed to, creating a gap of perception between behavior (e.g., use of fossil fuels) and their effects on the environment. It is essentially an intergenerational problem.
3. The people who are most responsible for emissions (particularly the wealthy in developed countries) also have the most resiliency and will suffer the least. It is therefore, to some degree, "someone else's problem."
4. It is unlike any problem we as a species have encountered before. For most of human history, our actions have had local, small-scale environmental effects but not global, large-scale impacts of this magnitude. The Earth has, by and large, seemed unchangeable—and its resources virtually infinite.
5. The visible signs of climate change are intermittent; and on most days its effects are largely unnoticeable. Climate change returns to our attention whenever there is a major event such as a hurricane or forest fire, but then fades as other, more-immediate issues arise. However the increasing frequency of extreme events may be changing this.
6. The greenhouse gases that cause climate change are not detectable by human senses. CO_2 has been described as the perfect killer because it is invisible, odorless, and inherently not necessarily a bad thing. If CO_2 gas were detectable by our senses we would be more likely to address the problem.
7. Climate change is a perfect "tragedy of the commons" problem. Greenhouse gases released locally affect everyone globally. The atmosphere has been described as the "great sewer in the sky." Solutions, therefore, to a large extent require global cooperation.
8. Psychologically it is difficult to think that we—as individuals and collectively—are part of the problem. No one likes to think of themselves as "being bad" or causing harm to others. This can lead to defense mechanisms, such as apathy ("Sure it's a problem, but not my concern."), denial ("It's a hoax."), or placing blame elsewhere ("China is the problem.")
9. The effects of one's individual actions are small, leading to people's perception that any action they could take would be futile.
10. Climate change is too big a problem to comprehend. Thus people protect their mental health by adopting coping strategies rather than trying to address the problem. This is in itself a subtle form of climate denialism—we

are refusing to acknowledge it to ourselves because of the implications of what it means about who we are, how we live our lives, and the changes we might need to make.

11. Conflicting messaging from different groups (e.g., scientists, the media, and interest groups) can create an environment of uncertainty that—in combination with the above psychological effects—encourages a "wait and see" mindset.

Other psychological factors are also important. In particular, personal beliefs and values impact how people interpret information. Known as "motivated reasoning," people seek out and remember knowledge that supports their worldview while dismissing information that is contradictory (Corner, Markowitz, & Pidgeon 2014). The "Dunning-Kruger" effect (Kruger & Dunning 1999) describes how people with a small amount of knowledge about a subject are likely to overestimate the depth of their understanding, leading to excessive confidence. Combined with motivated reasoning, it can lead skeptical people to find enough misinformation to be convinced of an incorrect understanding.

12.3.2 Emotional Barriers

When thinking about climate change people can feel a wide range of emotions—including anger, anxiety, fear, guilt, and hopelessness (Pihkala 2022 and references therein). These feelings are widespread; e.g., a 2019 poll by the American Psychological Association revealed that 68% of adults in the U.S. are experiencing at least a little anxiety about climate change.[9]

Climate grief can be defined as the intense feelings people suffer from the climate-related losses to valued species, ecosystems, and landscapes that carry personal or collective value—not only economically but as part of one's identity, culture, and way of life; e.g., the loss of a family farm or traditional hunting ground. Kübler-Ross (1969) made famous the "five stages of grief"—denial, anger, bargaining, depression, and acceptance—that one goes through while we mourn. Mourning is the process of coming to terms with a profound loss. With the passing of a loved one, we mourn not only their loss but also the loss of our previous life. We face the challenge of building a new life without them. With climate change the problem is similar, but more complex—in part because the loss is occurring gradually but also because most cultures lack the words and structure to process it (Cunsolo & Ellis 2018). Climate grief is related to the concept of "solastalgia," the distress produced by environmental change impacting one's home environment (Albrecht et al. 2007). Climate grief can also be a form of "disenfranchised grief," defined as a loss that is not openly acknowledged or recognized socially (Cesur-Soysal & Arı 2022) because of our general reluctance to talk about climate change.

Climate anxiety can be defined as the fear of future climate impacts, and is especially felt among youth. Hickman et al. (2021) surveyed people aged 16–25 years

[9] https://www.apa.org/news/press/releases/2020/02/climate-change

in several countries. The majority were worried about climate change; 59% were very or extremely worried and 84% were at least moderately worried. More than 50% reported one or more of the following emotions: sad, anxious, angry, powerless, helpless, and guilty. More than 45% said their feelings about climate change negatively affected their daily life and functioning, and many reported a high number of negative thoughts about climate change; e.g., 75% said that they think the future is frightening and 83% said that they think people have failed to take care of the planet. Climate anxiety and distress were correlated with perceived inadequate government response and associated feelings of betrayal.

Climate grief and anxiety are not pathological conditions, but an expected, adaptive response to a real and enormous threat. Hurley et al. (2022) argue that it is unfair to ask a generation of young people to develop enhanced psychological stamina to face climate change. But because of decades of inaction on the part of previous generations, they are being asked to do so.

Rising anxiety about climate change has led to feelings of hopelessness among many people, especially youth. Known as "climate doomism," there is a growing sense among many that climate change is too big of a problem to solve and/or it is already too late. Unfortunately tapping into the doomer zeitgeist is becoming a common strategy for disinformation campaigns. If people can be convinced that it's hopeless, then they won't make any effort to change—which has the same outcome as climate denialism.

The problem with the doomer perspective is that it presents only two options—either salvation or extermination. But climate change isn't like an asteroid that either does or does not hit the Earth. It's a matter of degree—literally! While the goal is to keep the temperature anomaly below 2.0°C, and preferably 1.5°C, it is not as though the world will be destroyed if we fail to achieve either benchmark. A lower temperature anomaly will always be better, no matter what the scenario—so it will always be worth working to reduce emissions. In other words, it's never too late to do the right thing.

To counter doomism we need to demonstrate that the situation is not hopeless, and in fact we can change in ways that will actually make our lives better. It turns out that hope is not only a pleasant feeling but can also work as a motivational force (Ojala 2012). It is especially helpful for improving attitudes toward climate change policy and advocacy (Nabi et al. 2018). Optimism is important in countering the negative emotions that many feel about climate change, but it alone is not enough. There also needs to be a sense of urgency and agency—that is, the problem is fixable and we need to do it now. Passive optimism—that is, the belief that the problem of climate change will resolve itself without one's participation—is just as problematic as doomism because both lead to inaction. Ritchie (2023) argues that people need to feel that the world is malleable, and optimistic that we can do so; i.e., "the future can be better if we work hard to change it."

Part of the challenge of climate change is acknowledging to ourselves that we're culpable—or to put it more precisely, the way we live our lives is causing the problem. Belonging to a social group that is labeled as responsible for systemic harm can induce complex reactions, such as guilt, rationalization, and denial.

In fact, many forms of denialism are an emotional response to protect oneself from feeling pain (Moser 2007). It can be a way of processing the fear and helplessness we feel when faced with such a scary problem whose solution is daunting and unclear. To overcome the guilt we need to acknowledge our role in the problem, but also understand that it's not exactly our fault. We didn't choose the system into which we were born, but we do get to choose whether or not we work to change it. This is true not only for the societies in which we live, but also astronomy as a profession.

As scientists and astronomers we also need to recognize that climate change is having an emotional impact on us. How can we cope with the implications of what we know? One of the challenges is that scientists are trained to be rational and impartial; i.e., let's let the facts take us to the correct conclusions. But we cannot help but be impacted, and it's not healthy for us to pretend that we aren't. Attempts to persuade others are inherently tied to emotions (Rocklage et al. 2018), so it can be useful for us to share with others how our knowledge of climate change makes us feel. For example, the Is This How You Feel campaign presented an opportunity for climate scientists to share their thoughts and feelings in an open, candid way—which can be therapeutic for them and inspirational for others.

What can people (and scientists) do to help with their feelings? Beth Ellwood[10] offers some tips. First is to acknowledge that the feelings are there, and that they're natural. These feelings are a rational response to an irrational world. Next is to recognize that you're not alone in feeling this way. Talk with others who share your concerns. One of the best things we can do is work with others, so join a group that's working on a solution for climate change that you'd like to be part of.[11] Schwartz et al. (2022) find that climate change activism improves mental health. While it's not your job to solve the problem, you can play a part. Recognize the limitations of what you can do, but also know that small actions on your part can lead to bigger changes nonlinearly. Also recognize that solutions to climate change are already underway. And who doesn't want to be part of the winning team?

12.3.3 Social Barriers

As discussed in Chapter 6, social norms are an important motivator that can lead to social change. People are more likely to engage in a behavior if they see others in their peer group do the same (Goldstein et al. 2008). Personalizing and localizing the threat of climate change, and enhancing the norm that most people support action, can therefore be effective in overcoming doubt and inaction (Ballew et al. 2022).

One of the biggest problems with climate change is our general reluctance to talk about it. Elisabeth Noelle-Neumann[12] coined the phrase "spiral of silence" to refer to the reluctance of individuals to express opinions that might be rejected by their

[10] https://www.psycom.net/anxiety/coping-climate-grief-anxiety

[11] Like "Astronomers for Planet Earth"

[12] Noelle-Neumann, E. (1993). The spiral of silence: Public opinion–our social skin, 2nd Edition. University of Chicago Press. Chicago, IL.

peer group. As such, beliefs may not be expressed if individuals perceive that others don't agree—even though in reality their views may be widely accepted by members of their group. Maibach et al. (2016) find evidence for a climate change spiral of silence, wherein more than half of those who are interested in global warming or think that the issue is important "rarely" or "never" talk about it with family and friends. Why is that? Many feel that they don't "know the science" well enough to speak about it. (In fact, this is perhaps the most common concern we hear from astronomers.) Many are also afraid it will be controversial, or will start an argument. Sparkman et al. (2022) found that 80–90% of Americans underestimate the prevalence of climate concern and support for mitigation policies. While 66–80% Americans support these policies, Americans estimate the prevalence to only be between 37–43%. Thus, supporters of climate policies outnumber opponents two to one, while Americans falsely perceive nearly the opposite to be true (Ballew et al. 2019). In other words, most Americans are concerned about climate change but don't talk about it because they think that most *other* Americans are not. We need to help people break the silence and establish concern for climate change as a social norm, which is why Katherine Hayhoe has said that, "the most important thing you can do to fight climate change is talk about it." Lasting engagement on the issue is therefore more likely to arise from (1) active thinking about climate change via deep discussion with close members of one's social network, (2) revised mental models that make it easier to understand, think about, and feel motivated to act, and (3) reinforcement of pro-climate social norms so they become internalized over time (Goldberg et al. 2020).

12.3.4 Barriers for the Alarmed and Concerned

The majority of people are concerned about climate change, so what's holding them up? As discussed earlier there is a strong correlation between beliefs about climate change and political identity–in the U.S. about 90% of liberals think that climate change is happening while less than half of conservatives do (YPCC and Mason 4C 2022 and Ballew et al. 2019). While it's good news that overall the majority of people are concerned, it hasn't led to widespread action. Many may be engaging in individual behavior, but few are engaged in collective action; e.g., only 1% of Americans say they are currently participating in a campaign to convince elected officials to pass climate legislation (Leiserowitz et al. 2021).

Much of the research on climate change communication focuses on ways to change attitudes among the more skeptical. However, it is perhaps more important to help motivate the Alarmed and Concerned–the majority of Americans—into action. While Leiserowitz et al. (2021) found that only 4% of the Alarmed and 1% of the Concerned say they are currently participating in a campaign to convince elected officials, they also found that 58% of the Alarmed and 35% of the Concerned say they either definitely or probably would participate in such a campaign (Figure 12.4).

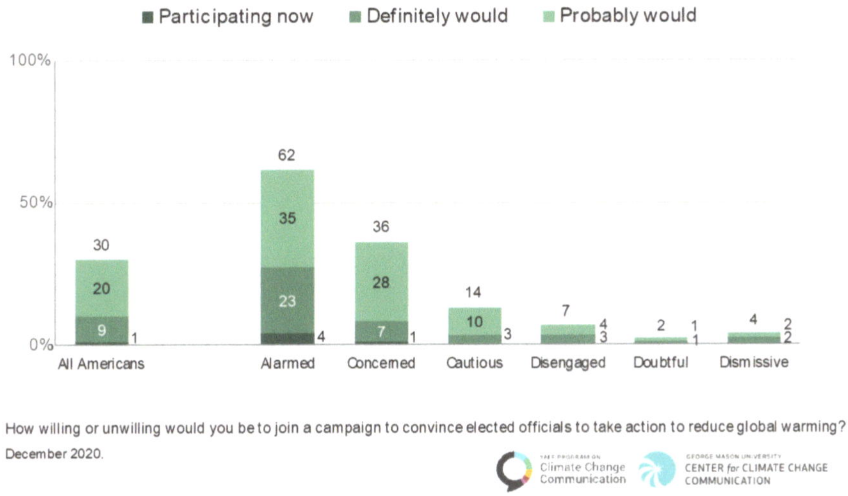

Figure 12.4. Leiserowitz et al. (2021) find that even among the Alarmed and Concerned, very few are actually participating in a campaign to motivate elected officials, but many said they would be willing to do so.

Why is there such a gap between the number of people expressing concern and the number engaged in activism? One problem is that for most people it isn't seen as a top priority. For example, in April 2022 a CBS news poll found climate change to be the lowest "high priority" issue for Americans—behind the economy, inflation, crime, Russia, and immigration.[13] Even among the most liberal leaning, climate change often lags behind other topics, such as health care and social justice. Since half of the Alarmed, and a third of the Concerned identify as liberal (Roser-Renouf et al. 2015), this likely applies to them as well. Climate change, of course, is *also* a health and social justice issue, so it may help to frame it as such for these audiences.

Latkin et al. (2022) find that the most frequent reasons people were not involved in climate change activism were that they felt others are better at it, hadn't been trained, hadn't been asked, didn't know how to get involved, didn't like letter writing, were too busy, were afraid organizations would ask them for money, or weren't encouraged to become involved. Overcoming these barriers was surprisingly simple. The most consistent reason people decided to engage in activism was because they had talked with someone about climate change in the prior month, reinforcing the importance of finding ways to discuss it.

There are ways to promote climate-friendly behavior. For example, "nudges" are a means of encouraging certain choices without restricting options (Thaler & Sunstein 2008). Descriptive norms have been demonstrated as effective nudges for promoting sustainable behavior (Demarque et al. 2015). An example of such a descriptive norm is, "Almost 75% of guests who are asked to participate in our new resource savings program help by using their towels more than once." When expressing a descriptive norm, avoid negative terms. Emphasizing that too few

[13] https://www.cbsnews.com/news/fewer-americans-see-climate-change-as-priority-opinion-poll-2022-04-22/

people or organizations are responding to climate change disincentivizes doing so; i.e., people might think, "Why should I do anything if no one else is?" You can however use dynamic descriptive norms even if the number is small; e.g., "Citizens are increasingly choosing to install solar panels."

In summary, the Alarmed and Concerned are largely ready to engage in climate advocacy as well as change individual behavior. One of the simplest and most effective ways is to simply talk about it—bring it to the front of their attention. Next, give them a simple "on ramp" activity they can engage in. In particular it helps to get them involved in a group working towards a common goal that resonates with their values. As mentioned before, climate change activism also improves mental health (Schwartz et al. 2022), which can help to overcome the anxiety which may be impeding action.

12.3.5 Barriers for Skeptics

As mentioned in the demographics section, in the U.S. the highest percentage of those skeptical of climate change identify as conservative. This skepticism is driven by social norms, perceived clashes with identity and values, and messaging from trusted leaders. Among some conservatives there is concern that environmental issues in general are a conspiracy to promote communism (Hoffarth & Hodson 2016). Lockwood (2018) argues that conservative resistance to climate science is largely driven by right-wing populism, wherein cultural and intellectual elites are perceived to be enemies of the common man. Former U.S. Vice President and climate advocate Al Gore serves as a symbol of such elitism.

Much of this skepticism is driven by fear of proposed solutions that do not resonate with their values. In general, those who identify as conservative are suspicious of "big government," and are therefore reluctant to endorse legislated emission-control measures. Gillis et al. (2021) found that conservatives and moderates are more supportive of climate change mitigation when informed about mitigation actions taken by the private sector. These results suggest that private-sector initiatives may be a way to bolster support for climate action across the political spectrum. When presented with solutions that align with their value system, e.g., free-market solutions, they are more likely to accept that climate change is happening. It should be noted that liberals also show "solution aversion"; i.e., skepticism when proposed solutions don't align with their values (Campbell & Kay 2014).

Conservatives also tend to value "property rights"—defined as the exclusive authority to determine how a resource is used, whether it be owned by an individual, group, or the government. Many conservatives have argued that the risks of climate change policy—and its threat to property rights—are greater than the risks of climate change itself (e.g., Adler 1997). However some are now recognizing the threat that climate change poses to property rights. In a change of viewpoint from his earlier works, Adler (2009) argued that "free-market environmentalism" requires that property rights must be safeguarded against harm by others, including harm caused by environmental pollution. For example, if you cause a flood that destroys someone else's home, it's considered a property rights violation—regardless if the cause was a broken dam or climate change. Framing the issue this way could inspire more support for climate action, especially among groups that prioritize individual property rights.[14]

Barker et al. (2022) found that conservatives were more likely to exhibit an anti-intellectual attitude, wherein they distrusted academic "book smarts" and preferred "common sense." This suggests that it may be more successful to convey information about climate change to conservatives by using "real world," everyday examples. For example, when talking with hunters and fishermen ask them about their recent experiences. Maybe they've noticed it was warmer, or that water levels were lower. Maybe there weren't fish in their favorite spots. These may be the result of climate change; e.g., it may be affecting the timing of the seasons where warmer temperatures are causing seasonal foliage to appear sooner and die off later, leading to different behavior for fish and game. For many the effects of climate change are plainly visible but are nonetheless difficult to acknowledge. For example, in this interview after a severe flooding event in 2019, a farmer in Nebraska said:

> "I'm not a climate change guy as far as, you know—climate change, global warming, whatever—any of that stuff. Has the climate changed? Yes, so those two words go together. But am I believer? I mean, I haven't seen all the facts I guess, just to make an assumption on my own. But have I seen the weather change in my 20-year farming career? Absolutely."

In effect he's acknowledging the reality of climate change but also asserting that his identity isn't aligned with believing in it. The challenge then is to help skeptics understand that solutions to climate change *can* be aligned with their identity and values. To avoid defensiveness it may not be necessary to use the term "climate change" when talking about adaptation strategies. Hine et al. (2016) found that omitting any mention of climate change and emphasizing local impacts increased adaptation intentions in dismissive audiences. But for conversations about mitigation it is important for people to understand that fossil fuels are the primary cause.

As mentioned before, many who are religious are skeptical, especially those who identify as evangelical. The moral foundations for protection of the Earth's climate can be found in all of the major religions, however it manifests itself in different ways and there will be a different focus for different groups within each religion; e.g. for evangelical Christians there is a greater emphasis on scripture, and so care for the climate should be tied to that (Marshall et al. 2016). A common value among evangelicals is children's health; e.g., pollution as well as food and water insecurity. Some religious groups are opposed to action on climate change because they believe it would hurt the poor by decreasing their

[14] https://yaleclimateconnections.org/2019/10/why-climate-change-is-a-property-rights-issue/

access to energy. However, climate change is hurting the poor the most; and renewable energies now represent the best route for poor communities to "leapfrog" the energy poverty gap (e.g., Szabó et al. 2013). Another argument is that climate change (if it is happening at all) is either not dangerous or is part of "God's plan." Many branches of Christianity assert that God doesn't stop bad things, such as murder, from happening. These events don't come from God but originate with man's sin. It is therefore mankind's responsibility to address the problem.

For many, skepticism is driven by concerns about how their employment will be impacted by the transition to greener energy sources. It's important to not blame industries or their workers, and to acknowledge their contributions to our modern society. Fossil fuels, despite their inherent problems, have provided the energy that enables the quality of life we now enjoy. It is also helpful to be honest about the challenges workers will face as they transition to new jobs.

Messaging about climate change, regardless of one's identity, is best received when it comes from trusted sources—for example farmers and fire fighters who are seen as being on the front lines of the problem. Americans tend to trust the military more than other national institutions.[15] It can herefore help to quote the military; e.g., the first line of Climate Action 2030,[16] the climate strategy of the U.S. Navy and Marine Corps, states, "Climate change is one of the most destabilizing forces of our time."

It is important to note that U.S. conservatives have not always been resistant to the problem of climate change, nor is their resistance fundamentally a result of their values or worldview. Before it became highly politicized, many conservative leaders—in the U.S. and worldwide—expressed concern about climate change. "Those who think we are powerless to do anything about the greenhouse effect forget about the 'White House effect,'" George H.W. Bush said in a 1988 campaign speech. "As president, I intend to do something about it." In 1989 British Prime Minister Margaret Thatcher stated that adding greenhouse gases to the atmosphere was "changing the environment of our planet in damaging and dangerous ways." In 2008 Republican U.S. Senator John McCain expressed concern about climate change in this campaign advertisement, where he states, "It's not just a greenhouse gas issue, it's a national security issue. We have an obligation to future generations to take action and fix it." This phraseology uses two framings that resonate with conservatives: national security and children. This rhetoric didn't necessarily lead to action, though. Bush did help to establish the U.N. Framework Convention on Climate Change in 1992, but overall Republican presidents have worked to oppose environmental regulations (Fredrickson et al. 2018).

In recent years there have been few prominent Republicans who have publicly stated a belief in climate change. Many Republican leaders understand that it is a

[15] https://www.pewresearch.org/fact-tank/2018/09/04/trust-in-the-military-exceeds-trust-in-other-institutions-in-western-europe-and-u-s/

[16] https://www.navy.mil/Portals/1/Documents/Department%20of%20the%20Navy%20Climate%20Action%202030.pdf

serious problem, but are fearful to acknowledge it publicly because of the risk of backlash among their supporters. This is not an unfounded concern. For example, former Republican congressman Bob Inglis from South Carolina was unseated by another Republican in his state's 2010 primary election—in large part because Inglis supported action on climate change. Since then he has founded republicEn, an organization that promotes conservative solutions to climate change. Debbie Dooley, a Tea Party conservative and founder of the Conservatives for Energy Freedom and Green Tea Coalition, in her own words is, "advocating for policy that removes regulatory barriers that prevents solar and other clean energy from competing in the marketplace." She argues that conservatives don't resonate with climate change, but do care about energy freedom, energy choice, and national security. She also appeals to people's real-world experience by stating, "If you think fossil fuel is not damaging the environment, pull your car into the garage, start up your engine, and inhale the exhaust fumes for a few minutes and see what happens."[17]

It can also help to show leaders in the business world who take climate change seriously. For example, Seth Klarman—one of the world's most successful hedge-fund managers—has made climate change a priority:[18]

> "The forces contributing to climate change, he noted, 'might be irreversible sooner than the damage from climate change has become fully apparent. You can't say it's far off and wait when, if you had acted sooner, you might have dealt with it better and at less cost. We have to act now.'"

There are conservatives who support clean energy because of job creation, economic benefits, and consumer choice—in large part because of the declining costs of renewable energy. Transition to renewable energy sources also aligns with other conservative values, such national security, small government, and free markets. Many conservatives are uncomfortable with subsidies for fossil fuels, the risks to American soldiers from defending access to foreign energy, as well as funding terrorism through foreign energy purchases (Hess & Brown 2017). The falling costs of renewables has helped shift attitudes as well. Many Republican politicians are beginning to support wind and solar purely for economic reasons. Amanda Ormond, a consultant who served as director of the Arizona Energy Office, said the state's politics have tracked the economics. She said, "I think the single biggest factor is clean energy cost reductions have completely flipped the economics of the utility such that the clean stuff is now the cheap stuff."[19]

Conservatives are also more likely to be nostalgic for the past, whereas liberals are more focused on a desirable future. Thus, conservatives are more likely to support solutions for climate change if they are connected to a fondly remembered

[17] https://www.youtube.com/watch?v=nbmt_WeNBck
[18] https://www.nytimes.com/interactive/2022/10/28/opinion/climate-change-bret-stephens.html
[19] https://www.scientificamerican.com/article/with-gop-support-arizona-mandates-cleaner-energy/

time in history (Lammers & Baldwin 2018). For example, one can compare the timeline of climate change to that of the space race, pointing out that in twelve years (1957–1969) the U.S. went from the failed "kaputnick" launch of Vanguard TV-3 to landing on the Moon. The goal is to connect our country's history and tradition of overcoming adversity to the problems we face now.

12.4 Framing of Messages

"Framing" refers to the way in which information is presented and the effect it has on how people understand and respond. The term comes from a metaphor: It's like holding up a picture frame on an issue. You can never see the whole issue in one picture, so we make choices–what's in the picture, what's out of it, what's in focus, what's not? Although there is no one precise definition of framing in media and communications research, the essential idea is that *how* you say something is as important as *what* you say–sometimes even more important. Words have meaning—but that meaning is impacted by factors such as the context in which the message is delivered, who is delivering the message, the audience receiving the message, and the order in which information is delivered (e.g., whether a news story quotes a scientist or a lobbyist first). In this way, message framing helps to determine what is relevant, what deserves attention, what can be discarded or ignored, and what ideas are deemed more reasonable or acceptable than others.

Framing is used by people who seek to influence the public—politicians, public relations practitioners, lobbyists and the like—to "frame" the messages in a way that makes their ideas more acceptable. Scientists also use frames—intentionally or not—although they are not always aware of the values associated with their decisions and how the public interprets them.

The elements of framing refer to the way different components are presented. They can vary depending on the issue and means of communication, but tend to include the following:

Frame-setting: this refers to the initial context in which the issue is presented, which can influence how it is perceived. For example, if a news story about a medical study begins with a headline that emphasizes the potential risks, it may frame the issue in a negative light from the start.

Emphasis: this refers to the aspects of the issue that are highlighted in the framing. For example, a news story about the risks of climate change to coastal communities might emphasize the personal impacts (e.g., tell stories of flooding in people's homes); or it could portray the issue through an economic lens (e.g., talk about the financial losses endured by beachfront businesses.)

Exclusion: this refers to information that is left out of the frame–accidentally or intentionally—that can also influence how people perceive an issue. For example, a lobbyist against climate solutions might talk only about the economic costs of climate change mitigation efforts and omit the potential costs of *not* reducing greenhouse gas emissions.

Tone: this refers to the emotional or affective aspects of the frame, which can shape how people feel about the issue. For example, a news story that uses emotive language to describe the impacts of climate change might elicit feelings of fear or anxiety in the audience; e.g., it can provoke a strong reaction when the media talks about "climate refugees"—with imagery of people fleeing their homes because of drought or hurricane devastation. Another example is an opinion piece by David Wallace-Wells in 2017 titled "The Uninhabitable Earth,"[20] which was criticized by many for portraying solutions to climate change as hopeless (e.g., Riederer 2019).

Metaphor: metaphors can be a powerful tool for conveying complex ideas in ways that are relatable, by connecting an established understanding of a known concept to an unknown one. For example, describing greenhouse gases as blankets that cover the Earth is a metaphor that simply but effectively conveys how they warm the planet. People may not know about radiative transfer but they of course understand what a blanket does. Metaphors are also more emotionally engaging and can resonate with people if they reflect their values and experiences (Citron & Goldberg 2014).

Storytelling: storytelling can make complex concepts easier to comprehend; and audiences find them more engaging and more memorable than traditional logical-scientific communication (Dahlstrom 2014). Stories are inherently personal and relatable and can help to make a connection with your audience. One of my (T.A.R.) favorite stories is of a visit to a third-grade classroom to talk about the solar system. After my presentation a boy asked me, "How old are the planets?" I couldn't think of a good way to explain it, so I simply said that it was around four and half billion years old. To which he replied, "Wow! That's older than my parents!"

Language: this refers to the wording that is employed to frame an issue in a particular way. For example, some news organizations have transitioned from using "climate change" to "climate crisis" or "climate emergency" to better communicate a sense of urgency.[21] Others (e.g., Depoux & Gemenne 2020) argue against using these phrases because they imply that climate change is temporary, short-term, and reversible.

12.4.1 Framing of Climate Change

So how can we frame climate change effectively? For the general public, Anthony Leiserowitz breaks down messaging of five key ideas about climate change into ten simple words: "It's real. It's us. Scientists agree. It's bad. There's hope."[22] An additional important idea is "You can help." Our task is to convey these six ideas in a way that resonates with our audience.

The message that scientists agree is particularly important because the consensus on climate change has been described as a "gateway belief." van der Linden et al. (2019)

[20] https://nymag.com/intelligencer/2017/07/climate-change-earth-too-hot-for-humans.html

[21] Stewart, T. (5 November 2021). Scientific American. Climate change is creating new vocabulary from, eco-anxiety to kaitiakitanga. https://www.scientificamerican.com/article/climate-change-is-creating-new-vocabulary-from-eco-anxiety-to-kaitiakitanga/.

[22] https://portal.ct.gov/-/media/DEEP/climatechange/GC3/GC3-2020-agendas-and-minutes/GC3_Forests_slides_AnthonyLeiserowitz_052120.pdf

found that once people understand that there is an overwhelming scientific consensus they are more likely to believe climate change is happening, be worried about it, and support action to address it. The "consensus gap" is the difference between the perceived and actual scientific consensus on climate change. Whereas 99% of climate scientists are convinced that climate change is real and human caused (Lynas et al. 2021), only 57% of Americans believe that such a consensus exists and 23% think there is still considerable disagreement (Yale Climate Opinion Maps 2021; see Howe et al. 2015 for data and methodology). It therefore helps to emphasize the consensus not only among scientists but also among other groups (e.g., professional societies like the American Medical Association[23]). Finally, we want people to understand the severity of the problem, but also to have an understanding of the solutions–and optimism that these solutions can be effective.

Many different framings are used when talking about climate change, for example:

- Environmental: "Climate change is a threat to plants and animals."
- Moral: "It's our obligation to future generations to stop climate change from getting worse."
- Economic: "Climate change will cause trillions of dollars in property damage and lost economic opportunity."
- Identity: "Climate change is a threat to our way of life."
- Personal security: "Climate change is making us more vulnerable to dangerous weather events such as firestorms and hurricanes."
- National security: "Climate change is an urgent and growing threat to our national security, contributing to increased natural disasters, refugee flows, and conflicts over basic resources like food and water."[24]
- Social norms: "75% of residents in your community have reduced their energy costs by installing solar panels."
- Austerity: "We must reduce our energy usage to minimize the impacts."
- Opportunity: "Solutions to climate change like wind and solar are creating high-paying jobs."
- Cost savings: "Switching to renewable energy will save you money."

Nisbet (2009) provides additional examples.

Naturally some framings are more effective than others, and a framing may resonate with one audience and backfire with another. For example, Sapiains et al. (2016) found that framing mitigation as an identity issue; i.e., in terms of preserving one's lifestyle and culture, was more effective than environmental and economic framings. Bain et al. (2012) argue that framing mitigation efforts as a means to promote a better society is more effective than focusing on the reality of climate change and averting its risks.

[23] https://www.ama-assn.org/press-center/press-releases/ama-adopts-new-policy-declaring-climate-change-public-health-crisis

[24] https://obamawhitehouse.archives.gov/sites/default/files/docs/National_Security_Implications_of_Changing_Climate_Final_051915.pdf

Gustafson et al. (2022) tested three renewable energy framings (mitigation of climate change, cost savings, economy and job growth). The cost-savings frame was found to be the most effective of the three for Democrats *and* Republicans. For Democrats, the climate-change and economic benefits framings worked equally well. For Republicans, the economic framing worked almost as well as cost savings; and the climate-change framing did not have a lasting impact. This reflects actual behavior; e.g., nearly all Democratic (91%) and Republican (92%) homeowners who have thought about—or have already installed—solar panels say that saving on utilities was a reason for doing so. However Democratic homeowners are more likely than Republicans to say that helping the environment (95% vs. 59%) was a motivator.[25] It turns out that, regardless of our political stripe, everyone is pro-money.

For most audiences, austerity messaging isn't effective and can backfire. This may be due to "loss aversion"–the psychological effect where losses are overvalued when compared to gains. Palm et al. (2020) found that recommendations for reduction-focused behavioral changes (e.g., reducing beef consumption or air travel) reduced an individual's willingness to take personal actions as well as their willingness to support pro-climate candidates. It also reduced their belief in the science of climate change, and decreased their trust of climate scientists. It is worth noting that opponents of climate solutions are aware that austerity messaging is unpopular and therefore often use it as a framing; e.g., by claiming (often falsely) that a climate policy will "take away" something of value.

As mentioned earlier, it also matters from where the messaging originates—it must be delivered by someone who is trusted by that individual or group; i.e., someone who can "bless the facts" (Callison 2014). Fielding et al. (2020) found that Democrats and Republicans both responded to climate change messaging more positively when it was endorsed by members of their in-group; e.g., Americans who identify as politically conservative were more likely to respond positively to an evangelical Christian, military leader, or Republican politician (Goldberg et al. 2021). Interestingly, children are especially influential in fostering climate change concern among their parents (Lawson et al. 2019).

The outcomes of framing can vary in complex ways. For example, Myers et al. (2012) found that framing climate change as a threat to public health was effective for all audiences, whereas framing it as a national security issue worked well for those who were concerned about climate change but that it elicited anger among those who were doubtful or dismissive. The source of the anger is unclear—it is possible they perceived it as an attempt to make a link between an issue they may care about (national security) and an issue that they tend to dismiss (climate change), thereby producing a negative reaction. Alternatively, it may be that they were experiencing anger toward the experimenters for having presented claims about global warming and national security that they didn't

[25] https://www.pewresearch.org/science/2022/03/01/americans-largely-favor-u-s-taking-steps-to-become-carbon-neutral-by-2050/

perceive as authentic or credible, suggesting that such information needs to come from a trusted source.

12.5 Astronomers as Communicators

The good news is that scientists are generally highly regarded by the public, consistently ranking among the top professions—as high as medical scientists and the military.[26] Of course not everyone trusts us. Trust in science in the U.S. has always been polarized to some degree by political affiliation, but in recent decades that trust has flipped. Over the last fifty years the General Social Survey[27] has tracked Americans' attitudes towards science. Until the start of the 21st century, those who identified as Republican had a greater trust in science—believing it was an economic driver and important for defense—whereas Democrats were more likely to blame science for social ills such as pollution. Since the 1990s, as concern for climate change has grown, belief in science among Democrats has increased while for Republicans it has declined. In large part because of the COVID-19 pandemic, this divide jumped in 2021 to a 30-point gap—wherein 64% of Democrats said they had "a great deal" of confidence in the scientific community compared to only 34% of Republicans. This is largely a result of Republican leadership expressing doubt about scientific research on COVID-19, much as they have about climate change.

Naturally, we perceive science as being for the benefit of humanity. Science of course has created knowledge that has been used for both good and bad. But it's important to recognize that not everyone has benefited from science equally. In particular, minoritized groups have often been mistreated. For example, the Tuskegee Study of untreated syphilis in black men knowingly caused harm to its participants (Baker et al. 2005). So be aware that the scientist's perspective is not always trusted, particularly if there is a perceived conflict of interest (e.g., Besley et al. 2017). Nor is our perspective the only one that's relevant.

12.5.1 Why Should They Listen to You?

Astronomy is fortunate in that we don't have a "dark side" to our profession in the same way that other fields of natural science do; e.g., physics and atomic weapons, geology and resource extraction, biology and experimentation on animals, as well as chemistry and the environmental damage from some of its products. As such, astronomers are often more highly trusted but also less accustomed to talking about controversial topics. That's not to say astronomy is completely without controversy, and sometimes astronomers struggle to understand the viewpoints and concerns of other stakeholders. The controversy over the construction of telescopes on Maunakea is a good example (Kahanamoku et al. 2020). There are of course conspiracy theories that are astronomically related; e.g., flat Earth and the Moon

[26] https://www.pewresearch.org/science/2022/10/25/americans-value-u-s-role-as-scientific-leader-but-38-say-country-is-losing-ground-globally/

[27] https://apnorc.org/projects/amidst-the-pandemic-confidence-in-the-scientific-community-becomes-increasingly-polarized/

landings. In a sense they don't matter—the Earth is round no matter what you think. However it *does* matter what people believe about climate change, as it influences personal and collective decision making.

So we need to think about how to connect with our audience, acknowledging that how we see ourselves and the world is not the same perspective as it is for others. When talking with a group, start by establishing your bona fides—in other words, why should they listen to you? This isn't just a summary of your CV, which only academics care about. You need to think: What is your relationship to your audience? Where do your interests overlap? Where do your values overlap? Will this group see you as an outsider, or as "one of them"? The most effective conversations begin with what you have in common.

Which raises an interesting question—what *are* the values of astronomers? We of course are a diverse group drawn from many cultures and identities. But there are perhaps some characteristics that apply to most of us. As academics and scientists we tend to value knowledge above all else—justifying (and even celebrating) antisocial behavior from individuals as long as they are scientifically productive. But many groups place a greater value on consensus—that is, what the group accepts as truth is more important than the actual truth. Astronomers highly value our advanced degrees, and tend to exclude those who don't have them. But among those with the appropriate credentials the power structure is relatively egalitarian when it comes to the exchange of ideas. However in other groups the flow of information is often one way, from the top down. Understanding these differences can inform how you decide to interact with different groups. In particular, expounding upon a body of knowledge—something we value—may not be effective in and of itself. How is what you know relevant to them? Take the time to understand their values and adjust your presentation or conversation accordingly.

It's also important to establish why they should be listening to an astronomer about climate change. People are often surprised and *curious* to learn about the close connection between our work and climate science, so use that curiosity as a way to engage them. For example, a compelling opening line might be, "I'm studying how climate change is affecting astronomy." This will probably pique their curiosity—how *is* astronomy being affected? Why would there be a connection? It helps to explain the overlap between astronomy and climate science, but keep in mind we're not trying to be climate scientists by proxy—we have our own unique knowledge and perspective to share. As discussed in Chapter 8, there are many connections between the science of astronomy and climate change. Be sure to tell people why they should be listening to *you*. What is your story? How does your knowledge of astronomy guide your thoughts and feelings about climate change?

Also keep in mind that your conversation may be one of the few times people have had an opportunity to talk with a scientist about climate change. Roser-Renouf et al. (2015) surveyed the Six Americas to see what questions they would like to ask an expert. The Alarmed and Concerned were most interested in hearing about what we—collectively and as individuals—can do to address the problem; i.e., they want to hear about solutions. The Cautious and Dismissive would most like to know what are the effects of climate change and why should they care; i.e., they want to know about the consequences, particularly to them. Finally, the Doubtful and Dismissive would like to ask how scientists know that climate change is real; i.e., they want to hear about the science. So it's important that your presentation address all three: causes, consequences, and solutions, so that every group in the Six Americas is hearing about what's relevant to them. Which of these three you choose to emphasize may depend on the audience.

12.5.2 How to Talk about It

What should you talk about? First off, don't treat climate change as a stand-alone subject. Talk about things people care about—and explain how climate change is affecting them; e.g. hunting or fishing, job security (or opportunity), or the welfare of children. Choose a framing that will resonate with your audience. As mentioned earlier, personalizing and localizing the threat of climate change, and enhancing the norm that most people support action, can be effective in overcoming doubt and inaction (Ballew et al. 2022).

Next, let's talk about *how we talk*. It's important to keep in mind that many words that scientists use have different meanings for the public. For example, scientists use the word "uncertainty" when discussing the range of plausible values or outcomes. However for many people the word suggests that scientists are unsure of their results as a whole (Kopp et al. 2023). Likewise, the phrase "positive feedback" can be interpreted as a good thing when it seldom is. It would be better to use a more familiar phrase like "vicious cycle." Also consider the order in which you present information. Somerville & Hassol (2011) list many terms that are often misunderstood—along with suggestions for alternatives. Scientists can also communicate more effectively with the public by inverting the order of their usual presentations to colleagues. Rather than start with the background, supporting details, and conclusions do just the opposite. That is, start with the "bottom line" and tell people why they should care (Somerville & Hassol 2011).

Keep in mind that language can leave people feeling left out. All communities—professional and social—use words that indicate membership and excludes others, just like slang among teenagers. Science is no different. Using a word or phrase that the listener doesn't understand can be belittling or alienating, even if not intended. So be sure to talk in a way that will connect with and be clear for your audience.

Also consider that your audience already has an awareness and understanding of the problem, but it is often flawed—particularly as a result of exposure to misinformation. Chapter 8 gives a list of common misconceptions. To overcome these misconceptions, it is effective to draw upon your audience's personal experiences. Alexander et al. (1991) found that most knowledge structure theories can be encapsulated within the broader realm of "schema theory." A schema is a piece of knowledge based on past learning or experience. As a whole they constitute the knowledge an individual uses to interpret and understand new information and experiences. Connecting information about climate change to their existing schema can be quite powerful. For example, most people implicitly understand that cold air holds less water vapor, hence the reason we need to use lotion in the wintertime for dry, cracked skin. By using this real-world example, you can reinforce their "warm = wet" schema. This can then help them understand that warmer air—which can hold more water vapor—can create wetter hurricanes that cause greater flooding. This schema can also explain increased snowfall in some winter storms. This is at first counter-intuitive, as the "snow=cold" schema would suggest that more snow means it is colder. But in fact extremely cold places, such as Antarctica and the interior of Alaska, experience little snowfall in the wintertime. Warmer air can therefore result in greater snowfall–within a certain temperature range. The "snow=cold" schema can therefore be countered by drawing upon the "warm=wet" schema.

Metaphors can also be more effective if they connect to existing schema. We've already talked about thinking of greenhouse gases as blankets on the Earth. Another example is using a bathtub analogy for the atmospheric levels of CO_2. People understand that in a bathtub the water level will go up if the faucet adds water at a rate faster than the drain can remove it. Likewise, atmospheric CO_2 levels go up if their sources (e.g., from fossil fuels) exceed that of sinks (e.g., plant life and oceans). However metaphors can fail; e.g., when you turn off the water a bathtub can drain quickly, whereas CO_2 added to the atmosphere can remain for thousands of years (Armstrong et al. 2018). Other metaphors include describing our situation as a "house on fire," or to "declare war" on climate change. Both convey ideas of threat and urgency but may fail to maintain motivation in the long term. After all, when is the fire put out? And when is the war won?

The National Network for Ocean and Climate Change Interpretation (NNOCCI) provides additional examples of framings and metaphors. As discussed in Chapter 8, one key framing they developed is to differentiate between "regular" CO_2 (i.e., the levels of CO_2 in the atmosphere from natural processes) and "rampant" CO_2 (i.e., that emitted from the use of fossil fuels). Raimi et al. (2017) find that medical analogies for climate change are also helpful. For example, using cancer as a metaphor: It's never a good idea to ignore cancer and always better to act sooner than later. Here is an example of how you might explain why a temperature anomaly of just a few degrees is a problem using the "temperature=fever" schema:

"Why is it dangerous if the global temperature changes by a few degrees? Here's an analogy: Think about the temperature of your body. For a typical adult, it can be anywhere from 97°F to 99°F when they're healthy. Think about how you feel if you have a fever of just a few degrees. For the elderly or immunocompromised a fever of just 101°F is cause for concern. For healthy adults a fever of 103°F or higher can be dangerous. And a fever over 106°F can be deadly. It's similar for the Earth. Ecosystems rely on the average temperature to be within a certain range. And more fragile ecosystems are disrupted first when the temperature increases, even by just a little."

Corner et al. (2018) offer guidelines to IPCC authors on how to communicate, which are largely relevant to scientists in general. For example, they recommend avoiding talking about the "big numbers" of climate change (i.e., global average temperature targets and concentrations of atmospheric CO_2) because they don't relate to people's day-to-day experiences. They also recommend using a narrative structure—with anecdotes and stories. Here is an example of a story that I (T.A.R.) often tell when talking about greenhouse gases:

"My grandmother grew up in South Dakota during the depression. When I was a kid she was always terrified I would freeze to death. Whenever I stayed at her house I would go to bed with a single blanket on me, nice and tucked in. But I would later wake up drenched in sweat because she'd sneak in and put another blanket or two on top of me. That's basically what we're doing to the Earth. It has a nice blanket of greenhouse gases that keep us warm. But we're adding more of these gases to the atmosphere. And, like my well-meaning grandmother, we're making things uncomfortably hot."

This is a true story,[28] and it's a simple example of using storytelling to convey information. It has elements that are relatable—we all know that feeling of waking up too hot in bed, and most people can relate to having an overbearing, if well intentioned, parent or grandparent. Many people in the audience laugh when they hear my story because they *are* that grandparent. This is a story about a metaphor, but you can also share stories of your personal experiences with climate change itself; e.g., experiencing the smoke from a massive fire. Gustafson et al. (2020) find that sharing personal stories of how climate change is already harming people is effective at engaging diverse and even skeptical audiences.

12.5.3 What to Show

Visuals can often convey ideas better than words. If giving an electronic presentation with visuals what should you show? In general statistics and graphs should be avoided, as they are impersonal and often hard to interpret. Sometimes the use of graphs is unavoidable, in which case Painter et al. (2023) offer insights on how to make them more clear and visually appealing. Judicious use of images can have a big impact. Corner et al. (2015) provide insights on what visuals are most effective for different audiences. Their recommendations include:

- Show images of real people in authentic situations instead of staged photographs.
- Avoid pictures of politicians.
- Avoid "classic" images of climate change, such as smokestacks and polar bears, as these prompt cynicism and fatigue in many people. Less familiar (and more thought-provoking) images can help tell new stories that can resonate with the jaded.
- Avoid focusing on individual behavior; e.g., show a congested highway rather than a single driver.
- Images of individuals, with identifiable emotions, suffering from climate impacts are emotionally powerful—especially if they're local to the audience.

[28] And it's mine—so use your own!

Figure 12.5. A comparison of the Aral Sea in 1989 (left) and 2014 (right). Once the fourth largest sea on Earth, it has shrunk to less than 10% of its size in the 1960s due to drainage for irrigation. Photos: NASA.

But such images can also be overwhelming. Coupling images of climate impacts with concrete behavioral actions can help to overcome this.

- Images of solutions (e.g., wind turbines under construction), while less emotionally powerful than impacts, have broad appeal. They can promote feelings of self-efficacy (i.e., belief that we can take action that makes a difference).
- Images of protests (or protestors) are less effective, as they reinforce an "us versus them" mindset.

Visuals can be used to effectively convey concepts. For example, there is often skepticism that human activity can have such a significant impact on nature or the Earth. The photos of the Aral sea below are a visceral example of how humans—outside of the context of climate change—can dramatically change the Earth for the worse (Figure 12.5).

12.5.4 Putting It All Together

With these communication tips in mind, let's unpack an example of ineffective communication. Imagine that scientists make the statement, "Models indicate that, if we don't reduce carbon emissions, by the year 2100 the global average temperature could increase by as much as 8°C.” While scientifically accurate, it is unlikely to motivate people because:

1. Most people don’t understand what a “model” is. They might think you mean a fashion model (and why would they know anything about climate change?)
2. The year 2100 is too far off in the future to care about. It also implied that there will not be any significant effects before then.
3. Why should we care about an 8°C temperature increase? What are the implications, especially locally?
4. Also, the temperature can vary that much from day to day in any location, so it may not seem significant.
5. There is no personal connection.
6. For Americans, temperatures are usually measured in Fahrenheit.

Instead, the messaging should use wording that is familiar. It should also convey how climate change is impacting, or will impact, people’s lives and livelihoods. Including imagery, but preferably not graphs, that illustrate these impacts will also be helpful. For example, if talking to a group of Alaskans you could say that, “Climate change is already deeply impacting Alaska’s storied fisheries. Record high temperatures in 2019 caused over one hundred salmon die-offs throughout the state, including salmon runs in the Yukon River.” (von Biela et al. 2022). This example makes it clear that climate change is happening now, and having severe impacts on something that most Alaskans care deeply about—salmon. It could be combined with an emotionally impactful photo or include a personal story, such as this evocative quote by Peter Westley, “A river that is usually teeming with life felt like a tomb." Of course this messaging is specific to Alaska and would be less effective elsewhere.

Regardless of the audience, it is critical to talk about solutions. But when discussing them, you need to acknowledge that every solution has strengths and weaknesses, and to be as honest as possible about the pros and cons for each of them. For example, wind and solar require more mineral resources for construction than fossil fuels. Lithium, nickel, cobalt, manganese and graphite are essential for current battery technologies. Rare earth elements are essential for permanent magnets that are vital for wind turbines and EV motors. And electricity networks need large amounts of copper and aluminum. The shift to clean energy will drive a huge increase in the requirements for these minerals (IEA 2021), and their extraction have environmental and social impacts that need to be addressed. However, it is also worth noting that these technologies are rapidly evolving. For example, new battery technologies that require less rare minerals are being developed.

Finally, we need to think about how we want people to feel after hearing our talk. The affective responses people experience toward climate change are consistently found to be among the strongest predictors of how seriously they believe the problem to be, and how supportive they are of solutions (Brosch 2021). We want people to feel more concerned, but also more optimistic and more engaged. To do this, we need to avoid language that causes despair or implies that the situation is hopeless. We also need to avoid phraseology, like "before it's too late," that imply there are only two possible outcomes. Instead emphasize that right now is the best time to be working on the problem because effective solutions like wind and solar are now ready to go, and we're still in a good position to prevent the worst potential consequences. Finish your conversation on a high note that leaves people excited for what their future can be. This can be done by focusing on success stories, as discussed below.

Note that there is often little point in debating whether or not climate change is happening or if it is caused by human activities. If you believe the question is sincere then of course you should do your best to answer it—especially if you're in an audience that would benefit from your response. But those who wish to engage in that debate usually have no intention of changing their minds, and would prefer to bog down the discussion so as to avoid talking about solutions. Dismissives tend to already have strongly held beliefs and attitudes, meaning they are likely to critically scrutinize the science of climate change while uncritically accept information that casts doubt. If needed, you can point to the consensus among scientists or that even oil companies acknowledge that their products are contributing to climate change.

In summary, you should do the following when preparing for a public presentation in which you anticipate talking about climate change:

Learn about your audience: if giving a presentation to a specific group, take the time to learn about their values and interests. Many organizations have a mission statement on their website that can be useful. You can also ask those who invited you what topics they think their group will want to hear about.

Tailor your presentation: It's good to have a few established talks for different audiences (e.g., K-12), but you should take the time to customize your presentation to match what you know about your audience. What framings and visuals will work best with them? Again, the key idea is to talk about what interests them, not what interests you. Also think about what is realistic to accomplish with the time you have with them. As mentioned above, each talk should include discussion of causes, consequences, and solutions. But the balance should be tailored to your audience and their interests. Again, always finish on a high note.

If appropriate, ask for permission: if you've been asked to speak with a group about a topic other than climate change, they may be surprised—and unpleasantly so—if you bring it up unannounced. So it's worth telling the organizers that you wish to talk about climate change as part of your presentation, and see if they think it will be well received.

Learn the latest misconceptions: finally, you can expect questions related to common misconceptions and recent skeptic/denier talking points. As discussed in Chapter 13, the Skeptical Science website offers useful tips on how to address both. This is a worthwhile resource to look at before every talk you give.

12.6 Interpersonal Communication

As scientists and astronomers we are accustomed to being the center of attention, and often the voice of authority in the room. We are in the habit of giving lectures and talks that are mostly "one way." Whether it be answering a question from the audience, or having a one-on-one conversation, it is also important to be able to have a successful dialog. There are many styles of conversation, but an effective strategy—known as "radical civility"—can be broken down into these elements:

Listen and reflect on what they said: "Deep listening" is the practice of taking an interest in another person and having them feel understood. Ask questions that help you to get to know them. Use follow-up questions that explore their ideas and express a desire to better understand what they think and feel. Try to learn about a person's core values and beliefs, and how they might feel threatened by climate change.

The goal isn't to change their mind, it's to understand them and to develop empathy. If you've ever thought in frustration, "how can anyone think that way?", this could be a chance to learn. And it can create an opportunity for them to learn from you, as people will trust you more if they feel like they've been heard. Try to focus on what they're saying, not on your response. When appropriate, summarize what they've said to show you're listening.

Don't try to win: this isn't a debate, nor a court of law. Ask meaningful, open-ended questions that allow people to explore their ideas and for you to understand them. Don't ask rhetorical questions (i.e., questions that don't invite an answer) or leading questions that are designed to trap them. This sort of interaction builds resentment and causes people to shut down.

Agree where possible, but also respectfully explain where you disagree: for example, imagine a skeptic says to you, "Humans aren't changing the climate. The Earth's climate has always changed. There were times when the Earth was hotter than it is now." The first sentence is wrong, but the next two are true. You could acknowledge this by saying, "It's true that the Earth's climate changes naturally, but that doesn't mean we aren't changing it now. It's also true that millions of years ago the Earth was hotter, but humans weren't here then. And neither were the plants and animals that we rely on." This is a respectful way of responding that may help them to see the difference.

Try to help people, not change them: when talking with someone who is skeptical or in denial about climate change, the goal shouldn't be to change their mind but rather create a space where you can both discuss and explore the issue. Rather than being focused on conveying the knowledge you have, think about what you can learn from them. What are their concerns, ideas, and possible misconceptions? The focus should be on having a conversation that leaves them feeling that they were respected and heard. That doesn't mean you have to agree with them on everything, and it can be helpful to point out where you don't (respectfully!)

Tell your personal story: what have you experienced? Why do you care? What events caused you to become active in addressing climate change? It is also helpful to acknowledge your own struggles and apparent hypocrisy. For example, I'm a vegetarian so in a sense it would be easy for me to suggest to others that they reduce their meat consumption. But I (T.A.R.) *love* cheese, and while I now eat less of it it would be very hard for me to give it up completely.

Recognize that their personal story is different: there are numerous factors that influence a person's view on climate change, including perceptions of social norms, portrayal of the problem by the media, cues from trusted leaders, as well as personal and vicarious experiences with extreme climate and weather events. And, as discussed in Chapter 13, they likely have been exposed to misinformation and disinformation.

Try to understand where they are coming from and see the value in their perspective. They may not have the knowledge or experiences that you've had—nor have you had theirs. Even for people who "should know better" there likely are motivations that go beyond what is visible. For example, it's easy to underestimate how hard it is to go against the social norms of the groups to which they may belong. Could accepting climate change put them at risk of rejection from their social circles? Remember that every person is fighting their own personal battles, often unseen. An act of kindness can go a long way.

Keep calm: once a person's identity or self value is at risk, the argument can get emotional. People will often argue back if only out of pride. If the conversation gets difficult, try not to take it personally. Try to remain upbeat, don't raise your voice, and don't be defensive or attack. If you feel yourself or the other person getting frustrated, find ways to diffuse the situation.

Find common ground: turn them from an adversary into a partner. What are things you share in common? It could be as simple as your favorite sports team or restaurant. Identify your shared values and explore ways to build upon them collaboratively. For example, you could ask them, "What's your opinion on how we could find solutions that we both want?"

Treat others as you would like to be treated: at the risk of saying the obvious, treat everyone with kindness and respect. Everyone wants to feel appreciated. But if they feel disrespected or humiliated they will further deepen their resistance, and be less willing to engage in the future. As Dan Shapiro said, "There is nothing more in the world that we like than to feel appreciated. Recognize your power to appreciate them."[29]

Build a foundation of trust: at the end of the conversation how does the person feel? Do they feel like they were heard? Respected? It's ok if the conversation ends without being in agreement. If you have a long-term relationship with them, the goal should be to strengthen your connection so that you can have more conversations in the future. If you are unlikely to see this person again, the hope is that they'll have had a positive experience with you—a scientist—and be more open to trusting scientists in the future. Keeping conversations about climate change going are important because we are regularly distracted by other crises in the short term. It can also be easy to forget about; e.g., it may be hard to reconcile with the severity of the problem on a beautiful day.

The hope is that each conversation will increase an individual's awareness and concern about climate change, even if just by a small amount. For those who are

[29] https://bigthink.com/the-learning-curve/harvard-negotiator-how-to-argue/

skeptical, it might soften their resistance and make them more willing to explore the issue. For those who are already concerned, it may increase their level of engagement. And for those who aren't already involved in advocacy it may inspire them to do so.

12.7 Success Stories

As mentioned earlier, one of the primary concerns people have is whether or not we can solve climate change, so it can help to share examples of success stories. Talk about the wins, show how we're making progress. As discussed in detail in Chapter 7, perhaps the biggest story in regards to solutions is the plummeting prices of renewable energy sources such as wind and solar—to the point where they are now in most cases less expensive than fossil fuels. It can also help to talk about how these successes have changed the equation; e.g., I (T.A.R.) often say, "Five years ago I would have told you that we need to switch to renewable energy even though they were more expensive because of the damage caused by fossil fuels. But now that wind and solar are less expensive there's no major reason not to switch that can't be overcome."

Share success stories of advocacy, such as the decision in 2021 by Harvard University to stop investing its endowment in fossil fuels[30]—a decision that came after years of pressure from students, faculty, alumni, and other stakeholders. Also share local success stories. For example, in Alaska we talk about how Kodiak Island has transitioned to 100% renewable energy, which has dropped the price of electricity and has grown the local economy.[31]

Also share other stories where science and policy have worked together to solve major problems. The hole in the ozone layer is a good example. Stratospheric ozone was depleted about 5% between the 1970s[32] and the mid-1990s,[33] the largest decrease occurring over Antarctica. Most of the ozone loss was associated with chlorofluorocarbons (CFCs), which were used in refrigeration, air-conditioning, insulating foams, and aerosol propellants. Starting in 1989, the Montreal Protocol and its revisions have phased out the use of CFCs internationally. As a result of this agreement, the decline of ozone has abated, and the hole over Antarctica is slowly recovering (Myhre et al. 2013 and references therein).[34] Remarkably, only fourteen years elapsed between the discovery of the CFC-induced destruction of ozone (Molina & Rowland 1974) and the signing of the protocol. Despite resistance from industry (e.g., DuPont; Mullin 2002) who tried to downplay the problem, the rapid response was in large part due to the clear messaging about the problem—the "hole"

---

[30] https://www.npr.org/2021/09/10/1035901596/harvard-university-end-investment-fossil-fuel-industry-climate-change-activism

[31] https://www.ktoo.org/2017/09/15/can-kodiak-teach-world-renewable-energy-lot/

[32] The greatest decade for music. Not the greatest decade for everything else.

[33] Truly an awful decade for music.

[34] Unfortunately the banning of CFCs led to a large increase in the use of hydrofluorocarbons (HFCs). While HFCs do not impact the ozone layer, they have just as strong of a greenhouse effect as CFCs—thousands of times stronger than CO_2. In 2016 the Kigali amendment was signed to phase out the use of HFCs over the next 30 years.

in the ozone layer was a simple concept to understand. The protocol is expected to prevent approximately 443 million cases of skin cancer, 2.3 million skin cancer deaths, and 63 million cataract cases for people in the U.S. born in the years 1890–2100 (EPA 2020).

There are many other examples as well. The problem of acid rain reached its peak in the 1980s. Fortunately, enacted laws have greatly reduced emissions of the gases that cause this phenomenon, and this is an example of successful science-based policy making (Grennfelt 2020). Asbestos and leaded gasoline are also good examples of how the process of science first led to their development but also to an understanding of their health risks. Most notably, the book *Silent Spring* (Carson 2002) about the impacts of DDT led to its ban and is credited with the start of the environmental movement. As is discussed in Chapter 13, the science on the health risks of all of these were met with stiff resistance from industry in much the same way that the fossil fuel industry has fought against the science of climate change. Despite this resistance, science-informed policies were eventually implemented to address the issues.

When thinking about climate change it's easy to focus on what is lost; e.g., believing that our children won't have the same positive experiences or opportunities we had when we were young. But we also need to look forward to what can be gained. History is full of examples. From air pollution to soda can "pull tabs" littering beaches, there is much about our past that we should be grateful we've left behind. Just as many youth are shocked to learn that people once smoked on airplanes, future generations will look back in dismay to think that we drove cars that would kill us if they were turned on inside the family garage. Success stories give us a glimpse into what we can look forward to.

References

Adler, J. H. 1997, The Costs of Kyoto: Climate Change Policy and Its Implications, ed. J. H. Adler (Washington, DC: Competitive Enterprise Institute)

Adler, J. H. 2009, Taking Property Rights Seriously: The Case of Climate Change (Cleveland, OH: Faculty Publications) 30, https://scholarlycommons.law.case.edu/faculty_publications/30

Albrecht, G., et al. 2007, Australas Psychiatry, 15, S95

Alexander, P. A., Schallert, D. L., & Hare, V. C. 1991, Rev. Educ. Res., 61, 315

Armstrong, A. K., Krasny, M. E., & Schuldt, J. P. 2018, Communicating Climate Change, A Guide for Educators (New York: Cornell Univ. Press) 70

Bain, P. G., Hornsey, M. J., Bongiorno, R., & Jeffries, C. 2012, NatCC, 2, 600

Baker, S. M., Brawley, O. W., & Marks, L. S. 2005, Urology, 65, 1259

Ballew, M. T., et al. 2019, Environ. Sci. Policy., 61, 4

Ballew, M. T., et al. 2022, ClCh, 173, 19

Ballew, M., et al. 2023, Global Warming's Six Americas across age, race/ethnicity, and gender. Yale University and George Mason University. (New Haven, CT: Yale Program on Climate Change Communication)

Barker, D. C., & Bearce, D. H. 2013, PRQ, 66, 267

Barker, D. C., Detamble, R., & Marietta, M. 2022, APSR, 116, 38

Båtstrand, S. 2015, Policy Polit., 43, 538
Bergquist, M., Nilsson, A., & Schultz, P. W. 2019, FrPs, 10,
Besley, J. C., et al. 2017, PLoSO, 12, e0175643
Besley, J. C., & Dudo, A. 2022, Strategic Science Communication (Baltimore, MD: Johns Hopkins Univ. Press)
Brosch, T. 2021, Curr. Opin. Behav. Sci., 42, 15
Callison, C. 2014, How Climate Change Comes to Matter: The Communal Life of Facts (Durham, NC: Duke Univ. Press)
Campbell, T. H., & Kay, A. C. 2014, J. Pers. Soc. Psychol., 107, 809
Carson, R.L. 2002, Silent Spring (London: Penguin)
Cesur-Soysal, G., & Arı, E. 2022, Omega (Westport), 00302228221075203
Citron, F. M. M., & Goldberg, A. E. 2014, J. Cogn. Neurosci., 26, 2585
Corner, A., Markowitz, E., & Pidgeon, N. 2014, Wiley Interdiscip. Rev. Clim. Change, 5, 411
Corner, A., Webster, R., & Teriete, C. 2015, Climate Visuals: Seven principles for visual climate change communication (based on international social research) (Oxford: Climate Outreach) https://climatevisuals.org/evidence/
Corner, A., Shaw, C., & Clarke, J. 2018, Principles for effective communication and public engagement on climate change: A Handbook for IPCC authors (Oxford: Climate Outreach)
Cunsolo, A., & Ellis, N. R. 2018, NatCC, 8, 275
Dahlstrom, M. F. 2014, PNAS, 111, 13614
Demarque, C., Charalambides, L., Hilton, D. J., & Waroquier, L. 2015, J. Environ. Psychol., 43, 166
Depoux, A., & Gemenne, F. 2020, Research Handbook on Communicating Climate Change (Cheltenham: Edward Elgar Publishing) 272
EPA 2020, Updating the Atmospheric Health Effects Framework Model: Stratospheric Ozone Protection and Human Health Benefits, Report No. 430R20005, U.S. Environmental Protection Agency
Fielding, K. S., Head, B. W., Laffan, W., Western, M., & Hoegh-Guldberg, O. 2012, EnPo, 21, 712
Fielding, K. S., Hornsey, M. J., Thai, H. A., & Toh, L. L. 2020, ClCh, 158, 181
Fredrickson, L., et al. 2018, AJPH, 108, S95
Gillis, A., Vandenbergh, M., Raimi, K., Maki, A., & Wallston, K. 2021, ERSS, 73, 101947
Goldberg, M. H., Gustafson, A., Rosenthal, S. A., & Leiserowitz, A. 2021, NatCC, 11, 573
Goldberg, M. H., Gustafson, A., & van der Linden, S. 2020, OEart, 3, 314
Goldstein, N. J., Cialdini, R. B., & Griskevicius, V. 2008, J. Consum. Res., 35, 472
Grennfelt, P., et al. 2020, Ambio, 49, 849
Gustafson, A., et al. 2020, CR, 33, 121
Gustafson, A., et al. 2022, NatEn, 7, 1023
Hess, D. J., & Brown, K. P. 2017, Environ. Sociol., 3, 64
Hickman, C., et al. 2021, LaPH, 5, e863
Hine, D. W., Phillips, W. J., Cooksey, R., Reser, J. P., Nunn, P., Marks, A. D. G., Loi, N. M., & Watt, S. E. 2016, GEC, 36, 1
Hoffarth, M. R., & Hodson, G. 2016, J. Environ. Psychol., 45, 40
Hornsey, M. J., Harris, E. A., & Fielding, K. S. 2018, NatCC, 8, 614
Howe, P. D., Mildenberger, M., Marlon, J. R., & Leiserowitz, A. 2015, NatCC, 5, 596
Hurley, E. A., Dalglish, S. L., & Sacks, E. 2022, LaPH, 6, e190
IEA 2021, The Role of Critical Minerals in Clean Energy Transitions (Paris: IEA) https://www.iea.org/reports/the-role-of-critical-minerals-in-clean-energy-transitions

Kahanamoku, S., et al. 2020, A Native Hawaiian-led summary of the current impact of constructing the Thirty Meter Telescope on Maunakea. https://doi.org/10.6084/m9.figshare.c.4805619
Kaufmann, R. K., et al. 2017, PNAS, 114, 67
Klöckner, C. A. 2020, Research Handbook on Communicating Climate Change (Cheltenham: Edward Elgar Publishing) 116
Kopp, R. E., et al. 2023, NatCC, 13, 648
Kruger, J., & Dunning, D. 1999, J. Pers. Soc. Psychol., 77, 1121
Kübler-Ross, E. 1969, On Death and Dying (New York: Macmillan)
Lammers, J., & Baldwin, M. 2018, J. Pers. Soc. Psychol., 114, 599
Latkin, C., et al. 2022, J of Prevention,
Lawson, D. F., et al. 2019, NatCC, 9, 458
Leiserowitz, A., Maibach, E., & Roser-Renouf, C. 2009, Climate change in the American mind: Americans' climate change beliefs, attitudes, policy preferences, and actions (New Haven, CT: Yale University) Available online at: http://research.yale.edu/environment/climate
Leiserowitz, A., & Smith, N. 2017, Oxford Research Encyclopedia of Climate Science (Oxford: Oxford Univ. Press)
Leiserowitz, A., et al. 2020, Climate Change in the American Mind: National Survey Data on Public Opinion (2008–2022),
Leiserowitz, A., et al. 2021, Climate activism: A Six-Americas analysis, December 2020. Yale University and George Mason University (New Haven, CT: Yale Program on Climate Change Communication) https://climatecommunication.yale.edu/publications/climate-activism-a-six-americas-analysis-december-2020/
Leiserowitz, A., et al. 2023a, Global Warming's Four Indias, 2022: An Audience Segmentation Analysis (New Haven, CT: Yale Program on Climate Change Communication)
Leiserowitz, A., et al. 2023b, Global Warming's Six Americas, December 2022. Yale University and George Mason University, (New Haven, CT: Yale Program on Climate Change Communication) https://climatecommunication.yale.edu/publications/global-warmings-six--americas-december-2022/
Lockwood, M. 2018, EnPo, 27, 712
Lynas, M., Houlton, B. Z., & Perry, S. 2021, ERL, 16, 114005
Maibach, E. 2009, Global Warming's Six Americas 2009: An Audience Segmentation Analysis, Report, Yale Project on Climate Change
Maibach, E., Leiserowitz, A., Rosenthal, S., Roser-Renouf, C., & Cutler, M. 2016, Yale University and George Mason University (New Haven, CT: Yale Program on Climate Change Communication)
Markowitz, E. M., & Shariff, A. F. 2012, NatCC, 2, 243
Marlon, J., Rosenthal, S., Goldberg, M., Ballew, M., Maibach, E., Kotcher, J., & Leiserowitz, A. 2022, Younger Americans are growing more worried about global warming (New Haven, CT: Yale Program on Climate Change Communication)
Marshall, G., Corner, A., Roberts, O., & Clarke, J. 2016, Faith & Climate Change - A guide to talking with the five major faiths (Oxford: Climate Outreach)
McCright, A. M., Dunlap, R. E., & Marquart-Pyatt, S. T. 2016, EnPo, 25, 338
Molina, M. J., & Rowland, F. S. 1974, Natur, 249, 810
Moser, S. C. 2007, ed. L. Dilling, & S. C. Moser Creating a Climate for Change: Communicating Climate Change and Facilitating Social Change (Cambridge: Cambridge Univ. Press) 64

Mullin, R. P. 2002, J. Bus. Ethics, 40, 207
Myers, T. A., Nisbet, M. C., Maibach, E. W., & Leiserowitz, A. A. 2012, ClCh, 113, 1105
Myhre, G., et al. 2013, Climate Change 2013: The Physical Science Basis. Contribution of Working Group I to the Fifth Assessment Report of the Intergovernmental Panel on Climate Change, ed. T. F. Stocker, et al. (Cambridge: Cambridge Univ. Press) https://www.climatechange2013.org/images/report/WG1AR5_Chapter08_FINAL.pdf
Nabi, R. L., Gustafson, A., & Jensen, R. 2018, SciC, 40, 442
Nisbet, M. C. 2009, Environment, 51, 12
Noelle-Neumann, E. 1993, The spiral of silence: Public opinion—our social skin, 2nd Edition (Chicago, IL: Univ. Chicago Press)
Ojala, M. 2012, EnvEdR, 18, 625
Painter, E., Zwar, J., Carino, S., & Kermonde, Z. 2023, Climate Change Communication Research Hub, https://www.monash.edu/mcccrh/publications/reports/best-practice-data-visualisation-guidelines-and-case-study (accessed 7.13.23).
Palm, R., Bolsen, T., & Kingsland, J. T. 2020, WCS, 12, 827
Pearson, A. R., Schuldt, J. P., Romero-Canyas, R., Ballew, M. T., & Larson-Konar, D. 2018, PNAS, 115, 12429
Pihkala, P. 2022, FrCl, 3,
Raimi, K. T., Stern, P. C., & Maki, A. 2017, PLoSO, 12, e0171130
Richardson, L. M., Machin, F., & Williamson, L. 2022, Climate Change: Concern, Behaviour and the Six Australias, Monash Climate Change Communication Research Hub (Melbourne: Monash Univ.)
Riederer, R. 2019, The Other Kind of Climate Denialism (New York: The New Yorker)
Ritchie, H. 2023, We need the right kind of climate optimism (Vox) https://www.vox.com/the-highlight/23622511/climate-doomerism-optimism-progress-environmentalism (accessed 7.13.23).
Rocklage, M. D., Rucker, D. D., & Nordgren, L. F. 2018, Psychol Sci., 29, 749
Roser-Renouf, C., Stenhouse, N., Rolfe-Redding, J., Maibach, E., & Leiserowitz, A. 2015, Engaging diverse audiences with climate change: Message strategies for Global Warming's Six Americas, ed. R. Cox, & H. Anders (New Haven, CT: Yale School of the Environment)
Sapiains, R., Beeton, R. J. S., & Walker, I. A. 2016, J. Appl. Soc. Psychol., 46, 483
Schwartz, S. E. O., et al. 2022, Curr Psychol, 42, 16708
Somerville, R. C. J., & Hassol, S. J. 2011, PhT, 64, 48
Sparkman, G., Geiger, N., & Weber, E. U. 2022, NatCo, 13, 4779
Sturgis, P., & Allum, N. 2004, Public Underst Sci., 13, 55
Suldovsky, B. 2017, Oxford Research Encyclopedia of Climate Science (Oxford: Oxford Univ. Press)
Szabó, S., Bódis, K., Huld, T., & Moner-Girona, M. 2013, RSER, 28, 500
Thaler, R. H., & Sunstein, C. R. C.R. 2008, Nudge (New Haven, CT: Yale Univ. Press)
van der Linden, S., Leiserowitz, A., & Maibach, E. 2019, J. Appl. Soc. Psychol., 62, 49
von Biela, V. R., et al. 2022, Fisheries, 47, 157
YPCC & Mason 4C 2020,Climate Change in the American Mind: National survey data on public opinion (2008-2018) [Data file and codebook]. doi: 10.17605/OSF.IO/JW79P
Zaleha, B. D., & Szasz, A. 2015, BuAtS, 71, 19

Chapter 13

Media and Misinformation

P Banchero and T A Rector

> "That's not the way the world really works anymore. We're an empire now, and when we act, we create our own reality."
>
> —Anonymous senior White House official during the George W. Bush administration[1]

Scientists and science communicators often lament our difficulties with helping people understand the problem of climate change and motivating them into action. It can feel that our message is falling on "deaf ears"; and we often blame ourselves. While there's definitely room for improvement, we need to keep in mind that we are not the only ones communicating about climate change. Much of what people learn comes from sources other than scientists. Mass communications—such as news media, social media, and corporate strategic communications—are also sending messages about climate change, and their messages are often counter to our own. Unfortunately these forms of media are often ineffective in communicating the science, scientific consensus, and severity of the problem. They are also often used to spread misinformation and disinformation. The result is a media ecosystem where the public is often confused and lacking trust in their sources of information. In this chapter, we'll explain how different forms of media communicate about climate change, what the modern media audience looks like, and the tactics of those who spread disinformation. We'll also provide insights about how astronomers can reckon with the challenges and opportunities in communicating to the public via the news as well as strategies to counter misinformation through science media literacy.

[1] Suskind, Ron. (October 17, 2004). Faith, certainty and the Presidency of George W. Bush. The New York Times. https://www.nytimes.com/2004/10/17/magazine/faith-certainty-and-the-presidency-of-george-w-bush.html

doi:10.1088/2514-3433/acfcb6ch13

13.1 How the Public Learns about Climate Change

Think of when you first heard of climate change. Was it from a newspaper article? An episode of *Bill Nye the Science Guy*? Was it in a classroom? Or perhaps at the kitchen table while talking with family? Or maybe you heard a public figure express their opinion about it?

Much of what you know about climate change probably comes from your experience as a scientist—from classes, science talks, and research journals. However, most of what the public hears comes from a wide range of sources—whether it be from news reports, advertisements from fossil fuel companies, YouTube videos, or even popular films such as *The Day After Tomorrow*. In fact, most of what people know about climate change doesn't come directly from scientists. The media, including news and social media, are the public's primary source of information about science (National Science Foundation 2018).

Astronomers can benefit from knowing about how different media communicate about climate change, the ways professional communicators may help or hinder the messages we want to share, and what can be done about it. In this chapter, we'll explain how the news and social media work, how corporations use communications to articulate their positions, and what tactics climate denialists use to thwart action to address climate change. We'll also discuss what scientists can do to communicate more effectively and to instill science literacy skills in students and the public.

13.2 The Modern Media Ecosystem

If we think about the Latin root of *media*, which means "middle", we understand that media "mediate;" they are between the "speakers" of social and political realms and their audience. Media distribute content, such as text, still images, video, and sound, to the public via different technologies (McQuail 1983). This includes traditional news sources in the form of newspapers, magazines, and television news, as well as media created and distributed by other stakeholder organizations such as political parties, nongovernmental groups, and corporations. In "traditional media" the flow of information is largely in one direction, whereas on social media consumers are also content creators and distributors.

The media in all its forms play a significant part in how people come to know about climate change, and what they think about its causes, consequences, and possible solutions. The people you interact with every day—your students, neighbors, and colleagues—have likely read or viewed media about climate change. Knowing how and why people use media helps to understand how public opinion is shaped. And public opinion is central to the political sphere—how citizens interact with government and other powerful institutions (including academia and for-profit corporations). Decisions about how people vote, what issues move them to activism, and which candidates they see fit to support are linked to public opinion and therefore the media they consume.

13.2.1 The Shifting Demographics of News Consumption

How we get our news is changing. Forman-Katz & Matsa (2022) find that consumers are shifting away from print, television, and radio to digital media.

The majority of Americans (82%) say they sometimes or often get news from a smartphone, computer, or tablet. The public may be accessing news websites, apps or social media feeds on these devices. But they are falling away from traditional forms of media: television, radio and print. In 2022 the slice of Americans who often get news from television has decreased to only 31%. Radio and print publications trail significantly, with only 8% of audiences saying they get their news often from print publications and 13% from radio. There are variations in news consumption across platforms due to age, gender, race, ethnicity, educational attainment, and political leaning:

Age: a large majority of people between the ages of 18–29 say they get at least some of their news from digital devices (91%) compared to those 65 and older (67%). A large majority (85%) of people 65 and older say they get at least some of their news from TV. Three-fourths of young adults also get at least some news from social media, while only one-fourth of people 65 and older do.

Educational attainment: about half of those with a high school education or less use the internet for news consumption whereas 75% of college graduates reported that they do. High school graduates were also less likely than college graduates to do an internet search for at least some of their news (53% compared to 65%).

Political leanings: in the U.S., about two-thirds (67%) of Democrats use television for at least some news, a rate that was slightly lower for Republicans (60%). Generally speaking, Republicans perceive many news sources as untrustworthy. Meanwhile, Democrats rely on a much wider assortment of news outlets—sources largely considered to be "mainstream." Roughly three-quarters of Democrats (77%) say they have at least some trust in the information they get from national news outlets, compared to only 42% of Republicans (Liedke & Gottfried 2022). The slide in trust of traditional news sources has been much faster among Republicans than Democrats (Gottfried & Liedke 2021). The partisan differences in trust are not as pronounced for local news organizations and social media sites. However, Democrats are still more likely than Republicans to have at least some trust in the information they get from local news outlets (79% vs. 63%) and social media sites (38% vs. 27%).

Because they don't trust as many news outlets, conservatives tend to get their news from only a handful of sources. There is also a conservative media constellation of which Fox News and talk radio are lodestars. They create a self-protective enclave, shielding conservatives from other information sources and promoting highly negative views toward their political opponents (Jamieson & Cappella 2008). This is a problem because a person's knowledge of climate change usually benefits from the use of multiple sources (Kahlor & Rosenthal 2009).

13.2.2 The Blurred Lines of Media

Until the late 1990s, most people received their news from print media and television, either produced locally or nationally. A majority of Americans believed in the news media—particularly their local news and information outlets. With the invention of the World Wide Web,[2] distributing news, information, and entertainment to large audiences became much easier. The internet quickly became a place where established news organizations sought out online audiences while competing for attention with others who were publishing blogs and websites, making it less clear who is a trustworthy source of news. This had repercussions for news organizations and their audiences. Traditional news outlets struggled with how to exploit digital advertising revenue as online competitors—particularly social media startups such as Facebook—became more effective at it. Initially, news sources also put much of their content online for free, resulting in a degradation of the perceived value of their product. As a result, legacy news organizations cut staff. Their audiences responded by canceling their subscriptions or tuning out their once trusted news sources (Mitchell 2013). Report for America, a nonprofit concerned with the adequacy of news coverage nationwide, estimates that in 1990 there were 4,490 Americans per journalist and in 2019 there were 14,251—a threefold loss.[3]

Another development was the proliferation of online and cable news outlets tailored to partisan perspectives—either liberal or conservative (Newman et al. 2018). Coverage of complex scientific matters such as climate change tended to get warped by this trend. Over the last thirty years climate change articles from major newspapers have become more politicized, with politicians and political advocates increasingly highlighted in coverage and scientists less so. The content has also become more polarized, with liberal and conservative discourses becoming starkly different (Chinn et al. 2020).

What *purpose* the media serves has also been blurred, as marketing imperatives mold media. Long before the internet, media producers found it useful to make advertisements or commercials that look and sound like the content that surrounded them. The strategy could be as simple as asking a popular radio host to endorse a product. This technique is effective because, if the media content is of interest to the consumer, it is often considered to be "information" rather than "news" or "advertising" (Grotta et al. 1976). Ad agencies placed "advertorials" in newspapers and magazines, as they counted on the credibility of the news organization and the believability of content that is tailor-made to look like a news article when in proximity to actual news. This technique is often called "sponsored content" or "native ads." Even though most sponsored content is identified as such in some way, it is often subtly labeled so that many readers are not aware of its true nature. Sponsored content is now a major source of revenue for many respected news sources (Hardy 2021); e.g., "T Brand" is the sponsored content studio of *The New York Times* and *The Washington Post* runs BrandConnect—for which both newspapers have received criticism.

One of the outcomes of the internet has been an acceleration of the blurring of boundaries that once separated different types of media. A news article often lives

[2] The World Wide Web software went into the public domain on April 30, 1993. https://www.home.cern/science/computing/birth-web

[3] Waldman, S. (June 28, 2021). Report for America. https://www.reportforamerica.org/2021/06/28/the-journalist-population/

alongside opinion pieces about the same topic. And far too often advertising is so embedded in news websites as to be virtually indistinguishable from the content that surrounds it.

Today's media environment includes content that defies easy classification. Historically theorists have categorized media in order to study *what* is said to *whom* and with what *effect*. Lasswell (1948) proposed three functions of media. The first is to survey the environment, such as reporting the news of a city council meeting or an armed conflict. The second function is to analyze and interpret the information we receive, including value judgments about the news of the day. Finally, the media play a role in transmitting social heritage from one generation to another; e.g., the sharing of customs associated with different holidays. Later, theorists added an entertainment function to this understanding (Wright 1986). Programs like *Oprah* and *The View* are examples of "infotainment." Many also rely on satirical shows, such as the comedy show *Last Week Tonight with John Oliver*, as sources of news (Gottfried & Anderson 2014).

13.2.3 Climate Change in Traditional News Media

News organizations express what they think is important by what they cover, how they cover it and often, by what they *don't* cover. Individual journalists and news outlets weigh factors such as timeliness, novelty, prominence, impact, and conflict. Journalism also prioritizes stories that can engage and hold the audience's attention. That often means a story with strong visual impact, whether it be in the form of still images, graphics, or video.

News media have been covering climate change for more than a century, with articles about the subject appearing as long ago as 1902 (Figure 13.1).[4] Coverage accelerated after 1988, when climate scientist James Hansen testified before Congress. The Media and Climate Change Observatory has tracked the level of attention by the news media globally since 2004. While 2021 had the highest amount of coverage of climate change—increasing 55% from 2020 (Pearman et al. 2022), in the following year media attention decreased 11% (Katzung et al. 2023). There are numerous gaps and oversights in media attention, but climate change coverage has become more sophisticated and comprehensive—particularly among legacy national media organizations; e.g., in 2022 NPR announced the launch of its Climate Desk. Many have reporting teams dedicated to climate and environmental issues, and they publish comprehensive explanatory and investigative work.[5] For example, in 2022 *The New York Times* published varied content related to climate change, such as an interactive feature about how South Asian monsoons are getting stronger and an investigation into then West Virginia Senator Joe Manchin's connections to the coal industry, exploring the political dimensions of climate change and why policy

[4] N.A. (October 15, 1902). "Hint to Coal Consumers" *The Selma Morning Times*. Selma, Alabama, US. p. 4. https://commons.wikimedia.org/wiki/File:19021015_Hint_to_Coal_Consumers_-_Svante_Arrhenius_-_The_Selma_Morning_Times_-_Global_warming.jpg.

[5] https://www.washingtonpost.com/pr/2022/11/28/visual-storytelling-global-reporting-insights-drive-washington-posts-expanded-climate-environment-coverage/

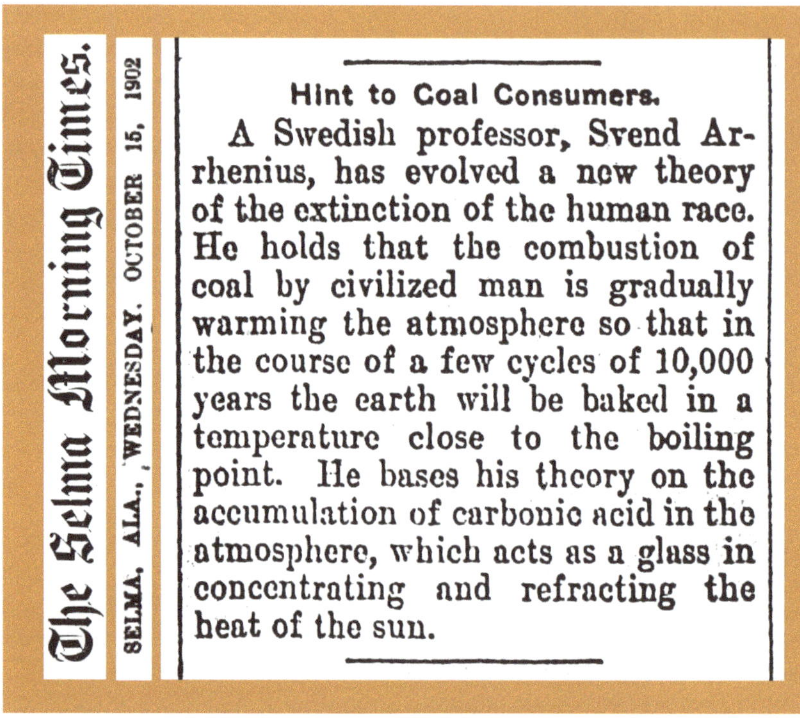

The Selma Morning Times.

SELMA, ALA., WEDNESDAY. OCTOBER 15, 1902

Hint to Coal Consumers.

A Swedish professor, Svend Arrhenius, has evolved a new theory of the extinction of the human race. He holds that the combustion of coal by civilized man is gradually warming the atmosphere so that in the course of a few cycles of 10,000 years the earth will be baked in a temperature close to the boiling point. He bases his theory on the accumulation of carbonic acid in the atmosphere, which acts as a glass in concentrating and refracting the heat of the sun.

Figure 13.1. [6]This brief appeared in the Selma (Alabama) Morning Times on October 15, 1902. It mentions Svend Arrhenius's theory about the warming effects of carbon dioxide.

proposals can be inhibited by powerful lawmakers. The uptick in coverage is a good thing because greater levels of exposure to media reporting on climate change have been found to be associated with increased certainty about the issue and support for mitigation policies (Feldman et al. 2014).

Recent coverage of climate change in traditional news media shows a variety of storylines and approaches, e.g.:

- "Is It Hot Enough for Politicians To Take Real Action?"
- "Things Don't Always Change in a Nice, Gradual Way"
- "The planet is getting hotter fast. This is what happens to your body in extreme heat"

These were three stories in national news media in July 2023, when the average global temperature reached a new high.[7] And they came after the hottest June in history (Hersher 2023).

[6] https://commons.wikimedia.org/wiki/File:19021015_Hint_to_Coal_Consumers_-_Svante_Arrhenius_-_The_Selma_Morning_Times_-_Global_warming.jpg

[7] https://climatereanalyzer.org/clim/t2_daily/

Although these articles all focus on extreme heat, they illustrate how the media differ in the ways they cover climate issues. The type of media platform (e.g., text-based versus video-based, daily news versus analysis- or commentary-driven) can determine how the story is told. Other factors include audience demographics, whether and to what degree partisanship factors into the media's agenda, and the outlet's ownership. The first two examples above, *The New Yorker* and *The Atlantic*, are both text-based news organizations with highly educated audiences.[8] Both offer analysis about the news, while CNN's business is focused on daily coverage of current events. Its audience is looking for quicker updates online and on its broadcasts.

Television news has, by and large, been slower to respond to the climate crisis and offers less robust coverage. In 2022 nightly news shows aired nearly 7 hours of climate coverage—a 17% increase from 2021. However, this was still only 1.3% of the broadcast TV coverage during that time (Macdonald 2023). The major networks have in recent years begun to invest more in climate coverage, deploying some of their best-known talent to Alaska, Greenland, and other destinations perceived to be on the front lines of climate change. For example, the news magazine program *60 Minutes* aired a segment in January 2023 about the sixth extinction event currently underway. But for print media audiences, one of the first mass media articles about the subject appeared in *The New Yorker* in 2009. Although television often comes to climate stories late, the medium has the potential to sway audiences because of its visual and often emotional narratives (Tompkins 2017).

It's also important to consider the partisan orientation of the news outlet. A content analysis of climate change coverage on Fox News, CNN, and MSNBC during 2007–8 showed that Fox News takes a more dismissive tone. Fox also interviewed a greater fraction of climate change doubters (Feldman et al. 2012). Exposure to these news sources influences perceptions of climate change and proposed solutions, particularly among those who identify as conservative (Feldman et al. 2014). Dubbed the "Fox News effect," Gustafson et al. (2019) found that viewing the channel was a significant predictor of both familiarity with, and opposition to, the "Green New Deal"—a climate change and economic renewal proposal introduced by Democrats in 2019. Another factor affecting coverage may be the ownership of the news source. For example, Pierce (2022) found that once a television station is purchased by Sinclair, Inc.—a conservative-leaning media corporation (Blankenship & Vargo 2021)—its volume of climate change reporting drops, and the remaining coverage deemphasizes climate science. In media markets with more political polarization, such as the U.S. and Australia, fewer said they have a need for climate change coverage (Robertson 2022).

Several forces have worked to keep journalists from delving deeply into the causes and consequences of climate change, or calling out the peril of maintaining the status quo. They include the financial calculations news organizations have to consider when corporations sponsor news. Although ethical news organizations try to

[8] https://www.pewresearch.org/politics/2012/09/27/section-4-demographics-and-political-views-of-news-audiences/

insulate editorial content from bottom-line concerns, there can be tension between news coverage and corporate interests. For example, the digital news source *Semafor* has been criticized for accepting sponsorship from Chevron for its climate newsletter (Legum 2022).

Another force that keeps journalists from delivering more complete coverage is the influence of strategic communications by corporate interests (as discussed in Section 13.5). Public relations practitioners depend on journalists to report information, and in return journalists count on public relations sources to provide information (Koch et al. 2020). But the ratio of public relations professionals to journalists is about six to one in the U.S., and a journalist's median salary is almost 20% less than for public relations professionals, according to the U.S. Bureau of Labor Statistics.[9] As newsrooms cut resources, public relations specialists are eager to step in and provide material that is consistent with their messaging.

13.2.4 Climate Change in Local News

National news organizations are a vital source of climate news, but they are not the only source. Historically the U.S. has had strong regional and local news outlets. Excluding the four national newspapers of *The New York Times*, *The Washington Post*, *USAToday* and *The Wall Street Journal*, the total combined print and digital circulation for locally focused U.S. daily newspapers in 2020 was 8.3 million for weekday (Monday–Friday) and 15.4 million for Sunday. Perhaps no sector of the news media has been under greater pressure in the digital age than local news. Total weekday circulation is down 40% since 2015 (Matsa & Worden 2022). Sunday circulation has fallen even more.

A web of organizations is working to reinvest in local news, including the Local Media Association and its Climate Collaborative. There is recognition that climate is not a stand-alone subject, but relates to other issues, such as transportation, health, and economic development.[10] Despite the economic challenges of the news business, local news outlets are investing in climate reporters, who are sometimes supported with grant money. For example, Colorado Public Radio employs two journalists dedicated to covering climate change in a newsroom of approximately 48 employees.[11]

Daily weather forecasts is one area where the investment in local journalism has paid off. Americans tend to trust television weather reporters as a source of climate change news reporting (Leiserowitz et al. 2016). Armed with that insight, in 2010 the George Mason Center for Climate Change Communication launched Climate Matters, a weekly source of climate materials for local media markets that are used by over 3,000 weather and climate reporters across the U.S. Television news viewers who watch climate reports are more likely to understand that climate change is happening, is human-caused, and is causing harm to their community. They also

---

[9] https://www.bls.gov/oes/current/oes_stru.htm

[10] Hopke, J. (November 1, 2022). Everyone is a climate reporter now. Nieman Reports. https://niemanreports.org/articles/climate-change-journalism-education/

[11] Colorado Public Radio. Retrieved August 16, 2023. https://www.cpr.org/about/staff/

feel that climate change is personally relevant, are more alarmed about it, and want to learn more (Feygina et al. 2020). Fortunately more meteorologists are using nightly broadcasts to deliver short, engaging climate lessons.

13.3 Frames Used in Climate Change Coverage

Chapter 12 describes "framing"—the way in which information is presented and the effect it has on how people understand and respond. The news media often use these frames in their coverage of climate change:

13.3.1 Economic Consequences

Climate change's economic impacts often make headlines, either focusing on the costs of mitigation and adaptation—e.g., shifting to renewable energy sources or reducing energy use—or the costs of climate change consequences; e.g., natural disasters and agricultural disruption. There is also increasing coverage on the potential opportunities of green businesses.

Stecula & Merkley (2019) conducted a content analysis of many of the most influential news media outlets organized around three frames: economic costs and benefits associated with climate mitigation, appeals to conservative and free market values and principles, and uncertainties and risk surrounding climate change. They found that around 30% of coverage had reference to possible economic costs of climate mitigation. Frames focusing on economic benefits of climate policy were less frequent. Kenny (2018) found that people's beliefs in anthropogenic climate change don't shift when they hear about economic impacts, and they are less likely to support urgent action if they have a pessimistic view of the economy.

13.3.2 Natural Disasters

Coverage of climate change often occurs in conjunction with natural disasters, such as floods, wildfires, and heat waves. Climate change tends to surface in analysis pieces or reports on longer-term trends rather than in breaking news. For example, reporter Seth Borenstein explained how the frequency of natural disasters is accelerating:

> *"This past year has seen a horrific flood that submerged one-third of Pakistan, one of the three costliest U.S. hurricanes on record, devastating droughts in Europe and China, a drought-triggered famine in Africa and deadly heat waves all over.... Weather disasters, many but not all of them turbocharged by human-caused climate change, are happening so frequently that this year's onslaught, which 20 years ago would have smashed records by far, now in some financial measures seems a bit of a break from recent years."*

The article paraphrases an NOAA applied economist who says that in the 1980s, the U.S. would average a billion-dollar weather disaster every 82 days. Now it is every 18 days (NOAA 2023).

Reports like the one above are increasingly common, in which journalists relay how climate change is contributing to extreme weather events. But daily journalism has a mixed record; e.g., these recent news reports about wildfires and heat waves lack any mention of climate change. Examples of thorough coverage seem to be increasing as the quantity of coverage does. Many national news organizations are doing more to put such events into context; e.g., this episode of PBS *NewsHour* explains how extreme heat in Rome, drought in Baghdad, intensifying monsoons in New Delhi, massive flooding in South Korea, and sweltering heat in Tokyo can be attributed to climate change. The correspondents describe what governments and citizens are or are not doing to respond.

News sources are also getting better at using effective wording when describing the role of climate change in natural disasters. For example, in October 2022 on this episode of *Face the Nation*, reporter Ben Tracy described the role of warmer water in powering stronger hurricanes: "The ocean waters are much warmer than they used to be, and warm ocean water is rocket fuel for a hurricane. So when one of these storms passes over this warm water, it just turbocharges [it]." The terms "rocket fuel" and "turbocharge" are memorable colloquialisms that have intuitive meaning to most Americans. Examples of well-conceived data visualization in climate change journalism are also on the rise, such as this explanation of the era of megafires, produced by Reuters.

13.3.3 Political Winners and Losers

The news media also tends to focus on what is gained or lost politically if specific public policies are enacted. This is in part because news audiences gravitate toward conflict—and politics are a natural area of tension. In the last 35 years, major newspapers have featured increasingly political coverage of climate change, with political actors featured more heavily than scientists as Democratic and Republican discourses about the issue have diverged (Chinn et al. 2020). On television, audiences have also seen stark partisan differences, which contributes to polarization on the subject. Use of this framing is common because it can be highly successful. For example, Fox News often uses political frames to cover most events, climate related or otherwise.[12]

The coverage of the Biden administration's March 2023 approval of Willow, a $8 billion oil and gas project in the National Petroleum Reserve-Alaska, demonstrated how politics can drive the narrative. The administration's decision to approve the project, led by oil giant ConocoPhillips, amounted to a broken campaign promise to move the nation away from fossil fuels. Willow could become the largest oil drilling project on U.S. public lands. It is expected to produce roughly 600 million barrels of oil over 30 years, generating more than 9 million metric tons of CO_2 a year (Friedman 2023). An analysis from Media Matters found that major TV cable news networks—CNN, Fox News, and MSNBC—aired nearly 90 minutes of total coverage about the approval. But Fox News spent more time than CNN and MSNBC combined. Only

[12] http://www.journalism.org/2014/10/21/political-polarization-media-habits/.

about one-fourth of the cable news segments discussed Willow's climate implications (Fisher 2023). Samuelson & Atkin (2023) analyzed 30 national news stories and found 75% featured primarily political framing, describing the Willow story as a loss for environmental activists and discounting the scientific position that new drilling must be stopped if nations are to meet the goals of the Paris Agreement.

13.3.4 Scientific "Consensus" and False Balance

A fundamental tenet of ethical journalism is to report information fairly and accurately. To appear neutral, journalists often present two or more sides of an issue. "False balance" refers disparagingly to the practice of journalists who, in their desire to be fair, present each side of a debate as equally credible—even when the factual evidence is stacked heavily on one side, as is the case for climate change. Sulzberger (2023) states, "In cases in which the facts have been established beyond reasonable dispute, journalists should not quote a fringe position to check a box or shield their work from accusations of bias. There is, for instance, no serious debate in the scientific community about the reality of climate change. The world is warming, with devastating consequences."

Unfortunately journalists, by relying on their traditional norms of balance, can introduce a false equivalence into their coverage of climate change. There is often a wide disparity between the scientific consensus about the causes and impacts of anthropomorphic climate change and what is reported (Boykoff & Boykoff 2004). Earlier research showed that approximately 70% of television coverage about climate change on major networks between 1996 and 2004 was falsely balanced (Boykoff 2008) and one-third of climate change articles in U.S. newspapers published in 2009–2010 included skeptical perspectives (Painter & Ashe 2012).

A textbook example of false equivalence aired on CNN in 2012 when host Piers Morgan invited Bill Nye "the science guy" to debate Marc Morano on whether or not climate change was happening, even though at that time 97% of scientific research papers endorsed the view that climate change is real and the result of anthropogenic emissions (Cook et al. 2013). While the debate shouldn't have happened at all, it also shouldn't have been between these two—as neither is a climate scientist. Nye was the host of a science-related television program and Morano was the director of communications at the Committee For A Constructive Tomorrow (CFACT), a conservative Washington, D.C. think tank. He was also a director of communications for U.S. senator James Inhofe.[13] Bill Nye is often called upon to defend climate science on television against climate-denying opponents, likely because he is a known personality. Satirist John Oliver showed the folly in such debates by bringing on a throng of climate scientists to illustrate the overwhelming scientific consensus about the subject, creating a memorable visual moment.

Fortunately, more recent research indicates journalists are moving away from false balance and toward more accurate coverage of scientific consensus. (Boykoff et al. 2021; McAllister et al. 2021). Criticisms of "both-sidesism" may be less relevant today

[13] https://www.desmog.com/marc-morano/

because top journalists are more transparent, more readily calling out lies, and producing deeper reporting, even when they may be criticized for showing bias.

13.3.5 Solutions Journalism

Often the news media creates stories about the *problems* caused by climate change, but have little emphasis on what scientists, communities, and organizations are doing for *solutions*. There is a tendency for news organizations to focus on "doom and gloom" (Arnold 2018). For example, the PBS roundup of extreme weather events mentioned above briefly discusses what governments are considering to alleviate climate change impacts, but daily journalism can leave the audience questioning what actions groups and individuals can take.

"Solutions journalism" is a movement to add a more complete view of social, economic, and political issues by expanding the range of coverage beyond a mere diagnosis of a problem, but also examining how people and systems are working to solve it. The movement aims to provide people with model responses to problems that are newsworthy (McIntyre 2019). It also provides evidence of results. For example, this story explores how well a rebuilding of housing and infrastructure worked to prevent damage during a subsequent flooding event. As discussed in Chapter 9, climate change is a multi-faceted issue—complex in its impact and manifold in the solutions needed. Solutions journalism is not just about reframing problems, but looking for examples of positive responses to those problems and explaining how those responses can work in other contexts.[14] It also seeks to provide insights that might be replicable in other communities or other circumstances. It also discusses failures and limitations. Solutions journalism, by its explanatory nature, does not lend itself to breaking stories. The Solutions Journalism Network trains and supports reporters who are working on such stories and expands the reach of such news.

13.4 Social Media

Social media can be defined as websites and applications that enable users to create and share content, or to participate in social networking. Examples include Facebook, YouTube, Twitter, and TikTok. Social media has become the new public square and a battleground on which public opinion is decided. While people often associate social media with personal updates or cat memes, it is increasingly where the public turns to for news and information. In fact, for people under 30 social media is considered to be as reliable as national news organizations (Liedke & Gottfried 2022). While traditional media has a presence on social media, much of the information spread is from influencers,[15] trolls,[16] and bad actors who profit by

[14] N.A. (December 17, 2018). How to sell solutions journalism. The Whole Story: Ideas and Dispatches from the Solutions Journalism Network. https://thewholestory.solutionsjournalism.org/how-to-sell-solutions-journalism-46b2861c5609

[15] An "influencer" is a person with a large following on social media who profits by promoting a product or service, often without revealing that they're receiving compensation to do so.

[16] A "troll" is someone who intentionally antagonizes others online by posting inflammatory, irrelevant, or offensive comments or other disruptive content.

disseminating propaganda, "fake news," misinformation, and disinformation. Note that misinformation is defined as factually inaccurate information that is unintentionally spread, whereas disinformation is created and disseminated with an intent to confuse and deceive.

The best aspects of social media live alongside the worst. Ideally it is a personalized forum where people can publish content that ranges from the banal to the intellectual. The information is available instantaneously and worldwide, with low barriers to entry. Anyone with access to the internet can post their own content, as well as share and comment on others.

However, there are multiple downsides. One problem is that users can unintentionally share misinformation and disinformation, amplifying its impact. Fake news diffuses more than six times faster than the truth, penetrating social media significantly farther, deeper, and more broadly because it is often more novel or compelling than true news (Vosoughi et al. 2018). As the saying goes, "a lie can travel around the world while the truth is lacing up its boots." Shared misinformation can have an even greater impact, as the source of the shared news—e.g., a friend or celebrity—can more heavily influence its perceived value than the credibility of the original news organization (Kang et al. 2011; Jang 2023). What's more, many of the posts on social media aren't made by humans but by "bots," autonomous programs that mimic real people, often spreading disinformation and inflammatory content (e.g., Figure 13.2).

Another problem is simply how content is disseminated on social media. Because it is impossible to share all of the material available, social media platforms use algorithms to tailor the content served to match individual users' tastes and preferences. Unfortunately, this targeted content narrows the scope of information an individual receives, to topics and viewpoints that often confirm and reinforce one another. In other words, you shape the algorithm—and the algorithm shapes you. Because social media companies organize people into groups that share tastes, preferences, and demographic characteristics, users can find themselves in "filter bubbles" or "echo chambers" that align with societal divisions—be they ideological, geographic, ethnic, linguistic, socioeconomic, religious or otherwise. Personalization therefore increases polarization. In this environment, social media amplifies emotions, and users can find themselves overriding their rationality as rumors or personal opinions are spread as if they are credible facts.

Like other news stories, social media is also the place where many people learn about climate change, particularly for young audiences. In a recent survey, over half of 14- to 18-year-olds said they learn "some" or "a lot" about climate change from social media (Prothero 2023). However social media is also a common way for climate misinformation and disinformation to spread. For example, after the U.S. withdrew from the Paris Agreement in 2017, a quarter of posts on Twitter related to climate change were made by bots, most of which spread misinformation (Marlow et al. 2021).

Social media companies have taken some steps to address misinformation about climate change on their platforms. Meta, which owns Facebook and Instagram, has added fact-checking labels to climate change content. Google, which also runs YouTube, pledged in 2021 to stop placing ads that deny the existence and causes of climate change. The goal was to stop climate deniers from making money on its

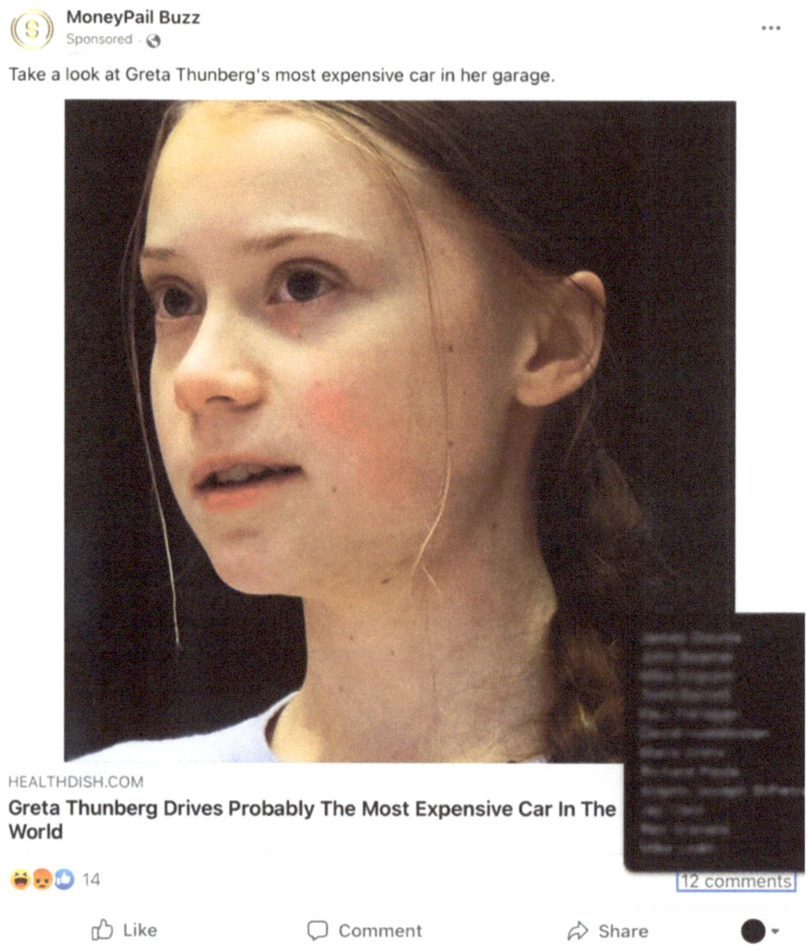

Figure 13.2. An example of misinformation shared on Facebook that falsely claims that climate activist Greta Thunberg owns an expensive car, apparently in an attempt to portray her as hypocritical. The origin of this claim is likely an altercation on social media between Thunberg and former kickboxer Andrew Tate, who bragged about his own car collection and its "enormous emissions." This misinformation was shared by an account that is likely a bot.

platforms (Grant & Myers 2023). Yet the Climate Action Against Disinformation coalition found that Google had failed to enforce its policy systematically, resulting in more than 71 million views of misinformation. The coalition also faulted Google for using too narrow a definition of climate disinformation.

Messages in the digital age—whether they are disinformation or legitimate discourse—are often spread widely by a few people who have outsized influence; i.e., influencers and celebrities. Social media enables the public audience to develop parasocial[17] relationships with influencers that can feel very personal. Industry uses

[17] Defined as a one-way or imagined connection between a regular person and a fictional character, celebrity, or other public figure.

these relationships to persuade the public about products or issues (Balaban et al. 2022). This is also true for climate change. For example, marketers affiliated with the American Gas Association and the American Public Gas Association have created social media campaigns that feature food and wellness influencers about the supposed virtues of cooking with gas stoves (Leber 2020). But if this communications strategy can work for industry, it can also work for positive climate change messages. The Reuters Institute's recent research about climate change in the news found that survey respondents under 35 in many global markets were more likely to say they pay attention to celebrities, social media personalities, or activists for climate change news than people over 35. Some of these climate change influencers have gained followers by drawing attention to the connections between climate change and social justice issues.[18]

13.5 Strategic Communications: Campaign Strategies from Industry

Strategic communications is the use of effective messaging to promote an organization's goals. It is used by corporations to communicate with stakeholders such as shareholders, politicians, and the public. This is often done via advertising and by maintaining a presence on social media. Sometimes this takes the form of "crisis communications" in response to an event; e.g., from a petroleum company in response to an oil spill. Strategic communication is a robust practice used by most corporations. It is also an emerging area of study in the social sciences (e.g., Thomas & Stephens 2015).

While not inherently nefarious, strategic communication is often used so. On the subject of climate change, usually the communications from industry—particularly those tied to the production and use of fossil fuels—run counter to the messaging from scientists. Unfortunately these messages often get more attention because the resources put into corporate communications far exceed that of science communications. Brulle & Downie (2022) found that U.S. trade associations engaged on climate change spent $3.4 billion on political activities between 2008 and 2017. Almost $2.2 billion of it went toward advertising and public relations, with the rest devoted to lobbying, grants, and political contributions. Trade groups in the oil and gas sector were the largest spenders. The American Petroleum Institute alone spent over six times more than four prominent renewable energy trade associations combined (Quinn & Young 2015). Considering that the estimated value of existing fossil fuel deposits is more than $10 *trillion* (e.g., Rempel & Gupta 2021), such investments in strategic communications to slow the transition from fossil fuels is a shrewd strategy.

These communications are in response to the pressure corporations are feeling. Known as "corporate social responsibility," companies are now increasingly expected to produce products that don't harm the environment. However, the messaging from corporations doesn't always match their performance—and is known as "greenwashing." While seldom involving outright lies it can take many

[18] https://reutersinstitute.politics.ox.ac.uk/digital-news-report/2022/how-people-access-and-think-about-climate-change-news

misleading forms, such as reporting environmentally friendly behaviors that are insufficient, irrelevant, or cannot be verified (de Jong et al. 2020). For example, for fifteen years ExxonMobil heavily touted the potential of algae biofuels—spending half as much on advertising as they did for research—before quietly discontinuing its funding (Westervelt 2023). Greenwashing can also involve setting goals in the distant future that sound impressive at the time but later will be quietly forgotten. For example, of fifty climate targets set by the aviation industry since 2000, only one had been met (Beevor & Alexander 2022). Oil companies also have a long history of the practice (Brulle et al. 2020).

There has been pushback on greenwashing, including calls for the banning of fossil fuel advertising (e.g., Spindle 2022), as has been done for other dangerous products such as tobacco. In 2022, France passed a law that bans all fossil fuel advertising. Likewise the UK's Advertising Standards Authority has banned ads from the petroleum companies Shell, Repsol, and Petronas as well as the airlines Lufthansa and Etihad Airways for being misleading.[19] An example of activists countering greenwashing messages is "brandalism"[20] or "badvertising," where greenwashing advertisements are parodied. Another strategy is "greentrolling," where critical (and often humorous) comments are posted in reply to social media attempts at greenwashing.[21]

While fossil fuel companies have increased their discourse related to the need for transition to clean energy sources, overall this has not been reflected in significant changes to their business models (Li et al. 2022). For example, in 2000 British Petroleum changed their logo to a green and yellow sunflower and rebranded itself as "BP." They adopted the tagline "Beyond Petroleum" and touted their investments in renewable energies. The advertising, marketing, and public relations agency Ogilvy & Mather created a campaign to promote BP's cleaner fuels. The aim of the ads was to demonstrate BP's efforts to make its fuels more environmentally friendly (MarketingWeek 2001). However their advertising didn't reflect their actual investments. In 1999, BP did spend \$45 million to purchase a solar energy company called Solarex, making BP the largest manufacturer of photovoltaics at that time. At the same time they also spent \$26.5 *billion* to buy ARCO to expand their oil extraction portfolio. In recent years BP investments have improved, but not near the levels necessary. Since the Paris Agreement was adopted at the end of 2015, BP has spent about \$3.2 billion on clean energy, and \$84 billion on oil and gas exploration and development (Ambrose 2022).

The framing that the fossil fuel industry uses for their messaging about their products and climate change is also highly developed. Supran & Oreskes (2021a) outline the "fossil fuel savior" frame, wherein:

- Everything about anthropogenic climate change is uncertain: a "risk," as contrasted with a reality.

[19] https://www.theguardian.com/business/2023/jun/07/shells-green-ad-campaign-banned-in-uk-for-being-likely-to-mislead

[20] http://brandalism.ch/activists-target-airline-advertising-over-its-climate-impact/.

[21] https://www.washingtonpost.com/business/2021/07/30/greentrolling-big-oil-greenwashing/

- Fossil fuel companies are passive suppliers responding to consumer energy demand.
- Continued fossil fuel dominance is inevitable, given the insufficiency of low-carbon technologies; and reasonable and responsible, because fossil fuels lead to profound, explicit benefits and only ambiguous, uncertain climate "risk(s)."
- Customers are to blame for demanding fossil fuels, whose "risk(s)" were common knowledge. Customers knowingly chose to value the benefits of fossil fuels above their risks.

Communications from the fossil fuel industry—e.g., from corporate executives and public relations specialists—often use phraseology such that their company is working to "meet the energy demands" of the public. This language transfers the blame to the consumer; i.e., it's not the company's fault that people use it—after all, the public is demanding it! They don't make people buy it, so they're not responsible, legally or morally, for the consequences of its use. This phrasing isn't mere semantics. As of 2022, at least twenty cities in the U.S. are suing oil and gas producers such as BP, Shell, Chevron, and ConocoPhillips. Such lawsuits are seen as essential tools for compelling corporations and governments to undertake more ambitious goals to address climate change (UNEP 2023). One of the arguments used in these suits is that these companies knowingly contributed to climate change by extracting and selling fossil fuels, obscuring the science of climate change, and fighting policies aimed at mitigating climate change (Burger & Wentz 2018). Recently, many of the cases center on the argument that companies failed to disclose what they knew about the role of their products in causing climate change. The "customer is to blame" framing helps to counter that argument. However, this framing overlooks that in most parts of the world people don't have viable options for other sources of energy; and that the fossil fuel industry has lobbied heavily for policies that make people reliant on their products (Basseches et al. 2022). There is also evidence that the fossil fuel industry increases its advertising in response to greater scrutiny from the public or government.[22]

One of the most notable reframings is the invention of "carbon calculators," which BP created as part of an advertising campaign in 2004. While they allow a person to estimate the emissions associated with their activities, they also reframe the problem as the responsibility of the individual to reduce their emissions. Not only do carbon calculators shift the blame, they don't work. Büchs et al. (2018) found that using a carbon calculator didn't reduce personal energy use in the long term.

There are other examples of how framing and wording are used by politicians and industry. In recent decades there has been a push to promote "natural gas" as a cleaner energy source and a "bridge" to a renewable energy future. Industry often describes natural gas/methane as "green" and/or "renewable," neither of which are true. The American public perceives "natural gas" much more favorably (76%

[22] https://heated.world/p/misleading-climate-ads-from-big-oil

favorable) than other fossil fuels like oil (51%) or coal (39%). People also respond more favorably to the term "natural gas" than "methane" because it contains the word "natural" (Lacroix et al. 2020). For this reason it is recommended that scientists refer to it as methane.

The coal industry did much the same thing with a $60 million advertising campaign that ran with the oxymoronic term "clean coal," as coal is among the most polluting fossil fuels for the Earth (Conniff 2008). One entity behind the campaign is the American Coalition for Clean Coal Electricity. So effective was this strategy that the 2008 Obama campaign made a point of backing "clean coal" projects, a promise President Obama continued with $3.4 billion in support for coal projects once he was elected.[23] Note that the phrase "clean coal" originally referred to the use of technologies that mitigate pollutants such as sulfur oxides (SO_x), nitrogen oxides (NO_x), particulates, and heavy metals (e.g., mercury and lead) but do not necessarily reduce CO_2 emissions.

Another form of strategic communications is the sponsorship of local organizations and events (e.g., Figure 13.3). Such sponsorship can increase brand visibility as well as improve public perception that the corporation is a positive member of the

Figure 13.3. Petroleum producer ConocoPhillips often sponsors nordic ski races in Alaska, including the U.S. Cross Country Championships and the Alaska High School State Championships. Winter sports such as skiing and snowmobiling are being negatively impacted by climate change (Wobus et al. 2017). Photo: T. A. Rector.

[23] https://www.scientificamerican.com/article/obama-and-clean-coal/ https://www.scientificamerican.com/article/obama-and-clean-coal/

community. Whether intended or not, such sponsorship can also stifle criticism. Organizations are less likely to speak out against a major funder, even if the corporation's product is harmful to the activities of the organization.

13.6 How to Talk with the News Media

Scientists should also communicate strategically. We are often called upon by news media to talk about scientific research or issues that the public finds intriguing. By and large, Americans express interest in news about science at higher levels than other topics, such as business and finance, sports, and entertainment (Saks & Tyson 2022). So there is an opportunity for scientists to seize on this interest to talk about climate change. Finding an opportunity to talk with local news organizations can be particularly fruitful, as they are the most trusted (Liedke & Gottfried 2022). It is also an underutilized venue for scientists (Howarth & Anderson 2019).

Before engaging with a journalist, either reactively (e.g., responding to an interview request) or proactively, it's important to know your messaging goals—which may differ from the journalist's perceived story. For example, journalists are often looking for a narrative about "discovery." Jamieson (2018) explains that journalists use the quest narrative because it is so familiar and enduring to audiences. Storytellers from *Gilgamesh* to *The Lord of the Rings* to reporters describing the sequencing of the human genome have used the structure. From the journalist's standpoint as well as the audience's, radical changes in science make for a more compelling story than that of incremental change. In fact, audiences not only crave quest narratives in science, but they put more trust in science after exposure to such structures in journalism (Ophir & Jamieson 2021). Conversely, when audiences were exposed to stories highlighting problems in the scientific process, it reduced their trust in scientists.

What this means is that scientists need to be strategic in our interactions with the media, knowing that public trust in science can be damaged by the way media narratives take shape. We need to help journalists understand that science is not just a body of knowledge, but also the process in which new knowledge is created. We need to be clear about what is firmly established (e.g., that anthropogenic emissions of CO_2 are warming the Earth) and what is less certain (e.g., the temperature anomaly at which a particular tipping point will be triggered). When discussing their results scientists tend to focus on uncertainty whereas the media want a clearer narrative, one that has a beginning and an end.

Keep these in mind when you engage with the media, whether it be on climate change or your own science:

13.6.1 Before the interview

- Develop relationships with journalists before you want to communicate a story. If your local newspaper has a science reporter, have an informal conversation about your research interests, climate change, and where these two might overlap. The journalist won't necessarily use this information for a story in the immediate future, but you can establish yourself to be a worthwhile source. You can also help the reporter to track down other sources.

- If contacted by an unknown reporter for an interview do an internet search on them and the news source. If you feel that they are unreliable, think carefully about whether or not to respond. Your words may be taken out of context—accidentally or intentionally.
- If a journalist or news source appears to be legitimate, respond to their request as quickly as possible. Journalists are often on a tight deadline and will have to move on to another source if they don't hear from you promptly. If you are unavailable suggest a reliable colleague.
- Respond to requests quickly, but don't allow yourself to be interviewed right away. Schedule an interview so that you have time to prepare.
- Ask journalists about the premise of the story they are working on. If you disagree with the premise or the line of questioning, you can say so; e.g., you can say, "That is a small part of the issue, but the real question is…" or "I'd like to direct your attention to this bigger issue…"
- Know your audience. You may decide to answer questions differently for a general news source than a specialty news outlet (e.g., *Wired* magazine). If speaking to local news, talk about the regional implications.
- Feel free to ask for the questions in advance so you can be prepared. But don't be surprised if they decline.

13.6.2 During the interview

- Prepare talking points ahead of time and have them in front of you during the interview. Every effective communicator, from CEOs to presidential candidates, recognizes that you can impart at most three or four key messages. An interview may only be able to capture one or two of them. So be sure that you are able to clearly articulate these concepts. Keep them brief and understandable. Whenever possible direct your answers toward these talking points, even at the risk of repeating yourself.
- Present your information in the opposite order you would for a science talk; i.e., start with the "bottom line" and tell people why they should care before presenting details (Somerville & Hassol 2011).
- Keep your answers short and focused. It's easy to ramble on and to drift into unnecessary detail. Try to answer each question in two to three sentences, or talk for no more than 15–30 seconds. This is particularly important for TV and radio where your answers will likely be edited for brevity.
- Avoid jargon, acronyms, or words that might be misinterpreted. As discussed in Chapter 12, Somerville & Hassol (2011) give examples of words and phrases that have different meanings to scientists than the public.
- Use anecdotes, analogies, and visual examples. Personal stories are particularly memorable.
- Scientists are best served to say, "I don't know" when necessary. It's better to say, "That's beyond the scope of my area of expertise," than to say something

that is inaccurate. If possible, direct them to another source that could answer their question.

- Don't ask to go "off the record" or to be a "confidential source" unless you feel comfortable with the implications of doing so. Note that anything you say to a journalist is "on the record" and can be used as part of a story, even if you think the interview hasn't started or has concluded.
- Use the "truth sandwich" model when asked questions about the claims of skeptics or deniers. As described by George Lakoff:

  A. Start with the truth. The first frame gets the advantage.
  B. Indicate the lie. Avoid amplifying the specific language if possible.
  C. Return to the truth. Always repeat truths more than lies.
- Try to keep the conversation lively and fun. It will make for a more compelling interview and story.

13.6.3 After the interview

- You can ask but don't expect to be able to proofread a story before publication. Not all journalists will grant you this request, often because they have a short timeline. Unless asked to do so, do not edit the journalist's work. It's fine to offer suggestions or clarifications. Most journalists want to be accurate and truthful, and will rewrite content to make it so.
- Follow up. If you had a good experience with a reporter, further develop that relationship. A quick email to acknowledge a good experience or a tip for another story can pave the way for future fruitful interactions.

Additional resources and advice on how to interact with the news media are available through the AAAS and the AGU.

13.7 Climate Denialism

While the majority of Americans—and people worldwide—are concerned about climate change, there are many who remain dubious. As discussed in Chapter 12, the Six Americas survey (Maibach et al. 2009) identifies them as being in the Doubtful and Dismissive categories—and they make up about 20% of the U.S. population. While the Doubtful are skeptical about climate change, the Dismissive are largely convinced that it isn't happening, isn't a major problem, or that solutions are worse than the problem itself.

Climate denialism is different from skepticism in that no amount of evidence can change the denialist mindset. Deniers are often perceived as being "anti-science" but are not necessarily so. They still claim to value evidence (i.e., "Do your own research"), but discount information that doesn't agree with their beliefs. Unfortunately they tend to fall victim to "motivated reasoning," seeking out and remembering knowledge that supports their worldview while dismissing information that is contradictory (Corner et al. 2014). Many denialists are products of the

"Dunning–Kruger effect" (Kruger & Dunning 1999), where they overestimate their abilities to understand climate science.

As discussed in Chapter 12, Kübler-Ross (1969) described the "five stages of grief"—denial, anger, bargaining, depression, acceptance—that one goes through while we mourn. Note that the first stage is denial—in fact, many forms of climate denialism are an emotional response to protect oneself from feeling pain (Moser 2007). It can be a way of processing the fear, guilt, and helplessness many people feel when they think about climate change. It can also be a way to avoid acknowledging that, if the problem is real, then we need to make life changes that they don't want to make. This is why denial is a common response to austerity messaging.

Conspiracy thinking can also play a role. A conspiracy can be defined as a plot carried out in secret, usually by a powerful group with a sinister goal. Uscinski & Olivella (2017) find that conspiratorial thinking can lead to climate denialism. This denialism can take the form of belief in theories claiming that climate change is a hoax, communist plot, or precursor to totalitarianism. Unfortunately evidence to the contrary of these beliefs is also likely to be rejected as being part of the conspiracy. Climate change conspiracy theorists are, therefore, less likely to participate politically or take actions that could alleviate their carbon footprint (Uscinski et al. 2017).

13.7.1 Tactics

Climate skeptic campaigns are organized and funded by groups that are threatened by solutions to climate change. It is often couched in sensationalist terms so as to increase the likeliness it will be shared and spread. Such efforts have been undertaken for decades; e.g., the American Petroleum Institute (API) was publicly downplaying the threat of climate change as early as 1980 (Franta 2021). The now-defunct Greening Earth Society—created in 1998 by the Western Fuels Association–argued that CO_2 is "plant food" and fossil fuels were "one of the Creator's greatest gifts to humankind."[24] Denialism can come from many sources; e.g., America's four biggest rail companies—which move the majority of the country's coal—have spent millions to deny climate science and block climate policy.[25] All of this was done while many of these organizations privately acknowledged that climate change was real and a problem. Supran & Oreskes (2017, 2020) assessed whether ExxonMobil has in the past misled the general public about climate change. They found that while 80% of their internal documents acknowledged that climate change is real and human-caused, only 12% of their advertorials did so, with 81% instead expressing doubt. Supran & Oreskes (2021b) gives examples of the many different kinds of misleading advertising fossil fuel corporations and groups have run over the years.

This is not the first time that the fossil fuel industry has used denial tactics. Starting in the 1920s "leaded gasoline" was created by adding the compound

---

[24] https://www.sfgate.com/business/article/Energy-Debate-Heats-Up-2709569.php

[25] https://www.theatlantic.com/science/archive/2019/12/freight-railroads-funded-climate-denial-decades/603559/

tetraethyllead (TEL). It was added to counteract "knocking," which is caused by the uneven combustion of fuel. Despite health concerns already known at that time, the use of TEL soon became widespread. Several decades later, Patterson (1956) used U-Pb radiometric dating to obtain the first accurate measurement of the age of meteorites and the Earth. In the process he also discovered widespread lead contamination of the environment from the use of leaded fuel, later determining that humans had about 100 times the natural level of lead in their bodies (Patterson 1965). The work of Patterson and other scientists was challenged by the fossil fuel industry, which relied on testimony from industry-funded scientists—such as Robert Kehoe—who argued that TEL was safe. Their denial of the science behind the health risks delayed the banning of leaded fuels, which eventually started to be phased out in the 1970s[26] and finally discontinued in the U.S.—at least for automotive use–in 1986 (Kovarik 2005).[27] The history of leaded gasoline demonstrates how difficult it is for science to resolve an issue with strong economic and social implications (Nriagu 1990).

It is also important to note that these denial strategies are not unique to fossil fuels. Perhaps the most obvious comparison is the tobacco industry's reaction to the medical science linking smoking to lung cancer (e.g., Proctor 2012). Beginning in the 1950s, the tobacco industry used sophisticated public relations strategies to undermine and distort the emerging science (Figure 13.4). Their strategy of producing scientific uncertainty undercut public health efforts and regulatory interventions

Figure 13.4. Examples of cigarette advertising that present trusted experts (i.e., physicians) and supposed scientific evidence that smoking is not dangerous. The messaging ran counter to the established science.

[26] Again, the greatest decade for music. Not the greatest decade for anything else.
[27] https://doi.org/10.1179/oeh.2005.11.4.384

designed to reduce the harms of smoking (Brandt 2012). Oreskes & Conway (2011) detail how the same strategies, and in many cases the same scientists, that were used to cast doubt on the scientific consensus about the dangers of tobacco were also used to obfuscate the certainty of the threat of climate change. These strategies have also been used on many other issues, including asbestos (Egilman, Bird, & Lee 2014), industrial pollution (Markowitz & Rosner 2013), and opioids (Michaels 2020). Likewise, economists have been employed to raise doubts about the benefits of proposed climate policies (Franta 2021). The goal in all of these cases has been to generate confusion about the certainty of the science, the seriousness of the problem, or the effectiveness of solutions, so as to prevent or delay policy changes that could damage the profitability of a corporation or industry.

Dunlap & Brulle (2020) give a comprehensive list of sources and amplifiers of climate denial. This includes groups that are resistant to climate change on economic and ideological grounds, such as corporations and trade associations that not only produce fossil fuels but also rely upon them (e.g., auto manufacturers), as well as conservative media, think tanks, and advocates on social media. This also includes politicians and contrarian scientists, many of whom have received funding from corporations and trade associations (Hoggan & Littlemore 2009). Farrell (2016) finds that language similarities between these groups reveal a well-coordinated network with ties to corporate benefactors.

Much of the language used by denialists originated with political strategist Frank Luntz (2002), who outlined several framing techniques that have become core elements of strategy to stall action on climate change. Luntz recognized the importance of sowing doubt about the scientific consensus, making it possible to argue that deniers are committed to "sound science" and delay policy until "all the facts are in hand." He also recommended, as a fallback position, arguing that "technology and innovation" can solve the problem without requiring regulation.

It's important to note that denialism can take many forms. Nuccitelli (2013) describes five stages of climate denial:

1. It's not happening.
2. It's happening, but it's natural (i.e., not anthropogenic).
3. It's not that bad, and may be beneficial (e.g., CO_2 is good for plants)
4. It's too difficult or expensive to solve.
5. It's hopeless (i.e., "climate doomism")

On the surface these arguments seem unrelated or contradictory; e.g., how can a problem be both hopeless *and* nonexistent? But the common thread is that they all argue against taking action to address climate change. Coan et al. (2021) outline a taxonomy of climate denial claims, with lists of sub-claims that are common talking points. It is similar to Nuccitelli's stages of denial but also includes claims that climate science is unreliable, scientists are overreacting (i.e., "alarmist"), and/or conspiring to hide the truth.

Cook (2020) describes rhetorical strategies used by climate deniers with the "FLICC" acronym: Fake experts, Logical fallacies, Impossible expectations, Cherry picking, and Conspiracy theories. Fake experts can include individuals who lack expertise, whose opinions are often amplified in traditional and social media. Logical fallacies include *ad hominem* attacks, "straw man" arguments, and misrepresentation. Impossible expectation arguments are set up to be unattainable, thereby stalling any action. Cherry picking involves selectively choosing information or data that leads to a different conclusion. Finally, as mentioned before, conspiracy theories are a convenient way to discredit climate scientists, because any source of evidence contrary to their beliefs can be discounted as part of the conspiracy.

An example of cherry picking is related to polar bears—a symbolic species of climate change. As discussed in Chapter 4, the global population of about 26,000 polar bears in the Arctic is divided into 19 subpopulations, each of which is being affected differently by climate change. While most populations are being negatively impacted, some have been reported as stable or productive in the short term (Regehr et al. 2016). Evidence of stability among some polar bear populations is cited by climate deniers as evidence that polar bears as a whole are not at risk—and therefore climate change either isn't happening or isn't a threat.

Following the advice of Luntz, one of the most common climate denial tactics is to attack the scientific consensus (Elsasser & Dunlop 2013) even though several studies (e.g., Cook et al. 2013 and Lynas, Houlton, & Perry 2021) find that greater than 99% of peer-reviewed scientific articles endorse the view that climate change is real and the result of anthropogenic emissions. As discussed in Chapter 12, understanding that there is an overwhelming scientific consensus is particularly important. It has been described as a "gateway belief" in that people are unlikely to accept other elements of climate change until they first understand the consensus, after which time they are more likely to believe climate change is happening, be worried about it, and support action to address it (van der Linden et al. 2019). It is likely for this reason that climate deniers focus on attacking the consensus. As part of this strategy, climate deniers also engage in misrepresentation of science results (e.g., Zeppetello 2020) in an attempt to give the impression that scientific research does not agree with the consensus.

A notable example of fake experts was the Global Warming Petition Project, which claimed to have signatures from more than 30,000 American scientists who don't think climate change is real or a problem. However, analysis of the signatures shows that many of them were fake. And less than 1% of the real signatories have research experience in climate science. Of these, many said they would not sign it now and/or were misled because the petition was formatted to look like it came from the National Academy of Sciences (Anderson 2011).

Often multiple techniques are at play. For example, there was an unusually strong "El Niño" year in 1998. For a few years afterward deniers cherry picked data to argue that global warming had stopped, choosing to ignore the larger trend. They then claimed a conspiracy among climate scientists, who had allegedly switched from using the phrase "global warming" to "climate change" because they supposedly knew the warming had stopped. However, "global warming" is largely used to refer to one element of climate change—and has been used as such since well before 1998. There is evidence that these disinformation campaigns have affected how scientists themselves present their work (Lewandowsky et al. 2015).

Many of the strategies used by climate deniers have taken on their own names, e.g.:

- "Paltering" is the use of truthful statements out of context to create a false impression. Paltering differs from lying by omission—the passive omission of relevant information—and lying by commission—the active use of false statements (Rogers et al. 2017). Here is an example of scientists reacting to the paltering in advertising by fossil fuel companies.
- "Climate delay" talking points accept the existence of climate change, but justify inaction or inadequate efforts. They focus attention on the negative social effects of climate policies and raise doubt that mitigation is possible (Lamb et al. 2020).
- "Woke-washing" is opposition to climate change solutions by framing them as environmentally irresponsible. For example, opposition to renewable energy sources such as wind and solar are discussed in terms of their environmental impact while neglecting to fairly discuss the consequences of fossil fuels; e.g., resistance to off-shore wind turbines by conservative politicians because of the alleged threats to whales. Likewise, climate doomer messaging, i.e., that the problem is insurmountable, may be presented as if it is coming from environmentally concerned individuals or groups, with the purpose of deflating hope and stifling action.
- "Whataboutism" is the tactic of blaming other groups. Those advancing this discourse often deploy statistics demonstrating their own small contribution to global emissions, or they point to large emitters such as China (Lamb et al. 2020). Related is the "free rider" argument, which posits that there will be no significant impact if our group changes its behavior but others do not.
- "Astroturfing" is the practice by corporations and advocacy groups of creating fake "grassroots" social movements. Astroturf groups also operate to oppose climate change legislation, but do so in a way that appears to be coming from everyday citizens (Cho et al. 2011). For example, in 2014 the Western States Petroleum Association was discovered to have created and funded over a dozen organizations that opposed state-level climate legislation in California, Oregon, and Washington.[28]

A common talking point for climate deniers is an event that has become known as "climategate."[29] In 2009 a computer file server for the Climatic Research Unit at the University of East Anglia was hacked by an unknown attacker, who illegally downloaded thousands of emails and files. These files were distributed to climate denier groups, who took segments of some of the emails out of context to argue that the climate scientists involved were deliberately misrepresenting their research. For several months climategate became a popular talking point in denier blogs and

[28] https://www.ucsusa.org/resources/how-fossil-fuel-lobbyists-used-astroturf-front-groups-confuse-public

[29] It has become tradition in the U.S. to append the word "gate" to the end of conspiracies. It originates from the "Watergate" scandal in 1972-1974 involving U.S. President Nixon, named after the Watergate Office Building in which it occurred.

conservative media, arguing that climate change was a conspiracy. Several investigations (e.g., by the University of East Anglia, Penn State University, the UK government, and the National Science Foundation) found no evidence of fraud or misconduct. However the incident did have a significant effect on public beliefs in global warming and trust in scientists. The loss of trust, however, was primarily among individuals with a politically conservative ideology (Leiserowitz et al. 2013).

Another strategy is to make accusations of hypocrisy to discredit the messenger (Gunster et al. 2018). For example, "If Al Gore or climate scientists really believe climate change is real why do they fly so much?" For example, this article from the French news source *Le Point* claims that climate scientists fly more than other researchers.

Climate deniers also engage in crisis communications, using events as opportunities to sow discord about climate solutions—even if the events themselves were caused or amplified by climate change. For example, in February 2021 a polar vortex caused a rapid drop in temperature across the central U.S. One of the results was severe disruptions in electricity and natural gas service. Texas Governor Greg Abbott and others blamed the problem on wind and solar, but in reality it was primarily the result of the failure of natural gas and coal power plants. Wind turbines did fail during the storm, but it was because their winter preparedness was inadequate (King et al. 2021). This is not the fault of the technology. Wind turbines that operate in cold climates have de-icing equipment that the Texas turbines did not.

Over the years climate denialism has evolved in response to changes in public attitudes. An analysis of contrarian claims about climate change, from blogs and think tanks, over the last twenty years shows a transition from arguing that climate change is either not happening or the science is unreliable to arguing that proposed solutions won't work (Coan et al. 2021). For example, common misconceptions spread about wind turbines:

- Construction requires more energy than generated during their lifetime;
- Disposal of turbine blades produces too much waste;
- Too dangerous for birds;
- Only affordable because of government subsidies;
- Unreliable and must be backed up by other energy sources;
- Unattractive;
- Too noisy;
- Cause cancer.

Some of these are based upon partial truths and others are complete nonsense. Some are a matter of opinion and others are demonstrably false. As discussed above, social media enables the proliferation of fake news, so belief in the outrageous claims is more common than you might think.

As discussed in Chapter 12, austerity framing is an ineffective way to motivate people to address climate change. Denialists are aware of this and often use austerity messaging as a way to scare people, falsely suggesting that they will have to give up popular items such as meat, pizza, and football.

13.7.2 Visuals

As discussed in Chapter 12, visuals are a powerful way to convey information—and they are often used as an effective tool by climate deniers. Former Oklahoma Senator James Inhofe understood this when he brought a snowball to the U.S. Senate after a massive snowstorm in Washington, D.C. as a prop to argue that climate change wasn't real. His argument was structured on a simple schema: it must be cold for it to snow, so more snow must mean it is colder. In other words, how can global warming be happening if there's so much snow outside?[30]

Climate change visuals are often spread as "memes"—images or videos that are copied and shared via social media, often with slight variations. Memes are typically humorous, but not always so. Climate change themed memes from conservative sources often discuss the hypocrisy of climate advocates, poke fun at prominent activists (e.g., Greta Thunberg and Alexandria Ocasio-Cortez), and liken the urgency of climate change to fear-mongering. In contrast, liberals are more likely to discuss racial and economic disparities related to the climate crisis, and the need for America to lead mitigation efforts (Wong et al. 2022).

Because of their success, wind turbines have in recent years become popular targets for climate denialism. Their imposing sizes make them visually dramatic, which is often exploited in memes. These images are often altered to make them appear faulty or dangerous (Figure 13.5).

Because of their size, the disposal of end-of-life wind turbine blades provides a dramatic visual that draws into question the environmental value of wind power. However, this needs to be compared to the waste produced by other energy sources. For example, the mass of solid waste from a megawatt-hour of coal electricity is 200

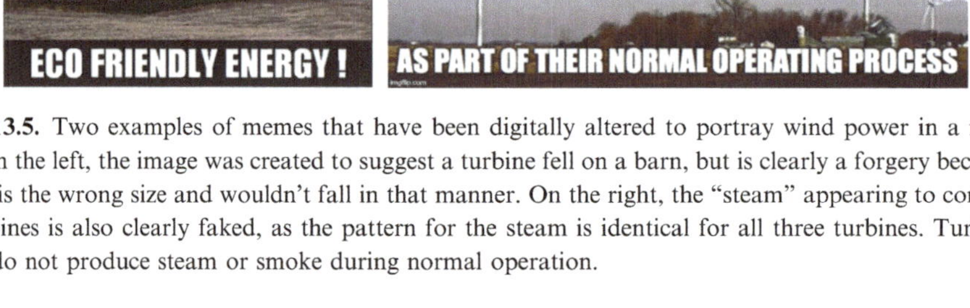

Figure 13.5. Two examples of memes that have been digitally altered to portray wind power in a negative light. On the left, the image was created to suggest a turbine fell on a barn, but is clearly a forgery because the turbine is the wrong size and wouldn't fall in that manner. On the right, the "steam" appearing to come from the turbines is also clearly faked, as the pattern for the steam is identical for all three turbines. Turbines of course do not produce steam or smoke during normal operation.

[30] As discussed in a previous chapter, warmer air can hold more water vapor—meaning that under the right conditions climate change is causing winter storms with more snow.

times the amount from a megawatt-hour of wind electricity, and the solid waste produced from wind or coal pales in comparison to one's daily trash output. To put things in perspective, If an individual U.S. resident gets all of their household electricity from wind energy, over 20 years their share of non-recyclable wind turbine blade waste will be 15 kg. That same mass of solid waste is produced by one person's share of a coal-fired power plant in 40 days, and it is just a week of municipal waste (Barnes 2020).

13.7.3 Countering Denialism

In Chapter 12 we discuss ways to talk with groups and individuals about climate change. In both scenarios you can expect questions related to common misconceptions and recent skeptic talking points. The Skeptical Science website offers useful tips on how to address both. This is a worthwhile resource to look at before every talk you give about climate change. It is possible to pre-emptively protect ("inoculate") against misinformation by either explaining how the misinformation is wrong (van der Linden et al. 2017) or the flawed argumentation techniques used (Cook et al. 2017) before they hear it elsewhere. Such strategies proved to be effective against the Global Warming Petition Project. Lewandowsky et al. (2020) have written a debunking handbook that—among other strategies—recommends the use of "truth sandwiches" to counter misinformation.

The most effective climate denial talking points are often based upon partial truths. When countering such misinformation it helps to acknowledge the truth and then explain the larger picture. For example, it is true that birds are killed by wind turbines. However fossil-fuel power stations kill about 30 times as many birds than wind turbines per GWh (Sovacool 2012). The U.S. Fish and Wildlife Service[31] estimates that about 250,000 birds die annually in collisions with wind turbines, but in comparison about 600 million birds die from collisions with buildings and about 2.4 billion are killed by cats. Also note that there are efforts underway to reduce avian deaths further; e.g., painting one blade of a wind turbine rotor black resulted in 70% fewer bird collisions (May et al. 2020).

Often denialists take statements by scientists and other experts out of context in a way that misrepresents their views. It can be particularly effective to point out examples of when the "quoted" person explicitly rebuts the denialist claim. For example, in this blog post Thomas Homer-Dixon addresses a meme about wind turbines that attributes a quote to him. He explains how the meme is fraudulent: the quote is not his, but based upon text from a book he co-authored. However, the quote on the meme selectively omits text in a way that clearly changes the meaning, demonstrating that it is misleading.

If misinformation is brought up, think about the context in which it arises—what is your relationship to the person? Are you alone or in a group? Do you think the misconception was brought up in good faith? Scientists are often reluctant to engage with deniers, worrying they'll do more harm than good. Known as the "backfire

[31] https://www.fws.gov/library/collections/threats-birds

effect," there is evidence that presenting information that refutes a misconception can more deeply entrench the belief. It's possible that repetition of the myth while debunking it strengthens the misconception (Pluviano et al. 2017). However, the backfire effect may only apply in certain circumstances; e.g., when the information challenges a fundamental worldview. Leaving denial talking points unanswered can have negative consequences. In fact, rebutting science denialism in public discussions doesn't necessarily backfire, but *not* responding to science deniers can (Schmid & Betsch 2019). If you engage deniers, try to do it on your terms, not theirs—as their intent is often to sidetrack the conversation. For example, if it comes up as part of a public talk you may say that you don't have the time to discuss it here but would be happy to talk with them later privately. Even if you think someone is not discussing in good faith, do not call them a denier. True or not, it is an *ad hominem* attack. In response you're likely to be labeled an "alarmist" and the conversation is effectively over—with them more deeply entrenched.

When explaining climate denialism it can help to show visual examples that have clear intentions to deceive. For example, a common denier talking point is based upon an erroneous claim that back in the 1970s[32] scientists believed that we were entering a new ice age. The purpose of this argument is to claim that scientists changed their mind and are therefore untrustworthy. In other words, "If they were wrong then, why should we believe they're right now?" Like many denier talking points, this one is based upon a kernel of truth taken out of context. As discussed in Chapter 8, scientists already recognized back then that aerosols from pollution (i.e., smog) could *in some scenarios* lead to a net negative radiative forcing. However, the far majority of research papers of that time made it clear that the primary concern was about global warming (Peterson et al. 2008). The meme shown in Figure 13.6 is

Figure 13.6. The meme on the left apparently shows a comparison of two covers of Time magazine. The cover on the left is allegedly from 1977, but it is a modification of the actual cover of Time from April 9th, 2007 (shown on the right). Not only is the title changed but so are the details along the top—the Iraq war was replaced with the cold war, and The Sopranos television show was replaced with M.A.S.H.

[32] *The* greatest decade for music, not the greatest decade for anything else.

based upon this talking point, showing two covers of Time magazine that appear to argue contradictory points. However the cover of Time magazine on the left is altered. Comparing it to the original is a clear way to demonstrate its intent to deceive.

13.8 Science Media Literacy

So how can we help protect our students against misinformation and disinformation? A goal for many "Astro 101" educators is to develop a student's interest and ability to consume media about astronomy after the course is over. As discussed above, traditional and social media aren't always reliable. So it is important to develop a student's "media literacy," defined as the ability to critically analyze the media they consume and share. A major area of media literacy is the ability to analyze news stories to determine their accuracy or credibility. This is important in astronomy because some topics—such as quasars, supermassive black holes, and antimatter—are so amazing they can sound like science fiction. Because our field is a popular topic, it is often a subject of "click bait," defined as content whose main purpose is to attract attention and encourage visitors to click on a link so as to generate advertising revenue. Unfortunately click bait articles often contain false and misleading information. While less of an issue for astronomy-related topics, sponsored content is especially common for climate change—and students are often confused by it.

A common assumption is that young people naturally pick up the skills necessary to use today's technology in an informed manner. But the evidence suggests otherwise. For example, a national survey of U.S. high school students by Breakstone et al. (2021) revealed major weaknesses in students' ability to evaluate the credibility of online sources. Asked to investigate a site claiming to "disseminate factual reports" on climate science, 96% failed to identify the organization's ties to the fossil fuel industry. Students often base their evaluations of websites on surface-level features, such as the visual appearance of a website or the relevance of the information provided (McGrew et al. 2018). In short, students are not so much digital natives as digital novices, lacking the basic navigational tools that would ensure they are not deceived (Osborne et al. 2022).

A key element of media literacy is "lateral reading," defined as using other sources to confirm the trustworthiness of an article rather than trying to assess the accuracy of the content itself. Students who learn how to use lateral reading grow significantly in their ability to judge the credibility of digital content (Wineburg et al. 2015). We'd like to think that the science content knowledge they learn from our classes would be sufficient to assess an article's veracity, but that's not a realistic goal. In fact, the Dunning–Kruger effect might lead a student to overestimate the depth of their understanding, leading to excessive confidence in their ability to assess an article's accuracy. Instead, we need to develop their science media literacy skills. This includes:

- How to determine the reliability of a media source. Who's behind the information? Who paid for it?

- How to confirm the news report. What do other sources say?
- How to determine what kind of media it is. Is it a news report? Sponsored content? Is it meant to entertain, persuade, or inform?

Can I Trust It? is an activity developed by the authors to teach science media literacy as part of an introductory astronomy class. It is intended to be done near the start of the semester, with additional articles provided regularly by the instructor for analysis using the same methodologies. With practice it can become an ingrained habit.

13.9 Conclusions

We are swimming in so much information. News alerts, email newsletters, advertisements, social media posts, podcasts, and videos are everywhere. And when it comes to climate change, much of that information is incorrect—accidentally or intentionally so. It's no wonder people are confused. If we want to communicate effectively about climate change ourselves, we have to understand how these different media sources are influencing the conversation.

This chapter outlines a modern media landscape that is rapidly changing, and, in many cases, struggling with a decline in trust. Traditional gatekeeping functions of the news media are failing; and the onus is on the consumer to figure out which sources of news and information are credible. As scientists and educators, we can help our students develop science media literacy skills so they can successfully navigate traditional and social media. The importance of developing these skills cannot be overstated, as climate change is the defining story of our time.

This chapter also explores the ways political and corporate interests use strategic communications—such as public relations and marketing—to counter climate change awareness and mitigation efforts. It's easy to underestimate the sophistication of these campaigns. While they are often factually incorrect and can appear simplistic, they can be surprisingly effective.

Finally, this chapter describes the shift we as scientists can make to engage more strategically with the public about climate change (Dudo et al. 2021). While Chapter 12 focuses on the interpersonal communication aspect of this engagement, it is incumbent on us to realize our capabilities as media creators. Whenever we are interviewed by the news, write or share a social media post, or publish an opinion piece, we contribute to and create media. Most of us already are creating media without realizing it. If you are actively explaining climate change, hopefully this chapter has helped your understanding of what the media does, how we can help the media do a better job when covering climate change, and how we can improve our own communication through the media.

Scientists and climate change social movements also need to employ strategic communications (Hon 1997; Capozzi & Spector 2016). The need to respond to climate change will be no less transformative than social movements such as women's suffrage, the civil rights movement, or the fight against tobacco. The difference is that we have more knowledge about the kinds of evidence-based strategic communication efforts that are needed to achieve sustained change. All of those causes required communications to build support across society, and responding urgently to climate change will as well. Doing so is our highest challenge.

References

Ambrose, J. 2022, BP has ambitious plans to move beyond fossil fuels—but are they enough? (The Guardian) https://www.theguardian.com/business/2022/feb/10/bp-has-ambitious-plans-to-move-beyond-fossil-fuels-but-are-they-enough

Anderson, E. 2011, Episteme, 8, 144

Arnold, E. 2018, Shorestein Center on Media, Politics and Public Policy, https://shorenstein-center.org/media-disengagement-climate-change/

Balaban, D. C., Szambolics, J., & Chirică, M. 2022, Acta Psychologica, 230, 103731

Barnes, R. 2020, Adding Perspective to the Wind Turbine Waste Debate (Climate Conscious) https://medium.com/climate-conscious/wind-turbine-end-of-life-waste-perspective-a561913dcbd9

Basseches, J. A., et al. 2022, ClCh, 170, 32

Beevor, J., & Alexander, K. 2022, Missed Targets: A Brief History of Aviation Climate Targets (produced by Green Gumption for Possible) https://www.wearepossible.org/our-reports-1/missed-target-a-brief-history-of-aviation-climate-targets

Blankenship, J. C., & Vargo, C. J. 2021, Electronic News, 15, 139

Boykoff, M. T. 2008, ClCh, 86, 1

Boykoff, M. T., & Boykoff, J. M. 2004, GEC, 14, 125

Boykoff, M. T., Nacu-Schmidt, A., Fernández-Reyes, R., Katzung, J., & Pearman, O. 2021, Media and Climate Change Observatory Monthly Summary: Big Oil's day of reckoning on the climate is here - Issue 53, May 2021 https://scholar.colorado.edu/concern/articles/76537251p

Brandt, A. M. 2012, AJPH, 102, 63

Breakstone, J., et al. 2021, ER, 50 505

Brulle, R., & Downie, C. 2022, ClCh, 175, 11

Brulle, R. J., Aronczyk, M., & Carmichael, J. 2020, ClCh, 159, 87

Büchs, M., et al. 2018, EPo, 120, 284

Burger, M., & Wentz, J. 2018, BuAtS, 74, 397

Capozzi, L., & Spector, S. 2016, Public Relations for the Public Good: How PR Has Shaped America's Social Movements (Hampton, NJ: Business Expert Press)

Chinn, S., Hart, P. S., & Soroka, S. 2020, Sci. Commun., 42, 112

Cho, C. H., Martens, M. L., Kim, H., & Rodrigue, M. 2011, J Bus Ethics, 104, 571

Coan, T. G., Boussalis, C., Cook, J., & Nanko, M. O. 2021, NatSR, 11, 22320

Cook, J. 2020, 62 ed. D. C. Holmes, & L. M. Richardson Research Handbook on Communicating Climate Change (Cheltenham: Edward Elgar)

Cook, J., Lewandowsky, S., & Ecker, U. K. H. 2017, PLoSO, 12, e0175799

Cook, J., et al. 2013, ERL, 8, 024024

Conniff, R. 2008, The Myth of Clean Coal. Yale E360 https://e360.yale.edu/features/the_myth_of_clean_coal

Corner, A., Markowitz, E., & Pidgeon, N. 2014, Wiley Interdiscip. Rev. Clim. Change, 5, 411

De Jong, M. D. T., Huluba, G., & Beldad, A. D. 2020, J. Bus. Tech. Commun., 34, 38

Dudo, A., Besley, J. C., & Yuan, S. 2021, Sci. Commun., 43, 33

Dunlap, R. E., & Brulle, R. J. 2020, Research Handbook on Communicating Climate Change (Cheltenham: Edward Elgar Publishing) 49

Egilman, D., Bird, T., & Lee, C. 2014, IJOEH, 20, 115

Elsasser, S. W., & Dunlap, R. E. 2013, Am. Behav. Sci., 57, 754

Farrell, J. 2016, NatCC, 6, 370

Feldman, L., Maibach, E. W., Roser-Renouf, C., & Leiserowitz, A. 2012, IJPP, 17, 3

Feldman, L., Myers, T. A., Hmielowski, J. D., & Leiserowitz, A. 2014, JCom, 64, 590

Feygina, I., et al. 2020, BAMS, 101, E1092

Fisher, A. 2023, Study: Fox News dominated cable news coverage of Willow Project announcement (Media Matters) https://www.mediamatters.org/fox-news/study-fox-news-dominated-cable-news-coverage-willow-project-announcement

Forman-Katz, N., & Matsa, K. V. 2022, News Platform Fact Sheet (Pew Research Center) https://www.pewresearch.org/journalism/fact-sheet/news-platform-fact-sheet/?tabId=tab-5a0b8b87-38bc-42d6-ba8d-2e666200e534

Franta, B. 2021, EnPo, 30, 663

Friedman, L. 2023, Biden administration approves huge Alaska oil project (The New York Times) https://www.nytimes.com/2023/03/12/climate/biden-willow-arctic-drilling-restrictions.html

Gottfried, J., & Liedke, J. 2021, Partisan divides in media trust widen, driven by a decline among Republicans (Pew Research Center) https://www.pewresearch.org/short-reads/2021/08/30/partisan-divides-in-media-trust-widen-driven-by-a-decline-among-republicans/

Gottfried, J., & Anderson, M. 2014, For some, the satiric 'Colbert Report' is a trusted source of political news (Pew Research Center) https://www.pewresearch.org/short-reads/2014/12/12/for-some-the-satiric-colbert-report-is-a-trusted-source-of-political-news/

Grant, N., & Myers, S. L. 2023, Google promised to defund climate lies, but the ads keep coming (The New York Times) https://www.nytimes.com/2023/05/02/technology/google-youtube-disinformation-climate-change.html

Grotta, G. L., Larkin, E. F., & Carrell, B. J. 1976, Journal Q., 53, 448

Gunster, S., Fleet, D., Paterson, M., & Saurette, P. 2018, FrC, 3,

Gustafson, A., et al. 2019, NatCC, 9, 940

Hardy, J. 2021, Digit. Journal., 9, 865

Hersher, R. 2023, Last month was the hottest June ever recorded on Earth (NPR) https://www.npr.org/2023/07/13/1187530636/last-month-was-the-hottest-june-ever-recorded-on-earth

Hoggan, J., & Littlemore, R. D. 2009, Climate cover-up: the crusade to deny global warming (Vancouver; Berkeley: Greystone Books)

Hon, L. C. 1997, JPRR, 9, 163

Hopke, J. 2022, Nieman Reports, https://niemanreports.org/articles/climate-change-journalism-education/

Howarth, C., & Anderson, A. 2019, Environ. Commun., 13, 713

Jamieson, K. H. 2018, PNAS, 115, 2620

Jamieson, K. H., & Cappella, J. N. 2008, Echo Chamber: Rush Limbaugh and the Conservative Media Establishment (New York: Oxford Univ. Press) https://global.oup.com/academic/product/echo-chamber-9780195398601

Jang, J. 2023, Soc. Sci. Comput. Rev., 41, 201

Kahlor, L., & Rosenthal, S. 2009, Sci. Commun., 30, 380

Kang, H., Bae, K., Zhang, S., & Sundar, S. S. 2011, JMCQ, 88, 719

Katzung, J., Nacu-Schmidt, A., Boykoff, M., Pearman, O., & Fernández-Reyes, R. 2023, Media and Climate Change Observatory, 2022, 1

Kenny, J. 2018, CliPo, 18, 581

King, C. W., et al. 2021, The Timeline and Events of the February 2021 Texas Electric Grid Blackouts: A report by a committee of faculty and staff at The University of Texas at Austin, https://energy.utexas.edu/sites/default/files/UTAustin%20%282021%29%20EventsFebruary 2021TexasBlackout%2020210714.pdf
Koch, T., Obermaier, M., & Riesmeyer, C. 2020, Journalism, 21, 1573
Kovarik, W. 2005, IJOEH, 11, 384
Kruger, J., & Dunning, D. 1999, J Pers. Soc. Psychol., 77, 1121
Kübler-Ross, E. 1969, On Death and Dying (New York: Macmillan)
Lacroix, K., Goldberg, M. H., Gustafson, A., Rosenthal, S. A., & Leiserowitz, A. 2020, Should it be called "natural gas" or "methane"? (New Haven, CT: Yale Program on Climate Change Communication) https://climatecommunication.yale.edu/publications/should-it-be-called-natural-gas-or-methane/
Lamb, W. F., et al. 2020, Glob. Sustain., 3, e17
Lasswell, H. D. 1948, ed. B. Lymon The Communication of Ideas (New York: Institute for Religious and Social Studies) 37
Leber, R. 2020, The gas industry is paying Instagram influencers to gush over gas stoves (Mother Jones) https://www.motherjones.com/environment/2020/06/gas-industry-influencers-stoves/
Legum, J. 2022, Sponsoring misinformation. Popular Information., https://popular.info/p/sponsoring-misinformation
Leiserowitz, A. A., Maibach, E. W., Roser-Renouf, C., Smith, N., & Dawson, E. 2013, Am. Behav. Sci., 57, 818
Leiserowitz, A., Maibach, E., & Roser-Renouf, C. 2016, Climate Change In the American Mind, https://www.climatechangecommunication.org/wp-content/uploads/2016/03/2009-Americans-Climate-Change-Beliefs-Attitudes-Policy-Preferences-and-Actions.pdf
Lewandowsky, S., Oreskes, N., Risbey, J.S., Newell, B.R., & Smithson, M. 2015, GEC, 33, 1
Lewandowsky, S., Cook, J., & Lombardi, D. 2020, Debunking Handbook 2020
Li, M., Trencher, G., & Asuka, J. 2022, PLoSO, 17, e0263596
Liedke, J., & Gottfried, J. 2022, U.S. adults under 30 now trust information from social media almost as much as from national news outlets (Pew Research Center) https://pewrsr.ch/3DF4dn1
Luntz, F. 2002, The Environment: A Cleaner, Safer, Healthier America (Alexandria: Luntz Research) https://www.sourcewatch.org/images/4/45/LuntzResearch.Memo.pdf
Lynas, M., Houlton, B. Z., & Perry, S. 2021, ERL, 16, 114005
Macdonald, T., Fisher, A., & Cooper, E. 2023, Media Matters for America https://www.mediamatters.org/abc/how-broadcast-tv-networks-covered-climate-change-2022
Maibach, E. 2009, Global Warming's Six Americas 2009: An Audience Segmentation Analysis,
Marketing Week 2001, Ogilvy & Mather creates global campaign https://www.marketingweek.com/ogilvy-mather-creates-global-campaign/
Markowitz, G., & Rosner, D. 2013, Deceit and denial: The deadly politics of industrial pollution (Berkeley, CA: Univ. of California Press) vol 6
Marlow, T., Miller, S., & Roberts, J. T. 2021, CliPo, 21, 765
Matsa, K. E., & Worden, K. 2022, Local Newspapers Fact Sheet. Pew Research Center's Journalism Project https://www.pewresearch.org/journalism/fact-sheet/local-newspapers/
May, R., et al. 2020, Ecol. Evol., 10, 8927
McAllister, L., et al. 2021, ERL, 16, 094008
McGrew, S., Breakstone, J., Ortega, T., Smith, M., & Wineburg, S. 2018, TRSE, 46, 165 DOI: 10.1080/00933104.2017.1416320

McIntyre, K. 2019, Journalism Pract., 13, 16
McQuail, D. 1983, McQuail's mass communication theory (London: Sage)
Michaels, D. 2020, The triumph of doubt: dark money and the science of deception (Oxford: Oxford Univ. Press)
Mitchell, A. 2013, Pew Research Center https://www.pewresearch.org/journalism/2013/03/17/americans-show-signs-of-leaving-a-news-outlet-citing-less-information/
Moser, S. C. 2007, ed. L. Dilling, & S. C. Moser Creating a Climate for Change: Communicating Climate Change and Facilitating Social Change (Cambridge: Cambridge Univ. Press) 64
National Science Foundation 2018, Chapter 7: Science and technology: Public attitudes and understanding https://www.nsf.gov/statistics/2018/nsb20181/assets/404/science-and-technology-public-attitudes-and-understanding.pdf
Newman, T. P., Nisbet, E. C., & Nisbet, M. C. 2018, Public Underst Sci., 27, 985
NOAA 2023, National Centers for Environmental Information (NCEI) U.S. Billion-Dollar Weather and Climate Disasters
Nriagu, J. O. 1990, ScTEn, 92, 13
Nuccitelli, D. 2013, The 5 stages of climate denial are on display ahead of the IPCC report (The Guardian)
Ophir, Y., & Jamieson, K. H. 2021, Public Underst Sci, 30, 1008
Oreskes, N., & Conway, E. M. 2011, Merchants of doubt: How a handful of scientists obscured the truth on issues from tobacco smoke to global warming (New York: Bloomsbury Publishing USA)
Osborne, J., et al. 2022, Science Education in an Age of Misinformation (Stanford, CA: Stanford Univ.) https://sciedandmisinfo.stanford.edu/
Painter, J., & Ashe, T. 2012, ERL, 7, 044005
Patterson, C. C. 1956, GeCoA, 10, 230
Patterson, C. C. 1965, Arch. Environ. Health, 11, 344
Pearman, O., Boykoff, M., Nacu-Schmidt, A., & Katzung, J. 2022, Media and Climate Change Observatory Special Issue 2021: A Review of Media Coverage of Climate Change and Global Warming in 2021 https://doi.org/10.25810/3vaz-2z04
Peterson, T. C., Connolley, W. M., & Fleck, J. 2008, BAMS, 89, 1325
Pierce, H. 2022, Media Ownership and Agenda-Cutting: The Effect of Sinclair Broadcasting on Public Awareness of Climate Change (Bloomington, IN: Indiana Univ.)
Pluviano, S., Watt, C., & Sala, S. D. 2017, PLoSO, 12, e0181640
Proctor, R. N. 2012, Tobacco Control, 21, 87
Prothero, A. 2023, Most Teens Learn About Climate Change From Social Media. Why Schools Should Care (Education Week) https://www.edweek.org/technology/most-teens-learn-about-climate-change-from-social-media-why-schools-should-care/2023/01
Quinn, E., & Young, C. 2015, Who needs lobbyists? See what big business spends to win American minds (Center for Public Integrity) http://publicintegrity.org/politics/who-needs-lobbyists-see-what-big-business-spends-to-win-american-minds/
Regehr, E. V., et al. 2016, Biol. Lett., 12, 20160556
Rempel, A., & Gupta, J. 2021, World Dev., 146, 105608
Robertson, C. T. 2022, How people think and access climate change news https://reutersinstitute.politics.ox.ac.uk/digital-news-report/2022/how-people-access-and-think-about-climate-change-news

Rogers, T., Zeckhauser, R., Gino, F., Norton, M. I., & Schweitzer, M. E. 2017, J Pers. Soc. Psychol., 112, 456

Saks, E., & Tyson, A. 2022, Americans report more engagement with science news than in 2017 (Pew Research Center) https://www.pewresearch.org/short-reads/2022/11/10/americans-report-more-engagement-with-science-news-than-in-2017/

Samuelson, A., & Atkin, E. 2023, Willow is not just an "environmentalist" concern https://heated.world/p/willow-is-not-just-an-environmentalist

Schmid, P., & Betsch, C. 2019, Nat Hum Behav, 3, 931

Somerville, R. C. J., & Hassol, S. J. 2011, PhT, 64, 48

Sovacool, B. K. 2012, J. Integr. Environ. Sci., vol 9, 255 Vermont Law School Research paper No. 04-13 https://ssrn.com/abstract=2198024

Spindle, B. 2022, My case against fossil-fuel advertising (Columbia Journalism Review) https://www.cjr.org/first_person/my-case-against-fossil-fuel-advertising.php

Stecula, D. A., & Merkley, E. 2019, FrC, 4,

Sulzberger, A. G. 2023, Columbia Journalism Review, https://www.cjr.org/special_report/ag-sulzberger-new-york-times-journalisms-essential-value-objectivity-independence.php/

Supran, G., & Oreskes, N. 2017, ERL, 12, 084019

Supran, G., & Oreskes, N. 2020, ERL, 15, 118002

Supran, G., & Oreskes, N. 2021a, OEart, 4, 696

Supran, G., & Oreskes, N. 2021b, The forgotten oil ads that told us climate change was nothing (The Guardian) https://www.theguardian.com/environment/2021/nov/18/the-forgotten-oil-ads-that-told-us-climate-change-was-nothing

Suskind, R. 2004, Faith, certainty and the Presidency of George W. Bush. (The New York Times) https://www.nytimes.com/2004/10/17/magazine/faith-certainty-and-the-presidency-of-george-w-bush.html

Thomas, G. F., & Stephens, K. J. 2015, Int. J. Bus. Commun., 52, 3

Tompkins, A. 2017, Aim for the heart (New York: Sage Publishing) 2

UNEP 2023, Global Climate Litigation Report: 2023 Status Review (United Nations Environment Programme) https://www.unep.org/resources/report/global-climate-litigation-report-2023-status-review

Uscinski, J. E., Douglas, K., & Lewandowsky, S. 2017, Oxford Research Encyclopedia of Climate Science,

Uscinski, J. E., & Olivella, S. 2017, RAP, 4, 2053168017743105

van der Linden, S., Leiserowitz, A., & Maibach, E. 2019, J. Environ. Psychol., 62, 49

van der Linden, S., Leiserowitz, A., Rosenthal, S., & Maibach, E. 2017, Global Chall., 1, 1600008

Vosoughi, S., Roy, D., & Aral, S. 2018, Sci., 359, 1146

Waldman, S. 2021, The Journalist Population (Report for America) https://www.reportforamerica.org/2021/06/28/the-journalist-population/

Westervelt, A. 2023, Big oil firms touted algae as climate solution. Now all have pulled funding (The Guardian) https://www.theguardian.com/environment/2023/mar/17/big-oil-algae-biofuel-funding-cut-exxonmobil

Wineburg, S., Breakstone, J., McGrew, S., Smith, M., & Ortega, T. 2015, Lateral Reading on the Open Internet http://dx.doi.org/10.2139/ssrn.3936112

Wobus, C., et al. 2017, GEC, 45, 1

Wong, E., Holyoak, K., & Priniski, H. 2022, Proceedings of the Annual Meeting of the Cognitive Science Society 44 https://escholarship.org/uc/item/9hk7673j
Wright, C. R. 1986, Mass Communication: A Sociological Perspective (New York: Random House)
Zeppetello, L. V. 2020, Don't @ Me: What Happened When Climate Skeptics Misused My Work (EOS) http://eos.org/opinions/dont-at-me-what-happened-when-climate-skeptics-misused-my-work

Part VI

Engagement and Advocacy

Chapter 14

Community Engagement

Kathryn Williamson

> "The most important thing you can do to fight climate change is to talk about it."
>
> —Katharine Hayhoe

The purpose of this chapter is to provide concrete engagement strategies for empowering ourselves and our astronomy students to learn about and take action on climate change in our communities and with decision-makers. From simple interactive demos to generate conversation at outreach events to course-based assignments and teacher training, and even showing up at the state legislature, there are a variety of ways to build a climate change engagement community. A variety of funding opportunities can support this type of work within our current framework, and we can iteratively work to create new opportunities, events, and positions that formally recognize climate change engagement as scholarship. This work of connection across our communities can help to redefine our work as astronomers as we develop our identities as Earthlings who will come together to take care of our one and only home.

14.1 Foreword

My aim with this chapter is to provide concrete examples of how to do climate change community engagement. This work is based on my personal experience as an astronomy professor and outreach professional and informed by a variety of sources, including formal research articles, but also personal conversations and popular culture books and podcasts.

I want to acknowledge that getting started can feel daunting. As a student, I had not been taught about climate change in any of my astronomy or physics courses, and as a professor, I had not taught about it in my Astro 101 course. I thought, "I'm

doi:10.1088/2514-3433/acfcb6ch14

not a climate scientist, so I should stay in my lane of expertise," and, "the only thing I can really do is try to do better in my personal life." Plus, I live in "Coal Country." Some of the oil and gas and coal companies lead educational activity sessions at the state science teachers association meeting. These companies provide well-paying jobs to many people in the area with no need for background knowledge, skills, or education. They even help pay for schools, scholarships, and community infrastructure. Many of my students come from generations of coal miners, and I have seen some wearing "Friends of Coal" bracelets and stickers. There is great pride for coal because it has allowed families to put food on the table. So, I feared backlash from my community members, my students, and even my institution.

I also want to share that, despite my initial trepidation, engaging with teachers, students, and my community on climate change has turned out to be very inspiring! I have made new connections, both personally and professionally. I have found a myriad of groups around my state, like the Reclaiming Appalachia Coalition, doing the hard work of restoring abandoned coal mines, repurposing old industrial plants, and training a new generation of workers. Students have told me about how they used some of our climate change content to have conversations with their parents, aunts and uncles, and grandparents. Student-generated climate actions have inspired dozens of news stories and their communities have celebrated them. So, while I am still very much a learner, I hope my example of "diving in" encourages other astronomy educators to give it a try. You may be pleasantly surprised!

14.2 Guiding Principles

We know that communicating the science alone is not enough (Nisbet 2009, Ardoin et al. 2013). Implicit messages might include: "Only scientists can fix this," and, "This is too complicated to fix." Minimizing our personal "Carbon Footprint" matters, but we must recognize the fraught history of this phrase (Kaufman 2021, Solnit 2021) and be careful about overly burdening individuals. Also, a focus on individual actions can feel exclusive to those who are less privileged, e.g.: "I can't afford an electric car or solar panels," and "recycle more" or "eat less meat" can feel trite. We need to be highly intentional about our goals, audiences, and strategies (Forrest & Feder 2011). Focusing on larger-scale solutions (for example, from Project Drawdown) is better, but it is also not enough. We must connect the science content with discussions of social information to help influence people to collective action (Ardoin et al. 2013). We must work with a variety of stakeholders we might not be used to working with, such as politicians, faith-based groups, Indigenous communities, and practitioners from very different disciplines such as medicine or art (Whyte 2014, CCEP 2016). Also, there are compounding crises around the world, with climate change disproportionately harming people of color and poor and under-resourced communities. I have heard teachers say: "People in my community don't have enough jobs or food, and there is a drug problem. How can we possibly think about climate change on top of that? (and there is currently a global pandemic!)." As we engage with this work, we also find that we need to remain humble, listen, and honor the interconnections between climate change and a myriad of related issues. Additionally, unlike most other astronomy topics, climate change is an existential threat that is here and now. Engaging with it brings an element of stress and highly emotional responses. As Earthlings, we mourn all that is being lost and we fear a dark future. As educators, we need to help students understand the science, the solutions, *and* we need to help them embody their individual power and process the many emotions that come from the existential threats of watching our planet suffer (Hayden et al. 2011). We are *all* co-learners in how to talk about this with each other.

14.3 Empowering Engagement

Katharine Hayhoe's TED Talk, "The most important thing you can do to fight climate change is to talk about it" has become my motto for climate change community engagement and activism. It gave me an "aha" moment. I don't have to be an expert. *I can make space for climate conversations to happen. I can model climate engagement and learning, and I can empower others to use their voices.*

How can we tap into our students' and our community's values and motivations to inspire climate engagement? It can sometimes be difficult to translate big ideas into everyday actions, so I aim to provide concrete examples of specific outreach demonstrations, videos, and activities for astronomy educators. I provide these to give the reader "touch points" that serve to catch the attention of others, educate, and inspire opportunities for conversations and further learning. These are not meant to be *the* best or only climate change engagement resources, but rather resources that I have found useful in my work with students, the public, and decision makers.

14.3.1 Just Say It

We underestimate the power of just saying out loud, as scientists: "Climate change is happening. It's human-caused. We need to act now." You can put a sign on your office door and at the end of any presentation (Figure 14.1). I included a climate

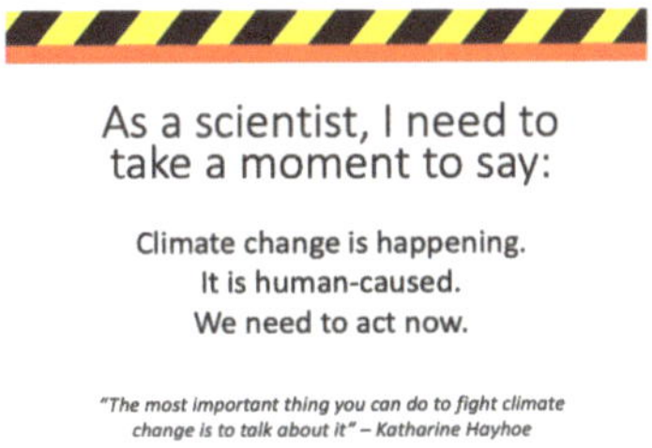

Example of a slide that can be included at the end of any presentation, even if it's not related to climate change.

Example of a sign you can hang on your door when out doing climate-related activities. Print one and offer it to all members of your department.

Figure 14.1. Visuals to include during any and all presentations and events, even those that are not focused on climate change.

slide after a colloquium I was giving about involving students in mapping the Milky Way, i.e. not a presentation about climate change at all. I checked with my colloquium host ahead of time to see his thoughts, i.e. to make sure he would be alright with it, and to see if he had an idea of how the audience might react. He thought the audience would be turned off by it, but he supported me in trying if I so desired. So I kept the slide, and after I reached the end of my presentation, I made the "game-time" decision to advance to this very last slide and said it out loud. The audience ended up being mostly college students, so perhaps more amenable to hearing about climate change, but they heartily applauded! Some students came up to me afterwards and thanked me for saying this and wanted to ask about climate change.

14.3.2 Interactive Demos

Using demonstrations to illustrate concepts in common practice for astronomers and physicists. Demonstrations are eye-catching, enjoyable, quick, and make the audience think, whether that audience is a classroom of students or passersby during an outreach event. What are the physical processes of climate change that can be illustrated with demonstration supplies you might already have or could easily purchase? My go-to climate change demonstrations, shown in Figure 14.2, include:

1. Infrared camera with plastic and/or a greenhouse gas pumped into a bag. The plastic bag shown in the photo below contains "canned air," like that used to clean dust off of keyboards, which contains a potent greenhouse gas called tetraflouroethane (CF_3CH_2F). Pass around the infrared camera as a touch point for audience engagement. You can also easily forego the tetraflourethane gas and just use a plastic or glass sheet as an analogy.
2. Purple cabbage juice, made by pouring boiling water over cabbage leaves, as a pH indicator to demonstrate ocean acidification. Give audience members straws to exhale CO_2 into the solution. CO_2 combines with the H_2O to create

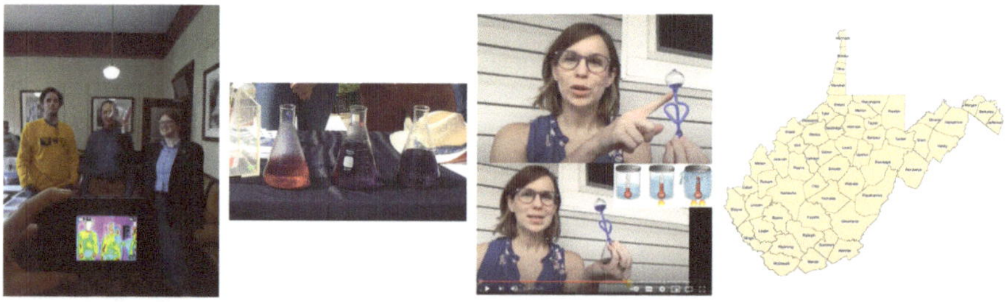

Figure 14.2. Useful climate change demonstrations: (a) Greenhouse effect with an infrared camera and plastic or gas. (b) Ocean acidification with purple cabbage juice as a pH indicator, (3) Sea level rise with a hand boiler showing thermal expansion, (4) Investigate your state's opinions by county with the Yale Climate Opinion Maps.

carbonic acid, slightly changing the color of the liquid. The effect is subtle and takes several repetitions, and it works better when the water is warm. Use a comparison vial with white vinegar to show the extreme color. (ex: see this activity from the American Chemical Society)

3. "Hand boilers" are common demonstrations in physics departments and are very inexpensive. Use this and pass around to illustrate sea level rise from thermal expansion.
4. Offer a computer with a browser set to the Yale Climate Opinion Maps. Show audiences how to navigate and let them explore for themselves. In my experience, most people are fascinated by these maps and they generate good conversations.

14.3.3 Getting Attention

Getting attention is an important goal in and of itself. News articles, social media, and TV clips can help change the broader culture of our communities, which is a goal that is even more important for climate change engagement than traditional astronomy outreach. We are trying to do more than teach people; we are trying to inspire them to connect to others in action.

The images below (Figure 14.3) are from a campus climate rally that was organized by the student Sierra Club, with intentional encouragement and planning by a core group of us faculty and community activists to coincide with the global climate strikes on September 20, 2019. We invited different groups to table out on the campus community lawn. Society of Physics Students (SPS) sold liquid nitrogen ice cream surrounded by signs saying, "There is no Planet B." The West Virginia Science Public Outreach Team (SPOT) hosted a booth using the cabbage juice ocean

WVU Society of Physics Students "There is No Planet B" booth with liquid nitrogen ice cream.

WV Science Public Outreach Team Ambassador being interviewed by the BBC with ocean acidification demo.

Figure 14.3. Images from the Campus Climate Rally, in coordination with global climate strikes, September 20, 2019.

acidification demonstration described above. Because this event coordinated with global events, and because climate change is a hot topic in coal country, several news groups reported on the event. Reporters from the BBC are shown in Figure 14.3 interviewing a student outreach ambassador, which turned out to require advanced communication skills. They wanted to take a political angle, but she expertly steered the conversation back to science and the importance of communicating climate change to younger students. In the resulting article the reporters did not accurately discuss the science that was actually featured at the booth, vaguely referring to “bright bubbling liquid to show how carbon emissions warm the atmosphere,” rather than the intended illustration of ocean acidification. As astronomers, we might not have as much experience navigating these types of interactions, but we should prepare ourselves and our students for the science to be taken out of context.

While in the example above, the reporters came to us, it is also possible to intentionally build in to education and outreach activities a significant dissemination component to share the artifacts that result from climate change actions, such as art (Jacobson, Seavey & Mueller 2016; Sommer & Klöckner 2021), music (Crawford 2013), writing (Fløttum & Gjerstad 2017), or videos (see the videos by Astronomers for Planet Earth). For example, in our West Virginia Climate Change Professional Development project (Williamson et al. 2023), we encourage teachers to help their students create climate change “Public Service Announcements” (PSAs), which, in contrast to a traditional commercial, aim to raise awareness rather than sell a product. We hosted a PSA Competition and aired the winning audio and video clips across the state for Earth Day in 2021, 2022, and 2023, reaching an estimated 40,000 West Virginians each year and generating roughly a dozen news stories (such as this one and this one). The students who participated in this project learned climate change science and communication in tandem with action, and the small fame they gained from winning led to feelings of pride by their peers, teachers, parents, and community members—pride that is associated with speaking up about climate change. Again, the power of this community engagement effort comes from the attention it generates, and the fact that it amplifies local youth voices, rather than the specific climate science facts shared.

14.3.4 Outreach via the Classroom

As shown in the examples above, empowering students can be a leverage point for community engagement. Youth have a powerful voice when it comes to climate change, with a particularly powerful impact of daughters influencing their fathers or conservative parents (Lawson et al. 2019), and while many students care deeply about the climate, they may not realize how important it is for them to speak up and generate conversations, even if they do not know what larger-scale actions their community needs to take.

Generating student-led climate conversations was the goal of a special topics one-credit book club I taught in the fall of 2019. Throughout the semester, we read three books on climate change science and action: *Two Mile Time Machine* (Alley 2002),

Earth in Human Hands (Grinspoon 2016), and *Renewable* (Flanagan 2015), and had reflection conversations each week, including one week with Grinspoon as a guest speaker. For their final project, students chose a climate change researcher or community member to interview for a podcast episode recorded at the local student news studio. The resulting "WVU Climate Conversations" podcast is now available on all streaming platforms and shared broadly and regularly. While most of the effort of this climate change work happened in the classroom and on campus, and with only eight students, the work was able to live on and help influence the community.

I also built the assignment to have a "Climate Conversation" in my Astro 101 course. For homework, students were required to use the lecture material to engage with a friend or family member, then share a reflection in our course discussion forum. Students were asked: "Was the person reluctant or excited to have this conversation? Were some topics easier to discuss than others? If so, which ones, and why? Did they already know some of the climate science and/or did anything surprise them? If so, what and why? What did you learn from this conversation? Would you try to have conversations like this in the future? Why or why not, and with whom? Did the conversation lead you to more questions? If so, what are they and how could you learn more about them?" From this one assignment, ~200 students in West Virginia had an informed climate change conversation with their friends and family. Students mentioned feeling nervous at first, but then being surprised that the person they talked to wanted to learn and/or actually agreed! Students mentioned wanting to keep the conversation going, that they had more confidence to speak about climate change with understanding and evidence. At the end of the semester, 36/125 (29%) of students mentioned climate change engagement as one of the most important lessons/skills they learned out of everything in the semester. A representative student comment is: "I learned about how to engage with people about serious topics … I can now conduct interviews on climate change."

Finally, I will mention the idea of hosting a "Teach in" as a way to make a statement in the classroom that can connect to the community. Each year, the Bard Center for Environmental Policy hosts climate solutions dialogues in each U.S. state and many countries. These are virtual panel discussions with local environmental organizations, policy makers, and researchers where each guest shares their one big, practical, idea for solving climate change in their area. The dialogue is broadcast live and recorded and shared across social media with the hashtag "#MakeClimateaClass." Teachers can use accompanying lesson plans for any discipline to generate further conversations with their students and communities. Readers could get involved in existing efforts to host these dialogues in their state and/or share the recordings for Teach-Ins in their own classrooms.

14.3.5 Civic Engagement

I view civic engagement to be an aspect of community engagement that specifically engages people in power, especially those who enact laws. In my personal life I engage in climate change activism by writing to my members of Congress about climate change and visiting their offices with my local activist group, and I have experimented with various ways of bringing this effort to my professional life. When it comes to legislation, we, as professionals, generally cannot lobby for a particular bill or policy, but we *can* share scientific results and consensus statistics (ex: from NASA and the Union of Concerned Scientists), tell our legislators about climate change research and educational efforts happening on our campus, and show demonstrations of the science of climate change.

In my college astronomy class, I have tried a few interventions to help empower students to use what we have learned in class for civic engagement. I have invited various guest speakers, such as from our WV Citizens Climate Lobby (they have chapters in every state), our local delegate who proposed a solar power bill and is a scientist, and a leader from an Appalachian revitalization group called Coalfield Development. These are place-based connections which most students often have not heard of, but they show that people in our state are taking action, change is possible, and that anyone can get involved. Additionally, I have assigned homework to write a letter to someone "in power," adapted from the "Writing a Position Statement" Activity from Climate Generation. Students could take any position they wanted as long as their arguments included scientific facts. While a few students wrote that climate change is not important to them, a vast majority expressed a desire for action on climate change. Students were not required to actually mail their letters, but those who did got some encouraging responses, such as from our University President's office and members of Congress. Some students were surprised at how easy it was to do civic engagement, and at least one student joined the Sierra Club afterwards. However, most of the letters were generally not as well-written as I had hoped, and some students commented that it felt like a childish exercise.

Perhaps my most exciting climate change civic engagement activity has been leading a group of WVU faculty and students to host a "Climate Education Day" at the West Virginia state capitol during the February 2020 legislative session (Figure 14.4). From the initial idea to the day-of, this effort taught me a lot about climate change civic engagement, so I want to provide some of the details here. This was an idea that came up during a meeting of a group of climate-motivated faculty in our college of arts and sciences. Simultaneously, our delegate invited us to come to the state capitol and have meetings with legislators. I sent out a Google Form to faculty and students to gauge interest in going for a "Climate Education Day." I asked the Assistant Dean to see if he could get us a few thousand dollars to cover the cost of a bus and meals, but we started planning the trip without the funding commitment. He later committed his discretionary funding from an honorarium for a research presentation. I had no idea what to expect. The bus could hold 50 people. Only 15 people attended, but the people who did attend were passionate and proactive, which was just what we needed for our first time trying this.

For the day at the state capitol, I made a 1-page handout (front and back) of my university's climate change research and education efforts to leave on every legislator's desk. Our delegate host introduced us during the floor session and escorted us to different offices. Again, this was fairly ad-hoc and fast-paced. We tried

WVU faculty and students at the West Virginia State Capitol during the February 2020 legislative session.

WV Delegate and students discussing the greenhouse effect demo.

Figure 14.4. Images of university faculty and students engaging legislators for a "Climate Education Day" at the state capitol, February 2020.

to respond to the moment. Fortunately, our delegate was also able to reserve a meeting room to invite legistors for a 2-hour drop-in session with all of the climate change activities and demos. We had graduate students showing their tree ring research. We had all the demonstrations from Figure 14.2, including a computer open with the Yale Climate Opinion maps to show legislators the opinions of their constituents in their districts.

This was the first time a group of faculty and students has ever showed up at the state capitol to talk about climate change. All legislators gained information about climate change research and education happening at our university. One student who grew up baling hay in a rural part of the state was able to connect with his representative about how the uncertain rains made things difficult for him and his community. One legislator called me later to get more information about the Yale Climate Opinion Maps. The students were "over-the-moon" proud of themselves and excited to do this again.

A big lesson learned is that being present matters. There was another activist group there the same day as us. They had matching t-shirts and over 100 people in their group. We saw them in every hall, everywhere we went. It made me realize that, in addition to content expertise, making a visual impact with large numbers of people is important. Additionally, I learned that, while coordinating these actions may feel a bit messy, some key university stakeholders and students will rise to the challenge. We can ask for help and rely on each other, and that is incredibly empowering.

14.4 Funding This Work

Most of the above efforts simply take time and effort, which can be difficult to find amidst the constant business of academia, but some efforts also take some amount of money. In the interest of transparency and practicality, I share the following list (Table 14.1) of examples of climate education and outreach efforts I have successfully and unsuccessfully funded with department/college funding commitments, and small grants to our state Space Grant Consortium, annual campus internal Community Engagement solicitation, and a regional community foundation.

Table 14.1. Activities, costs, and funding sources that have been requested for climate-related education and outreach. *indicates that it was successfully funded.

Activity	Cost	Source
Demonstration Supplies*	~$100–$500	Department
One-Day In-person Teacher Workshops, including travel stipends and meals*	~$400 per teacher (with 15–20 teachers)	Space Grant Consortium
Virtual Teacher Workshops*	~$100 gift card per person per workshop (generally 30–40 people)	Space Grant Consortium
Registrations & travel for annual state science teachers association conference*	~$1000 per person (generally 3–5 people)	Various: Space Grant Consortium, Department
Collaborator salaries/stipends, including for an environmental educator, a community activist and communications expert, and a Master Teacher*	$200–$1000 per person (for 5–7 people)	Various: Space Grant Consortium, Department, Community Foundation
PI and Co-I partial summer salaries	$2000–$4000 per person (for 2–3 people)	Campus Grants, Community Foundation
Purchase of a new climate-related planetarium show*	~$5000, one-time purchase	Department
Climate Education Day at the State Capitol*	~$1,500 bus, ~$20 per person for meals	College Dean Discretionary
Climate Conversations Podcast*	$100 production fee	College Teaching Fund
Public Service Announcement Competition*	$50–$200 per award for 3–6 winners ~$2,500 for airtime on TV/radio	Space Grant Consortium, Community Foundation
Graduate Teaching Assistant for education/outreach project*	~$15,000 per year	Department

Finally, it is critical to mention that I had one course-release built-in to my position for community outreach, giving me great freedom to pursue this kind of work, yet this is something I have not heard of at other universities. We, as a community of astronomy professionals, should work together to think creatively about how to redefine how we spend some of our time so that climate activism is integrated into what we do rather than over and above our regular job requirements. Our time and energy are some of our most valuable resources, and funding sources should recognize that. Most of us in faculty and/or research positions have some sort of "service" requirement to our job, which can capture climate engagement, but many people are already overloaded with service tasks. Intentionally engaging with climate change in our communities goes beyond service; it is scholarship and should be recognized as such. For any Department Chairs or Deans reading this chapter, I highly recommend advocating for and generating faculty positions with built-in outreach to formally value this work. We can all also work to create and support Graduate Assistantships for climate change education and outreach.

14.5 Conclusion

I hope the concrete examples of climate change community engagement in this chapter give the reader ideas and inspiration to speak up, reach out, and engage in whatever ways feel right for their unique professional position and community needs. Like the process of research that we are familiar with as scientists, the process of climate change community engagement can be non-linear, confusing, and full of highs and lows, but it can ultimately lead to new connections and progress. As astronomers who understand the fragile beauty of our home planet, we have the responsibility to help our students and community members develop their identities as Earthlings. We can help to make the connections between earth and sky to come together and protect our only home.

Finally, I will reiterate that I am a learner in this space, and I continue to find new bodies of knowledge, people, organizations, and methods of education and advocacy that push me to grow. Many of these resources have been shared in the Astronomers for Planet Earth collaboration spaces, which I encourage all readers to join. These resources include things like how to help ourselves and our students deal with climate emotions and eco-anxiety, how to listen to, learn from, and empower Indigenous groups, who sustainably manage key habitats and carbon reservoirs, and how to turn inward and change myself to be more nurturing toward my fellow Earthlings, including plants and animals. I look forward to building more community with those reading and with all those we touch.

References

Alley, R. 2002, Two Mile Time Machine: Ice Cores, Abrupt Climate Change, and Our Future (Princeton, NJ: Princeton Univ. Press)

Ardoin, N., Heimlich, J., Braus, J., & Merrick, C. 2013, Influencing Conservation Action: What Research Says About Environmental Literacy, Behavior, and Conservation Results (New York: National Audubon Society)

CCEP Climate Change Education Partnership 2016, *Climate Change Education: Effective practices for working with educators, scientists, decision makers and the public* https://ccepalliance.org/2016/12/ccep-effective-practices-guide/ [accessed 19 June 2023]

Crawford, D. 2013, A Song for Our Warming Planet (Institute for the Environment, Institute on the Environment, Univ. Minnesota) https://www.youtube.com/watch?v=5t08CLczdK4 accessed 19 June 2023

Flanagan, E. 2015, Renewable: One Woman's Search for Simplicity, Faithfulness, and Hope (She Writes Press)

Fløttum, K., & Gjerstad, Ø. 2017, Wiley Interdiscip. Rev. Clim. Change, 8, 429

Forrest, S., & Feder, M. A. 2011, Climate Change Education: Goals, Audiences, and Strategies: A Workshop Summary (Washington, D.C: The National Academies Press)

Grinspoon, D. 2016, Earth in Human Hands: Shaping Our Planet's Future (Grand Central Publishing)

Hayden, M., et al. 2011, Journal for Activism in Science & Technology Education, 3, 119

Jacobson, S. K., Seavey, J. R., & Mueller, R. C. 2016, Ecol. Soc., 21, 30

Kaufman, M. 2021, The carbon footprint sham: A 'successful, deceptive' PR campaign" Mashable https://mashable.com/feature/carbon-footprint-pr-campaign-sham [access 19 June 2023]

Lawson, D. F., et al. 2019, NatCC, 9, 458

Nisbet, M. 2009, Environment Magazine, 51, 2 12

Sommer, L. K., & Klöckner, C. A. 2021, Psychol. Aesthet. Creat. Arts, 15, 60

Solnit, R. 2021, Big oil coined 'carbon footprints' to blame us for their greed. Keep them on the hook (The Guardian) https://www.theguardian.com/commentisfree/2021/aug/23/big-oil-coined-carbon-footprints-to-blame-us-for-their-greed-keep-them-on-the-hook [accessed 19 June 2023]

Whyte, K. P. 2014, Hypatia, 29

Williamson, K., Shinn, J., Hemler, D., & Fallon, S. 2023, A case study for climate change professional development in West Virginia, Journal of Sustainability Education, https://www.susted.com/wordpress/content/a-case-study-for-climate-change-teacher-professional-development-in-west-virginia_2023_03/

Chapter 15

The Power of Activism

B Rodgers

> "When you realize that the 2020's will be remembered as the decade that determined the fate of human history, you will tap into an eternal evolutionary force that has transformed the world time and time again."
>
> —Joëlle Gergis

> "Another world is not only possible, she is on her way. On a quiet day, I can hear her breathing."
>
> —Arundhati Roy

Addressing the urgent crisis of climate change requires a global movement of engaged and energized citizens to promote and demand an energy transformation. This chapter is a call to be part of that movement. Activists play a critical role in building public momentum and pushing for systemic change. I argue that astronomers are well placed to do both. Taking action is worthwhile, rewarding, and even joyful, as I describe through my own journey and with practical tips and resources to kickstart, or ramp up, your own activism.

15.1 Introduction

If your house were on fire, would you be going to work today? If a loved one were in grave danger, would you be planning your next vacation? The answers are, most likely, no. And yet, the habitability of our collective home is in grave danger—it is (almost literally) on fire. That this is a slow-moving disaster makes it no less urgent or perilous. Most people are alarmed, or at least concerned, about the climate crisis, and yet, too many of us are sitting on our hands, or content with small personal changes. But this chapter is not about guilt or shame, it is about accelerating change.

Climate scientists tell us we have *less than a decade* to accelerate the transition away from fossil fuels in order to avoid the worst climate outcomes. And yet, in 2023 global greenhouse gas emissions continued to rise. It is small consolation that there are reasons for this disconnect between the problem and humanity's response, both

doi:10.1088/2514-3433/acfcb6ch15

Figure 15.1. Sign from a youth climate march in Portland, Oregon in 2021. *Image credit: B. Rodgers; author unknown*

psychological and diabolically intentional (Freedland 2023; see also Marshall 2015, Brulle 2023, Oreskes & Conway 2011).

Nevertheless, momentum is building and change is happening. Governments and corporations around the world, including the US, are ratcheting up climate action plans and initiatives. However, the critical issue is time, which is not on our side. Decades of inaction have put us far behind, and even current action is far too slow. What is required now is rapid and large-scale transformative change—nothing less than a worldwide paradigm shift. History suggests this kind of sweeping change is driven not by existing leadership but by the sustained pressure of a massive people-powered movement.

For this, civic activism is critical. Action commensurate with the scope and severity of this problem must go well beyond individual lifestyle choices (even while these are also important).

Collective action is needed, and activism–in a myriad of forms—helps drive society to rapid and dramatic change. Young people, and indigenous and frontline communities around the world already feeling the impacts, recognize the urgency of this crisis better than most, and many are leading the way (Figure 15.1), but these groups often lack political power and resources. Astronomers and other academics have power and agency to make a difference.

When I left my professional astronomy career nearly a decade ago, I had no idea what lay ahead. I only knew I could and, I decided, *must* commit more of my own time and energy to facing this crisis. One of the many things I have discovered along the way is that activism is far from joyless work–it is fulfilling, energizing, and rewarding, and well worth your time.

15.2 Why We Need More Activists

15.2.1 Progress Is Too Slow

In 2015, the nations of the world signed the historic Paris Climate Agreement,[1] agreeing to hold the increase in the global average temperature to "well below 2°C" and pursue efforts to limit it "to 1.5°C above pre-industrial levels," as well as increase the world's ability to adapt, and make "finance flows consistent with…climate-resilient development" (UNFCCC 2018). This agreement was a huge international achievement. Unfortunately, even with increasing action across all levels of society, the world is failing to meet these goals.

The synthesis report of the sixth assessment cycle (AR6) of the International Panel on Climate Change (IPCC) states it is now likely that global warming will exceed 1.5°C in the near term, despite the Paris Agreement goal, and the ***"modelled pathways that limit warming to 1.5°C (>50%) with no or limited overshoot…involve rapid and deep and, in most cases, immediate greenhouse gas emissions reductions in all sectors this decade."*** However, crucially, the report also emphasizes that ***"Deep, rapid, and sustained mitigation and accelerated implementation of adaptation actions in this decade would reduce future losses and damages related to climate change for humans and ecosystems"*** (IPCC 2023).

Figure 15.2 is part of figure SPM.5 in the Summary for Policymakers of the same report. It shows the current trajectory, labeled "Implemented policies" (in red), and

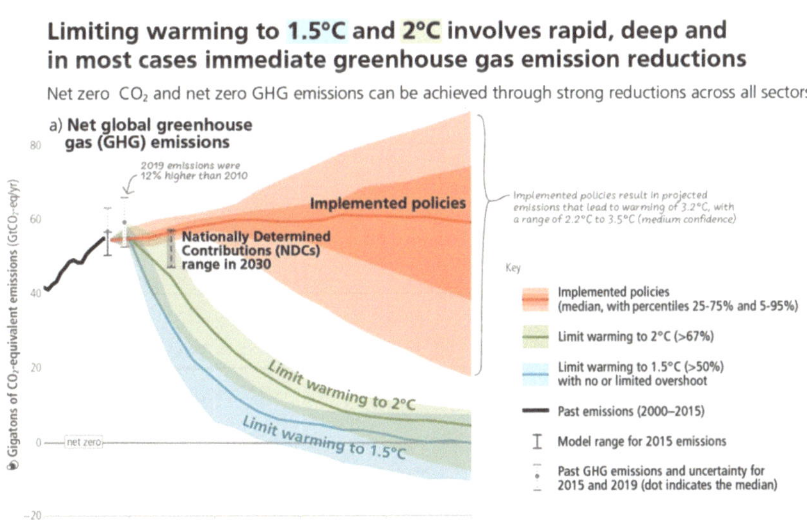

Figure 15.2. IPCC AR6 Synthesis Report Summary for Policymakers Figure SPM.5:[2] Global emissions pathways consistent with implemented policies and mitigation strategies (IPCC 2023). Only panel a of the original figure is shown.

[1] 195 of 198 nations are party to the agreement. The United States withdrew in November 2020 but rejoined on 20 January 2021 (UN Treaty Collection, 2016).
[2] https://www.ipcc.ch/report/ar6/syr/figures/summary-for-policymakers/figure-spm-5/

the required emissions reductions to limit warming to 2°C (green) and 1.5°C (blue). To meet the 1.5°C line in 2030 requires cutting emissions roughly in half by the end of *this* decade. The difference between projections and ambition is known as the Emissions Gap (UNEP 2022). Since the greenhouse effect depends on the cumulative greenhouse gases in the atmosphere, the longer we delay cutting emissions, the steeper the slope to limit warming. Hence, the urgency.

This is a well-defined problem scientifically and, although challenging in scope, technologically achievable. The measure of success is simple: *"Did that number [carbon emissions in the atmosphere] stop rising and start to drop?"* asked Sir David Attenborough at the start of the Glasgow Climate Conference (Attenborough 2021); and, is it dropping fast enough? In fact, emissions have leveled or dropped some in many wealthy countries in recent years, but they are rising in many developing countries as these countries understandably seek to raise their standard of living. It is therefore incumbent on the countries who have prospered from the fossil fuel economy–at the expense of our collective atmosphere—to reduce emissions more rapidly, and aid less wealthy countries to do the same (see Chapter 16.)

15.2.2 What Is Activism?

Civic activism, a fundamental tool of democracy, effectively accelerates societal change by drawing attention to an issue, swaying public opinion, and pressuring those in power to take action. By the broad definition in Figure 15.3, activism can encompass a far-ranging spectrum of activities, from basic civic duties like voting to civil disobedience.

The progression up the spectrum to ever more impactful actions is often called the *ladder of engagement.* While all actions are useful, the more impactful ones have, well, greater impact. Boycotts, sit-ins, blocking bridges and oil trains all are at the indigo-violet end of the activist spectrum. These actions get noticed, and hence can

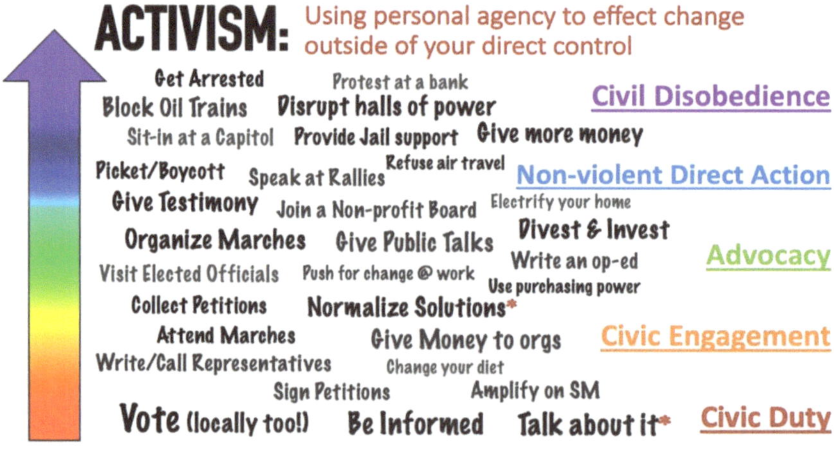

Figure 15.3. Examples of actions across the activism spectrum. Actions do not have to be 'radical' to be effective. *See text for more on these. *Image credit: B. Rodgers*

achieve results more quickly. Still, it is important to note that (1) actions at the bottom of the spectrum are important and necessary, and (2) most actions are not dangerous or illegal.

Climate scientist Katharine Hayhoe says one of the most important things we can do about the climate crisis is talk about it (2021a; see also Chapter 14). Constructive conversations "help people connect over shared values including family, community, health, and religion" and "can help people think more deeply about how a changing climate affects who and what they care about" (Ballew 2023). Studies show climate change is not a common topic of conversation for most people, and therefore people often shy away from it, creating a "spiral of silence" (Maibach 2016, Ettinger et al. 2023). Bringing this topic into regular conversation is itself a powerful act.

Talking about climate solutions, especially, not only breaks this silence but also counters the doom and helplessness too often associated with this issue, and normalizes the clean energy transformation already underway. Raising topics such as renewable energy, electrification, diet, wealth redistribution, limiting growth, and climate justice brings them into our daily discourse, and out of the realm of extremism or fantasy. After all, a clean and just energy transition is good for people and the planet–it creates jobs, saves money and improves health (Saha et al. 2022).

Most of the activities in Figure 15.3 are not illegal. The left side lists things typically associated with activist behavior, but most are actions anyone can take in everyday life. Some are well-suited to astronomers and other academics: giving public talks, engaging government officials, and requesting changes at work. A few are personal choices (e.g., change your diet, electrify your home); spreading the word about what you are doing and why brings these into the realm of activism. Broadening activities laterally as well as vertically is useful to increase your contribution; but again, moving "up the ladder" to higher energy actions is most effective. As Dr Christiana Figueres, architect of the Paris Agreement, says: "*Most of us have not been answering the call from science in this way. We've not been accessing the hidden inner strength, courage and endurance we need to do what is necessary*" (Figueres 2021).

15.2.3 Six Americas and 3.5%

Social progress is not linear. Research by political scientist Erica Chenoweth shows that social campaigns do not fail once they have "achieved the active and sustained participation of just 3.5% of the population" (Chenoweth & Stephan 2011).[3] Twenty million Americans, 10% of the population at the time, participated in the first Earth Day in 1970, and within one year the Environmental Protection Agency was created[4], followed quickly by landmark anti-pollution laws.[5] Once political 'tipping points' are reached, change can happen very fast.

[3] Based on a study of political campaigns for regime change. Not all agree this number is equally applicable to the climate movement (e.g., Matthews 2020).

[4] https://www.earthday.org/history/

[5] Although attention to environmental justice came much later, and is still an issue today.

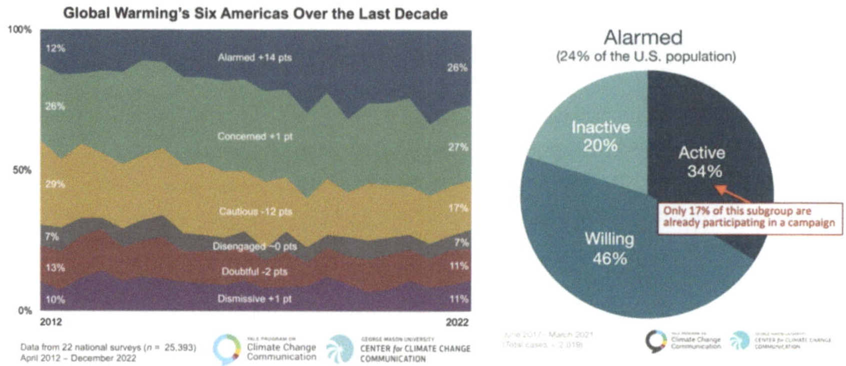

Figure 15.4. (a) Global Warming's Six Americas over time (*left*; Leiserowitz et al. 2023), and (b) Detailed breakdown within the Alarmed category, 2017-2021 (*right*; Goldberg et al. 2021). Red text box added.

The Yale Climate Change Communication program has defined "Six Americas" reflecting the level of concern about the climate crisis among Americans, based on decades of survey data.[6] Figure 15.4a shows their results over the last decade (Leiserowitz et al. 2023). Interestingly, the bottom three categories—the Dismissive, Doubtful and Disengaged—are stubbornly consistent. These folks are not likely to be persuaded. In contrast, the Cautious and Concerned have shrunk considerably as the Alarmed category has more than doubled. At 26%, this group alone is more than enough to meet the "3.5% rule" (worldwide, the number is even higher; UN 2021). So, where is everyone?

Unfortunately, detailed analysis of the Alarmed category reveals only 34% "exhibit high levels of climate activism"; and, although most of these "Active Alarmed" do discuss global warming with friends and family, *only 17% of them* are "most likely to already be participating in a campaign" (Goldberg et al. 2021; Figure 15.4b). The rest responded they 'definitely' or 'probably' would participate.[7] This suggests there are some 70 million people in the US alone who are already alarmed about this issue, but apparently waiting for the right catalyst to engage.

Granted, not everyone is in a position to act. The authors note "the reasons for action or inaction are not only driven by psychological factors…People face a variety of personal and structural barriers" (Goldberg et al. 2021). Nevertheless, energizing a significant fraction of this pool could have profound effects. Activist and author Emily Johnston says, *"Imagine if even 10% of the country started engaging deeply, even just one day a week. Our possibilities would be—will be—entirely different from what they are now"* (Johnston 2020).

[6] Take the short survey yourself at:https://climatecommunication.yale.edu/visualizations-data/sassy/.

[7] Calculating the percent of the US population actively engaged is left to the reader. (The answer is 1.5%.)

15.3 What's Stopping Us?

Although there are numerous reasons people fail to act, some of the most common can be overcome. Let's look at a few, before moving on to all the good reasons *to* act.

15.3.1 The "We Are the Enemy" Myth

This is perhaps the most common anti-activist trope: "Didn't you drive your gas-powered car to the rally? What a hypocrite." A high school physics teacher, talking about his reticence to do more to fight climate change, once told me: "I'm not sure what gate to storm, and even if I found it, I suspect it's a gate I pass through every day." Many people feel this way; paralyzed by what we might call "the enemy is us" guilt.

There is a villain in this story, but it is not most of us. Holding people personally responsible deflects attention from the systemic changes needed. This has been part of the fossil fuel industry's strategy for decades. In fact, fossil fuel companies invented the carbon footprint concept as a public relations campaign (Kaufman 2021, McFall-Johnsen 2021). The carbon footprint can be a useful tool (lifestyle choices are *part* of the solution), but unfortunately we live in a fossil-fuel based society. This is the problem. It was not our choice to build an energy system that destroys our home, and it requires energy to change it. "Hypocrisy is the price of admission in this battle," says Bill McKibben (2016). Burning fossil fuels is not a good reason to avoid fighting for the end of fossil fuels.

Perhaps a more useful concept which has arisen in recent years is the carbon *handprint*. As explained by Gregory Norris, co-director of SHINE,[8] "a contribution that causes positive change in the world—including reductions to your own or somebody else's footprint—is a 'handprint'" (Norris 2017). Rather than promoting guilt and shame by focusing on our negative impact, a carbon handprint emphasizes the good we can do, which is something we have more control over. While your carbon footprint can at best go to zero, your carbon handprint can grow indefinitely. Speaking out and influencing others through your actions can generate far greater results than only reducing your individual emissions.

15.3.2 Does It Do Any Good?

Scientists, among others, are often skeptical of the efficacy of activism, particularly more dramatic actions like civil disobedience. It is sometimes possible to draw a direct line from activist work to tangible results—like that first Earth Day—but more often this is not the case. Like many things worth doing, the immediate benefit is often not obvious, but taken collectively, the tide shifts. In hindsight, it is easy now to see the results of the Women Suffragettes and the Civil Rights movements, and

[8] The Sustainability and Health Initiative for Net-positive Enterprise at the Harvard School of Public Health's Center for Health and the Global Environment

yet these were long and hard fought battles whose eventual successes were not at all obvious at the time.

It is exciting when the impact of a certain tactic or campaign *is* obvious. In the Pacific Northwest of the US and Canada—once envisioned to become a major fossil fuel export hub—over three dozen fossil fuel infrastructure projects have been stopped[9] by relentless, organized citizen action across the spectrum: educating, petitioning, marching, lobbying and non-violent direct action (Cunningham 2022). An oil executive once said the Northwest is "where [fossil fuel] energy projects go to die" (de Place & Gruen 2015) –a reputation celebrated by the many local groups and tribes that made it so.

The climate group 350.org, taking a play from the apartheid playbook, launched the Divestment movement on college campuses in 2011 with this simple logic: if it's wrong to wreck the planet, it's wrong to make money off of it. With sustained public pressure, in ten years nearly 1500 pension funds, university endowments, and public and private investment portfolios have divested almost $40 *trillion* dollars from fossil fuel companies, much of it re-invested in clean energy and sustainable solutions (Lipman 2021).

Many campaigns may not achieve their immediate goals but nonetheless have great impact. Standing Rock is an example. The Dakota Access pipeline was completed (although the fight continues), and yet that movement undeniably raised awareness of climate injustice, indigenous rights, paramilitary policing in defense of corporate interests, and the sheer power and joy in solidarity, and galvanized countless people (Gunderson, 2021).

Youth activism has proven to be especially powerful. Greta Thunberg's now famous one-person school strike galvanized and consolidated a burgeoning youth movement, bringing to the fore the neglected work of thousands of youth activists around the world, and inspiring millions more (Parker 2020). Anyone even vaguely following climate news is aware of the impact young people—from middle schoolers to thirty year olds—are having on the political discourse. Thunberg, Vanessa Nakate, Xiye Bastida, and others like them around the world, couldn't know their initial small actions would start a tidal wave; with youthful clarity and determination they chose to act, and the rest followed.

15.3.3 Avoidance and Apathy

One does not have to be a Climate Scientist to speak out about the climate crisis. Astronomers know more than enough to understand this problem and its consequences. Carl Sagan was one of the first to bring climate change into popular context based on studies of Venus and its atmosphere; his Congressional testimony in 1985 preceded NASA Climate Scientist James Hansen by three years (C-SPAN 1985). Claiming "objectivity" as a reason not to engage publicly is little more than avoidance. There is no doubt that the science is clear. If the issue weren't so political,

[9] Of 55 projects proposed since 2012, 40 have been canceled, according to a Sightline Institute study (Moore 2022).

perhaps it would be easier for scientists to speak up for the truth, and in fact, this is happening more and more (Muir 2020; see more on Scientist Rebellion below). Still, too many are reticent, claiming that taking a position risks their reputation, or, concerned for their careers; however, supporting a false debate is a disservice to the public. We can debate solutions, but the basic science of the consequences of human-caused greenhouse gas emissions is no more debatable than the effect of gravity.

It is easy to become numb to the ever more alarming headlines when the impacts are not felt in our own lives. Many people *are* already experiencing life-changing impacts, but most of us are not. If the climate crisis feels distant in time and space (even as it becomes less so all the time), it is difficult to truly internalize the danger. Or perhaps we feel powerless or overwhelmed. But humanity created our fossil fueled society (relatively recently in fact), and we can dismantle it. The good news, according to IPCC author Joëlle Gergis, is "there is no evidence to support the notion that we are currently facing runaway climate change or the inevitability of an unlivable future" (Gergis 2023). Every metric ton of carbon pollution not emitted lessens future impacts. To say it is "too late" is another avoidance tactic we can not afford.

15.4 Reasons to Act

15.4.1 Scientists Make Excellent Activists

I have alluded to a few reasons why astronomers can be good activists, now let's examine this in more detail. Throughout history, scientists have chosen to be political on matters of great import. The climate crisis is arguably the biggest crisis humanity has ever faced, and many scientists, including several authors in this volume, have taken up public activism and are urging more scientists to join them.

In 1992, around the time of the first international climate conference (COP1), nearly two thousand scientists signed the first "Scientists' Warning to Humanity" about the ecological crisis. In 2017, a second notice was issued, garnering over 15,000 signatures (Ripple et al. 2017; also Gardner & Wordley 2019). A blog on their website laments the "almost total avoidance of realism about climate breakdown" in university curricula, and argues for university faculty to become "crusading scholars" (Read 2017).

Astronomers have specific skills and resources particularly useful for climate activism. These include relevant knowledge, intellectual authority, scientific agency and public interest.

Relevant knowledge: Global warming is the physical result of a planet out of balance. Astronomers know planets. We study, and teach, solar radiation, the greenhouse effect, radiation transfer, energy equilibrium. All of these are relevant to explaining the scientific basis for climate change, and debunking public misunderstandings. Astronomers are also fluent in the benefits and limitations of computational models, and the peer-review process, which can be helpful in addressing skepticism regarding predictions and scientific consensus, and deciphering the sometimes opaque language of technical literature (like the IPCC reports). We don't have to be climate scientists to provide clarity to the general public.

We know there is no Planet B. We can explain, for example, why the idea of colonizing another planet is not a preferable or even realistic alternative to repairing planet Earth.[10] As researchers and educators, we can use our platforms to share our knowledge and explain this issue to the public, our students, and policymakers. As public servants, we have a responsibility to do so.

Intellectual authority and scientific agency: Astronomers are trusted to comprehend a complex planetary-sized problem like climate. We have access to halls of power through universities and research facilities where we work. We can use our influence not only to inform, but to demonstrate the seriousness of the problem through our own words and actions. It is not often our field is relevant to a concrete problem of importance to all of humanity. When astronomers do not speak and act out, we send the wrong message–tacit approval of the status quo. If we do not seem alarmed, perhaps people less scientifically literate will assume they don't have to be. Let's show concern, use language that conveys the seriousness of the problem, and *act accordingly*. Insist our institutions take action. Make bold and impactful changes, commensurate with the emergency we face. Concern is contagious.

Activist organizations are sometimes dismissed as extremists, alarmists, radical environmentalists. When scientists join activist groups and events, speak at rallies or to elected representatives, we add intellectual credibility which helps to convey the gravity and urgency of this crisis. When we participate in non-violent direct action, we communicate that this *is* an emergency. Our actions carry a greater-than-average potential to motivate others.

Public Interest: Astronomy is a gateway science. People from all walks of life love to talk and hear about it. Strangers approach us to ask about black holes or the latest exoplanet in the news. Astronomers are natural science ambassadors by virtue of the fascinating field we have the privilege to work in; we can take advantage of this to talk about the climate crisis, not just in our classrooms but with family, friends, strangers, and in public talks. As we engage and enthrall people with fascinating and beautiful images of the distant Universe, let us also share beautiful images of this remarkable planet, and instill a sense of reverence, respect, and love, for our one-and-only home. Beyond economics and politics, all people share a desire to protect and care for what they love.

15.4.2 Moral Obligation

Not only do scientists have agency, we have privilege and with privilege comes responsibility. Many professional astronomers are public servants and reliant on public funds for our work. More importantly, our profession is committed to discovering and sharing scientific truths, and what could be more important than sharing that the scientific evidence says we face a global, catastrophic threat?

[10] "We humans have to go to space if we are going to continue to have a thriving civilization" Jeff Bezos told CBS Evening News in 2019 (cited from Pierre-Louis 2020).

Inspired by the Hippocratic oath of ethical conduct, two UK groups have created a scientific oath for the climate: a pledge of scrutiny, integrity and engagement.[11] Although targeting climate scientists specifically, the call to speak out truthfully, align personal behavior, and engage with institutions to "curtail and mitigate the worst outcomes" (Cairns 2020) is applicable to all scientists.

Impacts of the climate crisis are already being felt, most notably among poor nations of the Global South,[12] who have contributed the least to create this problem and have the fewest resources to address it. This climate injustice should give all of us who live in relative comfort pause. As the largest historical contributor, and one of the wealthiest nations, the US holds both a large responsibility, and a powerful opportunity, to act. Western scientists do too. To throw up our hands is itself a privilege many around the world, fighting for their very survival, cannot afford.

Scientists have a long history of activism, driven by their moral conviction, especially on issues of global significance. Carl Sagan was an ardent environmentalist throughout his astronomy career, expounding on the preciousness of our home planet from the start of the modern environmental movement until his death. See for example his now famous "Pale Blue Dot" quote (Sagan, 1994).

The climate scientist James Hansen, Director of NASA Goddard Institute for over 30 years, spoke to Congress about the certainty of global warming in 1988. Hansen later retired from NASA[13] specifically to be able to speak out more strongly, citing a moral obligation, and has been arrested multiple times (Gillis 2013). Hansen is now a public speaker and author, and a plaintiff in the *Juliana v. United States* lawsuit (Columbia Law School 2015). Recently, he wrote: *"We need activists… You don't need to lie down before a train or get arrested to make a difference… We can still make our democracies work, but it's hard work–it will require a lot of people"* (Hansen 2021).

Numerous astronomers and other scientists are heeding this call. Astronomers for Planet Earth has over a thousand members, and groups like Scientist Rebellion are growing.

Scientist Rebellion (SR) is an international scientists' environmentalist group that campaigns for degrowth, climate justice and more effective climate change mitigation.[14] Their website states:

> We are scientists and academics who believe we should expose the reality and severity of the climate and ecological emergency by engaging in non-violent civil disobedience.
>
> Unless those best placed to understand behave as if this is an emergency, we cannot expect the public to do so. Some believe that appearing

[11] Find the oath here: https://www.sgr.org.uk/projects/science-oath-climate-text-and-signing.

[12] The Global South is a term that broadly comprises countries in the regions of Africa, Latin America and the Caribbean, Asia (without Israel, Japan, and South Korea), and Oceania (without Australia and New Zealand), according to the United Nations Conference on Trade and Development (UNCTAD). (https://en.wikipedia.org/wiki/Global_North_and_Global_South)

[13] Dr. Hansen is now with Columbia University's Earth Institute.

[14] https://en.wikipedia.org/wiki/Scientist_Rebellion

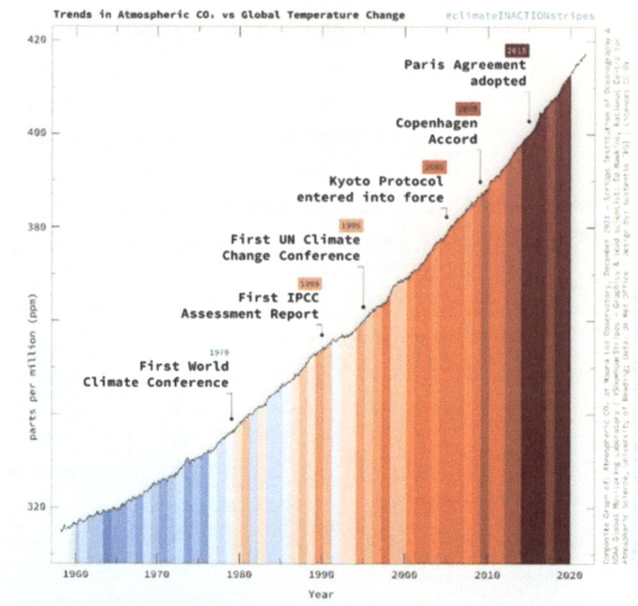

Figure 15.5. "Why we Rebel" from the Scientist Rebellion website, showing rising CO_2 in the atmosphere and increasing surface temperature (colored stripes), overlaid with relevant political milestones of the last 60 years. [Original source: Bergensia 2022.]

> "alarmist" is detrimental—but we are terrified by what we see, and believe it is both vital and right to express our fears openly.

SR advocates and organizes non-violent direct action, including civil disobedience, in order to challenge the status quo and point out that what we collectively accept through our apathy is not acceptable (Artico et al. 2023, Harvey 2022). "In times of emergency, the scientific community must act their part: we are among the most trusted members of society and we cannot let our immense prestige go to waste," says Dr. Fernando Racimo, an SR leader (Racimo et al. 2022).

Figure 15.5, also from the SR website, captures what they see as our collective failure, hence the need for rebellion. Speaking of scientists specifically, the website states:

> Scientists have spent decades writing papers, advising governments, briefing the press: all have failed. What is the point in documenting in ever greater detail the catastrophe we face, if we are not willing to do anything about it?
>
> Academics are perfectly placed to wage a rebellion: we exist in rich hubs of knowledge and expertise; we are well connected across the world, and to decision-makers; we have large platforms from which to inform, educate and rally others all over the world, and we have implicit

> authority and legitimacy, which is the basis of political power. We can make a difference. We must do what we can to halt the greatest destruction in human history.

Started in 2020 as an offshoot of Extinction Rebellion, SR has grown rapidly around the world, and their actions are gaining attention. For example, a recent target of Scientist Rebellion (and others) is private jet use, because of its outsized contribution to carbon pollution by a very small fraction of the population. There were a series of disruptive actions at airports around the world in Fall 2022, including a large demonstration which resulted in hundreds of arrests at Schipol airport in The Netherlands. It is likely this action played a role in the carbon reduction measures announced by the airport a few months later, including limiting private jet traffic (Jacob et al. 2023). While Scientist Rebellion is just one of many direct action groups, that it is an organization of scientists adds gravitas and credibility to its demands.

15.4.3 Cognitive Dissonance and Community

Finally, there are selfish reasons to increase your activism too. When you know humanity is facing an existential crisis and not doing enough about it, it is hard to go about your life as if everything is ok. The more you learn, the harder it is to ignore, and this cognitive dissonance takes a mental and emotional toll, even if, or especially if, we don't acknowledge it. To truly face the reality of the climate crisis is to recognize the imperative for action. This is necessary work, and so it feels good to do it. Perhaps Arundhati Roy summed this up best: "The trouble is that once you see it, you can't unsee it. And once you've seen it, keeping quiet, saying nothing, becomes as political an act as speaking out. There's no innocence. Either way, you're accountable" (Roy 2001).

Meeting like-minded people, and joining forces with them, is the joyous, rejuvenating part of this work. Facing an existential crisis is best not done alone. "Organizing is the antidote to despair" is the mantra of climate activism. While climate anxiety is a very real problem, people doing this work also talk of active hope and stubborn optimism. *"Hope is not a lottery ticket you can sit on the sofa and clutch, feeling lucky. It is an axe you break down doors with in an emergency*" (Solnit 2004). Changemakers recognize that the sooner we let go of the status quo, the sooner we can start to build something new, and that is strong motivation.

Winning feels good too. Public pressure does affect change, and sometimes those changes are tangible, especially at a local level. Stopping new fossil fuel infrastructure, adding vegetarian and vegan options in schools, creating city climate action plans, passing local ordinances that reduce emissions, addressing energy and food inequity, electing climate-conscious officials, getting your university to divest—these are just a few examples of real systemic change happening because people demand it. Being involved in a local battle and seeing it succeed is not only empowering, it is an excellent way to strengthen your local community. Conversely, not winning becomes less and less acceptable, which is motivation to try even harder.

While I have tried to highlight reasons astronomers make good activists, most of this section applies equally to anyone because, as Hayhoe (2021a) tells us, "to care about climate change, all we really have to be is a human living on planet Earth."

15.5 What You Can Do

The question "what can I do?" belies either an astonishing lack of imagination or, more likely, yet another type of paralysis or avoidance. There is endless advice in the literature and online (e.g., Wilkinson 2021), and probably right in your local community. A transformation this big affects all aspects of our lives, providing unlimited opportunities to participate. The "To-Do" list at the end of the book "The Future We Choose" (Figueres & Rivett-Carnac 2020) offers Step 1, to do *right now*:

> Take a deep breath and decide that collectively we can do this, and that you will play your part. You will be a hopeful visionary for humanity through these dark days. From this moment, despair ends and tactics begin.

Joëlle Gergis, in the powerful essay "A Climate Scientist's Take on Hope" (Gergis 2023) puts it another way:

> My answer is simply this: recognize that you are living through the most profound moment in human history. Averting planetary disaster is up to the people alive right now. When you realize that the 2020's will be remembered as the decade that determined the fate of human history, you will tap into an eternal evolutionary force that has transformed the world time and time again.

These seemingly abstract, but galvanizing, pronouncements are already a powerful start. They make inaction untenable. Looking for something more concrete? Everyone's path is unique, and no one can tell anyone else what their own climate commitment should look like (Heglar 2023). What I can do is tell you my story, as one example, and the simple, iterative recipe I follow: (1) Educate yourself; (2) Find affinity; (3) Go for it!

15.5.1 My Story

My climate epiphany moment—most climate activists have one—was a casual announcement at a science colloquium. The Principal Investigator of an instrument destined for the airborne SOFIA Observatory (now decommissioned) would not be using the instrument on-board out of concern for the carbon emissions associated with the long plane flights. This was 2009, and it was the first (but certainly not the last) time I'd heard of someone allowing climate change to affect their career. I was Head of Science Operations at Gemini South at the time, and squarely in the "Inactive Alarmed" climate bubble. I was reading a lot about this issue, making small personal changes, but had not let the climate crisis affect my life much. What I heard that day had a profound effect on me, and prompted a question I still ask myself today, "What more can I do?"

Together with a few like-minded colleagues, I organized the first sustainability meeting within the January 2010 AAS (Rodgers et al. 2010). The meeting was a big success and ultimately led to the formation of the AAS Sustainability Committee. (This was the start of my carbon "handprint.") I'll never forget something the keynote speaker, physicist-activist Joe Romm, told us that day: "Enjoy astronomy while you can, because in 20 years we're all going to be working on climate." Thirteen years later, at least a few of the authors in this book are testament to this prophecy.

From there, my current trajectory was launched. I began to get more involved online, and by 2014 when my family left Chile for Portland, Oregon, I left professional astronomy to look for climate work. I started volunteering while working as adjunct faculty at local colleges and universities. Within two years, I was board chair of the local Portland 350.org chapter (350PDX), and found myself getting arrested (Pesanti 2016) with 20 other people on a railroad track, blocking oil trains to protest a near-miss disaster in Moser, Oregon (KATU 2016). I spent 7 hours in jail, and a few more in court (grateful for pro-bono representation from the ACLU), which concluded with a \$60 ticket and no record. (Not all arrests go this way of course, but many do.) I was cited again in 2022 at a Scientist Rebellion sit-in and those charges were promptly dismissed.

Today I consider myself a full-time (pro bono) climate activist and a part-time science educator. I am board chair of a different organization now (SustainUS), as well as a Climate Reality leader, a public speaker, a climate educator, and a part of Scientist Rebellion, Extinction Rebellion and Astronomers for Planet Earth. I've criss-crossed the activist spectrum: giving talks, raising money, speaking to politicians and media, organizing and emceeing rallies, attending more marches, protests and trainings than I can count, and supporting and participating in civil disobedience (Figure 15.6).

None of this was planned. Along the way, each new opportunity has presented itself. Like a boulder rolling downhill, it is my experience that activism begets more activism. Mine has been a humbling and exhilarating journey so far, and even if I am occasionally nostalgic for the work and colleagues I left behind, I don't regret any of it. I am grateful for a supportive family and modest financial security that allow me to do what someone else might not be able to do. That's the virtuous side talking; the other truth is that the rewards in relationships, personal growth, and satisfaction at each small (and not so small) victory, are far greater than I could have imagined. There are so many incredible, courageous, life-affirming people doing this work.

My story is proof the "butterfly effect" is real. Many people have knowingly and unknowingly guided my path, and I know my own actions have influenced others in unexpected ways. As I also work to align my lifestyle with my values, I strive for a carbon handprint that well exceeds my carbon footprint. Like Dr. Leah Stokes (2020), "My offset plan is activism."

One last point from my experience: Young people experience this crisis in ways perhaps challenging for older generations to fully comprehend. My story does not

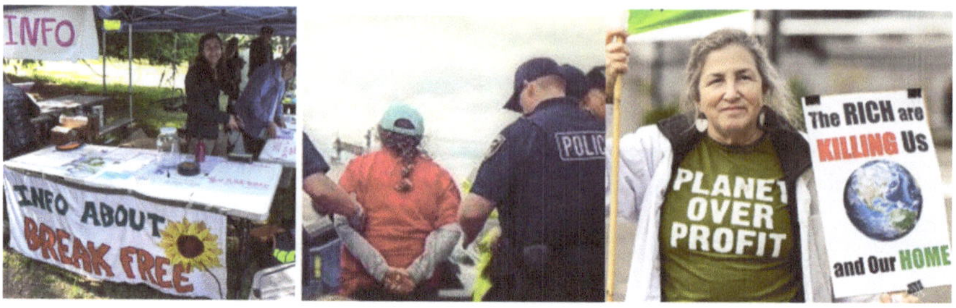

Figure 15.6. The author at a 3-day mass action in 2016 (left), getting arrested in Clark County, Washington in 2016 (center) *[Images by 350PDX]*, and at a Scientist Rebellion action in 2022 (right) *[Image courtesy of Mike Quinlan]*.

surprise them. Climate change has been with them their whole lives, if not directly impacting them, it is a shadow over every life choice they make (Hickman et al., 2021). This is both powerful motivation, and a sobering point to keep in mind, as you pursue your own path.

15.5.2 Educate Yourself

The published and online climate literature is vast. Do not limit yourself only to science and technology, nor overdose on doomism. Instead, reach outside your comfort zone to seek voices you may not normally hear. You may be surprised at the hopefulness you find. The climate crisis will not be solved by technology alone. For those of us educated and embedded in western science, and perhaps accustomed to a certain level of privilege, we have a lot to learn. Not only do the traditionally marginalized in our colonial and patriarchal society deserve to be heard, but they have wisdom to share that has too often been ignored or suppressed. To paraphrase Einstein, we cannot solve our problems with the same thinking that created them—new ideas and new leadership are needed.

Indigenous wisdom, or Traditional Ecological Knowledge (TEK), is particularly relevant to this problem. Activism for Native peoples around the world is integral to their way of life, to protect it from the extractive paradigm of "modern" society. Indigenous writers bring a completely different and valuable perspective to humans' place on the planet. They speak of the interconnectedness of all life, and of being in "right relationship" with the land, practicing reciprocity with Mother Nature (Kimmerer 2013). Chief Arvol Looking Horse, the Great Sioux/Lakota/Dakota Spiritual leader, tells us Mother Earth is "the Source of Life, not a resource" (Romero Inst 2015). This paradigm shift emphasizes reverence and respect for the planet and non-human life that goes beyond the strictly quantitative, economic, and human-centric calculations which typically dominate western climate solutions. Both approaches are valuable, and complement each other; TEK belongs on an equal footing with western science.

It is hard to select just a few recommendations from the ever-growing mountain of climate literature. However, here are a few recent favorites, ordered roughly from the more spiritual or holistic, to the more quantitative:

- *Braiding Sweetgrass—Indigenous Wisdom, Scientific Knowledge and the Teachings of Plants* by Robin Wall Kimmerer (2013). This beautifully written bestseller is as joyful to read as it is instructive. An excellent primer on indigenous thinking.
- *All We Can Save—Truth, Courage and Solutions for the Climate Crisis* edited by Drs. Ayana Johnson and Katherine Wilkinson (2021). An anthology of impactful essays and poems from a diverse who's who of women climate activists–a wide-ranging collection of wise voices.
- *Not Too Late—Changing the Climate Story from Despair to Possibility* edited by Rebecca Solnit and Thelma Young Lutunatabua (2023). Another excellent collection of essays; besides the title theme, a common thread shines through of joy and community through action.
- *The Future We Choose–Surviving the Climate Crisis*, by Christiana Figueres and Tom Rivett-Carnac (2020). These architects of the Paris climate agreement offer a parable for our time, and a pragmatic, well-informed, and highly readable call to action.
- *Less is More: How Degrowth Will Save the World* by Jason Hickel (2020). Hickel provides both a thorough historical perspective and a roadmap to break free of the capitalist and consumerist paradigm.
- *Drawdown: the Most Comprehensive Plan Ever Proposed to Reverse Global Warming* by Paul Hawken (2017). A detailed, quantitative analysis of the bounty of solutions available now (continually updated on the Project Drawdown website).
- News and Podcasts: *The Guardian/Climate Crisis, NYTimes/Climate Forward, Inside Climate News, Grist.org, Yes! magazine, Heated* (blog), *Outrage & Optimism* (podcast)
- For reading more related to rebellion and direct action, see Scientist Rebellion's Resources.
- For staying on top of the latest book club options, Yale Climate Connections regularly posts recommendations.

These are all fairly recent, as the political and social landscape is changing fast. And they are realistically optimistic choices, to inspire and motivate, because we cannot afford doom and despair.

15.5.3 Seek Affinity

Find your people and pursue your interests. You may find yourself spending a lot of time doing this, so by all means find work and people you enjoy. Foster relationships. Activism is hard work, but it can also be a lot of fun.

Nonprofits doing this work are plentiful and diverse. From local to national and international, from education to direct action to hands-on implementation, the possibilities are endless. There are organizations for many identity groups (e.g., women, youth, indigenous, BIPOC, LGBTQ+), small orgs with few or no staff (where your contribution can make a huge difference), to "big greens" like Sierra Club and others with large budgets and resources. There are opportunities in government and the private sector as well, but not all "solutions" are equal. Unfortunately, as addressing climate becomes more popular, so does "greenwashing"; read mission statements and look for accomplishments to determine if a particular opportunity is worth your time and money.

The next section provides a selective, very incomplete list of organizations, biased toward my own focus on ending fossil fuels. There are many other equally important and interconnected causes, including regenerative agriculture, education and health care for women and girls, stopping plastic pollution, protecting our oceans, addressing deforestation, land use and animal extinction.

Having selected one or more organizations to apply yourself to, the next question is: what will you actually do? People new to climate activism often feel understandably unsure how to step into the work. Certainly, it is wise to listen first, seek to understand before imposing. That said, if you demonstrate your willingness and commitment by showing up and pitching in, you will almost always find plenty to do.

A strategy for success is to find the intersection between your skills, your interests and the mission of the organization. Author-scientist-activist Dr. Ayana Johnson recommends drawing a Venn diagram answering these three questions: (1) What are you good at?; (2) What work needs doing? and (3) What brings you joy? The intersection of these is where you will find your climate action.

The (mostly outdated) image of a lone scientist toiling away does not apply here. Climate organizing is primarily community-based work with many hands, and roles, within and between organizations. A model for social movements defines four roles critical for success: citizens (also sometimes called "helpers"); changemakers (or advocates), rebels (to call out the bad) and reformers (to build the new) (Moyer 2001). Find the one that speaks to you.[15]

Astronomers tend to be good "big picture" thinkers, and particularly useful with quantitative tasks. These skills are typically in high demand, particularly for inward-facing jobs such as budgeting and bookkeeping, strategic planning, and grant writing. But as I hope I have made clear, we have an important outward-facing role to play too, and an important one to consider is civil disobedience.

15.5.4 Civil Disobedience

Non-violent direct action (NVDA) and civil disobedience are highly effective forms of activism (Chenoweth 2013). Dramatic images draw attention to an issue, attract news coverage, and demonstrate urgency, passion and conviction. Nearly 400 people were arrested on the White House lawn in 2014 to protest the now-canceled Keystone XL pipeline, the largest act of civil disobedience at the White House in

[15] Try the "What Kind of Changemaker Are You?" quiz: https://action.storyofstuff.org/survey/changemaker-quiz/

a generation (Henn 2014). The UK Parliament became the first national government to declare a Climate Change Emergency in 2019, less than a month after Extinction Rebellion led ten days of protests resulting in over 1000 arrests in London (BBC 2019).

This form of activism, by definition, is often illegal and therefore risky (and sometimes dangerous). Deciding to risk arrest is a personal decision, not to be taken lightly. My own experience notwithstanding, being arrested has consequences, and can be dangerous. One should consider their own physical and mental health, the political environment, and more—some situations have lower risk than others, and some people are at lower risk than others. Environmental activists around the world literally risk their lives, and many outside the U.S. die every year (Greenfield 2022). Systemic racism in the U.S. criminal justice system increases the risk for Black people, Indigenous people, and other people of color. White people, particularly healthy adults, may choose civil disobedience specifically because it is relatively lower risk for them. This is a good use of white privilege. You do not have to engage in NVDA to be an activist, but if you can, it is a powerful way to contribute.

If you are unable or unwilling to be arrested, there are many other ways to support civil disobedience. Jail support is an essential function that provides completely lawful, and very much appreciated, support to arrestees. Roles such as "legal observer" and "police liaison" are important lawful tasks during direct actions. Financial support, e.g., for bail and legal services, is always needed and extremely helpful, as are endorsing and amplifying actions.

15.5.5 Go for It!

Whatever you learn, whatever people you find, the final and most important step is to *do*, don't wait. "We must act to be hopeful, not the other way around" (Hayhoe 2021b). Be on the field, not on the sidelines. And be open. Pick something, then add, grow, change, go deeper or pick something else if where you find yourself is not a good fit. Nothing is static—the only constant is the need for rapid, global transformation—the solutions, the challenges, and the opportunities continue to evolve rapidly.

As of this writing, I am actively involved with no less than six organizations and I follow, and sometimes financially or otherwise support, several more. For me, the variety keeps it interesting, and at the same time, it is a lot—which is not for everyone. You might rather commit yourself fully to a single cause or group, or have only a small amount of time to offer but more to give financially, or choose to become more active where you already are, in your workplace or school or neighborhood group. It doesn't matter. There are countless ways to contribute.

The next section offers some organizations to consider. A quick Ecosia search will lead you to options in your area and interests. The ladder of engagement is multi-dimensional and never-ending. Reassess your position frequently—ask yourself: "What more can I do?"; "How can I have the most impact?" In answering, try to push the envelope just a bit further, taking inspiration from others. At the same time, know your limits and take care of yourself and others. Check in with yourself and your loved ones on a regular basis. Burnout is real, and this mission is a marathon, not a sprint. The climate crisis is not going away anytime soon; the movement to address it needs resilient warriors.

15.6 Resources and Organizations

There are literally thousands of non-profit organizations working to address climate. As the climate crisis touches (or will touch) every aspect of our lives, addressing this issue intersects with a multitude of communities, causes and fields of study. Some organizations focus on direct action or political lobbying or education, while others focus on specific solution areas such as agriculture or electrification or legal avenues, and others pay attention to process and analysis like the World Resources Institute, for example. Some work on climate exclusively but many work at the intersection of climate and other environmental or social justice issues.

Table 15.1 provides a small sampling of organizations, more to prompt your curiosity than make specific recommendations. This is an admittedly biased and arbitrary list from my own experience and knowledge, coarsely and imperfectly categorized.

Table 15.1. A sampling of organizations addressing climate from different perspectives, with links. Many, but not all, are US-centric. Bold type indicates organizations directly familiar to the author.

Education	Alliance for Climate Education, **Climate Reality Project**, Climate Interactive, **Global Optimism**, Yale Climate Change Communication
Faith-Based	Catholic Climate Covenant, Ecofaith.org, Green Muslims, Interfaith Power & Light, Jewish Climate Action, Plum Village, **Poor People's Campaign**
Indigenous & BIPOC-led	Climate Justice Alliance, Hip Hop Caucus, Honor the Earth, Indigenous Environmental Network (IEN), NDN Collective, Pachamama Alliance
Legal, Financial	EarthJustice, **Our Children's Trust**, STOP Ecocide international, Stop the Money Pipeline
Political/Direct Action	**350.org**, Citizens Climate Lobby, Climate Action Network, Climate Mobilization, **Extinction Rebellion (XR)**, Oil Change International, Third Act
Scientist-activists	**Astronomers for Planet Earth (A4PE), Scientist Rebellion**, Scientists Warning, Union of Concerned Scientists; Katherine Hayhoe, James Hansen
Solutions/analysis	Generation180, **Project Drawdown**, Rewiring America, Stockholm Resilience Centre, **World Resources Institute**
Women-led	Daughters for Earth, Women Donors Network, **Women's Earth Alliance**
Youth-led	Fridays for the Future, Movement Generation, **Sunrise Movement**, **SustainUS**

I have seen a pointed cartoon showing a young child sitting on an older person's lap asking, "What did you do to fix climate change, Granddad?" Whatever actions you choose, know that you are contributing to an epic task unlike any in human history. Like it or not, the climate crisis is happening on our watch, and what we do (and don't do) in the next decade will affect life on this planet for generations, perhaps millenia to come. It is up to us.

References

Artico, D., et al. 2023, Front. Sustain., 4, 1155897

Attenborough, D. 2021, Speech to the United Nations Conference of Parties (COP26), Glasgow, Scotland https://www.youtube.com/watch?v=o7EpiXViSIQ

Ballew, M., et al. 2023, Who is most likely to talk about climate change? Yale Program on Climate Change Communication (New Haven, CT: Yale Univ. and George Mason Univ.)

BBC 2019, UK Parliament declares climate change emergency *BBC News* https://www.bbc.com/news/uk-politics-48126677 [accessed 20 July 2023]

Bergensia, 2022, The Keeling Curve and 40 Years of Blah-blah-blah (Bergensia The Sustainable Gazette), https://bergensia.com/the-keeling-curve-and-40-years-of-blah-blah-blah/

Brulle, R. J. 2023, EnPo, 32, 185

C-SPAN, 1985, Greenhouse Effect *C-Span* https://www.c-span.org/video/?125856-1/greenhouse-effect [accessed 20 July 2023]

Cairns, S. 2020, "Take the Science Oath for Climate" (Scientists' Warning) https://scientistswarning.org/2020/11/11/the-science-oath-for-climate/ [accessed 18 Dec 2023]

Chenoweth, E. 2013, The Success of Nonviolent Resistance *TEDxBoulder* https://www.youtube.com/watch?v=YJSehRlU34w [accessed 20 July 2023]

Chenoweth, E., & Stephan, M. 2011, Why Civil Resistance Works: The Strategic Logic of Nonviolent Conflict (New York: Columbia Univ. Press)

Columbia Law School, 2015, Juliana v. United States *U.S. Climate Change Litigation database* Sabin Center for Climate Change Law at Columbia Law School http://climatecasechart.com/case/juliana-v-united-states/ [accessed 20 July 2023]

Cunningham, N. 2022, The Pacific Northwest has defeated dozens of fossil fuel projects — but the industry still wants to quietly expand (Nation of Change), https://www.nationofchange.org/2022/06/28/the-pacific-northwest-has-defeated-dozens-of-fossil-fuel-projects-but-the-industry-still-wants-to-quietly-expand [accessed 16 Jul 2023]

de Place, E., & Gruen, D. 2015, The Thin Green Line Is Stopping Coal and Oil in Their Tracks (Sightline Institute), https://www.sightline.org/2015/08/13/the-thin-green-line-is-stopping-coal-and-oil-in-their-tracks/ [accessed 16 July 2023]

Ettinger, J., McGivern, A., Spiegel, M. P., et al. 2023, ClCh, 176, 22

Figueres, C. 2021, 'The IPCC Sixth Assessment is a Call to Break the Mental Boundaries Holding Us Back' (Outrage and Optimism) https://www.outrageandoptimism.org/blog/the-ipcc-sixth-assessment-is-a-call-to-break-the-mental-boundaries-holding-us-back [accessed 20 July 2023]

Figueres, C., & Rivett-Carnac, T. 2020, The Future We Choose - Surviving the Climate Crisis (New York: Alfred A. Knopf)

Freedland, J. 2023, 'As heat records break, the climate movement has the right answers - but the words are all wrong' (The Guardian) https://www.theguardian.com/commentisfree/2023/jul/14/big-oil-climate-crisis-fossil-fuel-public [accessed 14 July 2023]

Gardner, C. J., & Wordley, R. 2019, Nat Ecol Evol, 3, 1271

Gergis, J. 2023, A Climate Scientist's Take on Hope Not Too Late, ed. R. Solnit, & T. Lutunatabua (LaVergne: Haymarket Books) pp 38

Gillis, J. 2013, Climate Maverick to Retire From NASA (New York Times) https://www.nytimes.com/2013/04/02/science/james-e-hansen-retiring-from-nasa-to-fight-global-warming.html [accessed 20 July 2023]

Goldberg, M., et al. 2021, Yale Program on Climate Change Communication (New Haven, CT: Yale Univ. and George Mason Univ.)

Greenfield, P. 2022, 'More than 1700 Environmental Activists Murdered in the Last Decade - Report' (The Guardian) https://www.theguardian.com/environment/2022/sep/29/global-witness-report-1700-activists-murdered-past-decade-aoe [accessed 20 July 2023]

Gunderson, D. 2021, 'I live with Standing Rock in my heart': Massive pipeline protest resonates 5 years later (MPR News) https://www.mprnews.org/story/2021/04/01/i-live-with-standing-rock-in-my-heart-massive-pipeline-protest-resonates-5-years-later [accessed 27 July 2023]

Hansen, J. 2021, "Dr James Hansen: Silent Forests" (Red Green and Blue) https://redgreenandblue.org/2021/06/21/dr-james-hansen-silent-forests/ [accessed 20 July 2023]

Harvey, C. 2022, Scientists Risk Arrest to Demand Climate Action Scientific American (E&E News) https://www.scientificamerican.com/article/scientists-risk-arrest-to-demand-climateaction/ [accessed 23 June 2023]

Hawken, P. 2017, Drawdown: the Most Comprehensive Plan Ever Proposed to Reverse Global Warming (New York: Penguin Books)

Hayhoe, K. 2021a, Saving Us: A Climate Scientist's Case for Hope and Healing in a Divided World (New York: One Signal Publishers/Atria Books)

Hayhoe, K. 2021b, In the Face of Climate Change, We Must Act So That We Can Feel Hopeful —Not the Other Way Around, (Time.com) https://time.com/6089999/climate-change-hope/ [accessed 20 July 2023]

Heglar, M. 2023, Here's Where You Come In Not Too Late, ed. R. Solnit, & T. Lutunatabua (LaVergne: Haymarket Books) 19

Henn, J. 2014, Keystone XL Protest at the White House Leads to Mass Arrests 350.org https://350.org/press-release/keystone-xl-protest-at-the-white-house-leads-to-mass-arrests/ [accessed 20 July 2023]

Hickel, J. 2020, Less is More: How Degrowth Will Save the World (New York: Penguin Random House)

Hickman, C. 2021, LaPH, 5, E863

IPCC Core Writing Team, Lee, H., & Romero, J. (ed) 2023, Climate Change 2023: Synthesis Report. A Report of the Intergovernmental Panel on Climate Change. Contribution of Working Groups I, II and III to the Sixth Assessment Report of the Intergovernmental Panel on Climate Change (Geneva, Switzerland: IPCC) 36 pages. (in press)

Jacob, S., Rudgard, O., & Roach, A. 2023, This Dutch Airport Wants to Ban Private Jets (Time Magazine) https://time.com/6276617/schiphol-amsterdam-airport-ban-private-jets/ [accessed 20 July 2023]

Johnson, A. E., & Wilkinson, K. K. 2021, All We Can Save, ed. A. Johnson, & K. Wilkinson (New York: One World)

Johnston, E. 2020, Loving a Vanishing World All We Can Save, ed. A. Johnson, & K. Wilkinson (New York: One World) pp 256

KATU 2016, 14 oil train cars derail in Columbia River Gorge, railcars erupt in flames; I-84 reopens (KATU News) https://katu.com/news/local/train-catches-fire-outside-mosier-in-the-columbia-river-gorge-oil-smoke-hood-river [accessed 20 July 2023]

Kaufman, M. 2021, The carbon footprint sham: A 'successful, deceptive' PR campaign" (Mashable) https://mashable.com/feature/carbon-footprint-pr-campaign-sham [accessed 22 June 2023]

Kimmerer, R. W. 2013, Braiding Sweetgrass: Indigenous Wisdom and the Teachings of Plants (Minneapolis, MN: Milkweed Editions)

Leiserowitz, A., et al. 2023, Global Warming's Six Americas Yale Program on Climate Change Communication (New Haven, CT: Yale Univ. and George Mason Univ.)

Lipman, J. 2021, Global Fossil Fuel Divestment Commitments Database, managed by Stand. Earth https://divestmentdatabase.org/report-invest-divest-2021/

Maibach, E., Leiserowitz, A., Rosenthal, S., Roser-Renouf, C., & Cutler, M. 2016, Yale Program on Climate Change Communication (New Haven, CT: Yale Univ. and George Mason Univ.)

Marshall, G. 2015, Don't even think about it: why our brains are wired to ignore climate change (London: Bloomsbury Publishing) https://www.bloomsbury.com/us/dont-even-think-about-it-9781632861023/

Matthews, K. R. 2020, Interface: a journal for and about social movements, 12, 591

McFall-Johnsen, M. 2021, The companies polluting the planet have spent millions to make you think carpooling and recycling will save us (Business Insider), https://www.businessinsider.com/fossil-fuel-companies-spend-millions-to-promote-individual-responsibility-2021-3 [accessed 22 June 2023]

McKibben, B. 2016, Embarrassing Photos of Me, Thanks to My Right-Wing Stalkers (New York Times) https://www.nytimes.com/2016/08/07/opinion/sunday/embarrassing-photos-of-me-thanks-to-my-right-wing-stalkers.html [accessed 20 July 2023]

Moore, E. 2022, A Decade of Successes Against Fossil Fuel Export Projects (Sightline Institute), https://www.sightline.org/wp-content/uploads/2022/06/A-Decade-of-Successes-Against-Fossil-Fuel-Export-Projects-in-Cascadia_Sightline-Institute-June-2022.pdf [accessed 16 July 2023]

Moyer, B. 2001, Doing Democracy: The MAP Model for Organizing Social Movements, 2001, pp 21-22 & pp 28-29 accessed from The Commons Social Change Library (https://commonslibrary.org/the-four-roles-of-social-activism/)

Muir, B. 2020, Should Scientists be Activists? (Chemistry World), https://www.chemistryworld.com/careers/should-scientists-be-activists/4011293.article [accessed 23 June 2023]

Norris, G. A. 2017, Introducing Handprints: A Net-Positive Approach to Sustainability (Harvard Extension School Blog), https://extension.harvard.edu/blog/introducing-handprints-a-net-positive-approach-to-sustainability/ [accessed 22 June 2023]

Oreskes, N., & Conway, E. M. 2011, Merchants of doubt: how a handful of scientists obscured the truth on issues from tobacco smoke to global warming (London: Bloomsbury Press)

Parker, L. 2020, National Geographic Magazine Earth Day Issue, https://www.nationalgeographic.com/magazine/article/greta-thunberg-wasnt-the-first-to-demand-climate-action-meet-more-young-activists-feature [accessed 20 July 2023]

Pesanti, D. 2016, The Columbian, https://www.columbian.com/news/2016/jun/18/21-arrested-in-vancouver-oil-train-protest/ [accessed 21 July 2023]

Pierre-Louis, K. 2020, All We Can Save, ed. A. Johnson, & K. Wilkinson (New York: One World) 138

Racimo, F., et al. 2022, eLife, 11, e83292

Read, R. 2017, Scientists' Warning, https://www.scientistswarning.org/2021/06/18/why-we-all-need-to-become-crusading-scholar [accessed 22 June 2023]

Ripple, W. J., et al. 2017, BioSc, 67, 1026

Rodgers, B., et al. 2010, American Astronomical Society, AAS Meeting #215, BAAS 42, 408
Romero Institute, 2015, The Romero Institute Report, https://romeroinstitute.wordpress.com/2015/07/08/mother-earth-is-a-spirit-the-source-of-life-not-a-resource/ [accessed 20 July 2023]
Roy, A. 2001, Power Politics (2 edn; Cambridge: South End Press)
Sagan, C. 1994, *Pale Blue Dot* excerpt cited by The Planetary Society https://www.planetary.org/worlds/pale-blue-dot [accessed 20 July 2023]
Saha, D., Shrestha, R., & Jordan, P. 2022, How a Clean Energy Economy Can Create Millions of Jobs in the US (Washington: World Resources Institute) https://www.wri.org/insights/us-jobs-clean-energy-growth [accessed 20 July 2023]
Solnit, R., & Lutunatabua, T. 2023, Not Too Late, ed. R. Solnit, & T. Lutunatabua (LaVergne: Haymarket Books)
Solnit, R. 2004, Hope in the Dark: Untold Histories, Wild Possibilities (LaVergne: Haymarket Books)
Stokes, L. 2020, All We Can Save, ed. A. Johnson, & K. Wilkinson (New York: One World) 337
UNEP, 2022, Emissions Gap Report 2022: The Closing Window—Climate crisis calls for rapid transformation of societies (United Nations Environment Programme) www.unep.org/resources/emissions-gap-report-2022 [accessed 20 July 2023]
UNFCCC, 2018, The Paris Agreement - Publication Paris Climate Change Conference November 2015, Session COP21
UN Treaty Collection, 2016, 7.d Paris Agreement, https://treaties.un.org/Pages/ViewDetails.aspx?src=TREATY&mtdsg_no=XXVII-7-d&chapter=27&clang=_en [accessed 29 June 2022]
UN, 2021, Climate change is a 'global emergency', people say in biggest ever climate poll *UN News* https://news.un.org/en/story/2021/01/1083062 [accessed 22 June 2023]
Wilkinson, K. 2021, The Climate Crisis is a Call to Action. These 5 Steps Helped Me Figure Out How to be of Use. (Time 2030), https://time.com/6071765/what-can-i-do-to-fight-climate-change/ [accessed 20 July 2023]

Chapter 16

Climate Justice

B Rodgers and R Mason

> "The biggest injustice of climate change is that the hardest hit are the least responsible for contributing to the problem."
>
> —Bali Principles of Climate Justice

Climate change is primarily caused by affluent nations and individuals, but it is disproportionately affecting vulnerable communities in the Global South,[1] as well as low-income people and people of color in the Global North. Supporting those least responsible for, and most affected by, climate change in adapting and responding to it is a moral imperative. On a practical level, fostering a diverse and inclusive climate movement allows a wider range of solutions to be envisioned, explored, and adopted. As global citizens and participants in a carbon-intensive scientific field, astronomers should actively amplify voices and solutions that are often overlooked in our institutions.

16.1 A Moral Imperative

Climate justice is perhaps easiest understood by its absence. "That's probably the simplest thing about climate change—the injustice," says Mary Annaïse Heglar in Not Too Late (Heglar 2023). Existing layers of injustice and inequity—racial, gender-based, generational, economic, and non-human—are amplified and multiplied by climate breakdown. Displacement of millions in Pakistan by unprecedented flooding; the devastating impact of hurricanes on Caribbean islands (Shultz et al. 2020); storm-related toxic discharges in poor communities of Louisiana's Cancer Alley; indigenous villages in Alaska and Pacific island nations losing their homes to sea level rise; food scarcity, drought and conflict across the Global South, leading to dangerous forced migration to unwelcoming countries. These are just a few

[1] Countries classified by the World Bank as low or middle income, primarily located in Africa, Asia, Latin America and the Caribbean.

doi:10.1088/2514-3433/acfcb6ch16

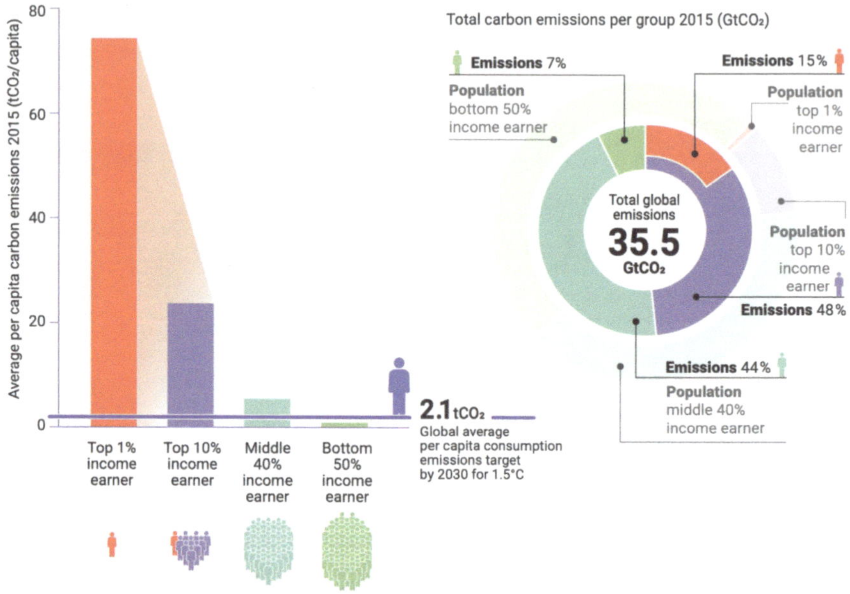

Figure 16.1. The disproportionate impact of high income earners.[2] Figure ES.8 from *Emissions Gap Report 2020* (UNEP 2020).

examples of climate injustice. As impacts worsen, a more pointed term perhaps is *climate apartheid*—the "differentiation in who pays the disproportionate price of climate breakdown, who is made expendable, and who is spared for now" (Sultana 2023).

Marginalized peoples across the world have always suffered the worst of environmental degradation. "Those who hold more power and privilege in society are less likely to be exposed to environmental injustice and hazards" (Thomas 2022). One example in the U.S. is the "air pollution burden": Black and Latinx people are exposed to "56% and 63% more pollution than is caused by their consumption, and which is linked to higher rates of cardiovascular and respiratory issues" while the "opposite is true for white Americans, who experience a "pollution advantage" in which they breathe in 17% less air pollution than they cause" (Thomas 2022, citing Rice 2019). It is no surprise then, that the communities being hit first and worst by climate breakdown are the ones that have contributed the least to cause the problem and have the fewest resources to defend themselves.

That wealthy nations, and the wealthiest individuals, have contributed most to create the climate crisis is also undeniable. See, for example, the 2020 UNEP Emissions Gap report (Figure 16.1). "The emissions of the richest 1 percent of the global population account for more than twice the combined share of the poorest 50 percent." (UNEP 2020).

[2] Check your own position in the global income ladder here.

This figure reports annual consumption emissions for 2015 only. Considering *cumulative* greenhouse gas emissions, the disparity is even more striking. Roughly 57% of historic cumulative emissions come from Developed countries (~25% from the United States alone), while nearly 50 of the Least Developed Countries contributed less than 0.5% of emissions for the same period (Dhakal et al. 2022). These data are from global studies by country, however, since emissions are well correlated with income, similar disparities exist between high- and low-income populations within countries as well.

That the U.S. and western Europe used fossil fuels to build wealthy societies is not the "fault" of the people living today who benefit from these societies. It does, however, place a greater responsibility on wealthy nations and individuals to mitigate the problem, and to aid low income countries and peoples with adaptation and resilient development.

Not only is there a clear moral imperative to address these unequal impacts, there is a clear *opportunity* to address inequity by transforming from an extractive and exploitative economy to a sustainable and regenerative economy that provides benefits for all levels of society. At a very practical level, the communities most impacted (which make up the majority of the world's people) are already developing solutions. Their hard-gained experience must be part of the global response.

16.2 Valuing Frontline Experience

Climate justice is not only about addressing the harm done to vulnerable communities; it is also recognizing the role these communities have in building a resilient future. Representation and inclusion allow the entire world to benefit from the experience of people doing that work now.

The Intentional Environmentalist website says it this way:

> The Earth and its ecosystems thrive on diversity and so does climate action. When communities of color, who are impacted the most by environmental injustices and are also leading the way in creating solutions, are included in environmental decision-making, movements, and educational systems, environmentalism will be brighter, more equitable, and more revolutionary for all.

Impacted communities, also called "frontline" communities, are experiencing climate impacts and other environmental crises *now*. The people in these communities are not victims, they are survivors. Mitigation and adaptation are existential requirements. They are the experts, by necessity. They have answers, and they are already implementing solutions, often with limited financial resources and negligible political power. Many climate solutions start at the local level, and some of these solutions can be replicated and scaled up worldwide (e.g., the UpLink Top Innovators who won the Climate Justice Challenge).

The damaging and inaccurate representation of environmentalists as white elites ignores the generations of struggle happening outside the dominant ruling class. People of all colors and creeds, and especially women, around the world have always—and continue to—implement sustainable solutions for their families and communities as a way of life. They are not waiting for anyone to tell them what to do: "We're giving directions, demanding action, and riding into battle to save our children and our country from the impending destruction" (Toney 2021; Schiffman 2023).

Indigenous communities worldwide constitute less than 5% of the global population and yet safeguard roughly 80% of the Earth's biodiversity (World Bank 2023). These are among the most economically impoverished peoples, having suffered deadly injustice at the hands of colonization for centuries, and yet indigenous wisdom is now recognized as a critical part of solving the climate crisis. Forest and fire management, as well as traditional agricultural and fishing techniques are just a few examples of "indigenous peoples' practices [that] enrich and accelerate collective progress towards achieving the goals of the Paris Agreement" (UNFCCC 2022). Indigenous wisdom provides "living models of sustainability that are rooted in ancient wisdom and that inform us how to live in balance with all of our relations on Mother Earth" (Mitchell 2021). Mitchell[3] believes that current mainstream ideologies are not capable of accommodating this holistic view. To benefit from the alignment of Indigenous knowledge with Western thought, she argues that "mainstream thinkers will have to be courageous enough to challenge the barriers that have prevented Indigenous knowledge from coming forward previously and begin to expand their sight to a more holistic vision" (Mitchell 2021). For a more concrete take on what this might require, see this opinion piece by Jeremy Jiménez (2023).

If everything already stated is not enough motivation, there is the obvious fact that frontline communities constitute the majority of humanity; "solving the climate crisis" must involve solutions developed by, with, and for those communities. For more on the need for, and progress toward, what they call "transformative adaptation," see the thorough review by Shi and Moser (2021, and references therein). To imagine a future without a just transition, consider that climate breakdown is already driving dangerous and unmanageable migration, and experts predict that climate-driven refugees will likely grow to the millions worldwide as conditions worsen (Cho 2021). Humanity cannot create a resilient and sustainable future without improving the lives of over half the world's population.

16.3 Uneven Progress

This reality has become obvious. As the impacts of climate chaos intensify, the injustice becomes harder to ignore. The need for, and benefits of, equitable solutions are becoming more pronounced across the world and at all levels of government (Berwyn 2022). At the same time, Diversity, Equity and Inclusion (DEI, or JEDI, adding Justice) programs that help to increase representation have become much more common across the Western world from nonprofits and businesses to academia and government. Impacted communities are demanding to be heard.

[3] Sherri Mitchell, JD, is a Native American attorney, teacher and activist, and author of *Sacred Instructions: Indigenous Wisdom for Living Spirit-based Change.*

The Synthesis of the United Nations' IPCC Sixth Assessment, AR6, states clearly that "Adaptation and mitigation actions, that prioritize equity, social justice, climate justice, rights-based approaches, and inclusivity, lead to more sustainable outcomes, reduce trade-offs, support transformative change and advance climate resilient development" (IPCC 2023). National and local government programs, like the US Government's Justice40 Initiative, work to provide much needed rebalancing of effort and resources. Countless non-governmental organizations from local to international are doing the hard work of implementing just solutions. (A few of these are listed at the end of this chapter.)

Some businesses also recognize their responsibility and are taking action. The B Lab is a nonprofit network committed to transforming the global economy (and best known for certifying "B corporations"). With others, they have created a Climate Justice Playbook to "help business leaders understand the intersection of climate action and social justice and advance a justice-centered approach to climate action" (B Corp Climate Collective et al. 2021). This helpful volume provides extensive background information, insights, case studies and more.

The youth of today offer another important voice, and they are rising up in impressive numbers and having an impact, despite almost no political power. Young people and future generations will experience a climate-changed world, and how bad it will be depends on the actions we take *this* decade. Many youth are taking their fight for justice to the courts. Around the world, governments are being sued for inaction. In the United States, Our Children's Trust has filed lawsuits on behalf of young plaintiffs in many states and federally, suing for a livable future. The federal case, officially *Juliana v. United States* and commonly called Youth v. Gov, asks the courts to recognize the atmosphere as part of a common trust and therefore protected under existing public trust law. The case also argues that the actions and inaction of the U.S. Government regarding climate change have violated the youngest generation's constitutional rights to life, liberty, and property. After years of delay, the case appears to now be on the path to trial (E360 Digest 2023). EU courts are hearing similar cases (Quell 2023).

Despite this progress, pursuit of short-term profits and the legacy of colonialism and racism are not easily overcome. Recent trends are at once hopeful and frustratingly inadequate. As the UN COP26 Conference closed in Glasgow in 2021, rich nations were congratulating themselves on weak progress while poorer nations expressed "disappointment." Reminding us all of the gravity of this crisis, Asad Rehman, speaking for non-governmental organizations, did not mince words: *"It is immoral for the rich to talk about their future children and grandchildren when the children of the Global South are dying now"* (UN 2021).

16.4 What Astronomers Can Do

Astronomers may feel that the topic of climate justice is not relevant to them. Maybe it is too 'political', or seemingly tangential to the core issue of reducing greenhouse gas emissions. We argue that this is not the case. Cutting emissions isn't simply a question of finding the right technology; it involves difficult decisions about who in society will benefit and sacrifice, and what will have to change. As scientists, we are committed to communicating truths, and our words carry weight and credibility. Engaging with climate justice allows us to take part in shaping a world in which certain people, places and species are no longer considered expendable. More selfishly, no one will be left untouched by the social upheaval that, without concerted efforts to the contrary, climate breakdown will provoke. Recognizing, and amplifying, the critical interdependence of these issues—the intersectionality of environmental destruction and social injustice—must be part of all of our work to address the climate crisis.

As academics, among the most important things we can do is educate ourselves and others, and use our platform to raise awareness and lift up voices and examples not always heard in mainstream media and in our institutions. Call out misinformation and highlight just solutions, and the *importance* of just solutions. We can also lend our own voices, our time, labor and money, to support any of the countless groups implementing and fighting for justice-based solutions whether in our local communities, nationally or internationally.

The intention of this brief chapter has been to highlight the importance of centering human (and non-human) rights and justice within the climate solutions conversation. The authors are not experts, and this short contribution is far from comprehensive. Many, many resources are available online; we list a few below.

Literature:

- Books with online resources:
  - The All We Can Save project accompanies the book (Johnson & Wilkinson 2021) which is a powerful anthology of feminine voices, many of whom are people of color. The website offers numerous resources including class lessons and assignments, facilitation guides for creating "Climate Circles," and accompanying TED talks.
  - The Intersectional Environmentalist is "a climate justice collective radically imagining a more equitable + diverse future of environmentalism" (from the website). The website includes a blog, a podcast, offerings for trainings and speakers, and other resources. The founder, Leah Thomas, widely credited with defining the term, is author of the book with the same title (Thomas 2022).
- Useful references/short articles:
  - Bali Principles of Climate Justice | corpwatch—Adopted at the Rio+10 United Nations Climate Summit in 2002, building upon the Environmental Justice principles developed at the First National People of Color Environmental Leadership Summit in Washington D.C. in 1991.
  - COP26 in Glasgow, Scotland—Climate Justice Alliance
  - Global Spotlight Report #48: Climate Justice
  - Two Kinds of Climate Justice: Avoiding Harm and Sharing Burdens (Caney, 2014)
  - We Can't Solve the Climate Crisis Unless Black Lives Matter | Time

- What is 'climate justice'? | Yale Climate Connections
- What is Climate Justice? | Global Witness
- What is Climate Justice? And what can we do [to] achieve it? | UNICEF Office of Global Insight & Policy

Organization Websites:

- International
  - Mary Robinson Foundation Climate Justice center
  - United Nations Local Communities and Indigenous Peoples Platform
- US-based government
  - Environmental Protection Agency—Environmental Justice
  - The White House Justice40 initiative
- Business-oriented
  - B Lab and the Climate Justice Playbook
  - Equitable & Just National Climate Platform
- Non-profits
  - Climate Justice Alliance
  - Global Citizen
  - Green 2.0
  - Grist: Climate. Justice. Solutions. (media)
  - The Solutions Project
  - Youth v. Gov

References

B Corp Climate Collective, 2021, B Lab, the Skoll Centre for Social Entrepreneurship at the University of Oxford, and the Climate Action Champions Team, The Climate Justice Playbook for Business, version 1.0, licensed under a Creative Commons Attribution 4.0 International License. Available at https://www.bcorpclimatecollective.org/climate-justice-playbook [accessed 31 July 2023]

Berwyn, B. 2022, Inside Climate News, https://insideclimatenews.org/news/11032022/ipcc-report-climate-justice/ [accessed 3 Aug 2023]

Caney, S. 2014, JOPP, 22, 125

Cho, R. 2021, Columbia Climate School State of the Planet, https://news.climate.columbia.edu/2021/05/13/climate-migration-an-impending-global-challenge/ [accessed 3 Aug 2023]

Dhakal, S., et al. 2022, Emissions Trends and Drivers IPCC, 2022: Climate Change 2022: Mitigation of Climate Change. Contribution of Working Group III to the Sixth Assessment Report of the Intergovernmental Panel on Climate Change, ed. P. R. Shukla, et al. (Cambridge: Cambridge Univ. Press)

E360 Digest, 2023, Yale Environment 360 (New Haven, CT: Yale School for the Environment) https://e360.yale.edu/digest/juliana-youth-climate-lawsuit-trial [accessed 3 Aug 2023]

Heglar, M. A. 2023, Not Too Late, ed. R. Solnit, & T. Lutunatabua LaVergne (Chicago, IL: Haymarket Books) 19

IPCC, 2023, Climate Change 2023: Synthesis Report. Contribution of Working Groups I, II and III to the Sixth Assessment Report of the Intergovernmental Panel on Climate Change, ed. H. Lee, & J. Romero (Geneva, Switzerland: IPCC) 1

Jiménez, J. 2023, Why the UN must rely more on indigenous wisdom and less on fossil fuels Resilience.org https://www.resilience.org/stories/2022-05-23/why-the-un-must-rely-more-on-indigenous-wisdom-and-less-on-fossil-fuels/ [accessed 3 Aug 2023]

Johnson, A. E., & Wilkinson, K. K. 2021, All We Can Save, ed. A. Johnson, & K. Wilkinson (New York: One World)

Mitchell, S. 2021, All We Can Save, ed. A. Johnson, & K. Wilkinson (New York: One World) 16

Quell, M. 2023, Top European rights court hears pair of high-profile climate cases (Courthouse News Service) https://www.courthousenews.com/top-european-rights-court-hears-pair-of-high-profile-climate-cases/ [accessed 3 Aug 2023]

Rice, D. 2019, Study finds a race gap in air pollution—whites largely cause it; blacks and Hispanics breathe it (USA Today) https://www.usatoday.com/story/news/nation/2019/03/11/air-pollution-inequality-minorities-breathe-air-polluted-whites/3130783002/ [accessed 30 July 2023]

Schiffman, R. 2023, The Color of Grass Roots: Diversifying the Climate Movement Yale Environment 360 (New Haven, CT: Yale School for the Environment) https://e360.yale.edu/features/heather-mcteer-toney-interview [accessed 30 July 2023]

Shi, L., & Moser, S. 2021, Sci, 372, 6549

Shultz, M., Sands, D. E., Kossin, J. P., & Galea, S. 2020, NEJM, 382, 1 [accessed 3 Aug 2023]

Sultana, 2023, Not Too Late, ed. R. Solnit, & T. Lutunatabua LaVergne (Chicago, IL: Haymarket Books) 58

Thomas, L. 2022, The Intersectional Environmentalist: How to Dismantle Systems of Oppression to Protect People + Planet Voracious (Boston, MA: Little, Brown and Company)

Toney, 2021, All We Can Save, ed. A. Johnson, & K. Wilkinson (New York: One World) 75

UNEP, 2020, Emissions Gap Report 2020. Nairobi https://www.unep.org/emissions-gap-report-2020

UN 2021, Asad Rehman speech at 3:11:20 in #COP26 LIVE - Closure of Negotiations ~ Climate Action ~ UNFCCC ~ United Nations *YouTube United Nations Channel* https://youtu.be/MtE-t3js1BM?t=11480 [accessed 31 July 2023]

UNFCCC, 2022, How Indigenous Peoples Enrich Climate Action (UN Climate Change News) https://unfccc.int/news/how-indigenous-peoples-enrich-climate-action [accessed 3 August 2023]

World Bank, 2023, The World Bank Understanding Poverty, https://www.worldbank.org/en/topic/indigenouspeoples [accessed 20 July 2023]

Part VII

Climate Change and Astronomy

Chapter 17

Impact of Climate Change on Astronomical Observations

Maaike van Kooten, Faustine Cantalloube and Travis A Rector

> "The wide world is all about you: you can fence yourselves in, but you cannot forever fence it out."
>
> —J. R. R. Tolkien, The Fellowship of the Ring

In this chapter we discuss the main impacts of climate change on astronomical observations, including operational activities and data quality. We look at how a changing climate affects observatories on large scales and how it poses difficulty in building and maintaining observatories. We discuss how changes in the atmosphere can affect how frequently one can observe as well as the quality of the final scientific data product. We then present the latest studies that look at climate change and astronomical sites, discussing important parameters the community should monitor to better understand how climate change is affecting a given astronomical site. We urge the community for active monitoring of the weather and climate at observatories to enable better preparation for the future of ground-based telescopes.

17.1 Introduction

As human beings it is important to understand the impact of climate change on our lives and the world we live in as well as understand our role in causing climate change. As astronomers it is important to acknowledge that climate change not only affects us personally but also our jobs and research. Likewise, our actions directly impact the climate. With changes to our climate, observatories are and will experience different weather and conditions compared to when they were first built. Climate change affects telescope facilities and data they produce.

With extreme weather such as floods and fires, the very existence of our observatories is threatened. Weather also affects when we can observe and the intrinsic quality of the data we collect. We need to adapt and ensure future observatories–such as the Square Kilometer Array (SKA) or the Extremely Large

doi:10.1088/2514-3433/acfcb6ch17

Telescope (ELT)–are prepared to operate and provide high quality data despite changes to their environment. This challenging task requires robustness at a basic facilities level as new technologies and instruments will not be able to account for decrease in atmospheric transparency, high winds, or the many other parameters that can affect observations.

Every wavelength of the electromagnetic spectrum observed from the ground will be affected by changes in weather and climate. In this chapter we aim to provide an overview of the different variables that affect the operations of astronomical facilities on the ground and their data quality. Our overview is biased towards shorter wavelengths, but many of the factors discussed below also have implications for submillimeter and radio astronomy.

17.2 Large Scale Impacts on Facilities and Infrastructure

Observatories are located at remote and extreme sites that have dry, dark, and stable conditions—usually at high altitudes and far away from major cities. Not only are they difficult to reach, having the necessary infrastructure (i.e., water, electricity, high-speed internet) is difficult and costly to build and maintain. Nature poses a great challenge for observatories with extreme weather events adding to nominal natural stresses affecting the facility. Storms, monsoons, floods, and fires have caused significant damage to facilities; and an increasing frequency of these events due to a changing climate increases the risk and cost for current and future observatories.

Sadly there are already many examples. Multiple storms over the years (including 2017 and 2020) in Puerto Rico caused damage to Arecibo Observatory, slowly eroding the telescope: destroying tiles (Drake 2017), instruments and finally a cable break causing the receiving platform to collapse and cause irreparable damage to the radio dish below. Winter storms on Maunakea with 191 mph wind caused minor damage to observatories (Burnett 2019). In North America, extreme drought conditions in California resulted in over 9000 fires during 2020 (Safford 2022). A fire nearly destroyed Lick Observatory on Mount Hamilton with burn marks visible on the road opposite the telescopes and requiring those working and living on the mountain to evacuate. The resulting ash around the mountain meant that once the fire had passed, science observations could not resume as normal due to the risk of ash being blown into the dome (Dickinson 2020). In Arizona, Mt Graham Observatory (location of Large Binocular Telescope and the Submillimeter Telescope) was saved in 2017 from fire damage thanks to firefighters' efforts (Gabbert 2017) and aerial images show fire damage on all sides of the site. While no telescopes were damaged in the 2022 Contreras fire on Kitt Peak (Figure 17.1) support buildings were lost. Power lines and its high speed network were also lost, requiring the physical transportation of data from the mountains, which was also impacted by monsoon-induced mudslides that damaged the access road (Hill 2022). Worsening trends in wildfires in Australia have resulted in more severe and extreme bushfire seasons. In 2003, explosive wildfires near the Australian capital of Canberra gutted the Mount Stromlo Observatory, destroying all eight of the research

Figure 17.1. Part of the Contreras Fire burning on the slopes of the Kitt Peak mountain on Friday early morning 17 June 2022. In the foreground is the NRAO's Very Long Baseline Array antenna. Credit: KPNO/NOIRLab/NSF/AURA

Figure 17.2. Remains of the Great Melbourne telescope at Mount Stromlo Observatory after the fire in January 2003. Credit: Enoch Lau (CC BY-SA 3.0)

telescopes at the site (Irion 2003). The remains of the Great Melbourne telescope are shown in Figure 17.2. In 2013, Siding Springs Observatory (home to 12 telescopes) located in New South Wales saw extensive damage to support facilities and homes in what was at the time the worst fire in the state's history (Kramer 2013). The 2019–2020 fire season was particularly devastating (Fasullo 2021). Between Oct 2019–Feb

2020 16.89 million hectares of land were burned with most of it being in New South Wales (Granwal 2022). With the SKA being built in Australia and South Africa, astronomers need to understand the increased risks of building in dry remote places and we need to adjust practices to keep personnel and facilities safe. These are just a few examples of fires coming physically close to observatories causing damage. Wildfires not only pose immediate threats to facilities, large fires have the potential to dramatically alter the climate through large emission of aerosols that affect albedo and potentially displace tropical deep convection that could influence El Niño and in turn the atmosphere above observatories (Fasullo 2021). Another effect preventing astronomical observations is the subsequent haze from wildfires that can last for a few days to several weeks. Depending mainly on wind conditions cloud of ashes can travel on several thousands of kilometers, increasing significantly the opacity and background light of the night sky.

17.3 Impacts on Daily Operations

For night-to-night operations of an astronomical observatory, data can only be taken under certain conditions and any threat to the infrastructure or optics will result in the night not being used for observations.

For radio astronomy, high winds vibrate the telescope dish itself, contaminating the data sufficiently that it is futile to observe. For short wavelengths, the dome housing the telescope itself can not be opened under high wind due to vibrations and shaking of the dome structure and the telescope. Van Kooten & Izett (2022), using in situ data on Maunakea, show a 2-fold increase in dome closure events driven mainly by increased wind speed but also temperature and humidity. In the submillimeter and into the optical, water plays an important role (Utsunomiya & Sekine 2005). For longer wavelengths, water vapor degrades data rapidly. In the optical and near-infrared (NIR), high humidity (>95%) prevents observing as there is a large risk for condensation to form on the primary mirror and other mirrors. In addition, high relative humidity (e.g. above 70%) may cause leaching, corrosion and prevents from using high-voltage components. This can happen when there is precipitation, low clouds, or fog but also under clear-sky conditions. Finally, in the optical and NIR ground-based observatories any precipitation prevents the dome from opening. When precipitation occurs there will be clouds that fully obscure the night sky for optical wavelengths, and there are also nights when cloud coverage on its own either makes the sky fully opaque or that partial cloud coverage attenuates light from all but the brightest sources making the night unusable for new and interesting science. Measuring, forecasting, and predicting cloud coverage is very difficult and with large uncertainties especially for astronomical observing (Haslebacher et al. 2022 & Ye 2011).

Recent studies have tried to assess the impact of a changing climate on different observatories and the ability to observe. For example, van Kooten & Izett (2022) used meteorological observations to conclude that the number of nights lost due to bad weather on Maunakea have doubled over the past 20 years. Further, Haslebacher et al. (2022) used climate projections to investigate changes in various parameters for eight different astronomical sites. In all cases, the temperature is rising, however, the impact on relative humidity and specific humidity (and precipitable water vapor) is less obvious with some locations and scenarios leading to drier conditions and others more humid conditions. While more humidity might have a negative impact on the daily operations and quality of data, and dry conditions might lead to better conditions for observing, they are also better conditions for fires. With such great uncertainty on how a specific location might be affected by climate change, observatories must be prepared for these worst case scenarios to optimize our data.

Particle pollution or particulate matter (PM) impacts daily operations in different ways depending on the PM size. For optical telescopes limits are set depending on danger to the telescope optics (for example Mt Hamilton Weather Limits sets 0.3 mm particle count to 17000 and 10 mm to 3).[1] Larger PM (10 mm) associated with ash have strict limits to prevent collection of ash in the dome and on the optics. In regions where fires have burned or are burning, even if the fires are not a direct threat to the facility, they could impact daily operations and cause dome closures depending on how far and where the PM travels. In the Canary Islands, the Calima causes a reduction in visibility due to the high concentration of dust in the atmosphere coming from the Saharan. While in some cases, the dome can be opened for observing (though with considerable extinction), in extreme cases and to prevent dust getting onto optics, no observing can be done.[2]

17.4 Impacts on Data Quality

When observing, there are a lot of variables that impact the quality of the data. Above, we focused on effects in the theme of daily operations (i.e., events that prevent observing completely or a specific type of observation such as faint objects) but in many cases, the milder versions of the above sources that affect operations affect data quality (e.g. high winds prevent the dome from opening but slightly slower fast wind speeds result in more fast evolving turbulence). Observatory sites are often selected for stable conditions after a site testing campaign (e.g. TMT site testing) but considerations about how conditions might change over long periods of time have not been accounted for or considered previously.

The transparency and chromaticity of the atmosphere has a large impact on observations. Extinction of light by the atmosphere can cause observations to not only take more time requiring long exposure times but can also have an impact on long baseline photometry depending on how the sky is changing (Steinbring 2022). Increased frequency in cloud coverage (i.e., cirrus clouds) and density, fog, increased precipitable water vapor (for NIR-IR), and aerosols, sand, and PM can all contribute to a decrease in sky transparency. For observatories close to urban areas, the concentration of PM (especially PM10 that comes from car and truck exhaust, industry, and fires) and the sky brightness on a cloudless night have a linear

[1] https://mthamilton.ucolick.org/techdocs/telescopes/Nickel/limits_weather/.

[2] https://www.ing.iac.es//Astronomy/telescopes/wht/weathloss.html.

relationship. If pollution continues to increase, the sky brightness could increase for typical 'good' nights (Ścieżor & Kubala 2014).

The ground wind speed and temperature all affect how air flows through a telescope dome. Temperature differences between the inside and outside of the dome and how well they can be equalized (which depends on wind speed) can cause turbulence inside the dome and other unique features (i.e., the low wind effect) from how atmospheric parameters interact with the observatory structure. Changes in wind speed can also change which/how vibrations get amplified. Changes in these variables can change the frequency and amplitude of these effects.

In the free atmosphere (atmosphere above the ground layer ~1 km above the observatory), the profiles of temperature, humidity, and wind speed all impact light. The index of refraction of air (n_{air}) changes with the above parameters. The modified Gladstone relationship (Masciadri et al. 2017) provides a way to convert from the above profiles to the n_{air} structure function (Cn2). Specific atmospheric profiling instruments such as MASS, DIMM, SCIDAR and others, use starlight to estimate the Cn2 profile. In the optical, temperature dominates the n_{air} while for longer wavelengths humidity/water vapor becomes more important (Birch, 1993). Optical turbulence or seeing is due to changes in n_{air} brought on by spatial and temporal changes in temperature. The spatial changes are often assumed to be statistically stationary processes, with the temporal changes approximated by the Frozen Flow Hypothesis which indicates that the temporal changes are driven by wind speed so that a spatially static phase screen is pushed across the telescope aperture by a constant wind speed. Therefore, the coherence time of turbulence is related to the wind profile. The spatial fluctuations also influence the gradient of the wind profile, the vertical wind shear profile. Enhanced (decreased) vertical shear will increase (decrease) vertical mixing, which also impacts the temperature and humidity profiles. Overall, changes in the wind profile impact both the coherence time and the seeing/Fried parameter. Humidity and temperature changes only influence the Fried parameter.

Chromaticity of the atmosphere has implications for the optical design of an instrument (e.g., atmospheric dispersion corrector; ADC). van den Born & Jellema (2020) show for example that their goal of 50 uas relative astrometric accuracy is possible with the current ADC design, accounting for both different atmospheric models of refractivity and sight-line effects from local weather (using reanalysis products—see next section) for H-band. These careful calculations highlight the relation between weather and temperature profiles with refractivity as well as relying on the current atmospheric models to be constant. An instrument at longer wavelengths might be more sensitive to these refractivity changes. It is also worth noting that telluric absorption features are also affected by changing concentrations of greenhouse gases, and in particular with water content. This is crucial for spectroscopic measurements—mostly considering the many upcoming instruments of high spectral resolving power. To avoid the use of standard star observations (doubling the observing time and overheads), one usually relies on radiative transfer models of the Earth's atmosphere, which take basic weather parameters (e.g., temperature, pressure, relative humidity) and greenhouse gas concentration as inputs.

17.5 How to Do Your Own Analysis

17.5.1 Data Sources

In situ observations are available at many observatories, often at a variety of different locations at/near the summit. These observations might be made with standard ground-based weather stations (e.g., measuring wind speed using an anemometer), or profiling instruments such as radiosondes. Radiosondes are balloons released from the surface that carry instrument platforms through the atmosphere, sampling meteorological parameters as they ascend. Depending on the location of the sensor providing the data, different sources of biases will exist (i.e., located in the dome or the sensor was changed 5 years ago, etc). The data themselves are not always easily accessible. Therefore, liaising with a support astronomer at a given observatory to obtain access to and understand the data is a good practice.

On Maunakea, publicly available data is shared through the Maunakea weather center, which includes meteorological observations and turbulence profiling measurements from MASS/DIMM instruments near the Canada–France Hawaii Telescope (CFHT). They also report meteorological data measured at CFHT. Other telescopes have their own data that is often piped to the header of the scientific files. On Paranal, the European Southern Observatory (ESO) provides data through the Astroclimatology of Paranal webpage[3] as well as Paranal Ambient Query Forms.[4] These pages also contain links to various previous data analysis, maps of the summit, manuals, and disclaimers on when new instruments were installed. The turbulence profiling data available is from a variety of sources and instruments including DIMM (now FASS), Stereo SCIDAR, and MASS/DIMM. A good overview of the various methodologies is provided in Osborn et al. (2018), Osborn (2010), and Le Louarn et al. (2021). Other smaller facilities will have meteorological data (to provide observers with current conditions) but with varying availability of historical data due to the infrastructure needed. For example, Schöck et al. (2009) provide an overview to the site testing data for the Thirty Meter Telescope. A summary of in situ data sources at the eight biggest observatory locations is provided by Haslebacher et al. (2022).

On a global scale, we can make use of reanalysis products developed, run, and delivered by (inter-) national agencies. Reanalysis products provide the most comprehensive view of the past weather and climate by running modern weather forecasting models over historical periods and assimilating available observations. They provided global estimates of meteorological fields such as wind, temperature, and humidity on multiple vertical levels and with hour(s)-level temporal resolution making them powerful tools for climate analysis. These models have spatial resolution of typically 30–100 km and vertical resolution of tens to hundreds of pressure levels to represent what is happening at different altitudes. Some well known up-to-date reanalysis products are listed below under the provider of the product.

[3] https://www.eso.org/gen-fac/pubs/astclim/paranal/index.html.

[4] http://archive.eso.org/cms/eso-data/ambient-conditions/paranal-ambient-query-forms.html.

1. European Centre for Medium-Range Weather Forecasts (ECMWF)
   a. ERA5
   b. ERA-20C
2. US National Center for Atmospheric Research (NCAR)
   a. NCEP/NCAR Reanalysis
3. NASA's Global Modeling and Assimilation Office
   a. MERRA-2
4. Japan Meteorological Agency
   a. JRA-55

Many of these data products allow for you to download a specific subregion of the global model domain, for example, that surround an observatory of interest. The reanalysis datasets can also be delivered in many different file formats, including the NetCDF format that stores multiple multidimensional variables, which are well supported by various programming languages. In Python, the NetCDF API provides support for reading in the file. We would recommend the use of Python's xarray[5] which combines numpy and pandas to support multi-dimensional time series and variables which maps well to NetCDF data and has specific tools to aid with NetCDF data analysis including the ability to read and concatenate data split over multiple NetCDF files.

Since the spatial resolution of the re-analysis products is tens to hundreds of kilometers, the data need to be interpolated (or similarly downscaled) to a specific location of interest such as an observatory. The vertical resolution of the datasets may require similar resampling to compare with in situ data. Once again these processes are easily done making use of Python's xarray package.

A word of caution when using reanalysis data. Incorporation of in situ data into the reanalysis can result in errors due to an instrument change; the assimilated data quality is not necessarily thoroughly checked. Specific regions might have more assimilated data than others, resulting in some regions being better represented than others while spatial sampling might not capture all the important physics for a specific site; specifically Haslebacher et al. (2022) found that ERA5 data for Maunakea is not as well matched to in situ data compared to other observatory locations, which is assumed to be a result of the grid size poorly capturing the Islands. In addition, the coarse spatial resolution of such re-analysis data usually cannot take into account detailed orography or interaction between land-oceans, which can result in a slightly biased results. Last, there are also regional reanalyses available from the above product providers that might better represent a specific location.

Projecting into the future, climate scientists make use of global climate models to capture the interactions between air, land, and ocean. These models are typically run as ensembles making use of specific scenarios outlined by agencies (and/or themselves) so that the outputs can be compared between different model providers to get a sense of the uncertainty in the projections. For example, a set of

[5] https://docs.xarray.dev/en/stable/.

latest scenarios is given by CMIP6 (Coupled model intercomparison project phase 6)[6] and for models that were run and meet the CMIP6 scenarios, the Intergovernmental Panel on Climate Change (IPCC) uses the results from all these models for various reports.[7] For climate projections, the ensemble provides insight on the errors within a model while comparing to different models (that might implement physics differently but follow the same scenario) to provide insight into the general error on predicting the climate. Therefore, the use of just one model output is not recommended and that is why studies focus on analyzing multiple climate data products. We highlight the Primavera project[8] which is a specific intercomparison European Union project that compares multiple different models from different groups with the aim of improving models and projects. The high-resolution models were used by Haslebacher et al. (2022) and were chosen due to their high spatial resolution that can better capture the mountainous geography where observatories are located. The availability and processing procedure for working with climate model output is similar to that of obtaining and using a re-analysis data set.

17.5.2 Important Parameters

When looking at all the available data it is necessary to parse out what parameters are important. Different data sources will have different temporal resolutions (e.g., ERA5 is every 3 hours while in situ data could be every few minutes). Running averages, yearly, and decadal averages along with their standard deviation provide useful insight. Transforming these into probability density functions and cumulative density functions over certain ranges can also be beneficial. We need to, however, understand normal weather cycles to parse out any long term trends. The solar cycle (approximately an 11 year period), El Niño/La Niña and the Southern Oscillation (historically ~4 years, though less regular recently), the Pacific Decadal Oscillation (every 25–45 years), the North Atlantic Oscillation (NAO), and the Atlantic Multi-decadal Oscillation (AMO) are all examples of natural climate cycles that could impact our ability to find long term trends in data. For in situ data, it is important to note when an instrument was upgraded or replaced as there might exist a bias or change in noise especially when given access to raw data.

Both re-analysis and climate projection data products include temperature, humidity, wind speed, and pressure profiles at a given location. Once we have resampled these to match the location of our observatory, we can look at specific features. These products, however, won't necessarily represent well what is happening at the summit where the observatory is located due to the spatial resolution they were run at. Instead, the data could be most useful to study the free atmosphere above the observatory. The specific humidity profile can be integrated to get the precipitable water vapor (van Kooten & Izett 2022). Cloud coverage as a function of

---

[6] https://pcmdi.llnl.gov/CMIP6/.

[7] https://climate-scenarios.canada.ca/?page=cmip6-overview-notes

[8] https://www.primavera-h2020.eu/.

altitude is also an available data product and can be integrated to get a ratio of coverage. Gradients of profiles can be important as well; the gradient of the wind profile is known as wind shear and gives insight into the underlying cause of turbulence in the atmosphere. Specifically, the modified Gladstone relationship (Osborn et al. 2018 and Masciadri et al. 2017) allows vertical structure function of the index of refraction, the Cn2 profile, to be calculated from the structure function of temperature (C2T), temperature profile, pressure profile, and the potential temperature (temperature profile weighted by pressure). The temperature structure function is related to the gradient of temperature and the largest energy scale of turbulent flow which can ultimately be related to the wind shear. The Cn2 profile can then be integrated to find the Fried parameter (Hardy, 1998). With the Fried parameter and the wind profile, the coherence time of the atmosphere can be calculated following Hardy (1998) where the effective wind velocity is determined from integrating the wind profile weighted by the Cn2 profile such that the atmospheric layers with the most turbulence contribute the most to the effective velocity.

Understanding the turbulence at the summit is more tricky and nuanced. In situ data can be used to determine how well different models (re-analysis or climate projects that include historical runs) represent what is happening on the summit. Using in situ data, the Fried parameter and coherence time can be found using measurement not corrupted by nearby structures (domes etc). However, the true impact on the data will depend on the dome the air is flowing over and through as well as nearby domes. A given observatory might have empirical equations or correlations relating image quality to temperature or wind speed at the summit. For example, the low wind effect occurs at certain observatories with specific wind speeds, degrading image quality. If we see a trend in time of the probability density function towards those wind speeds, then the low wind effect will happen more frequently. The full impact of a change in summit climate (i.e., increase in mean wind speed or temperature) might require a fluid dynamic model to fully understand and will depend greatly on the telescope structure but also operation details. For example, increasing temperatures might be fine if the dome can be cooled to have the same temperature inside as outside when the dome is open. In Cantalloube et al. (2020), the authors point out that the telescopes can only maintain a maximum temperature of 16°C; and with increasing temperatures the occurrences where the outside temperature is above 16°C is increasingly resulting in more dome turbulence and worse image quality. Understanding what is happening on the summit and its impact on image quality is extremely complex and requires not just an understanding of the available data but also of the site, geography, and the telescope operations as well.

Finally, previous site testing campaigns can provide insight into how various in-situ data can be reduced and analyzed. More recently, the Thirty Meter Telescope performed 5 year campaigns at various sites. A list of the associated papers is available along with access to data on the TMT website.[9] The papers cover general

[9] https://sitedata.tmt.org/references.html#TSTseries

site testing, turbulence profiling, meteorological parameters, cloud coverage and light pollution, and precipitable water vapor. Similar analysis can be done using long term in situ and climate models.

17.6 Future Monitoring of Sites

Current studies to date (Cantalloube et al. 2020; Seidel et al. 2023; van Kooten & Izett 2022, and Haslebacher et al. 2022) that have looked at comparing all the above data conclude that there should be improved infrastructure at observatories to have continuous monitoring of key parameters and that correlations between instrument performances and atmospheric conditions should be well documented. For sites that are home to multiple independent observatories, a shared analysis of all available data that is open to the public would be ideal along with the publication of dome open and closure events. This would enable the astronomy community to make better decisions about current and future telescopes and instrument upgrades. This is especially important in the case of hitting tipping points in our climate where runaway processes could change the climate quickly as predicted by many climate projections.

References

Birch, K. P., & Downs, M. J. 1993, Metrologia, 30, 155

Burnett, J. 2019, Winter storm causes minor damage to Maunakea observatories (West Hawaii Today) https://www.westhawaiitoday.com/2019/02/13/hawaii-news/winter-storm-causes-minor-damage-to-maunakea-observatories/ (accessed Dec 5 2022)

Cantalloube, F., Milli, J., Böhm, C., et al. 2020, NatAs, 4, 826

Dickinson, D. 2020, Historic lick observatory survives California fire (Sky & Telescope) https://skyandtelescope.org/astronomy-news/historic-lick-observatory-survives-california-fire/ (accessed Dec 5 2022)

Drake, N. 2017, Hurricane Damages Giant Radio Telescope—Why It Matters (National Geographic) https://www.nationalgeographic.com/science/article/arecibo-radio-telescope-damaged-puerto-rico-hurricane-maria-science (accessed Dec 5 2022)

Fasullo, J. T., et al. 2021, GeoRL, 48, e2021GL093841

Gabbert, B. 2017, Telescopes survive onslaught by Frye Fire (Wildfire Today) https://wildfirctoday.com/2017/06/19/telescopes-survive-onslaught-by-frye-fire/ (accessed Dec 5 2022)

Granwal, L. 2022, Total area burned by bushfires in Australia as of January 2020, by state (Statista) https://www.statista.com/statistics/1089996/australia-total-area-burned-by-bushfires-by-state/ (accessed Dec 5 2022)

Hardy, W. 1998, Adaptive Optics for Astronomical Telescopes 448 (Oxford: Oxford Univ. Press)

Haslebacher, C., Demory, M.-E., Demory, B.-O., Sarazin, M., & Vidale, P. L. 2022, A&A, 665, A149

Hill, S. 2022, Kitt Peak National Observatory mostly operational after Contreras fire (Astronomy Magazine) https://www.astronomy.com/science/kitt-peak-national-observatory-mostly-operational-after-contreras-fire/ (accessed Dec 5 2022)

Irion, R. 2003, *Australian Observatory Destroyed* | Science https://www.science.org/content/article/australian-observatory-destroyed (accessed Dec 5 2022).

Kramer, M. 2013, Huge Wildfire Damages Australia's Largest Optical Observatory (Space) https://www.space.com/19254-australia-siding-spring-observatory-wildfire.html (accessed Dec 5 2022)
Le Louarn, M., et al. 2021, Astronomical site monitoring: current status and ideas for the future (AO4ELT6) https://ao4elt6.copl.ulaval.ca/proceedings/401-LPP4-251.pdf (accessed Dec 13 2023)
Masciadri, E., et al. 2017, MNRAS, 466, 520
Osborn, J. 2010, Profiling the turbulent atmosphere and novel correction techniques for imaging and photometry in astronomy (Doctoral Thesis), Durham Univ.
Osborn, J., et al. 2018, MNRAS, 478, 825
Safford, H. D., Paulson, A. K., Steel, Z. L., Young, D. J. N., & Wayman, R. B. 2022, Glob. Ecol. Biogeogr, 31, 2005
Schöck, M., et al. 2009, PASP, 121, 384
Ścieżor, T., & Kubala, M. 2014, MNRAS, 3, 2487
Seidel, J.V., Otarola, A., & Théron, V. 2023, Atmosphere, 14, 1511
Steinbring, E. 2022, PASP, 134, 125003
van den Born, J. A., & Jellema, W. 2020, MNRAS, 496, 4266
van Kooten, M. A. M., & Izett, J. G. 2022, PASP, 134, 095001
Utsunomiya, T., & Sekine, M. 2005, Int J Infrared Milli Waves, 26, 905
Ye, Q.-Z. 2011, PASP, 123, 113

AAS | IOP Astronomy

Climate Change for Astronomers
Causes, consequences, and communication
Travis Rector

Chapter 18

The Carbon Footprint of Astronomy Research

Jürgen Knödlseder

> "All we have to do is to wake up and change."
>
> —Greta Thunberg

In order to limit global warming to well below 2°C, as agreed upon in the Paris Climate Agreement, rapid and deep greenhouse gas emission reductions are needed in all human activity sectors. In this context, astronomers are becoming increasingly aware and concerned about their own carbon footprint, seeking for ways to reduce the greenhouse gas emissions that are produced by their research activities. In this chapter I will summarize the current status of this new field of scientific research. I will start with presenting the methodology of carbon accounting, so that carbon footprint estimates can be compared and understood within a standardized framework. Published carbon footprint estimates are then discussed, identifying the principal sources of greenhouse gas emissions in astronomy. Actions proposed in the literature to reduce the carbon footprint that can be taken by individuals, research labs, organizations, and communities are then summarized.

18.1 Introduction

Not a single month passes now without headlines about record-breaking climate extremes, including heat waves, droughts, wildfires, heavy rainfall, thunderstorms, and widespread flooding that annually costs thousands of lives and creates millions of displacements. These events are examples of observed impacts from anthropogenic climate change, as summarized in the sixth assessment report of Working Group II of the Intergovernmental Panel on Climate Change (IPCC 2022a). According to the report, increased global warming will "cause unavoidable increases in multiple climate hazards and present multiple risks to ecosystems and humans." Recognizing these risks, 192 countries have signed the Paris Climate Agreement that aims at "holding the increase in the global average temperature to well below 2°C above pre-industrial levels and pursuing efforts to limit the temperature increase to

doi:10.1088/2514-3433/acfcb6ch18

1.5°C." In order to reach this goal, the agreement prescribes "to achieve a balance between anthropogenic emissions by sources and removals by sinks of greenhouse gasses in the second half of this century." In other terms, humanity needs to reach net-zero emissions by 2050, i.e. in less than 30 years from now.

Near-term greenhouse gas (GHG) emission reduction goals are country specific. For example, the European climate law creates a legal obligation to reduce by 2030 GHG emissions in the European Union by at least 55%, compared to 1990 levels (EC 2021). The United States has set a comparable target of 50–52% below 2005 levels in 2030 (US 2022). Other high-income countries have declared similar targets. Adjusting from the reference years' to today's GHG emissions, these targets translate into reductions of about ~40% between today and 2030, corresponding to reductions of about ~4–5% per year. This level is comparable to, but slightly larger than, the emission reductions that were reported due to COVID-19 lockdowns in 2020 (Jackson et al. 2022). This level needs to be achieved continuously in high-income countries over the next three decades, with each year's reduction coming on top of those achieved in the year before.

In their sixth assessment report, the Working Group III of the IPCC emphasizes that limiting warming to 1.5°C or 2°C "involve[s] rapid and deep and in most cases immediate GHG emission reductions in all sectors" (IPCC 2022b).

Astronomy, as we will see, is a sector that contributes to global warming, and consequently it is concerned by this imperative. The IPCC reminds that "doing less in one sector needs to be compensated by further reductions in other sectors if warming is to be limited" (IPCC 2022b). Eventually, policy makers and the society at large may set different reduction efforts for different activity sectors, yet until this is done, a reasonable assumption is that astronomy needs to reduce GHG emissions at a typical rate of ~7% per year over the next decades.

The following sections provide an overview of what is currently known about the carbon footprint of astronomy, and what actions can be taken to reduce it. Yet before diving into the results and actions, an introduction in carbon accounting is given that serves as a reference and guideline to readers that envision engaging into carbon footprint assessments.

18.2 Carbon Accounting

Carbon accounting, also referred to as GHG accounting, is a framework of methods that were developed for assessing the amount of GHG emissions attributable to the activity of an organization. GHGs are all gasses that, once released into the atmosphere, have an impact on the Earth energy balance, including in particular carbon dioxide (CO_2), methane (CH_4), nitrous oxide (N_2O), hydrofluorocarbons (HFCs), perfluorocarbons (PFCs), sulfur hexafluoride (SF_6) and nitrogen trifluoride (NF_3). The radiative effects of the different gasses are usually translated into CO_2 equivalents (CO_2e) assuming some emission metrics and timescale (usually 100 years) so that GHG emissions for different gasses can be combined into a single number, known as the carbon footprint.

Over the last two decades several standards were developed for GHG accounting, the most widely used being the GHG Protocol Corporate Standard, ISO 14064–1, and Bilan Carbone©. Rules and methodologies among these different standards are similar, and the standards are fully compliant. It is therefore not critical which standard is chosen for the carbon accounting, yet it is crucial to select one, and to apply it correctly and rigorously so that the resulting GHG inventory is accurate and reliable. In particular, all standards insist on the importance of completeness, including all GHG emission sources in the assessment, and requiring explicit disclosure and justifications of any specific exclusion.

To assure accurate and reliable carbon accounting, it is therefore important to first familiarize with the selected standard, which is best achieved through dedicated training. For example, the GHG protocol website[1] offers free online training in English, and certified trainings for the Bilan Carbone© method are provided for some fee by the Institut de Formation Carbone[2] in French and English. In the following I provide a summary of what will be learned during such training, adapted to an astrophysical research lab.

The first step in carbon accounting consists of the setting of the boundaries for the assessment. This starts with the determination of the organizational boundary for which different approaches exist: the equity share approach (accounting GHG emissions according to economic interest), or the control approach (accounting GHG emissions over which the organization has control). The latter can be done either based on financial or operational control, with the baseline being operational control that is also best suited for a research lab. The second boundary to consider is the operational boundary, which is defined by identifying the GHG emissions associated with the operations of the research lab. This is best done by drawing a map of all energy, material and service fluxes that are generated by the activities of the lab, including incoming and outgoing fluxes. To identify the fluxes that need to be included, the question to ask is: "Will the activity of the lab be impacted if the flux is removed?" If the answer to the question is yes, the flux needs to be included in the assessment. The third boundary is the temporal boundary, which is for example the reference year for which the carbon accounting will be done.

As an example, the GHG emissions that were identified within the operational boundary of the *Institut de Recherche en Astrophysique & Planétologie* (IRAP) that followed the Bilan Carbone© standard for carbon accounting, are illustrated in Figure 18.1, with GHG emissions classified by scope (Martin et al. 2022a, 2022b). Scope 1 comprises all direct emissions from sources owned or controlled by the lab straight into the atmosphere, such as emissions from gas heating facilities, the lab's vehicle fleet or the refrigerant gasses leaking from air conditioning devices. Scope 2 refers to indirect emissions arising from the consumption of purchased electricity, heat, steam or cooling, for example, emissions related to purchased electricity needed for lab operations. Scope 3 refers to indirect emissions from other activities not directly controlled by the lab, for example emissions related to employee

[1] https://ghgprotocol.org.

[2] https://www.if-carbone.com.

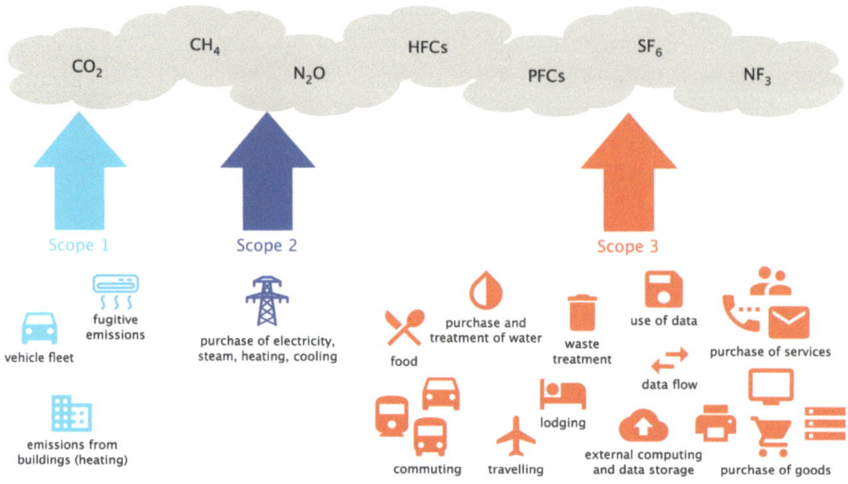

Figure 18.1. Sources of GHG emissions that typically arise in an astrophysics lab. The sources are split by emission scope (see text).

commuting and food, professional traveling and lodging, purchase of goods and services, use of data, data transfer and external computing, and water and waste treatment. For many organizations scope 3 emissions comprise the dominant part of the carbon footprint; for IRAP, they account for about 95% of the carbon footprint (Martin et al. 2022a, 2022b). Note that the same scope 3 emissions may be reported by different organizations if they fall within their operational boundaries, implying that some double counting would occur if their carbon footprints were summed up. Yet this is okay, since summing up organizational footprints is not the purpose of carbon accounting. The primary purpose of carbon accounting is the identification of lever arms for GHG reductions, and several organizations may take action to reduce the same emission source (see box).

After the definition of the boundaries follows the most time-consuming part of the carbon accounting, which is the collection of activity data and emission factors. Activity data describe how much of a given activity has contributed to the GHG emissions (for example how much electricity was consumed, how many km were traveled, etc.) while emission factors specify the GHG emissions in CO_2e per unit of activity data. While the first are usually collected from administrative sources, the latter are generally extracted from databases based on detailed life cycle analyses of various products and services. A list of relevant databases can be found at https://ghgprotocol.org/life-cycle-databases, some of which are freely available (such as Base Empreinte© database in France at https://base-empreinte.ademe.fr or the IPCC emission factor database at https://www.ipcc-nggip.iges.or.jp/EFDB/main.php) while others are available only for a fee (such as the ecoinvent database).

The accounting can be supported by using a carbon footprint calculation tool, such as the spreadsheets available at https://ghgprotocol.org/calculation-tools. Recently, the French initiative *Labos 1point5* has developed a dedicated webtool for the field of public research (https://apps.labos1point5.org) which was already used by almost 700 research laboratories in France, including many astrophysics labs (Mariette et al. 2022). While the tool so far has a limited perimeter and applies only to research labs in France, plans exist to extend its scope and applicability to other countries (e.g., Flagey 2022).

The next step is the reporting of the GHG emissions, with the report being as transparent, accurate, consistent and complete as possible. The report shall include the gross absolute GHG emissions, separate and independent of any GHG emission trades the lab may engage in (see box). The GHG protocol corporate standard only requires mandatory reporting of scope 1 and scope 2 emissions, yet given the general dominance of scope 3 emissions, reporting of all three emission scopes is strongly encouraged (in France, for example, reporting of scope 3 emissions is mandatory since 1st January 2023). An independent verification of the report is encouraged in order to strengthen its credibility.

The final and actually most important step in carbon accounting is the setting of GHG emission reduction targets and the definition of a credible action plan to reach the target. Obviously, this is only possible with the commitment by lab directors and a lab-wide buy-in into the objective of reducing GHG emissions. Target setting involves the definition of target boundaries, the selection of a base year and target completion date, and the tracking and reporting of GHG reduction progress. The GHG Protocol Corporate Standard emphasizes the need for credibility when including GHG emission offsets or credits in reduction targets (addressing of what would have happened in the absence of the offset project; whether the offset project has resulted in emission reductions or removals in addition to what would have happened in the absence of the project; that the project has no relevant secondary effects; that the offsets are not temporary but long lasting), and the need for the developing a Target Double Counting Policy when doing so (to make sure that offsets are not yet accounted for in someone else carbon accounting); also, when reporting on the target, the contribution of offsets to the target reduction needs to be clearly specified, and as already mentioned above, offsets should never be deduced from the lab's GHG emission (see box).

Frequent misconceptions
We list here a few misconceptions that may arise in carbon accounting.

Double counting. Scope 3 emissions are often shared by several organizations, which may lead to some double counting of the emissions when summing the carbon footprint of these organizations. An obvious example is the carbon footprint of some IT equipment, which will show up in the books of the manufacturer (comprising the footprints related to production and utilization) and of the research lab that buys the equipment. The GHG Protocol recognises that double counting in scope 3 emissions exists and is not avoidable, yet this actually is not a problem since the primary objective of carbon accounting is the identification of lever arms for GHG reductions, and several organizations may have an influence on the same emission source. In the example of the IT equipment, the manufacturer may have an impact on the carbon footprint of the production chain and the energy use of the equipment, while the customer controls the purchase (opting for example for a computer with a smaller screen or less energy use) and the use of the equipment.

Use of green certificates. Some labs purchase electricity with green certificates, and it is tempting to claim the accordingly lower GHG emissions as compared to conventional electricity grid emissions. This approach is referred to as market-based estimation, as opposed to the location-based method that makes use of electricity grid information. The latter is mandatory in all carbon accounting standards! While the Bilan Carbone© standard even explicitly prohibits using market-based estimates, the GHG Protocol standard allows for reporting of market-based estimates if they are reported together with location-based estimates. Such double reporting is however only allowed if certain quality criteria specified in the GHG Protocol Scope 2 Guidance document are fulfilled. On the problems behind market-based estimates and the rationale of only using location-based estimates the reader is referred to Brander et al. (2018).

Zero emissions. It is tempting to set the GHG emissions for renewable energy usage or electric car usage to zero, as they release no GHG into the atmosphere during operations. This however ignores the embedded carbon footprint arising from the construction of the renewable power plant or the electric cars, and that needs to be folded into the carbon accounting. This is done by choosing the applicable emission factors from appropriate databases. For example, the Base Empreinte© gives an emission factor of 25, 32 or 44 gCO_2e/kWh for electricity generated via solar panels, depending on whether the solar panels were produced in France, Europe or China, respectively; wind power generated electricity has an emission factor of 14 or 16 gCO_2e/kWh for either terrestrial or marine production. For cars, the Base Empreinte© gives an emission factor of 103 gCO_2e/km for a medium-class electric car in France as opposed to 232 gCO_2e/km for an equivalent petrol car. Switching to an electric car or renewable energies hence clearly reduces the related GHG emissions, yet the result is still well above zero GHG emissions, and needs to be accounted for properly.

Treatment of offsets and GHG credits. Purchased offsets or GHG credits shall never be subtracted from the GHG emission inventory. The rationale behind this is that there is currently no generally accepted methodology for quantifying GHG offsets. Organizations have no control over the effectiveness of offset or credit programs, and there is no guarantee that emission reductions will effectively take place. Instead, offsets or credits may be reported in an "optional information" section of the carbon accounting report. Appropriate information addressing the credibility of purchased or sold offsets or credits should be included. The same rationale applies to sequestering atmospheric carbon.

Treatment of Recycling. Similar to offsets, possible emission reductions that may arise from recycling shall never be subtracted from GHG emissions. Recycling is treated upstream in carbon accounting, i.e. purchasing a recycled product may have a lower carbon footprint than purchasing a new product. For example, the Base Empreinte© gives an emission factor of 2210 $kgCO_2e$ for purchasing one ton of new steel, while one ton of recycled steel has only a footprint of 938 $kgCO_2e$. On the other hand, there are products like paper or cardboard where recycled material has a larger footprint than new materials.

18.3 The Carbon Footprint of Astronomical Research

Now let's turn to what is known about the carbon footprint of astronomy. Several astronomical institutes and communities have assessed their carbon footprints with the aim to identify lever arms for GHG emission reductions. Stevens et al. (2020) estimated the GHG emissions of astronomical research in Australia, including powering of office buildings, business flights, supercomputer usage, and electricity consumption for operations of some ground-based observatories. Jahnke et al. (2020) estimated the GHG emissions of the *Max Planck Institute for Astronomy* (MPIA) in Germany, including business flights, commuting, electricity, heating, computer purchases, paper use, and cafeteria meat consumption. Van der Tak et al. (2021) estimated the GHG emissions arising from professional astronomy activities in the Netherlands, including business flights, commuting, electricity, heating and supercomputing. Simcoe et al. (2022) estimated the GHG emission of the *MIT Kavli Institute for Astrophysics and Space Research* (MKI), including heating, cooling and electricity for the office buildings, business flights, commuting and high-performance computing. Martin et al. (2022a, 2022b) estimated the GHG emissions of IRAP including running the office buildings (electricity, heating, water, air conditioning, waste), food, commuting, professional traveling, purchase of goods and services, external computing and use of observational data. The latter work is the most comprehensive carbon footprint estimate performed to date, and the only study that employed a formal carbon accounting methodology for the assessment.

The results of these studies, expressed as annual per capita GHG emissions for a person working at an astronomy institute, are summarized in Figure 18.2. The results were grouped by emission sources, where "Office" combines all GHG emissions related to running the office building (electricity, heating, water, air conditioning, waste) and "Employee" covers food and commuting. Jahnke et al. (2020) presented a similar figure to compare the MPIA carbon footprint to the one of Australia, and like for their comparison, the air travel footprint quoted by Stevens et al. (2020) for Australia was multiplied by a factor of two to account for non-CO_2

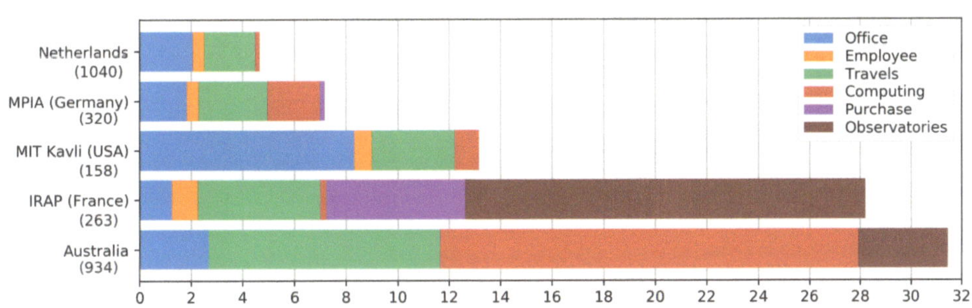

Figure 18.2. Estimates of the annual average carbon footprint of a person working in astronomy. The numbers indicate the number of employees used to compute the per employee footprint.

radiative forcing; the same was done for the air travel footprint quoted by Simcoe et al. (2022); the travel footprint quoted by Jahnke et al. (2020) was multiplied by ⅔ since their results were obtained for a Radiative Forcing Index (RFI) of 3. Hence all business flight footprints in Figure 18.2 are for RFI = 2.

The most striking feature of Figure 18.2 are the huge differences between the various carbon footprint estimates, which are partially explained by the carbon intensity of electricity production in a given country. For example, the largest contributor to the GHG emissions in Australia at that time (2020) was computing, with 16.2 metric tons (tCO_2e) per year and employee. Would the same amount of computing be performed in France, which has a carbon intensity for electricity production that is about 14 times lower than the one of Australia, computing would only account for 1.2 tCO_2e per year and employee. A further important factor is the energy performance of the institute buildings, which explains for example the large office-related footprint associated with MKI due to the poor performance of the Ronald McNair building (Simcoe et al. 2022).

Another reason for the large differences are the different boundaries employed for the studies. Most of the studies ignored some potentially important GHG emission sources, such as purchase of goods and services and use of observational data, or considered them only partially. For example, Jahnke et al. (2020) made an estimate for purchase, yet only included computers and paper in the assessment. Stevens et al. (2020) made an estimate of the carbon footprint of observatories, yet only covered national ground-based facilities in Australia and included only electricity use during operations in the assessment.

Only the IRAP estimate provides a comprehensive assessment of all astronomy-related GHG emissions (Martin et al. 2022a, 2022b), including a complete inventory of purchase of goods and services, and the footprint related to the usage of astronomical observatories by lab researchers. The latter covers construction and operation of the facilities, and involves attributing a fraction of their annual footprint to IRAP based on the size of the user community in the lab with respect to the world community (Knödlseder et al. 2022a).

Including all sources of GHG emissions, the IRAP assessment suggests a per-capita footprint of ~28 tCO_2e per year and employee. Since France has a rather low carbon intensity for electricity production, the IRAP estimate likely presents a lower limit to the average carbon footprint of a person working in astronomy. Adding the IRAP estimates for purchase and use of observational data on top of the estimates for "Office," "Travels" and "Computing" in Australia would lead to a per-capita footprint of about ~50 tCO_2e per year and employee, which may be typical for a country that has a large carbon intensity for electricity production and that relies heavily on air traveling.

So in summary, the average annual per employee carbon footprint for a person working in an astronomy lab is probably between 30 and 50 tCO_2e. Let's now take a more detailed look into the nature of the GHG emission sources that are behind these numbers.

18.3.1 Office-related Footprint

The office-related carbon footprint combines all GHG emissions related to running the office buildings of an institute, including electricity use, heating, water consumption and waste water treatment, fugitive gasses from air conditioning, and waste management. Figure 18.2 indicates that significant differences exist between the estimates of this sector, which can be explained by differences in the carbon intensity of electricity production, energy performance of the buildings, and differences in scope of the studies.

The estimate for IRAP results in the smallest office-related footprint despite the fact that it has the broadest scope, including also water and waste management and fugitive emissions from air conditioning that were excluded in all other studies. This can be explained by the low carbon intensity for electricity production in France (60 gCO_2e/kWh, compared to 415 gCO_2e/kWh for the Netherlands, 460 gCO_2e/kWh for Germany, 522 gCO_2e/kWh for the United States of America and 840 gCO_2e/kWh for Australia; all values are national grid averages extracted from Base Empreinte©), together with the fact that IRAP's main building is primarily heated using wood, with an effective carbon intensity of 51 gCO_2e/kWh (Martin et al. 2022a).

The estimates for MPIA and the Netherlands also exhibit relatively small office-related footprints, yet this is partially due to the accounting of green certificates for electricity purchase, putting the related GHG emissions to zero. This corresponds to the market-based approach, described in the methodology section above. If a location-based approach had been used, as done by Stevens et al. (2022), Martin et al. (2022a, 2022b) and Simcoe et al. (2022) in their studies, larger office-related footprints would have been found for MPIA and the Netherlands.

The largest office-related footprint was reported for MKI, which includes the footprints related to energy usage in three buildings (Ronald McNair, NW-17 and NW-22) for which GHG emissions were attributed to MKI according to the square footage managed by the institute. The footprint is dominated by the McNair building, for which Simcoe et al. (2022) quote an energy use intensity of 260 kBTU ft^{-2} yr^{-1}, corresponding to 820 kWh m^{-2} yr^{-1}. The authors mention that the McNair building ranks above the median of comparable existing labs in the same climate zone, which they attribute to the poor thermal insulation of the building, as evidenced by thermal imaging.

Office-related GHG emissions other than from energy use were estimated by Martin et al. (2022a, 2022b) for IRAP, revealing contributions of 17% from waste management, 8% from fugitive gasses from air conditioning, and 1% from water management. Hence, while energy use dominates the office-related GHG emissions, other sources are not negligible, in particular if the carbon-intensity of energy use is low, as expected in the future from an increased share of renewable energy sources in the electricity mix.

18.3.2 Employee-related Footprint

Home-to-office commuting is one of the GHG inventory categories of all carbon accounting methods and consequently needs to be included in the carbon footprint estimate of a research lab. Four out of the five studies summarized in Figure 18.2 include commuting in their estimates, suggesting 0.4 tCO_2e/employee for MPIA, 0.5 tCO_2e/employee for the Netherlands, and 0.7 tCO_2e/employee for IRAP and MKI. While the estimates for MPIA and the Netherlands include only commuting by car, the other studies also consider other transport means, including in particular public transport, which may explain the slightly larger estimates for IRAP and MKI.

Martin et al. (2022a, 2022b) also estimated the GHG emissions associated with meals taken at midday, as they are necessary to the employees to carry out the work, and hence need to be included in the carbon accounting. They find emissions of 0.3 tCO_2e/employee, which is about half of the commuting footprint at IRAP. They provide details on the type of meals taken by the employees, pointing out that the main factor that determines GHG emissions from meals is the content of animal products. Jahnke et al. (2020) also made an estimate for meals taken at the canteen, yet included only meat consumption, finding emissions of 0.1 tCO_2e/employee at MPIA.

Taken together, home-to-office commuting and midday meals eventually sum up to about 1 tCO_2e/employee, which for an institute that is powered using low-carbon energy sources (such as IRAP) can be of the same order as the office-related footprint. By definition, the employee-related footprint is fully determined by individual choices of the employees, hence it is the category of GHG emissions over which an individual has the largest control (see below).

18.3.3 Business Travels

Business flights are identified as an important contributor to the carbon footprint of academic research, as testified by the abundant literature on the subject and illustrated by the comprehensive book edited by Bjørkdahl et al. (2022) on the subject. Most of the studies of business travels in astronomy therefore focused on the GHG emissions of air traveling, with the only exception being the work of Martin et al. (2022a, 2022b) who estimated for IRAP also other transport means and assessed the impact of lodging that generally is associated with business travels. For transport, they found that 96% of the GHG emissions were indeed associated with air traveling, while the remainder was due to traveling by car or cab (4%) and train (0.1%). For lodging, they found a footprint of 0.3 tCO_2e/employee, which corresponds to 6% of the total footprint associated with business travels.

The footprint of air traveling varies significantly among the different studies, with the smallest footprint of about 2 tCO_2e/employee estimated for the Netherlands and the largest footprint of about 9 tCO_2e/employee estimated for Australia (recall that the quoted values are for RFI = 2). While the latter can plausibly be explained by the size and remote location of Australia (Stevens et al. 2020), the former may be related to the large number of PhD students that are among the employees in the Netherlands, as studies suggest that students travel less compared to senior scientists (Berné et al. 2022; see also below). Jahnke et al. (2020) investigated the contributions of different flight destinations to the GHG emissions of air traveling at MPIA, finding that more than 90% of the emissions stem from intercontinental flights that are dominated by destinations in the United States and Chile. Van der Tak et al. (2021) made a similar observation for the Netherlands. Flight emissions scale nearly linearly with distance, so intercontinental travel will dominate in most situations.

All studies report large inequalities in GHG emissions from air traveling among the institute employees, with a general trend that carbon footprint increases with seniority. For example, once corrected for a RFI of 2, Stevens et al. (2020) find an annual mean per employee footprint of 24 tCO_2e for senior scientists in Australia, while they estimate 6 tCO_2e for postdocs and 3 tCO_2e for PhD students. For MKI, Simcoe et al. (2022) estimate values of 6 tCO_2e for professors, 5 tCO_2e for postdocs and 3 tCO_2e for research scientists (again corrected for RFI = 2). On the other hand, Martin et al. (2022b) find a limited impact of seniority on the traveling footprint of employees at IRAP, but reveal a gender inequality, with a disproportionately large fraction of the emissions attributed to male employees.

There exist multiple reasons for business travels in astronomy, including but not limited to participation in conferences, workshops, seminars, observing, projects, management, and job interviews. None of the studies have attempted to break down the footprint of business travels by reason for traveling, but assessments exist for the carbon footprint of conferences in astronomy that may provide some insights. For example, Burtscher et al. (2020) estimated the total travel-related emissions of the 1,240 participants of the European Week of Astronomy and Space Science (EWASS2019) in Lyon, France, to 1,855 tCO2e, corresponding to 1.5 tCO_2e/participant. Réville et al. (2021) estimated the carbon footprint of the Cool Stars 2021 conference in Toulouse, France, to 848 tCO_2e of which 90% were attributed to traveling, 2% each to lodging and meals, and 6% to the conference venue, which for 535 in-person participants resulted in an average footprint of 1.6 tCO_2e/participant (and 1.4 tCO_2e/participant for traveling). For a conference outside Europe (though not in the field of astronomy), Klöwer et al. (2020) estimated traveling emissions to 80,000 tCO_2e for the 28,000 participants of the 2019 Fall Meeting of the American Geophysical Union (AGU) in San Francisco, USA, suggesting emissions of 2.9 tCO_2e/participant. Hence clearly, a few international conferences per year can easily explain the estimated business travel footprints quoted above. Gokus et al. (2023) made a complete census of all 2019 astronomy conferences worldwide and came to similar conclusions, albeit with a substantial spread.

18.3.4 Supercomputing

Figure 18.2 also shows a large spread in the GHG emission from supercomputing, which can partly be explained by the carbon intensity of electricity production (see above). At the same time, the contribution of supercomputing also depends on the specific research activities, with large-scale simulations or processing of large data volumes being significant consumers of computing resources. This is illustrated in Table 18.1 where data on supercomputing provided in the five studies are summarized. The studies quote supercomputing information either in units of annual million CPU core-hours (MCPUh) or as electricity consumption (MWh),

Table 18.1. Estimated supercomputing usage per employee and associated power consumption. For MPIA it was assumed that 82.5% of the electricity is used for supercomputing. For MKI a Power Utilization Effectiveness (PUE) of 1.7 was adopted. Numbers in parentheses were derived from primary data quoted in the publications assuming a power consumption of 53 W per CPU core.

Study	Employees	MCPUh	kCPUh/employee	MWh	kWh/employee
Netherlands	1040	(12)	(12)	620	595
MPIA	320	(53)	(166)	2805	8766
MKI	158	(11)	(69)	578	3658
IRAP	263	7	27	(371)	(1410)
Australia	934	400	428	(21,200)	(22,698)

and conversion between both quantities were performed using a net power requirement of 53 W per core, as derived by Stevens et al. (2020) for Australia.

The annual per employee supercomputing volumes (either measured using CPU core hours or associated electricity consumption) vary by almost a factor of 40, suggesting values as low as 12,000 CPU core hours per employee for the Netherlands, and values as high as 428,000 CPU core hours per employee for Australia. It remains to be seen whether this spread indeed reflects different supercomputing usage patterns in the different countries and institutes, or whether (at least) part of the differences can be explained by the methods of estimation and differences in scope. In all studies, estimates of GHG emissions from supercomputing are subject to large uncertainties, primarily due to absence of detailed data and the assumptions made to derive the results. Definitely more research is needed in the field of supercomputing to understand its contribution to the GHG emissions of astronomy.

18.3.5 Purchase of Goods and Services

Each astronomy lab will purchase during the year some goods and services that are necessary to support the lab activities. This may include basic services, such as phone, internet, mail, transport, building maintenance and consulting services, and goods such as general office supplies, consumables, computers and other IT equipment, laboratory equipment and eventually hardware needed for prototyping or instrument development. All these goods and services will come with an embodied carbon footprint that needs to be accounted for in the lab's scope 3 emissions (item 9 of the Bilan Carbone© and ISO 14046-1 standards, item 3-1 in the GHG protocol).

In general, suppliers will not communicate the carbon footprint of their products, although there exist exceptions, in particular in the IT sector. In the case that no information is available, carbon footprint estimates can be made based on the cost of the goods and services, making use of sector-specific monetary emission factors. An example for such an exercise is illustrated in Figure 18.3, which shows the results obtained by Martin et al. (2022a) for IRAP. The authors classified the lab's spending

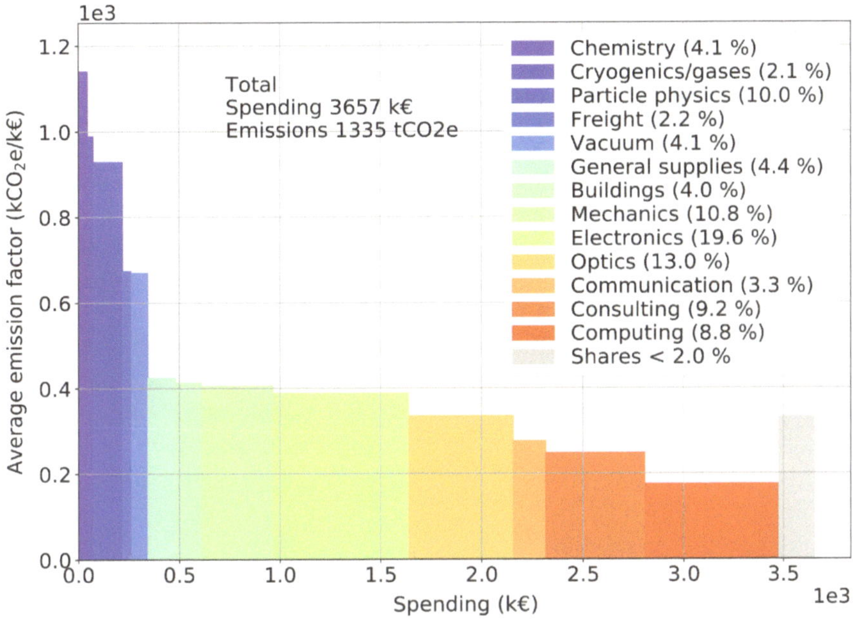

Figure 18.3. Average emission factors for the purchase of goods and services, organized by the main categories of expenditure, as a function of the cumulative expenditure. The size of the rectangles is proportional to the GHG emissions. The last gray block aggregates all shares that individually contribute less than 2% of the total GHG emissions. The fractional contribution to the GHG emissions for each category is shown in parentheses. Figure from Martin et al. (2022a).

using the French NACRES codes[3] into 35 broad activity sectors and associated each of them to an average monetary emission factor taken from the Base Empreinte© (an equivalent method is implemented in the *GES 1point5* tool that is applicable to research labs in France; Mariette et al. 2022). For a total spending of 3,657 k€ in the reference year 2019 they obtained a total carbon footprint of 1,335 tCO_2e, which corresponds to an average monetary emission factor of 365 gCO_2e per €. As illustrated in Figure 18.3, the specific activities of a lab will have an impact on the average emission factor, with labs using more chemicals, gasses, freight and vacuum technologies having a larger emission factor, while labs using more consulting and computing services (which also includes telecommunications) having a lower emission factor.

18.3.6 Use of Data from Astronomical Observatories

Astronomical observatories, be they located on the ground or aboard satellites in space, are the backbone of astronomical research. Consequently, the use of their data needs to be included in the scope 3 emissions of a research laboratory (item 9 of

[3] Equivalent codes exist in many countries, as well as at the level of the European Union (NACE) and the world (ISIC).

the Bilan Carbone© and ISO 14046-1 standards, item 3-1 in the GHG protocol). Knödlseder et al. (2022a) proposed an approach where for each observatory that is used a fraction of its annual carbon footprint is attributed to the lab on the basis of the user community. For example, if an observatory is used in a given year by 1,000 astronomers in the world, and ten of these astronomers work in the lab of interest, then 1% of the annual carbon footprint of the observatory will be attributed to the lab.[4] In that way, the carbon footprint of the world-fleet of astronomical observatories will be shared by the world-community of astronomers. As Knödlseder et al. (2022a) have shown, for the year 2019 this share corresponds to an average carbon footprint of 37 tCO_2e per astronomer.

So far there exists little information on the annual carbon footprint of astronomical observatories, hence Knödlseder et al. (2022a) used a monetary method to relate the full cost of a space mission and the construction and annual operating costs of a ground-based observatories to their carbon footprint. Based on carbon footprint assessments gathered from the literature, they determined emissions factors of 140 gCO_2e/€ for space missions, 240 gCO_2e/€ for ground-based observatory construction and 250 gCO_2e/€ for ground-based observatory operations. They also used an alternative approach for space missions based on the payload wet-mass, with an emission factor of 50 tCO_2e/kg, providing results that were comparable to the monetary estimates.

Recently, a few more carbon footprint assessments for ground-based astronomical observatories were published, including the W.M. Keck Observatory (WMKO) and the LOFAR radio observatory. For WMKO a carbon footprint of 1,994 tCO_2e was estimated for operations in 2019, including scope 1 and 2 emissions but only air traveling for scope 3 (McCann et al. 2022). Using quoted[5] annual operating costs of $16 M, which translate into 14.2 M€ for the year 2019, their result corresponds to a monetary emission factor of 140 gCO_2e/€, which is understandably lower than the value used by Knödlseder et al. (2022a) due to the exclusion of some scope 3 emissions. For LOFAR, Kruithof et al. (2023) estimated a carbon footprint of 1,867 tCO_2e for operations in 2020 that they converted into a monetary emission factor of 187 gCO_2e/€. They also estimated a construction carbon footprint of 19,904 tCO_2e that they translated using an initial investment of 200 M€ into a monetary emission factor of 100 gCO_2e/€ for construction. Kruithof et al. (2023) attributed their lower monetary emission factor with respect to Knödlseder et al. (2022a) to the fact that not all processes were included in their analysis.

In summary, the recent studies seem to support the carbon footprint estimates made by Knödlseder et al. (2022a) for astronomical observatories, and once more emphasize the importance of completeness in carbon footprint assessments. Still, more assessments are needed to get a comprehensive picture of the carbon footprint

[4] Specifically, Knödlseder et al. (2022a) counted the number of authors of refereed publications that cite a specific observatory in the reference year 2019 to estimate the fraction of the user community that resides at IRAP.

[5] https://www.ucolick.org/documents/KeckPRG_FINAL.pdf

of astronomical observatories, in particular in view of advising actions for carbon footprint reductions for the future.

18.4 What Can We Do?

As pointed out by the IPCC, limiting global warming to 1.5°C or 2°C "involve[s] rapid and deep and in most cases immediate GHG emission reductions in all sectors" (IPCC 2022b), hence carbon footprint estimates should not be seen as a goal in itself, but as a means to engage in a vigorous GHG emission reduction strategy. This engagement must be taken at all levels, including individual astronomers, labs, research and funding organizations, policy makers, scientific societies as well as the community at large. The following sections summarize the actions in the field of astronomical research that are currently discussed in the literature.

18.4.1 Acting as an Individual

Probably the biggest lever-arm to reduce GHG emissions that an astronomer has individually is to act on air traveling habits. While short-distance flights can be relatively easily replaced by train-traveling, the vast majority of GHG emissions arise from long-distance flights (e.g., Jahnke et al. 2020), hence reducing air traveling emissions ultimately means to travel less. For example, Martin et al. (2022a) calculated that using the train for all traveling in France would reduce IRAP's traveling carbon footprint by 20%, while limiting annual air traveling to two flights in Europe and one international flight would lead to reductions of 60%. For astronomers in the Netherlands, Van der Tak et al. (2021) suggest that air traveling needs to be diminished by a factor of 2 to 4 by 2030, and probably more beyond that, to reach the climate goals of the European Union. According to Simcoe et al. (2022), stopping traveling due to the COVID-19 lockdowns has led to overall GHG emission reductions of 15% at MKI, illustrating the order of magnitude that is achievable by this measure alone.

The COVID-19 lockdowns also illustrated what can be achieved by collective action, which is accompanied by community-wide efforts to change research habits (see also below). Stopping air traveling individually while the rest of the community continues with business-as-usual will of course have little effect, and likely isolate the exemplary astronomer. Changing research habits is probably also the solution to difficult-to-abate air traveling emissions, such as in-person collaborative visits and the installation and commissioning of hardware at astronomical observatories (Jahnke et al. 2020).

In the context of air traveling, carbon offsets are sometimes mentioned as a possibility to reduce GHG emissions (e.g., Stevens et al. 2020). Carbon offsets rely on the general idea that GHG emissions can be counterbalanced by GHG sinks, either in the form of emission reductions or carbon sequestration, for example in the form of newly planted forests. Financing new carbon sinks could therefore "offset" emissions, which is the rationale behind the offsets programs of airlines and other companies. Unfortunately, research shows that these schemes do not work (e.g., Coffield et al. 2022), and buying offsets is more "purchasing absolution of guilt than having a tangible impact on GHG concentrations in the atmosphere" (Stevens et al. 2020).

A more important lever arm for individuals is to reconsider computing habits. Portegies Zwart (2020) investigated the ecological impact of high-performance computing in astrophysics, and recommends avoiding the use of interpreted scripting languages such as Python for high-performance computing. Augier et al. (2021) argue that Python can be efficient and energy friendly if optimized and used with the right tools, pointing out the importance of human factors, such as time, work, knowledge and skills. Writing efficient code is therefore an interesting lever arm to reduce GHG emissions.

Little research exists on reducing computing as an efficient measure to reduce GHG emissions. Unfortunately, increased code (or computer) efficiency is in general not used to reduce computing, but to compute more, which is the classical rebound effect (also known as "Jevons paradox") that has been regularly observed since the 19th century when technological progress occurs. A bleak example of this effect is the exponential growth of the application of artificial intelligence and machine learning in astronomy (Rodríguez et al. 2022), which is worrisome in view of the related carbon footprint (Henderson et al. 2020). Questioning the nowadays widespread and quasi-systematic use of machine learning seems to be therefore an important avenue to be investigated.

Another category of GHG emissions over which an individual has significant control are employee-related emissions, i.e. home-to-office commuting and midday diet. Martin et al. (2022a) has shown that GHG emissions from home-to-office commuting can be significantly diminished by reducing car usage, switching to walking and/or cycling, public transport, and remote working. Replacing a car with a thermal explosion motor by a (light) electric car has also some benefits, depending on the carbon intensity of local electricity production. Simcoe et al. (2022) also point out the benefits of working from home as it reduces commuting needs, yet they warn about additional GHG emission from home heating and electricity consumption, emphasizing the need to engage with the employees on strategies to reduce their home emissions.

Concerning diets, Martin et al. (2022a) has estimated significant GHG emission reductions of up to 74% by moving from meals with a rich content of animal products towards vegetarian meals. Of course, the contribution of the midday meals to the total carbon footprint of an astronomer is rather small, yet any astronomer has full control over this emission source, and changing dietary habits will have no negative impact on research activities. The "only" change required is cultural, a change of habits and eventually values.

18.4.2 Acting as an Institute

The possible actions of an institute depend on the GHG emission sources over which the institute has some control, which include in general the sources listed under the office-related footprint above. The most significant emissions often come from energy-use of the buildings, either in form of electricity or heating/cooling. Hence the first action for any institute is to reduce energy use, for example through renovation of the buildings and/or the use of more energy-efficient technology. For example, Simcoe et al. (2022) suggest that replacing the single pane windows of the McNair building by double pane glass may reduce the annual GHG emissions by 150 tCO_2e, corresponding to 8% of the GHG emissions of MKI. Jahnke et al. (2020) propose to replace the existing oil heating system at MPIA by a system making use of ground heating in combination with electrically operated heat pumps, which is expected to reduce the heating related energy-demand by 50%.

Once energy-use has been minimized, the next lever-arm for an institute is the reduction of the carbon intensity of the energy sources. The most regularly mentioned option is the installation of solar panels on the institute buildings, as reported for the National Computing Infrastructure in Australia (Stevens et al. 2020) or proposed for the roof of MPIA (Jahnke et al. 2020). Yet the available roof space is rarely sufficient to satisfy the power needs of an institute (for MPIA it could cover ~10% of the needs; Jahnke 2023), and off-campus energy farms are proposed as a solution to reduce GHG emissions from electricity demand (Stevens et al. 2020). Using "green power" through purchase agreements is also proposed as a solution (Van der Tak et al. 2021), yet this will only lead to GHG emission reductions if such purchase agreements are directly linked to the installation of new renewable energy capacities. An example for such a direct link is the power purchase agreement concluded by MKI that allowed the construction of a 60 MW solar farm in North Carolina (Simcoe et al. 2022). Other examples are the agreements concluded by the European Southern Observatory (ESO) that led to the construction of solar farms at the La Silla, Paranal and Armazones Observatories (Filippi et al. 2022).

Another area of action for institutes is in the purchase sector of goods and services, which so far got little attention in the literature. Martin et al. (2022a) suggest reducing the related GHG emissions by modifying institute demand, including the introduction of clauses in purchase rules to disfavor or exclude suppliers that do not meet certain environmental standards and restructuring activities to favor less carbon-intensive purchases. Furthermore, they propose extending equipment lifetime, in particular in the IT sector, to reduce the overall institute demand. More research is definitely needed to better understand the GHG emission reduction potential in the purchase sector, which we recall is an important contributor to the carbon footprint of astronomical research, potentially comparable to professional traveling.

Finally, institutes also have an important lever arm by setting GHG emission goals and developing sustainability action plans, including, but not limited to, awareness raising, creating incentives, or setting rules for employees. For example, some institutes organize annual climate days or weeks with the purpose of raising awareness that can be the occasion to play serious games such as the one proposed by Gratiot et al. (2023) that is designed to build GHG emission reduction scenarios in the academic community, and for which a version exists for astronomers. Creating incentives can favor behavioral shifts, with an example being the reimbursement of public transport to reduce car usage for commuting (Van der Tak et al. 2021). Installation of decent video conferencing infrastructure and providing technical support personnel to facilitate the participation in and the organization of hybrid meetings can also help to curb GHG emissions from air traveling. For the same purpose, emissions quotas and/or carbon taxes are also regularly mentioned, for example by Van der Tak et al. (2021).

18.4.3 Acting as an Organization

There are a variety of organizations that support astronomical research, including universities, research organizations, funding agencies, academic societies, ground-based observatories and space agencies. They all have more or less important lever arms on the carbon footprint of the astronomical community and activating them has the potential to induce systemic changes that are urgently needed to curb GHG emissions.

As explained above, astronomical observatories are likely the dominant source of GHG emissions in astronomy, hence the organizations that build and operate observatories must be major actors in reducing the GHG emission of the field (Knödlseder et al. 2022a). For example, WMKO is aiming for net-zero emissions by 2035 by decarbonizing the vehicle fleet, efficiency improvements, a greener electricity grid, aviation footprint reductions and carbon offsets (McCann et al. 2022). While the net-zero emission goal is very ambitious, it remains to be seen whether substantial GHG emission reductions can indeed be achieved by the proposed actions. The U.S. National Science Foundation's NOIRLab aims to reduce by 2027 the organization's annual carbon footprint by 30% (or 2,500 tCO_2e) using travel reductions, and investment in energy-efficient equipment and renewable energy production (Miller et al. 2022). ESO has also set up a plan to decrease its carbon footprint, including the installation of solar arrays, divesting from air shipping, reducing business travels, extending lifetime of IT equipment, increasing the share of electric vehicles and low-carbon transport, integration of sustainability in design and procurement, with an estimated GHG emission reduction potential of up to 15%.[6]

Carbon footprint reductions get also increasing attention in the design phase of new observatories, as illustrated by Viole et al. (2022) who studied low-carbon power systems for the planned Atacama Large Aperture Submillimeter Telescope (AtLAST) at the Chajnator plateau in Chile. Sustainability is also a concern for large-scale projects such as the Square Kilometer Array Observatory (SKAO), which aims limiting power demand by design, using sustainably-generated electricity, minimizing non-essential travel, and considering environmental performance in the selection of prospective suppliers (Ball 2021). However, little quantitative information is available on the actual GHG emission reductions that will be achieved by these measures, and so far no environmental life cycle analysis of an astronomical observatory has been published to demonstrate that environmental burdens are not simply shifted from global warming to other areas of concern.

Also, new observatories are often added while the existing ones are kept in operations, leading to an accumulation of which the global carbon footprint is

[6] https://www.eso.org/public/france/about-eso/green/.

generally not quantified. To illustrate the impact of building new observatories, and based on data from Filippi et al. (2022), Knödlseder (2022b) has shown that the annual carbon footprint due to energy consumption at the ESO sites in Chile will increase due to the construction of the European Extremely Large Telescope (E-ELT), despite the organization's GHG emission reduction efforts mentioned above. Apparently, the reduction efforts are annihilated by the addition of new observatories, which is an issue of major concern. Knödlseder et al. (2022a) therefore argue that the only measure that will effectively curb the carbon footprint of observatories will be a slow-down in the deployment pace of new infrastructures.

Little attention has been paid so far to the role of academic funders in shaping sustainable research practices (Bousema et al. 2022), yet first positive signs appear in roadmaps and strategic documents. For example, in their 2020 environmental sustainability strategy, United Kingdom Research and Innovation (UKRI), the national funding agency investing in science and research in the UK,[7] aspires to be net-zero for their entire research undertakings. Among the actions to reach this goal is a commitment that all major investment and funding decisions will be directly informed by environmental sustainability. In the recently published roadmap for European Astronomy (ASTRONET[8]), covering the period 2022–2035, it is requested that all astronomy projects include environmental footprint assessments and reduction plans regarding construction and management of facilities, travel and computing, to follow (at least) the European timeline towards carbon-neutrality. At the last Council meeting of the European Space Agency (ESA) at ministerial level in November 2022, a resolution was endorsed that the agency should reduce its carbon footprint by 46% by 2030 as compared to the 2019 baseline.[9] Hence the direction is clear, and actions are identified, it's now up to the funding agencies to meet expectations.

Another way of how funding and research organizations can shape sustainable research practices is by directly supporting researchers that devote time and money to this activity. This is for example the case in France, where the CNRS and INRAE research organizations support the *Labos 1point5* Research Group that aims to better understand and reduce the carbon footprint of French public research.[10]

There is also a burden on funding and research organizations to enable systemic changes, for example by modifying evaluation criteria in hiring and promotion processes (ALLEA 2022). The current academic system promotes quantity over quality and rewards fast-paced science (Fischer et al. 2012) that directly relates to power structures and inequalities (Leite & Diele-Vigas 2022). Besides the associated large carbon footprint, this "publish or perish" culture affects mental health and creates burnout (Jaremka et al. 2020). Funding and research organizations can

[7] https://www.ukri.org/wp-content/uploads/2020/10/UKRI-050920-SustainabilityStrategy.pdf.
[8] https://www.astronet-eu.org/wp-content/uploads/2023/04/Astronet_RoadMap2022-2035_ExecSummary_interactive.pdf.
[9] https://esamultimedia.esa.int/docs/corporate/Resolution_1_CM22.pdf
[10] https://labos1point5.org

change this paradigm by promoting slow science and more generally sobriety as values (Berkowitz & Delacour 2020).

Another kind of organizations that should be important players for carbon footprint reductions are academic societies (Burtscher et al. 2022). On the one hand, they gather significant fractions of national, regional, or in the case of the International Astronomical Union (IAU), the world community of astronomers, enabling forums for community discussions that are essential for collective carbon footprint reductions. On the other hand, academic societies are often the organizers of large scientific meetings, contributing significantly to the community's carbon footprint and hence bearing responsibility for its reduction. Possible actions of academic societies can be exemplified by the Astronomical Society of Australia (ASA) which committed to reducing the climate impact of Australian astronomy by holding in-person meetings only when necessary, working with the community to improve the experience of remote meetings, advocating for sustainably-powered high-performance computing solutions, and supporting schemes to improve code efficiency. Many societies have swiftly adapted their annual meetings in response to the COVID-19 pandemic, demonstrating that significant carbon footprint reductions can be achieved by online, hybrid or hub-based formats without compromising scientific quality (Stevens & Moss 2023). Unfortunately, some societies returned immediately to business-as-usual once the pandemic was over, disappointing hopes of the community for more sustainable research practices (Antonova et al. 2022).

Astronomical societies can also create incentives, for example in the form of low-emission awards, akin to the Pleiades award that was created by the Astronomical Society of Australia for gender equity and diversity (Stevens et al. 2020). Similar awards can also be created by scientific collaborations to raise awareness of their climate impact, as illustrated by the CTA Consortium Travel Footprint Prize that is awarded to the consortium meeting participant who travels to the meeting with the lowest carbon intensity and highest travel duration (Biteau et al. 2023).

18.4.4 Acting as a Community

Finally, we can also act at the level of the community, where "community" designates any kind of grouping of people working in the field of astronomy, be it in the form of grass-roots movements, scientific collaborations, and regional, national or international groupings. An example for an international grass-roots movement is Astronomers for Planet Earth (A4E), with over 1,600 members from 75 countries, joined in the effort to combat the climate crises from an astronomical perspective (Burtscher el al. 2021).[11] Among the explicit goals of the movement is the establishment of a community to amplify individual voices and provide opportunities for meaningful action. A4E has organized several events and numerous public lectures, a webinar series that educates astronomers about the

[11] https://astronomersforplanet.earth

science of climate change, and campaigns to call for climate action within astronomy and beyond.

Astronomy conferences have been identified as having a large carbon footprint (Burtscher et al. 2020; Gokus et al. 2023), and the community bears the responsibility to organize conferences differently so that their GHG emissions can be slashed (Klöwer et al. 2020). Possible actions include, but are not limited to, selecting conference venues that minimize flying of attendees, creating regional hubs for large international conferences, enabling decent remote participation, decreasing the conference frequency, for example by switching annual meetings to biennial, and questioning the need for organizing ever more events. The changes to conference formats during the COVID-19 pandemic have clearly shown that conferences can be organized differently, now the community needs to take into account the lessons learned to design new ways of conferencing for the future (Moss et al. 2022).

18.5 Our Responsibility, Our Future

In this article I have tried to review our current knowledge about the carbon footprint of astronomy, and the avenues that are proposed to mitigate this footprint. Of course, as astronomers we will not fix climate change by ourselves, but since we understand the science and the threats involved, it is our responsibility to contribute our share. Our colleagues in climate and environmental sciences have been alerting the world of the dangers of continuing with business-as-usual for decades. To maintain our credibility, we have to act with responsibility, and lead by example (Ćuk et al. 2020).

We have to understand from where we speak: 89% of the world population is never flying (Gössling & Humpe 2020); more than half of the global world population has annual per capita carbon footprints below or within the target range of 1.6–2.8 tCO_2 that would keep global warming to below 1.5°C or 2°C (Bruckner et al. 2022). Compare that to the numbers of astronomy. Doing astronomy is obviously a privilege! And with great privilege comes great responsibility (Dotter 2019).

Now it is time to show up to the challenge. It's time to question the current state of affairs and to think beyond. Of course, technology will help us, for example to decarbonize astronomy, but this will not fix the problem. We have transgressed many planetary boundaries (Rockström et al. 2009), and climate change is only one of them. Decarbonization will not fix biodiversity loss, air, water and land pollution, excessive land use, scarcity of mineral resources, wide-spread micro plastics, exceeding biogeochemical flows, and freshwater scarcity, just to name a few of the problems. We have to move back into the safe operating range. And "we" concerns those who have transgressed!

Maybe this sounds overwhelming, but see it like this: how often do we have the chance to restart from scratch? And that's exactly what we need to do now, and it's gonna be exciting!

References

ALLEA 2022, *Toward Climate Sustainability of the Academic System in Europe and Beyond* https://allea.org/portfolio-item/towards-climate-sustainability-of-the-academic-system-in-europe-and-beyond

Antonova, A., et al. 2022, NatAs, 6, 765

Augier, P., Bolz-Tereick, C. F., Guelton, S., Mohanan, A. V., & Noûs, C. 2021, NatAs, 5, 334

Ball, L. 2021, *SKA Sustainability* https://zenodo.org/record/5159298

Berkowitz, H., & Delacour, H. 2020, M@n@gement, 23, 1

Berné, O., et al. 2022, ERL, 17, 124008

Biteau, J., Bulik, T., & Tibaldo, L. 2023, private communication

Bjørkdahl, K., Santiago, A., & Duharte, F. 2022, Academic Flying and the Means of Communication (London: Palgrave Macmillan)

Bousema, T., Burtscher, L., van Rij, R. P., Barret, D., & Whitfield, K. 2022, LaPH, 6, E4

Brander, M., Gillenwater, M., & Ascui, F. 2018, EPo, 112, 29

Burtscher, L., Barret, D., Borkar, A. P., et al. 2020, NatAs, 4, 823

Burtscher, L., et al. 2021, NatAs, 5, 857

Burtscher, L., et al. 2022, NatAs, 6, 764

Bruckner, B., Hubacek, K., Shan, Y., Zhong, H., & Feng, K. 2022, Nat. Sustain, 5, 311

Coffield, S. R., et al. 2022, GCBio, 28, 6789

Ćuk, M., Virkki, A. K., Kohout, T., Lellouch, E., & Lissauer, J. L. 2020, Pathways to Sustainable Planetary Science (Decaday Survey 2023-2032 White Paper) https://arxiv.org/abs/2009.04419

Dotter, A. 2019, JNCHC, 618, Online Archive https://digitalcommons.unl.edu/nchcjournal/618

EC 2021, Regulation (EU) 2021/1119 of the European Parliament and of the Council of 30 June 2021 establishing the framework for achieving climate neutrality and amending Regulations (EC) No 401/2009 and (EU) 2018/1999 ("European Climate Law") https://eur-lex.europa.eu/legal-content/EN/TXT/?uri=CELEX%3A32021R1119

Filippi, G., Scibior, P., van der Heyden, P., Arsenault, R., & Tamai, R. 2022, Proc. SPIE, 12182, id. 121823Z

Fischer, J., Ritchie, E. G., & Hanspach, J. 2012, Trends Ecol. Evol, 27, 473

Flagey, N. 2022, *Implementing an international version of the GES1p5 tool* A4E 2022 Symposium https://www.youtube.com/watch?v=7cqVDludfO4

Gokus, A., et al. 2023, in preparation

Gössling, S., & Humpe, A. 2020, Glob. Environ. Change, 65, 102194

Gratiot, N., et al. 2023, PLOS Sustain. Tranform, 2, e0000049

Henderson, P., et al. 2020, J. Mach. Learn. Res., 21, 1

IPCC 2022a, Contribution of Working Group II to the Sixth Assessment Report of the Intergovernmental Panel on Climate Change, ed. H.-O. Pörtner, et al. (Cambridge: Cambridge Univ. Press)

IPCC 2022b, Contribution of Working Group III to the Sixth Assessment Report of the Intergovernmental Panel on Climate Change, ed. P. R. Shukla, et al. (Cambridge: Cambridge Univ. Press)

Jackson, R. B., et al. 2022, ERL, 17, 031001

Jahnke, K., et al. 2020, NatAs, 4, 812

Jahnke, K. 2023, private communication

Jaremka, L. M., et al. 2020, Perspect. Psycol. Sci., 15, 519

Klöwer, M., Hopkins, D., Allen, M., & Higham, J. 2020, Natur, 583, 356

Knödlseder, J., et al. 2022a, NatAs, 6, 503
Knödlseder, J. 2022b, The carbon footprint of astronomical research infrastructures, Proc. SF2A2022 http://sf2a.eu/proceedings/2022/2022sf2a.conf..13K.pdf
Kruithof, G., et al. 2023, ExA, 56, 687
Leite, L., & Diele-Viegas, L. M. 2022, Nat. Hum. Behav, 5, 409
Mariette, J., et al. 2022, Environ. Res.: Infrastruct. Sustain, 2, 035008
Martin, P., et al. 2022a, *The carbon footprint of IRAP* https://arxiv.org/abs/2204.12362
Martin, P., et al. 2022b, NatAs, 6, 1219
McCann, K. L., Nance, C., Gavin, S., & Walawender, J. 2022, NatAs, 6, 1223
Miller, B., et al. 2022, A4E 2022 Symposium https://indico.icc.ub.edu/event/167/contributions/1506/attachments/611/1174/NOIRLab%20Sustainability%20Program%20-%20A4E%20Poster.pdf
Moss, V. A., et al. 2022, NatAs, 6, 1105
Portegies Zwart, S. 2020, NatAs, 4, 819
Réville, V., Brun, A. S., Bouvier, J., & Petit, P. 2021, *CS 21: A carbon footprint estimation,* Zenodo v1.0 https://zenodo.org/record/7467350
Rockström, J., et al. 2009, Natur, 461, 472
Rodríguez, J.-V., Rodríguez-Rodríguez, I., & Woo, W. L. 2022, WIREs Data Min. Knowl. Discov., 12, e1476
Simcoe, R. A., et al. 2022, *Carbon Footprint of the MIT Kavli Institute* https://space.mit.edu/wp-content/uploads/2022/09/MKI_Carbon_Footprint_revA.pdf
Stevens, A. R. H., Bellstedt, S., Elahi, P., Murphy, J., & M. T. 2020, NatAs, 4, 843
Stevens, A. R. H., & Moss, V. A. 2023, *Proc. CAP2022* https://arxiv.org/abs/2303.05259
US 2022, *The United States' Nationally Determined Contribution. Reducing Greenhouse Gases in the United States: A 2030 Emissions Target* https://unfccc.int/sites/default/files/NDC/2022-06/United%20States%20NDC%20April%2021%202021%20Final.pdf
Van der Tak, F., et al. 2021, NatAs, 5, 1195
Viole, I., Valenzuela-Venegas, G., Zeyrigner, M., & Sartori, S. 2022, *A renewable power system for an off-grid sustainable telescope fueled by solar power, batteries and green hydrogen* https://ssrn.com/abstract=4376384

Chapter 19

The Future of Meetings

Vanessa Moss, Glen Rees, Aidan Hotan, Emily Kerrison, Elizabeth Tasker, Rika Kobayashi, Claire Trenham, Ron Ekers and Travis Rector

> "In our view of the future, it is not so much distance that will be abolished, but rather our current concept of being there."
>
> —Jim Hollan & Scott Stornetta "*Beyond Being There*" (CHI, 1992)

Astronomy has a long history of innovative global collaboration, stretching back more than a century. Over the course of the last few decades, the discipline has embraced many technologies to better facilitate the dissemination of information. Yet the way in which we actually physically meet to partake in this dissemination has remained relatively fixed in format. During the COVID-19 pandemic we saw how our community could convene productively online. While virtual conferencing is very much in the "innovation stage" we also saw many immediate benefits, not only from drastically reduced carbon emissions but also by making it possible for many to participate who were previously excluded. With the rising impacts of climate change it is imperative—now more than ever—that we look to new ways of collaboration to improve the long-term impacts of our meetings on both our discipline and our planet. In this chapter, The Future of Meetings (TFOM) community presents some of the lessons we have learned in our quest for a "new normal" for academic meetings. We discuss some of the common challenges around online and hybrid interactions, and solutions to them. We conclude by looking to the future, whereby technology has the potential to dramatically improve how we work, collaborate, and socialize in an increasingly connected world.

19.1 Introduction

A century ago, there were no commercial flights, no computers, no internet and no smartphones. The world was far less connected than our contemporary one, especially when we consider the immediacy with which we can now spread information around the globe. Yet the way people back then met for conferences,

doi:10.1088/2514-3433/acfcb6ch19

meetings and collaboration actually looked remarkably similar to today, involving individuals or groups traveling sometimes great distances at often high cost to spend a relatively short period of time together.

Given then how much has changed about the world from the 1920s to the 2020s, it is curious that astronomy, academia and even society at large still consider the week-long conference as the gold standard for facilitating effective distributed collaboration, despite the fact that we can essentially pick up an electronic device of our choosing (telephone, tablet, laptop, mixed reality headset) and meet people from anywhere on Earth, as long as there is an internet connection bridging the two devices. Because the traditional in-person meeting format has existed for so long, we assume that it must be more effective, more useful and more beneficial than any other mechanism. So it has always seemed "worth it" to traverse the globe to meet and collaborate with colleagues—because that is how things are done, at least for those who have had the means to do so.

But we have a dilemma: as a global community, we recognise the need to reduce our carbon footprints urgently, by large amounts. An individual's carbon footprint is made up of a complex number of factors, with details outlined elsewhere in this book. But a significant component for an astronomer (and more importantly, one we have *direct control* over) is the extent to which we travel and the means by which we do it. Many discussions about sustainability in astronomy focus on all the things that might be lost when we change the way we work and collaborate, particularly when it comes to travel and conferences. There are commonly-raised objections to online and hybrid interaction, many of which have a basis in reality but the majority of which have solutions. It is not just sustainability which drives the need for change: considerations of accessibility, inclusivity and economics are major factors that encourage us to look for innovative solutions to distributed collaboration.

What we aim to show you in this chapter is that not only has astronomy always been an evolving field with a strong dependence on technological innovation for progress, but also that we gain much more than we potentially lose when we adopt enabling technologies to improve the ways in which we collaborate. It's early days still, and it remains to be seen which options we ultimately choose as a community. But our experience thus far has shown that the doors which open in a community that embraces online means of communication, and the seats at the table offered to those who have traditionally been denied in-person opportunities, are very much worth the effort, and they point to a brighter, more inclusive and more sustainable future for astronomy.

19.2 The Evolution of Meetings and Collaboration in Astronomy

Astronomy has a history of innovative global collaboration that goes back more than a century. This has been a natural requirement for a field under one sky that needs to share data from all parts of the world, crossing not only international lines but political and social barriers. Scientific conferences, in which astronomers from all over the world meet, have played a key role in the development of these collaborations as well as the agreements on conventions needed to work

Figure 19.1. Attendees of the first IAU assembly in 1922, in Italy. Credit: IAU.

harmoniously together. The International Astronomical Union (IAU) was formed *"to facilitate the relations between astronomers of different countries where international collaboration is necessary or useful."* It had its first meeting in Rome in 1922 (Figure 19.1). These conferences were also social events; and the interactions extended well beyond the meetings themselves, with participants often sharing travel by boat or train. There is almost no modern equivalent to the development of a collaboration while traveling together to distant lands.

In the following sections we look at the evolution of different aspects of astronomy conferences as examples of how changes have occurred in the past and as potential indicators of the changes we might expect in the future.

19.2.1 Poster Sessions

Poster sessions were introduced in the 1970s to solve the problem of too many participants needing to make oral presentations to get travel support. The excess of speaking participants was reducing the time available for discussion, and increasing the need for parallel sessions. Although initially disfavoured as lower status, posters slowly evolved to become an effective modern presentation platform, featuring "*sparkler talks*," direct contact with presenters, and the possibility for more complex presentations such as interactive computer models. Poster sessions can be readily translated into an online activity, with embedded links to powerful graphics and interactive displays. Proximity social media platforms such as Gather can facilitate direct interactions. Unfortunately these do not combine well with in-person interactions in a hybrid format, but online posters could effectively replace any need for in-person interactions and printed poster displays. Indeed, virtual exhibits and posters will probably become one of the first effective uses of virtual reality in academia.

19.2.2 Discussion Sessions

Questions and Answer (Q&A) sessions and discussions after talks are the essential difference between a meeting and a published paper. But there have been large and mostly negative changes since the introduction of digital technology. Up to about 1980, all discussions after presentations were recorded and published in conference proceedings. Meticulous processes were used to collect and verify the Q&A. This made it possible for those who couldn't attend to at least be aware of the content of the discussions, if not participate. It also provided material for historians seeking to understand the context of the scientific results being presented. This context is rarely captured in scientific publications. We have now lost some of this history because it was incorrectly assumed that new communication and recording technology would suffice. Any audio recorded at the conference is rarely transcribed accurately, the identity of the questioner is often not indicated, replaying recorded meetings is extremely time consuming, and Q&A discussions may not have been recorded or were lost. It is clear that modern technology offers great opportunities to retain the details of discussion sessions (e.g. with accurate automated transcription and AI-generated highlight summaries). But this requires active consideration and experimentation on the part of meeting organizers, as well as better standardized policies around the archiving of legacy-value meeting content.

19.2.3 Meeting Presentation Technology

Changing technology has had a big impact on how conference presentations are made, and we should expect this trend to continue. Few of the past changes could have been predicted and there is no reason to expect this will be different in future, but we can still look at past changes for examples of what can happen:

- 1960: Slides (3x4") glass—high quality but cumbersome.
- 1970-90: 35mm slides. Maintained high quality, enabled multiple projectors and remote control
- 1960-90: overhead projectors and transparent sheets—flexible and could be modified in real-time. At first these were hand drawn, then copiers were used, followed by computer generated transparencies. Some loss of quality but enabled real-time changes and more artistic personalized presentations.
- 1995: computer generated presentations using an LCD panel on overhead projectors. This was the beginning of computer generated graphics and it was to lay the path for future on-line presentations, although this was not the motivation at the time. Computer generated displays removed discrimination against presenters with poor writing skills.
- Mid 1990s: video presentations into multiple meeting spaces or lecture theaters.

- 2000: digital projectors. Initially inferior quality and brightness but still used because of convenience. Eventually they matched 35 mm slide quality. The introduction of digital projectors was very disruptive due to technical issues and immature support software, somewhat like the current technology for Virtual Reality.
- 2010: Digital projection technology finally stabilized and their use became ubiquitous.
- 2010: Video conferencing with individual participation using smartphone technology.
- 2020: COVID-19 rapidly accelerates the development and adoption of video conferencing tools. This was of course an unpredicted outcome.

However, until very recently technology has had little impact on the perceived need to meet in-person.

19.3 Why Do We Travel in Astronomy?

Digital technology is changing our research practices and society at large incredibly rapidly, but the uptake is not uniformly distributed. There also remains a widespread cultural impression that the most effective collaboration and learning occurs in person. In fact, we have built an academic system in which travel is widely accepted as necessary for career advancement. This increases astronomy's environmental impact significantly, and is highly exclusive given the many barriers to travel. Technological evolution offers a chance to improve our practices, reflect on the suitability of existing methods and shape the design of future interactions.

The need for effective communication with a global cohort seems like an ideal problem to solve with technology. Indeed, we have seen asynchronous communication platforms such as Slack become ubiquitous. However, the continuously connected nature of modern life can demand more of our attention and cause excessive context-switching. Instead of being the only way to collaborate, in-person meetings are now seen as rare opportunities to focus on a specific topic or task, with everyone in the same time zone. Here, we examine some of the most common contemporary forms of collaborative work with the intent to identify key factors.

19.3.1 Conferences and Workshops

While refereed journal publications remain the primary means of exchanging information with other astronomers, the sense of finality associated with publishing a result creates a need for other forums in which ideas can be exchanged less formally, particularly prior to publication. This role has typically been fulfilled by conferences of various structures and sizes, ranging from small, local workshops on specific topics, to large, international meetings covering the whole of astronomy and astrophysics. It has been argued that these meetings are necessary to spur ideas for new science and collaborations, by enabling people with shared research interests to meet.

The modern conference exhibits a poorly concealed duality in which its primary format (dominated by talks and other largely passive delivery of content) is often considered a prelude to the more free-form and productive discussions that take place in social settings around the periphery. Attendees often justify the use of institutional funds for travel and registration with the pretext of giving a talk or poster to disseminate their work. Yet in reality, it is this peripheral social element, the promise of meeting collaborators new and old, that is the strongest motivation for many to attend. Talks serve as an abbreviated form of introduction in which someone's current research interests are advertised to the assembled community. However, the growth of the astronomical community and proliferation of research topics has led to larger conferences needing multiple parallel sessions, such that no attendee can experience more than a small fraction of the formal content. Ironically, scientists who share research interests may be in attendance at the same conference but not meet simply because they chose to attend different parallel sessions. Some large conferences, winter meetings of the AAS for example, have upwards of thirty parallel streams. It is clear from this alone that the purpose and format of in-person meetings needs to be reassessed.

19.3.2 Education and Training

Education and training are also evolving with technology and many universities now offer online units. However, teaching complicated concepts at the postgraduate level is often done at schools that resemble conferences, with lectures on specific topics instead of research talks. Under ideal circumstances, in-person training has the advantage of rich feedback that can lead to a quicker or better understanding of the topic. However, in practice many training workshops inevitably involve periods of passive content-delivery which, as universities have shown, doesn't require face-to-face delivery to be effective in its purpose. Additionally, there is an increasing need for on-demand training that can be accessed when needed, rather than intermittent events. Finally as cohort sizes increase, physical limitations of scale become a problem. Technology can provide the tools to create effective online training material, circumventing these issues, but there are initial investment and re-training barriers to the development of quality content.

Community support is also a significant part of an astronomer's informal lifelong training, which naturally evolves alongside our social structures. It is now much easier to be connected to a larger community and therefore a wider pool of knowledge, but we still need to provide supportive forums in which questions are encouraged. How we best adapt the education and training of the next generation of astronomers based on their evolving needs is very much a work in progress across the field.

19.3.3 Observatory Visits

Technology has had the greatest impact on how astronomers build observatories and conduct our research. While it is important to support a wide range of facilities with varied roles, the largest and most data-intensive facilities tend to dominate the research landscape and reach the largest number of astronomers. As the complexity, scale and physical isolation of our observatories increases, it becomes less feasible to directly interact with them. In fact, many observatories even of small and medium scale have transitioned to more remote operations for reasons of efficiency and cost among others. For space-based observatories, observations are of course conducted remotely, but this is an increasingly standard observation mode for ground-based telescopes as well. Indeed, many astronomers now do their research entirely using online archives that make old and new data much more accessible, increasing the data's value and therefore observatory efficiency.

It is worth noting that the transition from "classical" observing—where the astronomer is physically present at the telescope at the time observations are completed—to remote, queue and robotic observations was also initially met with resistance; e.g., when Gemini Observatory was under construction in the 1990s and the idea of queue observing was proposed many astronomers felt that it was important to be at the telescope so as to be able to make quick decisions about the observing programme based upon weather conditions. As a concession Gemini did at first offer a classical observing mode, but it soon became clear that queue observing made the observatory more efficient and enabled kinds of science not possible under classical scheduling. Most astronomers eventually accepted queue observing, recognizing that it increased scientific productivity. Many modern telescopes, such as the CSIRO ASKAP radio telescope and the Vera C. Rubin observatory, are operated such that observations, data calibration and initial analysis are fully automated.

It is of course important for astronomers to understand how data is gathered so they can identify systematic errors and extract the best possible result. Previously, this information may have been gained by visiting an observatory or control center. With increasing decentralization of observatory operations, it is important for astronomers to have clear lines of communication with the observatories they use, but the format of these interactions must also evolve. From the educational perspective, it is also important and valuable to provide opportunities for the next generation of astronomers to interact meaningfully with observatories and their staff, and it is likely that at least some of these interactions will benefit strongly from taking place in person.

19.3.4 Collaboration Trips and Job Tours

For better or worse, it is widely acknowledged that building a successful career in astronomy requires developing a network of global contacts. There are usually several research groups from different parts of the world working on similar topics; and these groups may have expertise in different aspects such as theory, simulations or observations. It is common for an early career scientist to tour a variety of research groups in an effort to understand the totality of their field, build their professional network and create future employment options. Similarly, it is common for an established scientist to take a sabbatical, dedicating several months to pursuing a research goal in a location other than their main institution while setting other duties aside. Some aspects of these career-oriented trips are difficult to reproduce any other way, such as immersing oneself in the culture and surroundings of a potential future workplace. However with increasing freedom to work from anywhere, the need for such trips may diminish over time. Instead, we will likely need to develop a culture that supports setting aside normal duties to change focus without the need for physical separation from one's home institution.

19.3.5 Review Panels and Committees

For senior astronomers, it is typical to serve on various review panels, steering committees, resource allocation committees, and other esteemed assemblies. It has been common for such groups to meet in person, often in conjunction with a physical tour of a facility and social events. Such groups are responsible for making key decisions that impact the future of astronomy, so it is widely accepted that a significant devotion of time and travel is justified. However, since it is desirable for these panels to have diverse representation from around the world, senior astronomers tend to be responsible for disproportionately large carbon emissions. The goal of diverse representation would in fact be better served by relaxing the burden of travel and time required to participate in leadership activities. Technology can be used to create realistic virtual representations of facilities such as telescopes, virtual meeting spaces and easier access to ancillary information such as the technical feasibility of an observing proposal. Technology can also help increase the transparency of decision-making processes and organize input from a much wider range of people. Already virtual meetings are being used for review panels by many observatories and funding organizations; e.g., NASA and the NSF.

19.3.6 Building and Fortifying Distributed Communities

The experimental nature of research means that learning the process is just as important as understanding the background. While it is possible to teach the principles of the scientific method, knowing how to apply these principles has typically required hands-on experience. Knowledge is handed on from experienced astronomers to their postdocs and students. Research groups form as semi-autonomous units, then collaborations form between these groups. Typically, individual research groups have been co-located to facilitate rapid communication. However, the increasing scale of astronomy has led to groups and collaborations numbering hundreds of researchers, which means it is no longer feasible for everyone to be co-located. It is also increasingly difficult to know the skills and talents of everyone in the group, and yet without these sprawling groups with diverse skill sets it would not be possible to operate complex projects such as the Event Horizon Telescope and the LSST survey. Local interactions are still important, but to be an effective part of the global community, we must constantly seek to improve our ability to collaborate across large distances.

19.4 The Cost of "Business as Usual" Astronomy Practices

There will remain in the astronomy community an existing need for travel to some extent for the reasons listed above, but at what cost? It is time to rethink the continuation of these activities in person, as the disadvantages of "business as usual" far outweigh the benefits of alternative approaches. There is no doubt about the carbon footprint of travel. However, it has only recently been recognised how the disadvantaged are further marginalized by the way we do things now. We need to address problems with accessibility and inclusivity so the whole community can participate on a level playing field.

19.4.1 Sustainability

We live on a planet with finite resources, and in the current era, our travel is powered almost exclusively by fossil fuels. The fact that burning fossil fuels directly causes climate change cannot be ignored. Anthropogenic climate change is everybody's problem, but the responsibility to mitigate it lies particularly with those with the power, influence and ability to change our practices. As discussed in Chapter 18, travel to an in-person meeting averages 1–3 tons CO_2e per participant per meeting, with variation depending primarily on the distance traveled (Burtscher et al. 2020, Klöwer et al. 2022, Gokus et al. 2023). For perspective this can be more than an average individual in a developing country produces *per year* (IEA 2023). In order to reach the Paris Target limiting global warming to 1.5°C above the pre-industrial average, the global carbon budget needs to be just 2.3 tons of CO_2e per person per year by 2030 (Carbon Independent 2007). It is impossible to achieve this target while undertaking regular long-distance travel, and indeed most people with access to higher education enjoy lifestyles with carbon budgets many times this goal. As such, we have an urgent need to reduce emissions for the sake of future generations to have the opportunity to continue to engage in science and astronomy, in a world that remains habitable to humans. If the reasons we travel, as discussed above, can be at least partially replaced by online meetings and collaboration, then every bit of carbon not emitted into the atmosphere is doing good for the planet and all who live on it.

19.4.2 Accessibility and Inclusivity

There are many reasons to adopt virtual interaction beyond addressing climate change. In particular we need to acknowledge that in-person meetings as they are done now are exclusive in part because they are inaccessible to many. The term "accessibility" is frequently associated with the physically disabled, but there are more factors that make in-person conferences out of reach for many people; e.g., financial, geopolitical, family and job responsibilities. International travel in particular is skewed towards the privileged, those with the financial capacity to afford increasingly costly air travel, accommodation and associated expenses. Furthermore, many people are excluded through visa restrictions, bureaucracy and fees. Moving conferences online removes many of these barriers and attendance figures for virtual meetings have shown an increased participation of a far more diverse audience (e.g., Skiles et al. 2022).

But bringing people to a conference is only half the story. Once they are there they should feel welcome, as highlighted by the associated idea of inclusivity. Conferences often have a poor reputation regarding the safety of junior researchers—especially female—in traditionally male-dominated fields such as physics (Aycock et al. 2019). Online conferences can provide a safe environment for those who feel uncomfortable in such settings, but that does not mean they are necessarily welcoming. It is a frequent criticism that online conferences lack the human touch, and they can propagate existing inequalities; e.g., women who do not travel are less likely to be relieved of obligations at home during a virtual conference, which decreases their ability to fully participate in the event (NASEM 2021). Likewise, X (formerly Twitter) is a common communication tool for conferences, but users tend to interact more with people with similar stature or those from their region, mirroring aspects of inequality inherent to in-person conferences (Duncan & Shean 2023). But with good design social interactions can be improved, as will be discussed in the next section.

19.5 Technology as a Facilitator for Interaction

In 1985, the first network connecting institutes at the US and European space agencies was created across the Atlantic. The purpose was to rapidly share mission data from the International Cometary Explorer that had to be downloaded in Baltimore, USA, but analyzed in Darmstadt, Germany. The network reduced a task to a matter of hours that had previously taken months of recording data tapes and mailing internationally, and enabled the interception of a spacecraft with a comet to be broadcast live to a public audience.

The network was later extended to Japan and Russia, revolutionizing the ease in which the world's largest space agencies could work together. Tasks that previously required multiple long-distance trips to install instruments on joint missions could now be done remotely, making the collaboration process vastly simpler.

This early internet was very limited by today's standards, but it allowed real-time communication across huge distances. This meant people could interact as if their collaborators and equipment were in a single building. By the 2010s, message exchange was being supplemented by video conferencing, allowing regular meetings to be held easily with attendees from all over the world. The advent of smart phones similarly normalized distributed communications and increased the critical mass of people worldwide who relied on technology to stay connected with distributed family, friends and colleagues.

While incredibly useful at stripping away the restrictions of distance, online communication has also become widely used for even local teams due to the ease of supporting asynchronous discussion. Video call platforms are particularly successful for seminars and meetings where there is a one-way flow of information at a single time to a large group of people, or a live discussion between a small number of individuals. But for larger teams that may also need to communicate asynchronously over a long period of time, collaborative platforms aimed at creating online communities have grown in prominence.

Examples of this include Discord on the social front as well as Slack and Microsoft Teams for professionals, where groups of people can join a virtual space dedicated to their team, organization, or event and exchange messages, share media, and also host video meetings with other members. Slack now reports more than 10 million daily active users across more than 150 countries, Discord at 26 million active daily and 125 million a month (Curry 2023a), while Microsoft states a monthly figure of more than 270 million active users (Curry 2023b).

These kinds of spaces have proved particularly successful for group discussions, which are difficult to achieve via email as only the most recent contribution is easily seen, and also challenging for both in-person and video meetings where a single time must be agreed upon, and a small number of voices often dominate the discussion.

As a result, collaborative platforms such as these are used by organizations, local and disparate working groups, and also for conferences. The latter usage provides a place to ask questions about presented talks and posters to the speaker, and also see replies to questions from other attendees. This is useful both not only for on-line conferences, but also for in-person meetings where Q&A time is limited.

Through such advances, technology has been able to provide wide and varied ways of interacting online that come with added advantages and some pitfalls. By far the main advantage concerns accessibility and inclusivity. In addition to removing the barrier of mobility, the online environment can enhance the conference experience for people with a range of disabilities, such as through the use of closed captions for the hard of hearing and audio description for the visually impaired. Astronomy can go beyond the common senses (Vargas et al. 2022) through many accessibility tools enabling equal participation of physically and neurologically disadvantaged people (Tang 2021). Specifically for inclusivity, instant translation can break through language barriers for people from different nations.

While it should be recognised that online accessibility does not automatically translate to a welcoming experience, this is mostly an issue of implementation rather than the format itself. Too often online conferences are focused on the in-person component without taking time zones into account or giving presence to the online audience who are often treated as passive spectators. With good design, an online conference can be made to feel inclusive of the audience through structured sessions, such as "onboarding" to familiarize the participants with the online platform and tools or dedicated time to meet for discussion, while unstructured sessions such as games and shared breaks can be used to keep the online participants connected. The biggest issue of timezone can be mitigated by not being too reliant on synchronous content, with asynchronous viewing of recorded talks and the aforementioned tools, it is possible to keep questions and discussions going "after hours."

The result is that online technology greatly facilitates interaction and communication, regardless of whether physical distance is an issue. For situations where distances are great, these tools help remove the need to travel large distances and allow the easy, regular contact required to build a successful project.

Creative responses in the time of COVID leveraging technology to work more effectively online have shown that it is possible to move forward from "business as usual." Already we are seeing changes in the way we work as many companies embrace the flexibility of working from home. Similarly, the astronomy community should be pushing for better practice for a more sustainable, accessible, inclusive future.

19.6 Where Do We Go from Here?

So given the above, what comes next? It is clear that our communication methods and approaches have evolved over time, and there remain many reasons in astronomy for meeting and interacting. What remains to be seen is how we determine when these interactions need to happen in person, given considerations of the costs and benefits, and when they can be shifted to alternative modes that are less carbon-dependent and exclusive. As a global, distributed community, we are now faced with choices each time we meet: in-person, online or hybrid? In our experience, we continue to see many questions around the feasibility of non-traditional modes, particularly when it comes to the cost of hybrid and the effectiveness of online interaction.

For context, we currently operate as a community of practice known as "The Future of Meetings" (TFOM), stemming from the 2020 symposium of the same name (as outlined below). Our community includes scientists, researchers, educators and industry representatives, all of whom are united by a common interest in how future meetings and interactions may be optimized, with a central theme of leveraging technology for improved accessibility, inclusivity and sustainability. As TFOM, we continue to experiment with and explore new approaches (e.g., Figure 19.2) and to raise awareness of the many possibilities for effective hybrid and online interaction that have not yet become standard practice within the astronomy community or elsewhere.

The pandemic offered us an extreme opportunity to reset our assumptions about meeting and proved that even with several months of not meeting in person as freely as we once did, we were able to continue working and collaborating and, in the case of astronomy, keep progressing our exploration of the universe. There were clear limitations in many cases, but often these came back to a choice of design, approach or technology rather than a fundamental limitation of online interaction.

Now, for all of the reasons outlined above, it is critical that we find a way to step back, look at the broad picture of all the ways we meet and interact, and determine an effective balance between the familiar sociability of in-person interaction and the relatively unexplored possibilities of the online format. For us in TFOM, it has been clear that the most exciting element of using technology to go online is not the endless quest to reproduce in-person interaction, but all of the wide and unexpected possibilities that arise when you embrace the concept of going "beyond being there" (Hollan & Stornetta 1992).

Figure 19.2. Members of "The Future of Meetings" (TFOM) community of practice, gathered in Altspace at the conclusion of an event series run by TFOM in VR during 2022.

19.7 Addressing Common Concerns about Online and Hybrid

During the pandemic, the world experienced a near-complete shift to online formats for interaction which radically and abruptly displaced our expectations about meetings. Since mid-2021 with the return of international travel and the lessening of COVID restrictions, meetings and conferences have been evolving in many directions at once. We have seen the predominant return of the in-person only conference/meeting, as if it marks the triumph of "back to normal," alongside a variety of types of hybrid conferences and meetings, many of which have left people concluding that hybrid interaction is ineffective, or worse, impossible. Some meetings and conferences have stayed online-only, but these are in our experience the minority compared to the other two formats. In astronomy, it seems that conferences and workshops in particular have remained resistant to the reduction of in-person reliance, whereas more functional work practices (e.g. weekly or monthly meetings, colloquia) have shifted towards online or hybrid due to the intrinsically-distributed community.

As for why online formats aren't adopted more often as the primary format, or why hybrid is done either poorly or not at all, we have observed from the TFOM context that there are common assumptions being made across astronomy, academia, and society more broadly. In this section, we have collated a list of the most prevalent assumptions and criticisms being made of these formats, which we see as presenting barriers to the better and more widespread adoption of online formats. These barriers are not insurmountable because, in the majority of cases, they are embedded with misconceptions and there are straightforward ways to reduce the impact of these barriers.

The statements presented below were collated by the authors of this chapter, and then ordered based on the weighted consensus of their importance to include here and by how much an improvement can be made by using different methods and new tools, both of which we indicate for each statement. We aim with this section to help illuminate some of the current challenges about online and hybrid formats, and offer tangible solutions to address these for those who are organizing, attending and/or funding astronomy meetings and conferences.

19.7.1 "It's Fine for Those with Accessibility Needs to Be Online, Everyone Else Can Go to Meetings in Person"

When it comes to hybrid formats, there has been considerable discussion about how online participants are often treated like second-class citizens, because the experience and design of most conferences is focused too much on the in-person component. Some surveys (e.g., Remmel 2021) indicate that researchers want hybrid to be the future format of meetings and conferences, but when they are asked how they personally want to attend the majority choose in-person. The issue with this is that if we continue to put most of our effort into the in-person format because that's what we know and that's what the majority will choose, then we will never create innovative, equitable hybrid or online-only formats. This means that those who participate online will always have a second-class experience, and miss out on opportunities that are only offered to the in-person portion of the attendees. There are many reasons why someone might choose to attend online rather than in-person, and many more reasons which remove their ability to choose freely. It is an extreme stance to deliberately relegate disadvantaged communities to online formats with little funding, planning or support, while the more privileged can continue to gather in person. This is especially true since both hybrid and online interaction can (with some thought, effort and design) be done well.

Recommendations

- Ensure every meeting and conference has reasonable (even if basic) online access.
- Give attendees, speakers and stakeholders a choice in format from the beginning.
- Host online-only gatherings where possible to give people a more equal voice.
- If a meeting must happen in person, use accessibility grants to support both in-person and online attendance, to assist diverse attendance.
- Try to establish parity. Avoid situations where only a very small fraction of attendees are online-only.

19.7.2 "You Can't Reproduce the Organic and Serendipitous Interactions Online that You Have in Person"

The belief that people cannot have organic and serendipitous interactions online in the *same way* they have had in person may be somewhat true, but it is not true that equally effective interactions cannot be achieved online. Most online conferences currently do not allocate time or mechanisms for the side-chats that are part of attending talks, instead focusing too much on the functional structured elements of a conference program (e.g. talks, panels). It is also important to consider the exclusivity of in-person events. While meeting someone in person may remain the gold-standard for some, think of the future collaborator *you didn't meet* because they couldn't attend.

While in-person interaction continues to hold value in the future interactions of our communities, online interaction can be just as effective and impactful when it comes to serendipitous interactions. All it relies on is for organizers to think carefully about how best to facilitate these modes of interaction, whether it is by specific organized social events or by making use of technology that helps with introductions and asynchronous interaction. In particular, immersive technology (e.g. virtual reality) presents a much more natural way to interact that is closer to our traditional expectations of body language and shared "physical" spaces.

Recommendations

- Provide specific planned mechanisms and opportunities for people to interact online.
- Host structured interactive meet-ups such as poster sessions and social events.
- Be innovative in approach and experiment with new platforms/technologies.

19.7.3 "Hybrid Is Impossible because It Doubles the Cost of the Conference"

There are a few factors mixed into this misconception that are worth unpacking. The first is that there is only one way to hold a hybrid conference, when in fact hybrid meetings range from completely passive "fly-on-the-wall" style online attendance, to an experience that is maximally equivalent whether an attendee is in-person or online. The former is far too commonly accepted as "hybrid" and the latter is still to be realized in future with the rise of widespread mixed reality technology. For now, effective hybrid formats lie somewhere in between these two extremes, and the right approach (e.g. the effort/resources invested to bridge the gap between the two audiences) should be adopted based on the situation. That said, it can be relatively cheap and low-effort to enable a basic hybrid standard, such that online participants are granted some degree of active participation. This could include a video call platform, a text chat platform, a screen in the room showing online participants and cameras in the room showing both speakers and the room. It is worth noting that a key element in the above working well is active management by organizers of both audiences and by encouraging both in-person and online attendees to interact via the bridging platforms.

When it comes to why hybrid is thought to cost so much, this generally comes from the context of commercial conference venues and professional conference organizers (PCOs), especially in cases of parallel sessions and large numbers of attendees. But often the expensive solutions provided in these cases do not meet many of the basic needs of online attendees, and focus too much on the in-person experience alongside costly AV-heavy solutions. Fundamental questions that organizers need to ask in these cases is whether the conference format being adopted is truly the right approach, or whether the approach could be innovated upon to better suit the needs of the community and if they are truly getting value for money for their quoted hybrid conference? Is a content-overloaded conference with numerous parallel streams really an effective way to bring an audience together, regardless of their mode of attendance? Are traditional PCOs in their current form really the right solution to handle the organization of meetings, given the dramatic impact technology has had globally on interaction? Are "TV-style" broadcasts and post-production effects really needed for academic talks, do we really need a full production crew for setting up a Zoom broadcast? Or is learning to do it ourselves combined with good lighting, a well placed camera, and an effective mic enough? One hope we have on the TFOM side is to see the rise of flexible approaches in hybrid that better meet the needs of our communities, but this requires organizers to see past the current offering of commercial hybrid approaches and seek solutions that truly help both online and in-person attendees get the most out of meetings.

Recommendations

- Consider the needs of both online and in-person attendees from the beginning of conference design and organization.
- Explore and experiment with low-cost alternatives (including technology) to commercial hybrid provision.
- Request solutions from PCOs that employ more innovative approaches to hybrid while clearly justifying the costs involved.

19.7.4 "The Energy Cost from Running an Online Meeting Is Just as High as an In-person One"

This simply isn't true. When discussing the environmental impact of various meetings, this argument frequently arises as a criticism of remote formats, often without any more evidence than a hand-wavy sentiment that computers have carbon footprints too. While it is true that running personal computers, online event platforms, and remote data storage all come with an energy cost, these costs pale in comparison to those associated just with travel to and from conference venues, let alone all the other elements of in-person attendance including the production and waste of consumables (Tao et al. 2021). To put this in more concrete terms, for the online TFOM symposium hosted in 2020, the total estimated carbon budget for all 200 synchronous attendees combined was just 1420 kg CO_2e. This estimate aimed to be comprehensive, including pre-, during- and post-event estimates for all video processing, storage and streaming, device and server power usage, and even a fraction of attendees device manufacturing emissions. By comparison, this is less than the travel emissions alone of a single international attendee physically attending an in-person version of our event (Moss et al. 2020). This is worth repeating; the emissions for an entire 200 person online event were estimated as *less than half* the emissions of *a single one-way, economy-class flight* between Europe and Australia. Although meetings will always have an energy cost, online meetings for which the cost is dominated by electricity generation, can be made more sustainable by ensuring that the electricity is generated from renewable resources. At present, there is no carbon-neutral way to fly, and for many electric vehicles are still out of reach (Burtscher et al. 2020, Tao et al. 2021). This misconception may be the result of awareness of the carbon footprint of *supercomputing*, which is indeed quite large.

Recommendations

- Research the carbon footprint associated with different formats for a given event, and make sure the cost is balanced by the benefits in format choice.
- Consider options for minimizing or offsetting the energy cost of an event.
- Include sustainability and carbon cost as part of the budget of a meeting or conference so that it is considered during every step of the process.

19.7.5 "We Need to Go Back to In-person Formats for the Sake of the Early Career Researchers (ECRs)"

Though often made with the best intentions, this argument is overly simplistic and falls into many of the same traps as section 19.7.2 above, and section 19.7.9 below. It presumes that future scientists will work in the same manner as their seniors, even though every profession continues to evolve. It also presumes that meaningful connections, which so often grow into career opportunities for ECRs, cannot be forged online. While there is unequivocal value in in-person meetings *for the right reasons*, blanket statements like this one are often made by senior academics whose views are informed by their own ECR experience in which in-person connection trumped all, but the views of ECRs themselves tend to be more nuanced and balanced (Köhler et al. 2022). To be sure, in-person activities are invaluable for ECRs as they work to establish a name for themselves in their chosen field, and these should continue to be made available to all (in moderation). However, we must ensure that any such in-person activities are both effective and widely accessible, and they must be supplemented by other forms of interaction to provide equitable opportunities to all ECRs, whether they can attend an in-person event or not. Just as importantly, we need to ask ourselves what it is about in-person events that is so powerful for ECRs, and why this hasn't been emulated online; does it require different types of online interactions? A cultural shift amongst established academics in their attitudes to online events? Whatever the answer, taking steps to address this question will be integral to ensuring not only that ECRs survive in a digital-first world, but that they thrive.

Recommendations

- Make sure that changes to the meeting format doesn't disadvantage ECRs
- Clearly identify whether your event will cater specifically to ECRs, and if it will, what elements will help them derive tangible benefits from attending.
- Consider ways to maximize the value of the in-person events that do happen for ECRs through structured networking and dedicated social time.
- Encourage senior academics attending all events (but especially online ones) to engage with ECRs beyond just delivering their talk.
- Encourage the academic community to move away from a model which leads ECRs to believe that in-person meetings are crucial to their career.

19.7.6 "COVID Showed Us How Tedious, Fatiguing and Boring Doing Everything Online Is, and How Ineffective It Is as a Format"

All of us can attest to having a bad experience with an online conference. Often this is because the format of the meeting was essentially the same as what would have been done in person. This is understandable because many conferences were quickly moved online when the pandemic started, not allowing time to think out format changes. What COVID actually showed us was that if every interaction becomes a video call, then we find meetings and interactions very tedious, fatiguing and boring. At the same time, the pandemic did dramatically accelerate the future of work because it showed, globally, that many professions could be carried out productively and effectively even with a distributed workforce. That said, interpersonal connections did suffer during the lockdown period because humanity has spent thousands of years meeting in person, and less than fifty years using computers to interact. It was also true that new starters faced challenges in particular because it was difficult for them to integrate into already established teams and earn trust when the bandwidth of interaction was more limited. However, none of these challenges do anything to show that online itself is ineffective as a format, but instead showed that how online formats were used was ineffective in some ways. It is not the case that everything should shift to online formats, but there are certainly varied and creative ways to use online formats, platforms and technologies effectively beyond just the video call.

Recommendations

- Schedule breaks between/during online meetings to reduce the impact of fatigue.
- Adopt more interactive formats and tools for meetings to avoid passive attendance.
- Make use of other formats (e.g. asynchronous chat) to reduce live meetings.
- Adapt meetings to be less structured and more informal where possible.
- Make time for social interactions in professional settings.

19.7.7 "We Don't Need to Adapt, We Can Just Go Back to Normal Now because COVID Is over and We Don't Need Online Formats in the Future"

What is "normal" is always tricky to identify when the world is continuously changing. It goes without saying that the pandemic has had a huge impact on the world far beyond its medical implications. For months, even years, a significant percentage of the working population was forced to be more distributed than ever before because of lockdowns and virus spread prevention. This dramatically accelerated "future of work" trends that had been observed and predicted for the years leading up. Diversity in both attendance and active participation was seen to increase for online meetings, and even now we have seen a majority of organizations and companies adopt hybrid working formats (if not embracing complete flexibility). This alongside the two equally important trends of mixed reality and artificial intelligence point to a future of interaction that is both distributed and heavily dependent on technology. Not unlike the early 20th century declaration of Kelvin that the airplane would never be successful, the 1970s statement by Olsen that nobody would ever want a computer in their home or the pre-millennium prediction by Metcalfe that the internet would collapse before becoming useful, anyone who believes that COVID was a passing phase with no lasting impact and that we can go back to "normal" as if nothing changed seems disconnected from the reality of the trends across society. Even putting aside the context of the pandemic, a key point of this book is to prepare our communities for the rapidly rising impacts of climate change globally, and sustainability is a pivotal reason to accept the impetus to adapt that we have been offered. Combined with improving our accessibility and inclusivity as a global community, it seems obvious that online will be a critical part of our future. Those who adjust their ways of working sooner will be best placed in the years to come, especially with the changing economic factors that will equally demand changes in practice.

Recommendations

- Embrace hybrid ways of working as the future rather than a temporary situation.
- Provide flexibility in attendance as a default for all meetings, acknowledging that everyone's needs are different.
- Take active steps to ensure that in cases of hybrid attendance, both the in-person and online audience are able to engage as equally as possible.

19.7.8 "There Are Too Many Technical Issues When You Go Online and They Often Can't Be Solved"

While it is true that online meetings may run into technical issues unique to the medium, other meeting formats are not without their own pitfalls. It is not uncommon for small in-person meetings to be derailed by duplicate room bookings or an inadequate set up, and all meetings irrespective of format are susceptible to the dreaded AV problems. While we are quick to blame the technology if, say, a speaker cannot participate because their internet connection is down, we don't apply the same criticism if a speaker misses the conference because of a canceled flight. As for any meeting, technical issues in an online meeting can be minimized with appropriate preparation. Having contingency plans can go a long way to ensuring a smooth running event, and online this can be as simple as preparing a backup meeting link and checking that key meeting attendees (such as keynote speakers, session chairs, etc) are familiar with the platform before the event starts. Remember good preparation is always apparent to an audience, and typically your audience will be more understanding of technical difficulties when they can see you have taken steps to mitigate them beforehand.

Recommendations

- Do not hold your online meetings to a higher standard than in-person ones, and remember things go wrong all the time, regardless of format.

- Spend adequate time safeguarding against predictable technical issues early as part of your planning stage. These include things like difficulty navigating the platform for attendees, and access issues related to meeting materials.
- Provide low-tech options for those hampered with technology issues e.g. live streaming sessions of platforms people have difficulty accessing.

19.7.9 "You Can't Start New Projects or Form Effective Collaborations Online"

On the contrary, it is entirely possible to meet new collaborators and form effective collaborations online. However, just like with in-person events, the ease with which this can be done is heavily impacted by the attitude of both attendees and organizers. If no provisions are made for networking as part of an event, either in-person or online, then it will naturally be more difficult to meet new people than if there is a dedicated social space. Likewise, if attendees come to an event without any intention of engaging in networking or making introductions, then it will be very difficult for others to get to know them, regardless of the format. Since much of the recent shift to online was driven largely by the sudden emergence of the COVID-19 pandemic, many peoples' first experience of organizing an online meeting was hasty and fuelled by a need to produce a minimum viable product. As a result, many recent online meetings have focused purely on content delivery at the expense of providing space for networking, which is taken almost for granted as an element of in-person events, whether it be in the form of a networking hour, or tea and biscuits following presentations. Admittedly networking online often requires a different type of interaction to what is expected in-person, but already many younger researchers are more adept at navigating these online pathways, suggesting their prevalence has only to grow.

Furthermore, a number of world-leading collaborations in astronomy are globally distributed: EHT, Planck, SKA, and the list goes on—and international collaborations are by no means limited to the astronomy domain. If it were not possible to collaborate effectively online then many of these large-scale collaborations would quite rapidly cease to function. In fact, astronomy and academia more broadly have a very long history of remote collaborations, with first private letters, then scientific journals providing means for connecting researchers across the globe. When viewed through this lens, making new acquaintances and collaborations via online channels is merely the next logical step in this long history of remote collaboration.

Recommendations

- Deliberately build time for collaborative networking into your events where appropriate.
- Bear in mind that online networking may require a different skill set to the in-person version.
- Be proactive in your approach to online networking much as you would in-person.
- Use tools that facilitate multiple ongoing conversations at the same time, with user controlled join and leave options for each conversation.

19.7.10 "The Technological Requirements Needed to Go Online Excludes Regions of the World Without Good Internet"

Similar to section 19.7.8, online meetings have a high reliance on technology for success and are lambasted more for technical issues. While limited in many parts of the world, it is *far* easier to obtain access to the internet than to find the resources for travel. Wifi is much cheaper than a jumbo jet. As well as having contingency plans for technical hiccups, for global audiences where technology may be an issue, alternative low-tech options can be provided. For example, for VR events there are platforms that offer 2D access for those who cannot afford or are uncomfortable with headsets. Similarly, one can choose content and delivery to minimize bandwidth demands. For platforms that have high computer demand, alternative lightweight access through livestreaming the session or screen sharing Zoom may also help. The participant may not have the same interactivity with the platform but at least they can feel some involvement. The issue is also slowly being mitigated, the global pandemic and the reliance on technology for everyday life has driven a shake-up of the technological infrastructure of many countries, and indeed nearly all countries now have technical access in urban areas to internet speeds in the low tens of megabits per second (World Population Review 2023). The issue is therefore more related to financial and rural access to those speeds, but this is a problem that can be solved far easier and cheaper than visas, using accessibility grants, and will often yield longer lasting impacts in situ than providing travel and accommodation costs. Many initiatives along these fronts already exist, and indeed even large scale upgrades are possible in this manner, with projects like the African SKA expected to result in significant upgrades to the technological infrastructure of much of the continent (Bhogal 2018).

Recommendations

- Provide low-tech options for those hampered with technology issues.
- Offer accessibility grants where possible to support people who have less access to technology (e.g. internet, hardware, software).
- Actively engage with communities to determine what their accessibility needs are when it comes to technical requirements of participation.

19.7.11 "There Are Some Things You Really Just Can't Do Online and Have to Do in Person, Like Whiteboarding and Eating Together"

One of the challenges we regularly see in online interaction is that it is constantly being compared to, or trying to reproduce, the equivalent in-person interaction. In actuality, there are some activities that are better suited (currently, with contemporary technology) to in-person. Eating meals and drinking together is a good example of something that is much easier to do in person than, for example, in a VR headset. This may not stay true in the coming years if affordable and effective augmented reality options become widespread, since AR allows you to project virtual elements into a physical environment (and one could imagine, for example, eating a meal while sitting at a physical table, sitting next to a holographic projection of a far-away friend or colleague). In the case of whiteboarding, we have actually found that online methods (either in browsers like Miro/Mural or in VR) can be much more inclusive than gathering around a physical whiteboard, depending on how they are used, although in some cases it is true that the in-person version is more effective. A major part of moving to more sustainable meeting formats is learning to decipher which activities are best suited to in-person gatherings, and which can be accomplished just as (if not more) effectively online. Once you can do this, you can make an informed decision about the best format to accomplish the goals of your next meeting. The conclusion here is that we should be using in-person formats for when they are most effective, and looking to online formats for where they can be equivalent or even superior—but we should also maintain an open mind, because technology is evolving rapidly and there are increasingly fewer examples of activities where "only in-person can be effective."

Recommendations

- When choosing in-person formats, ensure these are activities that truly are most effective when carried out in person.
- Explore options for similar online formats to meet the same goals as the desired in-person activity.
- Regularly reassess the assumption that something best done in person has remained so in the face of advancing technology.

19.7.12 "You Can't Solve Time Zones or the Fact that We Have a Round Planet, Therefore You Can't Really Do Things Globally While Online"

This is a fundamental problem. For in-person meetings we experience it as jet lag. Who of us hasn't nodded off during a session because their biological clock was still in a different time zone? But there are ways to address it. One of the advantages of the virtual format is that interactions do not need to happen in a live or synchronous manner. Many organizations in both academia and industry have maintained a strong global presence for decades thanks to judicious use of both synchronous and asynchronous meeting formats. In fact, there is no compelling reason why much of the 'passive' content seen frequently in live meetings like conferences (talks, presentations, etc) could not be delivered asynchronously, with a reduced but more efficient live meeting scheduled for the interactive discussion that typically follows these events. At present we rely heavily on live meetings for many kinds of information transfer, but moving away from this and to a more flexible form of work will only strengthen our ability to form and maintain global networks.

On a more practical note, in reality it is often the case that meeting participants will hail from some subset of all global time zones. If a short or one-off event does require synchronous interactions, finding a suitable meeting time is usually fairly straightforward, and can be done simply if attendees' time zones are collected as part of the event registration (or during discussions prior to setting a time, for less formal interactions). For slightly more flexibility, it is worth considering collecting attendees' *preferred working hours*, rather than just their timezone, as some individuals will be naturally inclined to working late at night, and others will have a preference to start early. Obtaining this information makes it simple to accommodate attendees from a range of timezones with the help of a basic optimisation script (or a student happy to code one up for you). When an event has a truly global audience, it may also be useful to consider two repeat sessions separated by 8-12 hours, in order to allow all attendees the chance to participate.

Recommendations

- Schedule live meetings only for things that cannot be achieved asynchronously.
- Consider the actual time zones (or preferred working hours) of attendees when scheduling live meetings, which may include asking them directly via a survey.
- Very occasionally (and with enough notice), schedule a meeting which takes place as a single time to bring an entire community together (to avoid segregating regions), and rotate the timing in future to avoid locking in disadvantage.
- As an attendee, consider requesting funds for a local hotel or additional support for childcare, etc, even for a remote conference where you will not need to travel, especially if it is unsociable hours.

19.7.13 "There Are Too Many Tools Used Online and It's Too Overwhelming and Inefficient"

There indeed are many tools used online and it can be overwhelming. This is in part because virtual connectivity tools are still in the innovation stage, with new ideas and techniques continuously being developed. But this problem also occurs with in-person meetings, as there are also many different, evolving ways of facilitating them as well. The existence of these many tools is driven by different goals and purposes, and as such organizers too should be selective in their choices. From the point of view of an online attendee, it is true that being confronted by many tools, especially unfamiliar ones, can be confusing, but a counterpoint to this is that a reluctance to learn or try new things goes against the spirit of scientific experimentation and limits agility and innovation. That said, even excessive use of familiar tools can require multiple windows and tabs to be opened, which can be inefficient and so it is good to minimize the use of tools. When introducing new tools to an event, organizers should make sure they are clearly explained, allow time for troubleshooting and have plans for when certain institutes or countries refuse to allow a tool to be used. This does not mean however that we should always stay with the lowest common denominator of toolkits, as IT policies are partly shaped by demand, and many new tools have not been encountered enough for a policy to be put in place. Your department may be happy to do so given enough time and demand. Finally, depending on the complexity of the tool, organizers should also offer "onboarding" or familiarization sessions.

Because of the challenges of mixing and matching tools for conducting an event, it may be tempting to use a single monolithic platform approach instead. But there are downsides to this approach as well, namely that no one platform is good at every aspect. Some may provide excellent video streaming, others a better Q&A solution. So instead of overwhelming attendees with millions of tools, or stifling your event within the "walled-garden" of a single application, ensure instead that each additional tool used is backed by a clear, justifiable need and even more importantly, ensure that in all cases there is one (and preferably only one) tool for doing each thing, and that it is made obvious to the speakers, hosts and attendees.

Recommendations

- Carefully assess the need for a variety of tools and minimize the numbers necessary to meet the goals of a meeting or event.
- Organizers introducing new tools to their event should make sure they are clearly explained and that people are given time to gain familiarity.
- Balance the desire for an "all in one" platform against the need to use the right tool for a given approach, and experiment to find the best solution.

19.7.14 "Metrics for Career Progression Depend on Meeting In-person E.g. Number of Conference Presentations"

This is unfortunately probably quite true both directly, as in the case of conference presentations and indirectly through networking with the "right sort of people" to advance your career. Furthermore, in some disciplines conference proceedings are valued highly. Though reliance on in-person attendance for such conferences seems to be diminishing, there is still an inherent bias through favoring in-person presenters for accepting talks or high virtual registration fees ostensibly to subsidize in-person costs. Also, the number of publications arising from a conference has been cited as a measure of success of that conference and related to the various concerns about the importance of networking. Across academia there have been many discussions about diversifying the metrics for success to be more inclusive of the full contributions made by a given academic, and it is clear that this aspect falls under that umbrella of needing widespread improvement.

Recommendations

- There needs to be a continued overhaul of how academic performance is measured to something more equitable. ECRs should not be encouraged or required to travel to be successful.
- Conference proceedings, presentations and contributions delivered virtually should carry equal weight to those taking place in person.
- Build more global impactful networking initiatives that do not rely on in-person attendance.

19.7.15 "Online Would Be Better and More Effective If It Felt More Like Being in Person"

Early in the TFOM process, we were introduced to a famous human-computer interaction paper from 1992 titled "Beyond Being There" (Hollan & Stornetta 1992), in which many of the challenges we felt were unique to the pandemic era of 2020 were identified back in the early 1990s. In this paper, it was pointed out that one of the biggest impediments that we arbitrarily impose on online formats is the requirement to reproduce in-person interaction, as if that is the end goal that should determine success. However, one of the most prevalent and impactful elements of online interaction is semi-asynchronous text communication (e.g. being able to send instantaneous messages which someone may, or may not, reply to right away). This is an example of something completely impossible to do "in person," and something that really changes the way we interact with others both locally and globally. Based on this, we have adopted this sentiment in much of our work with online formats, constantly revisiting the assumption of online as a reproduction of in-person and asking ourselves: "How do we go *beyond being there* with online formats?" Online is not about just reproducing what we would do in person, but about opening up possibilities to do *more* than we could ever achieve with in-person only interaction.

Recommendations

- Consider what is possible online that might not be possible in person and make use of that as an advantage of the format.
- Ensure that the format choice for a meeting is driven by the goals of the meeting rather than by the established norms around how things typically take place.
- Adopt a blend of different formats to meet the different needs of a community (while constraining choices in the context of accessibility, inclusivity and sustainability).

19.7.16 "The Online COVID Version of the Conference Didn't Feature That Many New Diverse People, So It's Not True That Online Actually Improves the Diversity"

Actually, it is widely reported in the literature that diversity was measurably increased for many conferences and meetings that shifted online during the height of the pandemic (Skiles et al. 2022). The removal of traditional barriers of access meant that conferences which typically excluded many were suddenly available, particularly for marginalized groups. In many cases where the statistics of a meeting were analyzed, diversity factors such as gender, seniority, country of origin and country of residences changed significantly for the positive during the pandemic. In cases where the statistics did not, it is important to recognise that a traditionally-excluded community does not necessarily become instantly included just because of the door being open. In this sense, the difference between something being accessible and being inclusive is critical. Diversity starts with accessibility (e.g. ensuring that your meeting choices are such that everyone *can* take part), but it is only effectively achieved with inclusivity (e.g. the meeting is structured in such a way that diverse people feel empowered *to* take part).

Recommendations

- Define systematic and structured ways to measure and analyze meeting attendance (including consideration of ethics), such that diversity of participation can be tracked over time.

- Always include online attendance as a default to anything that takes place in person, and aim for online attendance which enables people to participate as much as possible.
- Actively support and engage under-represented people to ensure that they feel welcome and included to take part in the relevant meetings and conferences.

19.7.17 "Opportunities for Research and Collaboration Often Appear during Informal Networking Over Drinks or in Other Social Settings, So These Opportunities Won't Be Available in Online Meeting Formats"

It is true that, given thousands of years of interaction taking place in person, many people are generally pretty good at in-person socializing compared with the online equivalent. We are also more inclined to build relationships and trust people quicker with certain cues that are easier achieved in person (Kramer 2009, Kalia 2021). So given this, informal networking does play an important part in establishing new collaborations and forming relationships, and traditionally this has been well-achieved via in-person social settings. However, there are a few caveats to consider here. Firstly, most of the evidence for this is highly anecdotal, and has not been measured well either qualitatively or quantitatively for either in-person or online formats, and this is something that should be amended in the coming years. Secondly, this kind of statement does not acknowledge the dark side of relying on these kind of interactions to gain opportunities—are we saying that, as a community, we are happy with the idea that those who cannot take part in person (for all the reasons we know about when it comes to accessibility and inclusivity, including cultural differences around acceptable food and drinks) do not deserve the same kind of opportunities? And lastly, the world and opportunities within it are immense, and there are so many important and interesting challenges to solve together. It seems a great missed opportunity in itself if we choose to limit ourselves only to the opportunities that arise when we can physically gather in one location.

Recommendations

- Deliberately and considerately plan activities and events which aim to provide these kind of informal opportunities to people online as well as in-person
- When organizing in-person social events, consider groups and minorities who might be traditionally unable to take part and support their inclusion where possible
- Explore possibilities both online and in-person to provide ways in which these kind of opportunities may be realized in different or more direct ways (e.g. team-building, careers fair, social activities, games, etc)

19.8 Case Studies in Online and Hybrid Design

We have found in our activities that often the lack of widespread examples in how to approach online or hybrid interaction creates barriers for people designing meetings. In this section, we share case studies of some highlight events we have either orchestrated or been involved in, with the goal of sharing the context behind it, the goals, the unique elements of a particular case study and also the aspects which worked well or were challenges.

The case studies below were selected to cover a range of different scenarios, and we aim that they provide a useful insight into some of the broad possibilities in online and hybrid interaction, but we note that they are a small subset of the wide variety of creative and innovative approaches that have been documented. We encourage the reader to look beyond these case studies and seek further inspiration in the literature and online when it comes to the current state of best practice for online and hybrid meetings.

19.8.1 The Future of Meetings Symposium (TFOM, 2020)

"The Future of Meetings" symposium was held over three days (15th–17th September 2020) as a CSIRO Cutting Edge Science Symposium (CESS), originating from pre-pandemic discussions around sustainability and accessibility amongst astronomers and motivated by a desire to collaborate more effectively virtually (Figure 19.3). During the preparation of the proposal, COVID-19 happened and thus made the topic suddenly relevant to a much broader range of people, strengthening themes of the future of work and distributed collaboration. The lack of a precedent combined with being the first online-only CESS meant planning and design started from scratch, which fostered a culture of experimentation and innovation.

After considerable research into available technology, the Whova platform was chosen as a conference platform on the basis of features, cost and flexibility. Whova provided many desirable elements such as attendee management, timetabling including embedding recordings, streaming and Zoom meetings into the schedule, and extensive community features for keeping attendees connected. Most

Figure 19.3. The "expression of interest" page for the TFOM symposium in 2020. The current community of practice was formed by a subset of the organizing committee who stayed active after the symposium to continue to advocate for and experiment with new meeting technologies.

importantly, Whova as a platform gave quite a bit of control and autonomy to organizers, which was extremely important for oversight and management of the event. While the platform backend is slightly complicated, it includes many elements of automation to streamline workflows and provides an effective comprehensive solution that was well-suited to TFOM.

The three days were themed “Today, Tomorrow, Future” and made up of a combination of invited talks, panel discussions, workshops and social sessions. The majority of talks were presented as live-streamed recordings with the speakers available online for live Q&A via the Whova platform. Other live sessions such as panel discussions, extended live Q&A and social sessions were held through Zoom and Gather. Legacy recorded content can be found at the TFOM archive and lessons learned from planning and delivering the symposium have been made available in a publicly available report (Moss et al. 2020).

What worked well: This was aimed to be an innovative symposium in design and execution, and gained a wide audience (over 1100 attendees). It showed the possibility of making virtual conferences effective. The diversity of the audience introduced perspectives from beyond the normal circle of attendees giving rise to interesting and engaging discussions.

Limitations: This symposium was held in the early days of the pandemic and so there was a general lack of familiarity with online tools and format. Similarly, unfamiliarity meant there was not as much participation in the social sessions as hoped from people literally “switching off” after the formal content of the day.

19.8.2 International Conference on Women in Physics (ICWIP, 2021)

This was a fully virtual international conference, held 11–16 July 2021, but unlike the TFOM Symposium was part of a series that had previously been held only in-person and, furthermore, was designed to be accessible to participants from developing nations. The timetable for the conference was designed with a global audience in mind so that no specific time zone would be disadvantaged and the impact of time zones mitigated by making all content available for asynchronous viewing. The Whova platform was again chosen for this event and the community pages proved particularly popular, keeping attendees connected through sharing photos and discussion. The social aspect, which was traditionally a key part of the conference, was facilitated through a variety of social sessions from indigenous dance lessons in Zoom to trivia nights and Penguin Parade watch parties in Gather. These activities had been planned for the original in-person conference before it was postponed by the COVID-19 pandemic, and the participants appreciated that they could still have some form of local Australian experience. The Gather space was available throughout the conference and many of the participants made use of it for their personal social and work-related meet-ups.

What worked well: The social aspect where the community enthusiastically engaged in the organized activities and furthermore arranged their own. The well-attended poster sessions were by far one of the best demonstrations of the power of spatial-interaction platforms like Gather (Figure 19.4), with clusters of people

Figure 19.4. Attendees of the ICWIP 2021 conference in Gather, assembled for the conference photo.

scattered throughout the virtual poster room having conversations at the designated poster session times.

Limitations: ICWIP is traditionally aimed at developing nations so there were some issues with technology and familiarity with online platforms. As a conference series, there were many stakeholders involved in the structure and format which resulted in some constraints when it came to the approach taken.

19.8.3 Science Pathways 2021 (SP21, 2021)

"Science Pathways" is a national-level conference run by the Early and Mid Career Researcher (EMCR) Forum, which is a representative body that exists as part of the Australian Academy of Science. It is one of the only meetings that seeks to unite junior STEM researchers across Australia, and traditionally takes place every two years. The 2020 meeting was postponed to 2021 due to COVID, and the planning for the 2021 meeting took place while the pandemic was still very much active in Australia. The executive determined that SP2021 should take place in a mix of formats to meet the different needs of EMCRs, including hybrid, in-person only and online-only. As such, a hybrid launch half-day was designed (taking place in Melbourne in late November 2021), followed by a day of in-person workshops, and then in early December there were three online-only half days featuring panels, workshops and social activities. The hybrid launch day was designed to minimize the gap between in-person and online audiences where possible, such as having all questions asked via the conference platform only and ensuring all attendees were able to access and use the online platform. In addition, a member of the organizers was designated as the in-person representative of online attendees, and this member attended online while sitting in the room adjacent to the main conference venue (Figure 19.5). This meant that if any issues arose for the online audience, there was someone local aware and ready to take action immediately. For those in person, a visit to a local museum "behind the scenes" was arranged, and the workshops the next day were chosen with the goal of leveraging the in-person nature of attendees. The online-only days were scheduled to be a little spread out in time, meaning that there was the building of an online community between the launch day and the

Figure 19.5. To help bridge the gap between in-person and online attendees during the hybrid launch day, one member of the organising team attended "online" from the venue, acting as the in-person representative of the online audience.

online days, with discussions and interactions continuing beyond the "live" time of the conference via the platform.

What worked well: The mixed mode of formats meant that there were elements of the meeting that appealed to everyone. Having the conference spread out in time kept the community together, facilitated by the technical platform, which stayed active before and after the meeting and allowed people to interact and connect. The hybrid design, which aimed to minimize the degree to which in-person attendees were favored, meant that both audiences were active participants during the launch day.

Limitations: The in-person elements were made possible by generous sponsorship from the host institution, covering venue, catering and AV, which otherwise would have been extremely expensive compared to the other costs associated with running the conference. As with other online meetings, getting attendee buy-in for the online social activities was difficult and there was low turn-out in general for these, even though many people expressed desire for informal unstructured interaction online.

19.8.4 Infrared 2022 (IR2022, 2022)

"An Infrared Bright Future for Ground-based IR Observatories in the Era of JWST" (IR2022) conference was hosted by the JAXA Institute of Space and Astronautical Science and held online between February 14th–18th 2022. This was the first astronomy conference that the TFOM community partnered with,

Figure 19.6. TFOM worked with IR2022 organisers to build a VR-compatible add-on venue in Altspace, which was used to host posters, talks, livestreams of the sessions and a JAXA exhibition display.

using the gained experience in virtual reality to provide a social venue and exhibition space alongside the planned conference program which took part primarily in Zoom and Slack.

The conference organizers were keen to provide an area where attendees could interact, both as part of the conference poster session, and also for general discussion and networking. A 3D virtual space with multiple rooms was set-up using the Microsoft AltspaceVR platform (Figure 19.6), which provides a variety of ready-made conference environments that can be freely used. The venue could be accessed via a computer or VR headset, with visitors creating an avatar that allowed them to move around the area. As this was a 3D space, directional audio and audio fall-off allowed simultaneous overlapping conversations in the same room, as with an in-person venue.

Two rooms in the virtual venue were dedicated to posters, which were displayed as a large size and included embedded videos of the poster presenter summarizing their results. Another room included a stage where two live (in platform) talks were held, and the live research talks from the main zoom stream were also displayed for attendees who wanted to watch the presentations together. A fourth room allowed visitors to watch the pre-recorded talks together on-demand to encourage group watch parties, re-watching of key talk sections, and ongoing discussions. The final room was designed as an exhibit space for space missions from JAXA. This included videos, images and 3D models of spacecraft which could be handled by visitors. The exhibition space was particularly popular for networking, as attendees could explore the space and fall into conversation with others also examining the same objects or media.

While Zoom was still the main platform for the conference's presentations, the virtual space offered a natural venue to engage with other meeting participants, and the exhibition space was an attraction that could not have been easily implemented for an in-person meeting.

What worked well: The exhibition space was particularly successful at encouraging people to attend the virtual socials, and the models and exhibits were good ice breakers for seeding conversation.

Limitations: A few participants were not able to access the virtual space from their computer, due to struggles with installing the software. This could have been solved by stronger encouragement to install the software in advance, or adopting a platform that was browser-based.

19.8.5 Science for Public Good x TFOM (S4PGxTFOM, 2022)

Throughout 2022, TFOM hosted a series of events centred around the theme of "Science for the Public Good" funded by a grant from the Theo Murphy Initiative. As part of this series, four events were held solely in a virtual space built in Altspace, a VR platform compatible with both headsets and desktops (which has since been unfortunately retired). These events featured speakers drawn from a wide range of scientific disciplines, and touched on some of the many ways in which different areas of science, from astronomy to ecology, can build towards a greater public good. The events were held live and scheduled so that the series spanned all time zones, with at least one event falling during business hours for attendees from Oceania, Europe and the Americas, and it was intended from the outset that social connections and networking would be integrated into each event. This desire for social connection was in part what led to Altspace as the venue of choice, because unlike a simpler video conferencing platform, the 3D virtual venue could be easily transformed from presentation space to networking space. What is more, the spatial audio element allowed for organic conversation groups to form after the talks were concluded, with additional immersion possible for those accessing the events via VR headset thanks to cross-platform support.

Additionally, a fifth, in-person event was held in conjunction with the 8th Australia-Spain Research Forum "A Digital World" to introduce in-person attendees at the workshop to the possibilities of meeting and collaborating in VR spaces (Figure 19.7). Five VR headsets were set up in a room at the workshop venue and all connected to the same Altspace event venue, alongside a PC connected to the same venue and projected on a wall. This allowed dozens of conference attendees to explore the VR event space during breaks in the in-person workshop.

What worked well: The platform was easily accessible both in VR and via desktop, and the venue seamlessly transformed between presentation and socialization space. Coordinating an in-person event led to much greater VR exposure, and feedback on the VR immersive experience and ability to more naturally interact/socialize was very positive.

Limitations: The platform was not accessible on low-spec systems or sometimes blocked by institutional firewalls, creating a barrier for access. Although the series spanned time zones, events were synchronous so not all events could be attended by

Figure 19.7. During 2022, a series of events were run in VR as part of the "Science for Public Good x TFOM" collaboration, where attendees could join via virtual reality, desktop or livestream.

all attendees. Visitors from the in-person workshop only had a superficial introduction to VR events due to time constraints.

19.8.6 Astronomers for Planet Earth (A4E, 2022)

The Astronomers for Planet Earth held their inaugural symposium fully virtually on 28 Nov–2 Dec 2022 (Wagner et al. 2023). The program was designed to be inclusive of all time zones by allowing asynchronous access to recorded talks and offering two daily live discussion sessions in Gather (Figure 19.8). The opening and closing sessions, and a panel discussion with invited representatives from large astronomy and space consortia, were broadcast live via Zoom and YouTube, but recorded for those who couldn't attend. Furthermore, members of the A4E community were encouraged to organize independent workshops of which four were held. All recordings from the symposium are available on the A4E YouTube channel. The event was judged a success with live audiences of as many as 150 people and over 4,000 views of the recorded material. The level of engagement over the duration of the event showed a clear interest of people in sustainability and the online symposium allowed them to share their views with a wider like-minded community. This particular point was commented on by many who had been feeling increasingly isolated for holding such concerns.

What worked well: The discussion sessions were well-attended and generated lively discussion from a global audience. The "pre-watch" format with focus on discussion allowed for the consumption of the content at an attendee's own pace as well as some introspection before the open discussion.

Limitations: People still do not dedicate the same time and energy to online conferences as they would in-person events (especially social events and activities), and the global time zone scheduling inevitably inconvenience some.

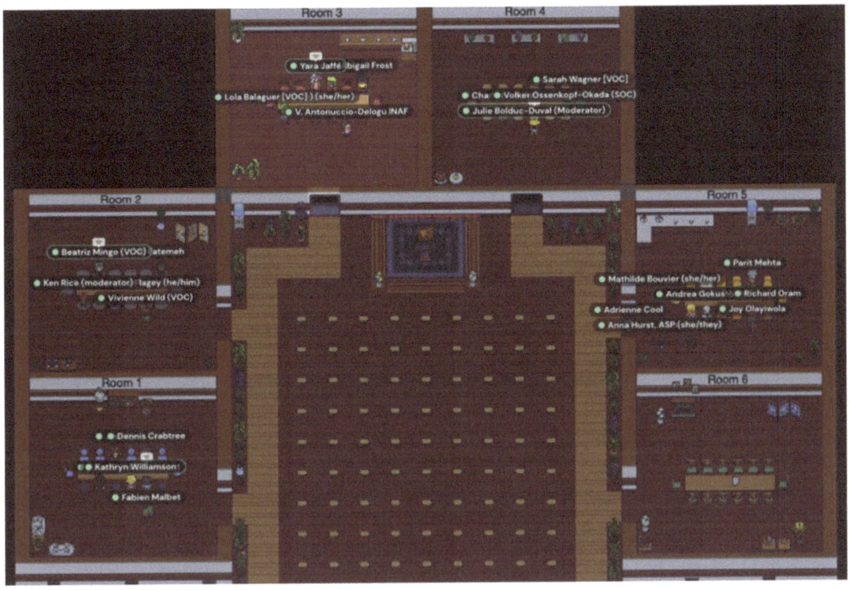

Figure 19.8. The A4E symposium aimed to maximise interaction time during the live sessions, with parallel discussion sessions taking place on a variety of topics.

19.9 Looking to the Future

As we have discussed above, the modern era has seen vast improvements to communications technology that have fundamentally changed how we work, live and socialize, making it easier than ever to interact across long distances without the monetary and climate costs of moving people. Yet despite these improvements we still travel for work far more than we should, with travel for conferences and meetings being two of the biggest drivers of academic carbon emissions to date.

Reducing these emissions should be a major priority. Many people still feel that in-person meetings bring a vital human connection, a sense of presence to collaborations that is just not reproducible remotely. This tension is an issue. While a return to purely in-person events would allow for that important in-person networking, it will also exclude many who are unable to travel, while driving up our current (post-pandemic) carbon emissions by several orders of magnitude. The solution then, is not to give up on remote collaboration. Instead we must find a way to bring the best parts of in-person meetings to our remote ones.

Virtual collaboration tools are very much in the "innovation stage," and like many of the technologies we take for granted now, e.g., CCD detectors and digital projectors, these tools are not yet fully ready—often frustratingly so. Luckily, communications technology has not stopped with the development of the video call, and in the decades since the video call's popularization (let alone creation) there are now many more advanced solutions available for remote work and social. The most promising of these, Extended Reality (XR), brings huge improvements in interaction quality, particularly in making users feel connected and making them feel like they

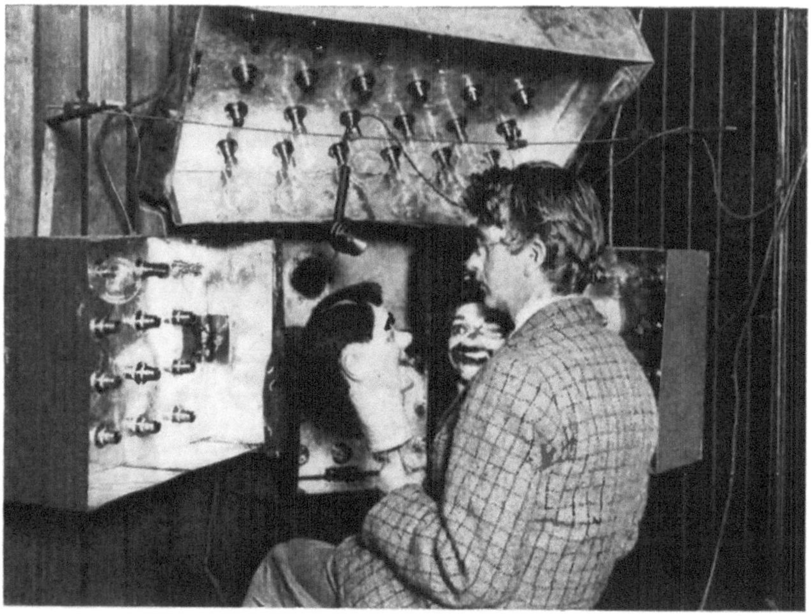

Figure 19.9. [1]British inventor John Logie Baird and his first publicly demonstrated television system, with which he transmitted moving pictures in 1925. In a few years is this what the early days of virtual conferencing will seem like?

are really 'there,' while also opening up opportunities to go beyond what is possible in person.

Virtual collaboration *formats* are also in this stage of innovation. There's no reason why we need to do conferences online as we do them in person—we're just doing them that way because that's what we know. A good analogy is the early days of television. When it was invented they didn't know how to do TV shows, but they did know how to do radio shows. So they did radio shows on TV. Over time people eventually recognized the power of the new technology, and new show formats developed. That's where we are now with virtual conferences (Figure 19.9). So keep that in mind if you've had a bad experience online.

It's exciting to think about what the future can bring. XR, encompassing both Virtual Reality (VR) and Augmented Reality (AR), is leaps and bounds ahead of what most people think, and the impact it will have on our world as it develops will be substantial. This is because at its core XR is not a game or a social platform, it is the next ubiquitous computing platform, subsuming laptops and smartphones to provide not only the functionality of both, but combining them with true spatial awareness and presence (e.g., Figure 19.10).

It is hard to overstate just how much our society might change when we can cast not just gigabytes of data around the world on a whim, but cast real-time digital

[1] https://en.wikipedia.org/wiki/John_Logie_Baird#/media/File:John_Logie_Baird_and_Stooky_Bill.png.

Figure 19.10. Interacting with digital versions of physical hardware (such as telescope components) is already easy to do in existing platforms like Spatial.

representations of both *ourselves and our surroundings* anywhere, to meet anyone, using nothing but a small headset and an internet connection.

Imagine the ramifications of being able to keep access to all of our current screen based technology while also enabling the feeling of true human connection that is so missed when staring through a screen. How will our world change, when we are able to walk around campus with a photoreal, 3D, real-time view of a colleague half the world away walking alongside us? What will happen when you can have a virtual meeting with eye contact, facial expressions, body language and spatial audio in a shared environment that seems real?

The applications of this in astronomy and beyond are vast. You could join a new colleague for dinner remotely (Figure 19.11), meet a potential collaborator for a round of virtual mini-golf (while discussing the latest paper of course), explore a 3D scan of an asteroid by standing on it together, whisper a question to a mentor while listening to a talk in a virtual lecture theater, or mingle with new and interesting people while exploring a virtual expo hall. Once this is possible, will we still regularly fly halfway around the world to sit in an office or lecture theater? Would our funding departments still approve travel?

The thing is, most of this already *is* possible with current XR headsets, and while it is yet to be rolled into 'one headset to rule them all', it is getting close. Current headsets are still forced to make the trade-off between graphical quality and form factor, convenience and cheapness, but this will not stay the case for long. Already one can have a decent fully-functioning standalone headset for a few hundred USD, or a spectacular headset with a supporting PC for a few thousand. Daunting though these numbers may seem, both are one-off costs well below those associated with attending a single international conference.

Figure 19.11. Virtual reality provides a new mechanism for organic distributed social interaction that has the potential to go well beyond what is possible in person, such as this example from "Walkabout Mini Golf." The immensely popular app offers convincing mini-golf physics and multi-user interaction in exotic and beautifully constructed worlds.

Indeed, with current headsets it is already possible to have a meeting, even a conference completely in XR (e.g., Figure 19.12). To have a true virtual stage with speakers, laser pointers, powerpoints and microphones, to have attendees stand up in the crowd to ask questions leading to a natural back and forth, and to have posters and pre-recorded on-demand video playback. Most importantly perhaps, the sheer sense of presence and spatial audio provided by being in XR enables that vital component of organic networking and social that is so often missing in current online conferences; that feeling of being together.

But why should we settle for mundanely reproducing an in-person conference? An XR conference can give every attendee a front row seat, or let the speaker speak in one language and be heard in the attendees' native tongue. It can provide conference badges that contain recent publications, interests and affiliations and provide posters that can be made interactive or even contain 3D visualizations, while still allowing the owner to stand in front of it to answer questions.

From an organizational perspective a XR lecture theater can be expanded or contracted on the fly to handle more guests or engender a more relaxed atmosphere

Figure 19.12. It is already possible using widely available platforms to have people gather in 3D immersive spaces, such as this space in Glue, where they can discuss, share data and collaborate on whiteboards together across VR headset and desktop devices.

(Figure 19.13), and in terms of accessibility the options for XR meetings (and more traditional online meetings) are far ahead of what is possible in-person. Whether it's something minor, such as allowing a speaker to reduce the apparent number of the attendees to deal with stage fright, to allowing vision impaired users to have built-in audio readers for the entire shared reality, XR meetings give us the opportunity to go beyond being there in supporting both attendees and speakers. Beyond all of this, however, simply by not requiring people to be away from their homes and duties for extended periods of time, XR meetings (like all online meetings) are just so much more accessible and inclusive than their in person counterparts, while still providing that all important "human" connection that we want in person for, at just a fraction of the financial and environmental costs.

So if this is all already possible, why has it not been done? The truth is it has, and *is* being done. There are many examples of effective conferences and meetings that are carried out in online VR or AR spaces by a wide range of academic, industry and government and social organizations. But the biggest issue with XR spaces is that most people's experience with them is via a backwards compatibility option (like a laptop or phone), and much like a video call where 90% of the attendees are dialed in by landline, the benefits of the better technology are not apparent unless the majority of users are connected using the new approach. The technology is also new to many, causing its own learning curve issues and triggering risk aversion strategies when doing larger events. In short, while XR conferences can solve many of the issues we have with online meetings, they unfortunately (much like video call conferences and meetings before the pandemic) currently suffer from a critical mass problem.

So at last we return to that all-important question: How can we help change our institutes, our governments and ourselves to help address climate change?

Figure 19.13. Members of TFOM gathered in virtual reality via headset and desktop to watch content together, despite being distributed across vast distances.

As astronomers we have a rather unique place as members of our truly worldwide research field. United by shared skies and international facilities, astronomy transcends borders and cultures in a way that just isn't that common in many academic or industry fields, and this means we are uniquely placed to push for better ways of doing things.

So we urge you, our readers, to push. Push for a better future. Advocate for your institutes to swap to renewable power supplies. Campaign for the time and resources to make effective use of more sustainable travel options, and speak out against inadequate organizational and political stances on climate, at all scales.

But add to your list making remote work, remote conferences and remote social events an integral part of your workplace. Not *just* because it is cheaper, but because every event you attend digitally is a plane not taken, a commute not driven, and a fellow scientist not excluded.

At the same time, acknowledge the current limitations of remote teams, and advocate for online meetings that make us feel more connected to each other. Use XR for what it's meant for; to spend time with all those you wish you could see more often, to share new experiences together, and to feel physically closer to your friends, family and colleagues, regardless of the miles and oceans between you.

Because by doing so, we have the potential to open the door to a better world. One brimming with new colleagues you might never otherwise have had the chance to meet, and old friends you might never normally have seen again. A world where remote conferences are as fun and productive as in-person, and where opportunities for scientists to share their work are available regardless of where they are born. A world that even with all of these improvements, is also hundreds of times more sustainable than our old.

So regardless of why you first picked up this book, whether you were looking for guidance on the state of our planet, seeking new colleagues who share your concerns, or simply wanting something that you can do personally to improve our future world, and our time in it. We urge you; strive for better remote collaboration. Not only because it *will* help us do better science, but because we hope that by embracing these fantastic new digital worlds, we might just get to keep our real one.

References

Aycock, L. M., et al. 2019, Phys. Rev. Phys. Educ. Res., 15, 010121

Bhogal, N. 2018, How the SKA telescope is boosting South Africa's knowledge economy. The Conversation. http://theconversation.com/how-the-ska-telescope-is-boosting-south-africas-knowledge-economy-96228

Burtscher, L., et al. 2020, NatAs, 4, 823

Carbon Independent 2007, *Aviation Emissions*, viewed 7 August 2023 at https://www.carbon-independent.org/22.html

Curry, D. 2023a, Discord Revenue and Usage Statistics (2023). https://www.businessofapps.com/data/discord-statistics/

Curry, D. 2023b, Microsoft Teams Revenue and Usage Statistics (2023). https://www.businessofapps.com/data/microsoft-teams-statistics/

Duncan, N. W., & Shean, R. 2023, Analysing the effectiveness of Twitter as an equitable community communication tool for international conferences PeerJ 11 e15270

Gokus, A., et al. 2023, submitted to PNAS Nexus

Hollan, J., & Stornetta, S. 1992, CHI '92: Proceedings of the SIGCHI Conference on Human Factors in Computing Systems 119

IEA 2023, The world's top 1% of emitters produce over 1000 times more CO2 than the bottom 1% (Paris: IEA) https://www.iea.org/commentaries/the-world-s-top-1-of-emitters-produce-over-1000-times-more-co2-than-the-bottom-1

Kalia, S. 2021, *The Swaddle* Available at: https://theswaddle.com/why-we-trust-the-people-we-trust/

Klöwer, M., Hopkins, D., Allen, M., & Higham, J. 2020, Natur, 583, 356

Köhler, J. K., et al. 2022, FrPs, 13, 906108

Kramer, R. M. 2009, *Harvard Business Review* https://hbr.org/2009/06/rethinking-trust

Moss, V. A., et al. 2020, The Future of Meetings: Outcomes and Recommendations, Zenodo (v1.0) https://zenodo.org/record/4345562

NASEM 2021, Engineering, and Medicine (Washington, DC: The National Academies Press)

Remmel, A. 2021, Natur, 591, 185

Skiles, M., et al. 2022, Nat. Sustain., 5, 149

Tang, J. 2021, Proceedings of the ACM on Human-Computer Interaction, 5,(CSCW1) 1

Tao, Y., Steckel, D., Klemeš, J. J., & You, F. 2021, NatCo, 12, 7324

Vargas, S., et al. 2022, Astronomy for Accessibility and Inclusion Revista Mexicana de Astronomía y Astrofísica. Serie de Conferencias Vol. 54, (Universidad Nacional Autónoma de México)

Wagner, S. M., et al. 2023, NatAs, 7, 244

World Population Review 2023, Internet speeds by country. https://worldpopulationreview.com/country-rankings/internet-speeds-by-country (accessed 15 August 2023)

Part VIII

Resources

Appendix A

Resources

T A Rector

> "If we could change ourselves, the tendencies in the world would also change. As a man changes his own nature, so does the attitude of the world change towards him."[1]
>
> —Mahatma Gandhi

Climate change is a huge subject. While we would like to think that this book has provided everything you will need, here are additional resources that may be useful. For brevity only two resources are provided per subtopic.

Learning the Science

Princeton Primers in Climate: A series of short books that explain climate science at a level appropriate for other scientists.

EdX.org: Offers courses and degree programs on climate change and related topics. Most courses are self paced and can be taken for free.

Current Status of the Climate

Intergovernmental Panel on Climate Change: The IPCC is the preeminent international organization for assessing the science related to climate change. Released in March 2023, the Sixth Assessment Report (AR6) is a summary of the most recent

[1] 1964, The Collected Works of Mahatma Gandhi, Volume XII, April 1913 to December 1914, chapter: General Knowledge About Health XXXII: Accidents Snake-Bite, (From Gujarati, Indian Opinion, 9-8-1913), Start Page 156, Quote Page 158, The Publications Division, Ministry of Information and Broadcasting, Government of India.

doi:10.1088/2514-3433/acfcb6ch20

understanding about climate change and its impacts as well as options to mitigate and adapt to its consequences.

GlobalChange.gov: The U.S. Global Change Research Program (USGCRP) comprises 14 federal agencies that conduct or use research on global change and its impacts on society. Like the IPCC, it releases regular reports on the status of climate change. Its reports assess current and future risks posed by climate change to ten distinct regions in the U.S. The Fifth National Climate Assessment (NCA5) report was released in 2023.

Educational Resources

National Network for Ocean and Climate Change Interpretation: NNOCCI is a network of individuals and organizations in formal and informal education, the social sciences, climate sciences, and public policy. They provide evidenced-based communications methods and the social and emotional support needed to engage as climate communicators.

Climate Change 101: This PowerPoint slide set is used by the author in most of his classes. It consists of five parts, each about 30 minutes long:

- Natural climate change;
- Anthropogenic climate change;
- Consequences;
- Solutions;
- Why the controversy?

You are welcome to modify them as you wish. Note that many of these slides, particularly the graphs and information about solutions, need regular updating.

Latest Data and Plots

Climate.nasa.gov: Contains information on the latest NASA-related climate science and an extensive collection of global warming resources, including before-and-after pictures of the Earth.

Climate.gov: A website from the National Oceanic and Atmospheric Administration that contains educational resources and the latest plots for a variety of climate indicators, such as ocean and surface temperatures, sea level rise, and ocean oscillation indices (e.g., El Niño).

Climate Communication

Yale Program on Climate Change Communication: Conducts scientific research on public climate change knowledge, attitudes, policy preferences, and behavior, and the underlying psychological, cultural, and political factors that influence them. They also engage the public in climate change science and solutions with a daily, national radio program.

George Mason Center for Climate Change Communication: They run several programs to study and disseminate information on people's understanding, including the "Climate Change in the American Mind" series of reports.

Climate Denialism

Skeptical Science: Provides resources to debunk climate misinformation by presenting peer-reviewed science and explaining the techniques of science denial, discourses of climate delay, and climate solutions denial.

Making Sense of Climate Denial: A free, self-paced EdX course on understanding climate denial arguments and ways to debunk them.

Science Media Literacy

Can I Trust It? is an activity developed by T. A. Rector and P. Banchero to teach science media literacy as part of an introductory astronomy class. It is intended to be done near the start of the semester, with additional articles provided weekly for analysis using the same methodologies.

Science Education in an Age of Misinformation: The Stanford report provides resources for teaching science media literacy.

Solutions

Project Drawdown: Offers comprehensive resources on the wide range of solutions available to address climate change, including their impact and ease of implementation. Can be used to give students an opportunity to explore solutions that resonate with their interests and values.

En-ROADS climate simulator: En-ROADS is a freely-available online simulator that provides the user with the ability to test and explore climate solutions. Can be used individually or as a group learning experience to demonstrate that, by implementing a wide range of solutions, it is possible to address climate change while keeping energy costs low and growing the economy.

Social Media

Global Warming Fact of the Day: A Facebook group that shares recent news and information on climate change.

Alaska Climate Action Network: An example of a Facebook group for a specific region. There likely are groups for where you live.

Advocacy

Climate Action Network: A global network of more than 1,900 civil society organizations in over 130 countries driving collective and sustainable action to fight the climate crisis and to achieve social and racial justice.

Scientist Rebellion: An extension of the Extinction Rebellion movement that works to get scientists involved in advocacy and non-violent protest.

Talking with the Media

SciLine: The American Association for the Advancement of Science (AAAS) SciLine program provides resources for scientists and journalists, including tips on how to interact with the media. If you have a PhD you can register with SciLine to become a contact for the media.

A Scientist's Guide to Working with the Media: The American Geophysical Union (AGU) provides a useful guide for getting your work in front of a broader audience and promote the value of scientific research.

Scientists Working for Change

Labos 1point5: An international, cross-disciplinary collective of academic researchers who share a common goal: to better understand and reduce the environmental impact of research, especially on the Earth's climate.

Astronomers for Planet Earth: A4E is an international grass-roots movement of astronomy students, educators, amateurs and scientists, working to address The Climate Crisis from an astronomical perspective.

www.ingramcontent.com/pod-product-compliance
Ingram Content Group UK Ltd.
Pitfield, Milton Keynes, MK11 3LW, UK
UKHW061829100926
8556IPUK00011B/926
* 9 7 8 0 7 5 0 3 3 7 2 5 0 *